Fermentation Processes Engineering in the Food Industry

Contemporary Food Engineering

Series Editor

Professor Da-Wen Sun, Director

Food Refrigeration & Computerized Food Technology
National University of Ireland, Dublin
(University College Dublin)
Dublin, Ireland
http://www.ucd.ie/sun/

Fermentation Processes Engineering in the Food Industry

Edited by
Carlos Ricardo Soccol
Ashok Pandey
Christian Larroche

CRC Press
Taylor & Francis Group
Boca Raton London New York

CRC Press is an imprint of the
Taylor & Francis Group, an **informa** business

CRC Press
Taylor & Francis Group
6000 Broken Sound Parkway NW, Suite 300
Boca Raton, FL 33487-2742

First issued in paperback 2016

© 2013 by Taylor & Francis Group, LLC
CRC Press is an imprint of Taylor & Francis Group, an Informa business

No claim to original U.S. Government works

Version Date: 20130220

ISBN 13: 978-1-138-19867-8 (pbk)
ISBN 13: 978-1-4398-8765-3 (hbk)

Visit the Taylor & Francis Web site at
http://www.taylorandfrancis.com

and the CRC Press Web site at
http://www.crcpress.com

Contents

Series Preface

Contemporary Food Engineering

Food engineering is the multidisciplinary field of applied physical sciences combined with the knowledge of product properties. Food engineers provide the technological knowledge transfer essential to the cost-effective production and commercialization of food products and services. In particular, food engineers develop and design processes and equipment to convert raw agricultural materials and ingredients into safe, convenient, and nutritious consumer food products. However, food engineering topics are continuously undergoing changes to meet diverse consumer demands, and the subject is being rapidly developed to reflect market needs.

In the development of food engineering, one of the many challenges is to employ modern tools and knowledge, such as computational materials science and nanotechnology, to develop new products and processes. Simultaneously, improving food quality, safety, and security continues to be a critical issue in food engineering study. New packaging materials and techniques are being developed to provide more protection to foods, and novel preservation technologies are emerging to enhance food security and defense. Additionally, process control and automation regularly appear among the top priorities identified in food engineering. Advanced monitoring and control systems are developed to facilitate automation and flexible food manufacturing. Furthermore, energy saving and minimization of environmental problems continue to be important food engineering issues, and significant progress is being made in waste management, efficient utilization of energy, and reduction of effluents and emissions in food production.

The *Contemporary Food Engineering Series*, consisting of edited books, attempts to address some of the recent developments in food engineering. The series covers advances in classical unit operations in engineering applied to food manufacturing as well as such topics as progress in the transport and storage of liquid and solid foods; heating, chilling, and freezing of foods; mass transfer in foods; chemical and biochemical aspects of food engineering and the use of kinetic analysis; dehydration, thermal processing, nonthermal processing, extrusion, liquid food concentration, membrane processes, and applications of membranes in food processing; shelf-life and electronic indicators in inventory management; sustainable technologies in food processing; and packaging, cleaning, and sanitation. These books are aimed at professional food scientists, academics researching food engineering problems, and graduate-level students.

The editors of these books are leading engineers and scientists from many parts of the world. All the editors were asked to present their books to address the market's need and pinpoint the cutting-edge technologies in food engineering.

All the contributions have been written by internationally renowned experts who have both academic and professional credentials. All the authors have attempted to provide critical, comprehensive, and readily accessible information on the art and science of a relevant topic in each chapter, with reference lists for further information. Therefore, each book can serve as an essential reference source to students and researchers in universities and research institutions.

Da-Wen Sun
Series Editor

Preface

Fermentation processes are some of the oldest technologies in food, having been developed with the aim of increasing the storage stability of foods and improving the organoleptic and textural properties of raw materials. Fermented foods remain very popular even today as a substantial percentage of daily consumed foods are fermented. Important examples include bread-making; dairy products, such as yogurts, cheeses, buttermilks, and sour milks; alcoholic drinks, such as wine, beer, and cider; fermented vegetables, such as sauerkraut and pickles; and fermented meats, such as sausages and salami. More recently, traditional products have started to include prebiotic and probiotic microorganisms and ingredients as a result of a better understanding of industrial microbiology and our physiology. With the advent of modern tools of molecular biology and genetic engineering and new skills in metabolic engineering and synthetic biology, the developments in fermentation technology for industrial applications have changed enormously.

This book comprises 17 chapters, providing state-of-the-art information concerning technological developments on engineering aspects of fermentation processes in the food industry. Chapter 1 by Dimitris Charalampopoulos and Colin Webb presents an overview about the lactic acid metabolism and the type of fermented foods that are produced using bacteria and yeast, including a presentation of the main processing steps for their manufacture; it also presents three examples of metabolic engineering of LAB with potential applications in the production of fermented foods and food ingredients. Chapter 2 by Syed G. Dastager describes the methods and techniques for isolation, improvement, and preservation of the microbial cultures used in the food fermentation industry. Chapter 3 by Vincenza Faraco and Antonella Amore discusses that fermentation processes are strongly affected by physical and chemical factors, such as temperature, pH, aeration, and medium composition. In this chapter, the authors present the effects of physical and chemical factors on fermentations in dairy and bakery products and alcoholic beverages.

Chapter 4 by Binod et al. presents the fundamentals of fermentation processes, modes of fermentation, and the principle operations of upstream processes that include the screening and selection of suitable organisms for a particular product, development of a suitable medium, and finally, the mass culturing of the organism for the product. The upstream processing costs make up approximately 20%–50% of the total; whereas the downstream processing costs about 50%–80%. Upstream processing is very crucial for getting the desired end products. The techniques of fermentation and fermentation parameters affect the product yield, and proper quality control measures should be taken during upstream operations for getting the desired product quality. Chapter 5 by Andre Lebert covers the use of mathematical models for analyzing the evolution of the pH and a_w of a process, and their relationship with the increase in the nutritional and safety qualities of the fermented products. Chapter 6 by Christophe Vial and Yussef Stirriba focuses on the effective use of computational methods for bioreactors, mainly using commercial solutions. These are now considered to be standard numerical tools, widely used within the industry although they are yet to be fully integrated into the design methodology. Conversely, the details of computational fluid dynamics (CFD) algorithm development remain beyond the scope of the discussion. The first section of this chapter describes briefly what can be expected from the characterization of bioreactors using CFD. The second section describes the main steps of CFD analysis. The third and the fourth parts address the physical modeling and numerical issues of CFD, focusing on the specificity of bioreactors. The last section analyzes the recent progresses and limitations of CFD, using the example of bubble column and airlift reactors so as to illustrate the steps and issues developed previously on bioreactors.

Chapter 7 by Talasi Satynarayana et al. provides an overview of various types of fermenters and bioreactors employed in submerged fermentations with special reference to their utility in food industries. Chapter 8 by José Angel Rodríguez-León et al. describes the evolution of the fermentation theory of solid-state fermentation that can be characterized by kinetic concepts. Several commonly employed SSF reactors are also described. Chapter 9 by Júlio Cesar de Carvalho et al. describes the most important operations for solid–liquid separation, concentration, and drying of fermented foods. Chapter 10 by Wilerson Sturm et al. describes automation systems, such as sensors, actuators, and controls, specifically in the food industry. Chapter 11 by Juliano De Dea Lindener et al. documents different aspects related to meat and vegetable fermentations, emphasizing the advantages of the fermentation processes and the benefits to health. Also, it presents some characteristics of fermented functional foods, including probiotics and prebiotics, and new trends for the fermentation of foods. Chapter 12 by Satinder Kaur Brar provides a brief outline on different varieties of alcoholic beverages, technical aspects of their production processes, and recent advances in this field. Chapter 13 by Luciana Porto de Souza Vandenberghe et al. presents general aspects about the manufacturing of dairy fermented products as well as technological challenges of the fermentative process and economic and functional importance of these products. Chapter 14 by Jean-Luc Tholozan and Jean-Luc Cayol describes new tendencies of dairy and nondairy probiotic products, including soy milk, fruits, vegetables, juices, and others beverages. Chapter 15 by Juliano Lemos Bicas et al. provides an insight about the main aspects involving the properties and biotechnological production methods of food additives, focusing on bio-aromas, bio-colorants, nondigestible oligossacharides, and biosurfactants. Chapter 16 by Jorge Alberto Vieira Costa and Michele Greque de Morais presents several applications of microalgae in the food sector, including supplementation of the products from biomass, specific biocompounds, polymers with potential for developing packaging, and the treatment of industrial effluents. Chapter 17 by Carlos Ricardo Soccol et al. presents the potentials of the application of a biorefinery concept to add value to the solid and liquid wastes from the food industries. Some results are presented about a case study on an integrated project in which the concept was applied for the integral exploitation of residues of the soybean industry for the production of bioethanol, xanthan gum, alpha-galactosidase, and lactic acid on laboratory- and pilot-scales.

Series Editor

Professor Da-Wen Sun, PhD, is a world authority on food engineering research and education; he is a member of the Royal Irish Academy, which is the highest academic honor in Ireland; he is also a member of Academia Europaea (The Academy of Europe) and a Fellow of International Academy of Food Science and Technology. His main research activities include cooling, drying, and refrigeration processes and systems; quality and safety of food products; bioprocess simulation and optimization; and computer vision technology. In particular, especially, his many scholarly works have become standard reference materials for researchers in the areas of computer vision, computational fluid dynamics modelling, vacuum cooling, etc. Results of his work have been published in more than 600 papers, including over 250 peer-reviewed journal papers (Web of Science *h*-index = 40; Google Scholar *h*-index = 47). He has also edited 13 authoritative books. According to Thomson Reuters's *Essential Science Indicators*[SM] updated as of July 1, 2010, based on data derived over a period of ten years and four months (January 1, 2000–April 30, 2010) from the ISI Web of Science, a total of 2554 scientists are among the top 1% of the most cited scientists in the category of agriculture sciences, and Professor Sun is listed at the top with a ranking of 31.

Dr. Sun received his first class BSc honors and his MSc in mechanical engineering, and his PhD in chemical engineering in China before working at various universities in Europe. He became the first Chinese national to be permanently employed in an Irish university when he was appointed a college lecturer at the National University of Ireland, Dublin (University College Dublin [UCD]), in 1995. He was then continuously promoted in the shortest possible time to the position of senior lecturer, associate professor, and full professor. Dr. Sun is now a professor of food and biosystems engineering and director of the Food Refrigeration and Computerized Food Technology Research Group at UCD.

As a leading educator in food engineering, Dr. Sun has contributed significantly to the field of food engineering. He has guided many PhD students who have made their own contributions to the industry and academia. He has also, on a regular basis, given lectures on the advances in food engineering at international academic institutions and delivered keynote speeches at international conferences. As a recognized authority in food engineering, Dr. Sun has been conferred adjunct/visiting/consulting professorships by over ten top universities in China, including Zhejiang University, Shanghai Jiaotong University, Harbin Institute of Technology, China Agricultural University, South China University of Technology, and Jiangnan University. In recognition of his significant contribution to food engineering worldwide, and for his outstanding leadership in the field, the International Commission of Agricultural and Biosystems Engineering (CIGR) awarded him the CIGR Merit Award in 2000 and again in 2006; the U.K.-based Institution of Mechanical Engineers named him Food Engineer of the Year 2004; in 2008, he was awarded the CIGR Recognition Award in recognition of his distinguished achievements as the top 1% of agricultural engineering scientists around the world; in 2007, he was presented with the only AFST(I) Fellow Award in that year by the Association of Food Scientists and Technologists (India); and in 2010, he was presented with the CIGR Fellow Award (the title of "Fellow" is the highest honor in CIGR and is conferred upon individuals who have made sustained, outstanding contributions worldwide).

Dr. Sun is a fellow of the Institution of Agricultural Engineers and a fellow of Engineers Ireland (the Institution of Engineers of Ireland). He has also received numerous awards for teaching and research excellence, including the President's Research Fellowship, and has received the President's Research Award from UCD on two occasions. He is also the editor in chief of *Food and Bioprocess Technology— An International Journal* (Springer) (2011 Impact Factor = 3.703, ranked at the fourth position among 128 ISI-listed food science and technology journals); series editor of the Contemporary Food Engineering Series (CRC Press/Taylor & Francis Group); former editor of *Journal of Food Engineering* (Elsevier); and an editorial board member of *Journal of Food Engineering* (Elsevier), *Journal of Food Process Engineering* (Blackwell), *Sensing and Instrumentation for Food Quality and Safety* (Springer), and *Journal of Ocean University of China*. Dr. Sun is also a chartered engineer.

On May 28, 2010, he was awarded membership to the Royal Irish Academy (RIA), which is the highest honor that can be attained by scholars and scientists working in Ireland. At the 51st CIGR General Assembly held during the CIGR World Congress in Quebec City, Canada, in June 2010, he was elected as incoming president of CIGR and will become CIGR president in 2013 to 2014. The term of the presidency is six years—two years each for serving as incoming president, president, and past president. On September 20, 2011, he was elected to Academia Europaea (The Academy of Europe), which is functioning as European Academy of Humanities, Letters and Sciences and is one of the most prestigious academies in the world; election to the Academia Europaea represents the highest academic distinction.

Editors

Carlos Ricardo Soccol is the research group leader of the **Department of Bioprocess Engineering and Biotechnology (DEBB)** at the Federal University of Paraná, Brazil, with 20 years of experience in biotechnological research and development of bioprocesses with industrial application. He received the bachelor's degree in chemical engineering from UFPR (1979), the master's degree in food technology from UFPR (1986), and the Ph.D. degree in *Genie Enzymatique, Microbiologie et Bioconversion* from the *Université de Technologie de Compiègne*, France (1992). He received his postdoctoral degree from the Institut ORSTOM/IRD (Montpellier, 1994 and 1997) and the *Université de Provence et de la Méditerranée* (Marseille, 2000). He is an HDR professor at the *Ecole d'Ingénieurs Supériure* of *Luminy*, Marseille, France. He has experience in the area of science and food technology, with emphasis on agroindustrial and agroalimentary biotechnology, acting in the following areas: bioprocess engineering and solid state fermentation, submerged fermentation, bioseparations, industrial bioprocesses, enzyme technology, tissue culture, bioindustrial projects, and bioproduction. He is currently a Coordinator of Master BIODEV-UNESCO, an Associate Editor of five international journals, and the Editor in Chief of the *Brazilian Archives of Biology and Technology Journal*. Professor Soccol received several national and international awards, including the Science and Technology award of the Government of Paraná (1996), the Scopus/Elsevier Award (2009), Dr. Honoris Causa, University Blaise Pascal-France (2010), and Outstanding Scientist—5th International Conference on Industrial Bioprocesses, Taipei, Taiwan (2012). He is a technical and scientific consultant of several companies, agencies, and scientific journals in Brazil and abroad. He has supervised and molded 88 master of science students, 35 Ph.D. students, and 12 postdoctorate students. He has 931 publications/communications that include 14 books, 87 book chapters, 250 original research papers, and 543 research communications in international and national conferences, and he has registered 37 patents. His research articles have so far been cited 4150 times (Scopus database) and 3950 times (Web of Science—ISI) with Index $h = 32$.

Ashok Pandey obtained his master's degree in chemistry in 1976 and his Ph.D. in 1979 in microbiology from the University of Allahabad. Professor Pandey was a scientist at the National Sugar Institute, Kanpur, during 1982 and 1985. During 1985 and 1986, he worked as a research scientist in Suddeutsche Zucker AG in Germany. In 1987, he joined the CSIR's National Institute for Interdisciplinary Science and Technology at Trivandrum as a scientist and since then continuing there. Currently, he is deputy director and head of the Center for Biofuels and Biotechnology Division there.

Professor Pandey's current main research focus is on bioprocesses and product development, mainly focused on agro-industrial solid waste utilization for the production of biofuels, biopolymers, industrial enzymes, etc. He has developed and transferred technologies on industrial enzymes to the industries and has completed several industrial consultancy projects. He has 900 publications/communications, which include 12 patents, 29 books, 23 special issues of journals as guest/special issue editor, 25 technical reports, 95 chapters in the books, 354 original and review papers, and 362 research communications in international and national conferences.

Professor Pandey is the recipient of many national and international awards and fellowships, which include Fellow of International Organization of Biotechnology and Bioengineering, Biotech Research Society of India and Association of Microbiologists of India; honorary doctorate degree from Univesite Blaise Pascal, France; Thomson Scientific India Citation Laureate Award; Lupin Visiting Fellowship, Visiting Professor in the University Blaise Pascal, France, Federal University of Parana, Brazil and EPFL, Switzerland; Best Scientific Work Achievement award, Government of Cuba; UNESCO Professor; Raman Research Fellowship Award, CSIR; GBF, Germany and CNRS, France Fellowship; Young Scientist Award, etc. He was chairman of the International Society of Food, Agriculture and Environment, Finland (Food and Health) during 2003 and 2004. He is founder president of Biotech Research Society of India and international coordinator of International Forum on Industrial Bioprocesses. He is editor-in-chief of *Bioresource Technology*, the Elsevier journal, and editorial board member of several international and Indian journals.

Christian Larroche is a graduate in biochemical engineering from the INSA, Toulouse (1979); docteur-ingénieur in organic chemistry from Paul Sabatier Toulouse 3 University (1982); and docteur ès sciences (Ph.D.) in biochemical engineering from Blaise Pascal University (1990). Professor Larroche has strong research interests in the areas of applied microbiology and biochemical engineering. His skills are related to the study and development of special processes for the use of microorganisms. This includes fungal spores produced by solid-state cultivation and their use as protein (enzyme) reservoirs in biotransformations. A special interest in phase-transfer phenomena coupled with metabolic engineering has to be noticed. It is applied to the design and optimization of biotransformations involving hydrophobic compounds and carried out in biphasic liquid–liquid media. These processes are related both to the food and environment (bioremediation) areas. His interests have recently been extended to bioenergy, and he is presently coordinator of two French research programs on biohydrogen production by nonphotosynthetic anaerobic microorganisms grown on complex media.

He is author of approximately 70 articles, two patents, and 12 book chapters. He has supervised 10 Ph.D. students and 20 MSc lab works. He is a member of the SFGP (French Society for Process Engineering) and chief international coordinator of the ICBF Forum, an international network entitled Food Bioprocessing: A

Global Approach for Advancing Sustainable Production of Value-Added Food. He is head of the department of the study and development of processes involving microorganisms of the platform for technological development biotechnology—material engineering of Blaise Pascal University and is in charge the research on solid-state fermentations—biotransformations of the team GePEB (process engineering, energetics, and biosystems) of the Institut Pascal in the same university. He has been vice president of the university in charge of research valorization and technology transfer (2008–2012) and is currently director of Polytech Clermont-Ferrand, a school of engineering of Blaise Pascal University.

Contributors

Antonella Amore
Department of Organic Chemistry and
    Biochemistry
University of Naples "Federico II"
Naples, Italy

Francisco Fábio Cavalcante Barros
Department of Food Science
University of Campinas
Campinas, Brazil

Juliano Lemos Bicas
Department of Chemistry
Biotechnology and Bioprocess
    Engineering
University of São João del-Rei
Ouro Branco, Brazil

Parameswaran Binod
Biotechnology Division
National Institute for Interdisciplinary
    Science and Technology
CSIR
Trivandrum, India

Satinder Kaur Brar
INRS-ETE
Université du Québec
Québec, Canada

Jean-Luc Cayol
Mediterranean Institute of
    Microbiology
Campus de Luminy
Marseille Cedex, France

Dimitris Charalampopoulos
Department of Food and Nutritional
    Sciences
University of Reading
Reading, United Kingdom

Jorge Alberto Vieira Costa
Laboratory of Biochemical Engineering
School of Chemistry and Food
Federal University of Rio Grande
Rio Grande, Brazil

Syed G. Dastager
NCIM Resource Center
CSIR-National Chemical Laboratory
    Pune
India

Júlio César de Carvalho
Bioprocess Engineering and
    Biotechnology Department
Federal University of Parana
Curitiba, Brazil

Juliano De Dea Lindner
Research and Development Department
    Incorpore Foods
Camboriú, Brazil
and
Departamento de Tecnologia de
    Alimentos
Universidade Católica de Santa
    Catarina
Jaraguá do Sul, Brazil

Alexandre Francisco de Moraes Filho
Instituto Federal do Paraná (IFPR)
Curitiba, Brazil

Michele Greque de Morais
Laboratory of Biochemical Engineering
    School of Chemistry and Food
Federal University of Rio Grande
Rio Grande, Brazil

Paula Fernandes de Siqueira
Bioprocess Engineering and
    Biotechnology Department
Federal University of Paraná
Curitiba, Brazil

**Luciana Porto de Souza
Vandenberghe**
Bioprocess Engineering and
    Biotechnology Department
Federal University of Parana
Curitiba, Brazil

Ivo Mottin Demiate
Departamento de Engenharia de
    Alimentos
Universidade Estadual de Ponta Grossa
Ponta Grossa, Brazil

Dario Eduardo Amaral Dergint
Universidade Tecnológica Federal do
    Paraná (UTFPR)
Curitiba, Brazil

Vincenza Faraco
Department of Organic Chemistry and
    Biochemistry
and
School of Biotechnological Sciences
University of Naples "Federico II"
Naples, Italy

Susan Grace Karp
Bioprocess Engineering and
    Biotechnology Department
Federal University of Paraná
and
Industrial Biotechnology Program
Positivo University
Curitiba, Brazil

Parvinder Kaur
Department of Microbiology
S.S.N. College
Alipur, New Delhi, India

André Lebert
Institut Pascal
Université Blaise Pascal
Clermont-Ferrand, France

Luiz Alberto Júnior Letti
Bioprocess Engineering and
    Biotechnology Department
Federal University of Paraná
Curitiba, Brazil

Mário Roberto Marostica, Junior
Department of Food and Nutrition
University of Campinas
Campinas, Brazil

Adriane Bianchi Pedroni Medeiros
Bioprocess Engineering and
    Biotechnology Department
Federal University of Parana
Curitiba, Brazil

Gustavo Molina
Department of Food Science
University of Campinas
Campinas, Brazil

Ashok Pandey
Biotechnology Division
National Institute for Interdisciplinary
    Science and Technology
CSIR
Trivandrum, India

José Luis Parada
Universidade Positivo
Curitiba, Brazil

Gláucia Maria Pastore
Department of Food Science
University of Campinas
Campinas, Brazil

Ana Lúcia Barretto Penna
Departamento de Engenharia e
    Tecnologia de Alimentos
Universidade Estadual Paulista
São José do Rio Preto, Brazil

Maria Rosa Machado Prado
Departamento de Engenharia de
    Bioprocessos e Biotecnologia
Universidade Federal do Paraná
Curitiba, Brazil

Cristine Rodrigues
Bioprocess Engineering and
    Biotechnology Department
Federal University of Parana
Curitiba, Brazil

Daniel Ernesto Rodríguez-Fernández
Bioprocess Engineering and
    Biotechnology Department
Federal University of Paraná
Curitiba, Brazil

José Angel Rodríguez-León
Industrial Technology Program Positivo
    University
Curitiba, Brazil

Cássia Tiemi Nemoto Sanada
Bioprocess Engineering and
    Biotechnology Department
Federal University of Paraná
Curitiba, Brazil

Saurabh Jyoti Sarma
INRS-ETE
Université du Québec
Québec, Canada

Tulasi Satyanarayana
University of Delhi South Campus
New Delhi, India

Alysson Hikaru Shirai
Universidade Tecnológica Federal do
    Paraná (UTFPR)
Curitiba, Brazil

Raveendran Sindhu
Biotechnology Division
National Institute for Interdisciplinary
    Science and Technology
CSIR
Trivandrum, India

Carlos Ricardo Soccol
Bioprocess Engineering and
    Biotechnology Department
Federal University of Paraná
Curitiba, Brazil

Michele Rigon Spier
Bioprocess Engineering and
    Biotechnology Department
Federal University of Parana
Curitiba, Brazil

Youssef Stiriba
Universitat Rovira i Virgila
Tarragona, Spain

Wilerson Sturm
Instituto Federal do Paraná (IFPR)
Curitiba, Brazil

Jean-Luc Tholozan
Mediterranean Institute of
    Microbiology
Campus de Luminy
Marseille Cedex, France

and

Attaché de Coopération Scientifique et
    Universitaire
Ambassade de France
Institut Français de Tunisie
Tunis, Tunisie

Vanete Thomaz-Soccol
Bioprocess Engineering and
    Biotechnology Department
Federal University of Paraná
and
Industrial Biotechnology Program
Positivo University
Curitiba, Brazil

Mausam Verma
Institut de Recherche et
    de Développement en
    Agroenvironnement Inc.
Québec, Canada

Christophe Vial
Polytech Clermont-Ferrand
Aubière Cedex, France

Ashima Vohra
Institute of Home Economics
New Delhi, India

Colin Webb
School of Chemical Engineering and
    Analytical Science
University of Manchester
Manchester, United Kingdom

Adenise Lorenci Woiciechowski
Bioprocess Engineering and
    Biotechnology Department
Federal University of Parana
Curitiba, Brazil

Caroline Tiemi Yamaguishi
Bioprocess Engineering and
    Biotechnology Department
Federal University of Parana
Curitiba, Brazil

1 Applications of Metabolic Engineering in the Production of Fermented Foods and Food Ingredients

Dimitris Charalampopoulos and Colin Webb

CONTENTS

1.1 INTRODUCTION

Fermented foods have been produced for many centuries with the basic aim of increasing the storage stability of processed foods and modifying the organoleptic and textural properties of raw materials. Fermented foods are very popular even today as a substantial percentage of daily-consumed foods are fermented. Notable examples include dairy products, such as yogurt, cheese, buttermilk, and sour milk; alcoholic drinks, such as wine, beer, and cider; fermented vegetables, such as sauerkraut and pickles; and fermented meats, such as sausages and salami. The main

1

reason for their popularity, however, is their specific organoleptic properties rather than their preservation stability.

A class of bacteria that has been associated with fermented foods is the lactic acid bacteria (LAB). LAB are generally associated with habitats rich in nutrients, such as various food commodities (vegetables, milk, meat), but some are also habitants of the normal flora of the oral cavity and the gastrointestinal and genitourinary tracts of animals and humans (Axelsson 1998). The genera of interest to fermented foods are *Lactobacillus, Leuconostoc, Pediococcus, Lactococcus,* and *Streptococcus. Lacto-cocci* are used as starter cultures for various types of dairy products, and *pedio-cocci* are used for sausage-making and as silage inoculants (Hammes et al. 1990; Hammes and Vogel 1995). *Leuconostoc* species have been used in dairy fermentation to produce the flavor compound diacetyl from citric acid and are often important in spontaneous vegetable fermentations, for example, sauerkraut (Daeschel et al. 1987). The only *Streptococcus* species that is associated with food technology is *Streptococcus thermophilus,* which is used as a starter culture in combination with *Lactobacillus delbrueckii* subspecies *bulgaricus* in the production of yogurt. The genus *Lactobacillus* is the largest in the LAB group. *Lactobacilli* are widespread in nature and have found applications in many areas of the food industry, such as fermented meats and vegetables, dairy products, wine, and cider production (Fleming et al. 1985; Hammes et al. 1990; Buckenhuskes 1993; Oberman and Libudzisz 1998; Salovaara 1998). In addition, certain species are members of the indigenous human microflora (Mikelsaar et al. 1998) and have been postulated to exert probiotic activities, making them ideal candidates for probiotic food products.

Driven by the advancement in the last 25 years in genome sequencing, comparative genomic analysis, and gene cloning, there has been a lot of interest in utilizing this knowledge to design metabolic pathways in bacteria used for the production of fermented foods and food ingredients, targeting the synthesis of specific compounds. Such compounds can have sensorial, textural, nutritional, and health attributes. LAB have received a lot of attention in this respect; as they are extensively used for the production of several fermented foods, they have a relatively simple metabolism and have limited biosynthetic abilities. Metabolic pathway engineering of LAB has been used for the synthesis of metabolites in bioreactors, which can then be used as food ingredients; the most notable example is lactic acid although other examples include bacteriocins (e.g., nisin), low-calorie sweeteners, aroma compounds (e.g., diacetyl), amino acids (e.g., alanine), exopolysaccharides, and B vitamins. These compounds are normally produced in large bioreactors, extracted from the fermentation broth, and purified by employing usually physicochemical (precipitation, liquid–liquid extraction, etc.) or chromatographic methods (Uteng et al. 2002; Goh et al. 2005; Leuchtenberger et al. 2005). Another application of metabolic engineering is to produce the compounds of interest in situ, specific components that can beneficially affect the food product, for example, to improve its nutritional value, its texture, or its sensory properties.

The first part of this chapter provides an overview of lactic acid metabolism and the types of fermented foods that are produced using bacteria and yeast, including a presentation of the main processing steps for their manufacture; in the second part, we present three examples of the metabolic engineering of LAB with potential applications in the production of fermented foods and food ingredients.

1.2 LACTIC ACID METABOLISM

Biochemically, LAB lack the Krebs cycle and a terminal electron transport system and obtain their energy only by carbohydrate fermentation coupled to substrate-level phosphorylation. The generated ATP is subsequently used for biosynthetic purposes. Unlike many anaerobes, most LAB are not sensitive to oxygen and can grow in its presence as well as in its absence (Condon 1987). The mechanisms against oxygen toxicity include enzymatic defense systems (flavoprotein-containing oxidases and peroxidases, superoxide dismutase) and a nonenzymatic manganese-catalyzed scavenging system of superoxide radicals (Condon 1987).

LAB are able to grow on simple sugars, such as glucose, sucrose, and lactose. LAB can be divided into two physiological groups, homofermentative or heterofermentative, depending on the hexose metabolic pathways used. Homofermentative LAB degrade hexoses via glycolysis (the Embden–Meyerhof pathway), producing lactic acid as the major end product, whereas the heterofermentative LAB use the 6-phosphogluconate pathway, producing lactic acid, CO_2, and acetic acid and/or ethanol (Kandler 1983) (Figure 1.1).

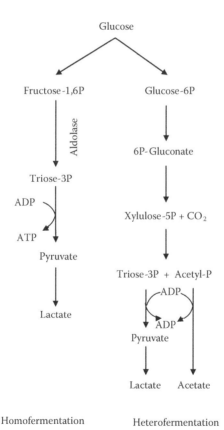

FIGURE 1.1 Schematic presentation of main pathways of hexose fermentation in LAB. (Adapted from Kandler, O., *Antonie van Leeuwenhoek*, 49:209–224, 1983.)

In *Streptococci, Pediococci, Lactococci,* and homofermentative *Lactobacilli,* glycolysis is characterized by the production of two triose phosphate moieties from the splitting of fructose 1,6-biphosphate by aldolase. These triose phosphate moieties are then converted to pyruvate by substrate-level phosphorylation at two sites. Under normal conditions, that is, with excess sugar and limited access to oxygen, pyruvate is then reduced to lactic acid (Axelsson 1998).

The presence or absence of fructose 1,6-biphosphate aldolase is responsible for the difference between homofermentative and heterofermentative LAB (Tseng and Montville 1993). Because of the lack of this enzyme in *Leuconostoc* and heterofermentative *Lactobacilli,* glucose is transformed into a pentose 5-phosphate via phosphorylation, dehydrogenation, and decarboxylation. Pentose 5-phosphate is then split by a phosphoketolase into a C-2 and C-3. The C-3 moiety is metabolized in the same way as for the glycolytic pathway to give one molecule of lactic acid, and the C-2 moiety is converted to acetic acid or ethanol depending on the oxidation–reduction potential of the system. Thus, equimolar amounts of CO_2, lactic acid, and acetic acid or ethanol are formed from hexose (Kandler 1983).

Homofermentative LAB grow significantly better than other bacteria present in the same ecological niche. This is a result of the fact that their growth is associated with the production of lactic acid as they lack the ability to produce other fermentation compounds in substantial amounts. As a result, homofermentative LAB are able to dominate in a wide range of environments, mainly those that are rich in nutrients. The reason for this is that LAB do not have significant biosynthetic capacities and, therefore, rely on the supply of nutrients, such as sugars, nitrogen, and vitamins, from the environment. It is not a surprise, therefore, that homofermentative LAB are able to grow in most food substrates, such as milk, vegetables, and meat, which are rich in nutrients (Hugenholtz 2008).

1.2.1 BYPRODUCT FORMATION BY HOMOFERMENTATIVE LAB

In theory, homolactic fermentation of glucose results in 2 mol of lactic acid and in a net gain of 2 ATP per mole of glucose consumed. Heterolactic fermentation produces 1 mol each of lactic acid, ethanol, and CO_2 and 1 ATP per 1 mol of glucose. In practice, these theoretical values are seldom obtained. Usually, a conversion factor of 0.9 from sugar to end-product carbon is obtained, which probably reflects an incorporation of sugar carbon into the biomass (Axelsson 1998). In addition, the use of complex fermentation substrates, as well as the environmental conditions (oxygen availability), may contribute to the formation of byproducts, such as acetic acid, and other small organic compounds (diacetyl, acetoin, acetaldehyde). In particular, it is well established that pyruvate, formed as an intermediate in all the pathways, may partly undergo several alternative conversions resulting in different end-product patterns than those seen with glucose fermentation under normal nonlimiting conditions (Kandler 1983).

Technologically, the most important alternative conversion of pyruvate is its conversion to acetoin and diacetyl, which proceeds via the mechanisms presented in Figure 1.2. These compounds contribute to the aroma of cottage cheese and buttermilk (Marshall 1987). Diacetyl is also present in cereal-based fermented

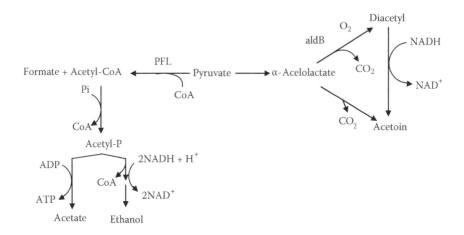

FIGURE 1.2 Schematic presentation of the alternative conversion pathways for pyruvate. Abbreviations: ADP, adenosine diphosphate; aldB, α-acetolactate decarboxylase; ATP, adenosine triphosphate; CoA, coenzyme A; NAD$^+$/NADH, nicotinamide adenine dinucleotide; PFL, pyruvate formate lyase; Pi, phosphate. (Adapted from Kandler, O., *Antonie van Leeuwenhoek*, 49:209–224, 1983; Axelsson, L., Lactic acid bacteria: Classification and physiology. In S. Salminen, A. von Wright (ed.) *Lactic Acid Bacteria: Microbiology and Functional Aspects*, New York: Marcel Dekker, pp. 1–72, 1998.)

products (Joshi et al. 1989), but it is undesirable for products such as fermented vegetables, wine, beer, and sausages (Hugenholtz 1993). The diacetyl/acetoin metabolism proceeds to a certain extent only if there is a surplus in pyruvate. This can be created if another source of pyruvate besides the fermentable carbohydrate, such as citric acid, exists or if a compound acting as an electron acceptor is present, thus sparing the pyruvate formed by carbohydrate fermentation (Kandler 1983). Usually, in dairy fermentations, the surplus of pyruvate originates from citric acid, which is present in or added to milk (Oberman and Libudzisz 1998). Citric acid is transported into the cells and cleaved to yield oxaloacetic acid and acetic acid, the former of which is subsequently oxidized to pyruvate and CO_2 (Hugenholtz 1993). This mechanism is mainly used by homofermentative LAB under anaerobic conditions. However, under hexose-limiting and aerobic conditions, homofermentative LAB can use oxygen as an electron acceptor and create a surplus of pyruvate, which is then available for metabolism through the diacetyl/acetoin pathway (Axelsson 1998). As shown in Figure 1.2, pyruvate can also be metabolized by a pyruvate formate lyase system, in which the enzyme pyruvate formate lyase catalyzes the reaction of pyruvate and CoA to formate and acetyl CoA; the latter can be used as electron acceptor leading to the formation of ethanol or can be used as a precursor for substrate-level phosphorylation leading to the production of acetate. This mechanism takes place under anaerobic and hexose-limiting conditions and results in a shift from homofermentative to heterofermentative metabolism (Axelsson 1998).

In general, knowledge of the metabolic pathways is necessary to control several characteristics of a fermentation process, such as the flavor, nutritional value, texture, and self-life of the product. Factors such as the intrinsic properties of the LAB species (homofermentative/heterofermentative), the activity of the particular strain (growing or nongrowing cells) (Palles et al. 1998), the experimental conditions (availability of oxygen, pH), and the composition of the raw material determine the formation of metabolic products. However, it must be emphasized that natural materials, such as cereals or vegetables, are very complex materials as they contain several fermentable organic compounds, such as hexoses, pentoses, disaccharides, oligosaccharides, citric acid, tartaric acid, malic acid, acetaldehyde, and various nitrogenous substances. As in Kandler (1983), in these cases, results may be confusing, and the control of the end-product distribution could be a very difficult task.

1.3 FERMENTED FOODS

Fermented foods have been produced for many centuries. Table 1.1 presents a list of the most common fermented foods and the type of fermentation taking place. In most cases, this is either lactic acid fermentation (e.g., cheese, yogurt) or ethanol fermentation (e.g., wine, beer, cider). Traditionally, fermented foods are produced either through natural fermentation or using a backslopping technique in which a successful batch of fermented product is used to inoculate a new batch. Nowadays, however, starter cultures are used, which consist either of a single strain or a mixture of strains, which are well defined and have specific, predetermined characteristics. As a result of the use of starter cultures, the production process is much more efficient and usually gives higher product yields, and the fermented products have consistent organoleptic properties (Leroy and De Vuyst 2004).

TABLE 1.1
Fermented Foods and Types of Biochemical Reactions

Food Product	Fermentation
Cheese	Enzymatic hydrolysis, lactic acid fermentation, mold fermentation
Yogurt	Lactic acid fermentation
Fermented milks	Lactic acid fermentation
Sauerkraut	Lactic acid fermentation
Olives	Lactic acid fermentation
Bread	Ethanol fermentation
Beer	Enzymatic hydrolysis, ethanol fermentation
Wine	Ethanol fermentation, malolactic fermentation
Cider	Ethanol fermentation, malolactic fermentation
Soy products	Enzymatic hydrolysis by molds, lactic acid fermentation, ethanol fermentation
Vinegar	Ethanol, acetic acid fermentation
Sausage	Lactic acid fermentation

1.3.1 CHEESE

More than 400 cheese varieties are produced throughout the world. This large diversity of cheeses is attributed to the variety of milk sources, the starter cultures used, and the production processes. Although the protocols for production differ, Figure 1.3 depicts a general processing scheme for unripened and ripened cheeses. Detailed information on the science and technology of cheese can be found elsewhere (Fox et al. 2000). Common processes for most varieties include the acidification of milk (resulting from lactic acid fermentation), the coagulation of milk proteins, and the curd dehydration and salting. LAB may play different roles in cheese-making: the starter LAB (SLAB) are involved in the fermentation of lactose and produce high concentrations of lactic acid, whereas the nonstarter LAB (NSLAB) are responsible for the ripening process. Typical SLAB include mesophilic starters (temperatures ~30°C), such as *Lactococcus lactis* (e.g., for Cheddar cheese) or thermophilic starters (temperatures of 38°C–45°C), such as *Lactobacillus helveticus, Lactobacillus delbrueckii* subsp. *bulgaricus, Streptococcus thermophilus* (e.g., for Emmental and Parmesan). The group of NSLAB is particularly heterogeneous with *Lactobacilli* being mostly represented, including obligately homofermentative species (e.g., *Lactobacillus farciminis*), facultatively heterofermentative species (e.g., *Lactobacillus casei, Lactobacillus paracasei,*

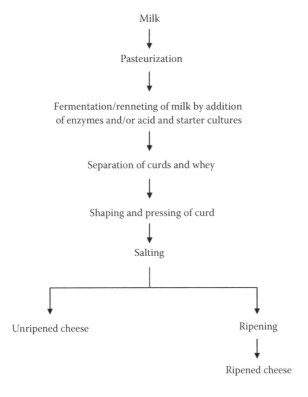

FIGURE 1.3 Outline of cheese-making process.

Lactobacillus plantarum, and *Lactobacillus rhamnosus*), and obligately heterofermentative (e.g., *Lactobacillus fermentum*, *Lactobacillus buchneri*, and *Lactobacillus brevis*) (Settanni and Moschetti 2010). The coagulation of casein is initiated through the addition of chymosin, a protease, and the high levels of lactic acid and results in the formation of the curd. The coagulated curd forms a gel that entraps fat, water molecules, enzymes, and starter bacteria (Boylston 2006). Subsequent steps include cutting and pressing of the curd, resulting in the release of whey, and salting, which aims to reduce the moisture content of the curd and inhibit the growth of spoilage microorganisms. Finally, ripened cheeses undergo a ripening period, which can last from a few weeks up to two years, and which contributes to the flavor and texture of the cheeses. The selection of starter and nonstarter cultures has a considerable effect on the quality of the final product as they are involved in the breakdown of cheese proteins to small peptides and amino acids, the formation of volatile compounds from lipids and amino acids, and the formation of diacetyl (buttery flavor) and can potentially have probiotic activities (Rodrigues et al. 2012; Smit et al. 2005).

1.3.2 YOGURT

The manufacturing protocols for yogurt and the raw materials vary widely depending on the country in which it is produced. In general, two types of yogurt are made: the stirred and the set type. In the former, the fermentation is carried out in a reactor, and then the yogurt is gently mixed and transferred to containers, whereas in the latter, the fermentation is carried out within individual containers (fermented-in-the-cup-style yogurt). Figure 1.4 depicts the generic process for the production of set-style

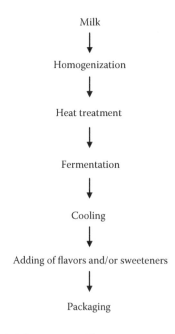

FIGURE 1.4 Outline of set-style yogurt-making process.

yogurt. Detailed information on the science and technology of yogurt-making can be found elsewhere (Tamime 2005). Briefly, the process starts with the standardization of milk to the desired fat and milk solids content (usually 12%–14%). The latter is achieved by adding nonfat milk powder and aims to reduce syneresis (i.e., the separation of water from the coagulated milk); to many consumers, syneresis is considered an indication of a less-than-good-quality yogurt. The milk is then homogenized in order to decrease the size of the fat globules, reduce syneresis, and increase the firmness of the yogurt. Pasteurization of the milk then follows, usually at 85°C for 30 min, in order to eliminate spoilage microorganisms; inactivate the milk enzymes; denature the whey proteins, thus exposing more amino acid residues; and reduce the levels of oxygen. Subsequently, the milk is cooled to 43°C–45°C and inoculated with the starter culture mixture, which consists of *Lactobacillus delbrueckii* subsp. *bulgaricus* and *Streptococcus thermophilus*. The inoculated milk is incubated at 40°C–45°C for 2–5 h, during which time lactic acid and, to a lesser extent, formic acid are produced through the metabolism of the starter cultures; this results in a drop in pH and, consequently, in the aggregation of the casein micelles and the formation of a gel. The two bacteria used for yogurt-making have a synergistic effect on each other's growth (Driessen et al. 1982; Oberman and Libudzisz 1998; Tamime 2005). *Lactobacillus delbrueckii* subsp. *bulgaricus* has protease activity and thus is able to break down the milk proteins to peptides and amino acids. These, in turn, can be taken up by *Streptococcus thermophilus,* which has limited proteolytic activity, and enhance its growth. On the other hand, *Streptococcus thermophilus* produces formic acid and carbon dioxide, which stimulate the growth of *Lactobacillus delbrueckii* subsp. *bulgaricus*. As a result, during the fermentation process, *Streptococcus thermophilus* grows faster initially because of the availability of nitrogenous compounds and then slows down as a result of the increased acidity. *Lactobacillus delbrueckii* subsp. *bulgaricus* dominates in the final stages of the fermentation as it is stimulated by formic acid.

1.3.3 FERMENTED MILKS

More than 400 different fermented milks are produced throughout the world. These include cultured buttermilks and sour creams, which are produced using mesophilic starter cultures, including *Lactococcus lactis* subsp. *lactis*, *Lactococcus lactis* biovar. *diacetylactis*, and *Leuconostoc mesenteroides* subsp. *cremonis*. These bacteria are able to produce diacetyl and contribute to the buttery flavor of the fermented products. The fermentations are carried out between 20°C and 30°C and are stopped by cooling (Oberman and Libudzisz 1998). Another fermented milk product is kefir, which is very popular in Eastern Europe. It is prepared by inoculating cow, goat, or sheep's milk with kefir grains; these are complex matrices of polysaccharides and proteins that contain LAB, yeast, and acetic acid bacteria. Kefir is sour, contains up to 2% ethanol, and is considered to have beneficial health effects (Thoreux and Schmucker 2001; Vinderola et al. 2005). Other types of fermented drinks with a considerable market share include fermented milks with probiotic bacteria, such as *Lactobacillus acidophilus*. This is called acidophilus milk. According to their definition, probiotics are live microorganisms, which, when administered in adequate amounts, confer a health benefit on the host (FAO/WHO 2001). Probiotic *Lactobacilli* have been associated with the prevention and treatment of gastrointestinal disorders, such as rotavirus diarrhea, antibiotic-associated diarrhea, and traveler's diarrhea, and have also been suggested as potential therapeutic agents against irritable bowel syndrome and inflammatory bowel disease (Lomax and Calder 2009; Parkes et al. 2009; Ruemmele et al. 2009).

1.3.4 Fermented Meats

Fermentation followed by drying of meats has been used since ancient times as a method to extend the shelf life of meat products. Nowadays, a wide variety of fermented sausages are produced, depending on the raw materials used, the starter cultures, and the processing conditions. Fermented sausages are classified on the basis of moisture content and range from dry (moisture content 25%–45%) to semidry (moisture content 40%–50%). The ingredients of dry and semidry sausages usually include the following (Adams and Moss 2008):

- Lean meat: 55%–70%
- Fat: 25%–40%
- Curing salts: 3%
- Fermentable carbohydrate: 0.4–2%
- Spices and flavorings: 0.5%
- Starter, ascorbic acid, etc.: 0.5%

Typically, the manufacturing process includes grinding and mixing the meat with fat at cold temperatures. Then, spices, flavorings, curing salts, carbohydrates, microbial starter cultures, and sodium ascorbate are added; the mixture is homogenized under vacuum and then stuffed into casings, which can be either natural (e.g., collagen-based) or synthetic (Lee 1996). The sausages are then placed in air-conditioned and temperature-controlled rooms. The temperature and time of fermentation vary depending on the type of sausage being made. High fermentation temperatures (35°C–40°C) are typical for sausages from the United States, intermediate temperatures (25°C–30°C) are for those from Northern European countries, and milder temperatures (18°C–24°C) are for those from Mediterranean countries. Fermentation times range between 20 and 60 h. During the fermentation, acids are produced as a result of microbial metabolism, and the pH drops to below 5.2. This promotes the coagulation of the meat proteins and the development of the desired texture and flavor. Fermentation is followed by a ripening/drying stage, during which the water activity decreases and the flavor is developed. This stage can take from 7 to 90 days, depending on the product's desired characteristics.

The fermentation characteristics depend, to a large extent, on the type of microorganisms that are present. Nowadays, in order to standardize the process, ensure the overall quality of the product, and improve its safety, most manufacturers use starter cultures. These are single or mixed cultures of well-characterized strains with defined attributes that are beneficial for sausage manufacturing. The main components of these starters are LAB, including *Lactobacillus*, *Lactococcus*, and

Pediococcus strains, more specifically *L. sakei, L. curvatus, L. plantarum, L. pentosus, L. casei, Pediococcus pentosaceus,* and *Pediococcus acidilactici* (Ammor and Mayo 2007), and nitrate-reducing bacteria, mainly *Micrococcus* species. Certain yeasts are also used, such as *Debaryomyces hansenii* and *Candida famata,* as well as molds, such as *Geotrichum candidum* and *Penicillium* species, in particular, *Penicillium camemberti* and *Penicillium nalgiovense* (Hammes and Hertel 1998; Adams and Moss 2008). During fermentation, the added carbohydrate is converted into lactic acid and, to a lesser extent, to acetic acid and acetoin.

1.3.5 FERMENTED VEGETABLES

Vegetables such as cabbage, cucumbers, and olives are used as substrates for the production of fermented products, such as sauerkraut, pickles, and fermented olives, respectively. Detailed information on the production of such products can be found elsewhere (Adams and Moss 2008). Typically, the process involves submerging the vegetables in brine (2%–6%) under anaerobic conditions and at temperatures between 18°C and 30°C, depending on the product (Lee 1996). This environment inhibits the growth of spoilage bacteria and promotes the growth of lactic acid bacteria. Starter cultures are available, but the above method of controlling the environmental conditions is preferred by manufacturers. During the fermentation process, the microbial population, which is naturally present in the raw material, changes, and as a result, a succession of species is observed. In the case of sauerkraut, in the beginning of the fermentation, LAB is less than 1% of the total microbial population, but at the end of it, they account for approximately 90% of the total population (Adams and Moss 2008). The fermentation is initiated by *Leuconostoc mesenteroides,* a heterofermentative LAB producing lactic acid, carbon dioxide, acetic acid, and ethanol. As the fermentation progresses, *Leuconostoc mesenteroides* is replaced by heterofermentative *Lactobacilli* and eventually by *Lactobacillus plantarum,* a homofermentative *Lactobacillus*; this particular species can survive high lactic acid concentrations. Overall, the fermentation takes between 4 and 8 weeks, and the final product contains 1.7%–2.3% lactic acid with a pH between 3.4 and 3.6.

1.4 METABOLIC PATHWAY ENGINEERING OF LAB

Over the last 15 years, in terms of LAB, 25 genomes have been sequenced and annotated (15 *Lactobacillus,* three *Lactococcus,* three *Spectrococcus,* two *Leuconostoc,* one *Pediococcus,* and one *Oenococcus*), and there are several ongoing sequencing projects (Mills et al. 2010; Pfeiler and Klaenhammer 2007). This extensive knowledge of the genetics and the physiology of LAB, as well as the large commercial interest in enhancing the production of specific metabolic products that have bioactive and functional activities either as food ingredients or within a fermented food, have increased the interest for applying metabolic engineering approaches in LAB. For a review of the bioinformatic tools used for genome data mining of LAB and their utilization in the exploration of their metabolic and regulatory networks, refer to Siezen et al. (2004). Obviously, the utilization of genetically modified microorganisms (GMOs) for food production is not widely applied and accepted; for example, in contrast to the United States, the current EU legislation limits the introduction of foods derived from GMOs. The safety issues associated with the use of GMOs in foods and, more specifically, with those of LAB are very important and are addressed in depth in a review by Sybesma et al. (2006).

Most work has been carried out with *Lactococcus lactis*, which has been the model species for LAB; however, *Lactobacilli* are used more and more for these studies as these species are used in a variety of fermented food products (cheese, yogurt, meat, etc.). The main gene expression system that has been successfully used for metabolic engineering of LAB is the NICE (nisin-controlled gene expression) system, which is based on the production of nisin, an antimicrobial compound. This system has been used to express other genes using expression plasmids that contain the controllable nisin promoter in a suitable host (Hugenholtz 2008). The most successful examples of metabolic engineering in LAB are associated with carbon metabolism and lead to increased amounts of ethanol or diacetyl and alanine production instead of lactic acid. Other examples include the production of vitamins (e.g., riboflavin, folic acid, B12), exopolysaccharide (EPS), low-calorie sweeteners, and flavor compounds. The rest of this chapter will present, in more detail, three examples of metabolic engineering, related to the production of diacetyl, low-calorie sweeteners, and EPS. Further reviews on the topic, covering, in detail, the production of flavor compounds and vitamins have been produced by Hugenholtz (2008), Kleerebezem and Hugenholtz (2003), and Teusink and Smid (2006).

1.4.1 Diacetyl

As mentioned previously, diacetyl is a compound that has a buttery aroma and is very important for fermented dairy foods. For this reason, it has been the main target of metabolic engineering for LAB thus far. A review on the topic of metabolic engineering of LAB has been produced by Kleerebezem et al. (2000). Normally, LAB produce small amounts of diacetyl; in *Lactococcus lactis*, which is the LAB species where diacetyl production has been mainly studied, diacetyl is produced from pyruvate through the synthesis of an unstable intermediate, α-acetolactate (Figure 1.2). To increase diacetyl production, various metabolic engineering strategies have been employed. One approach is to inactivate lactate dehydrogenase (LDH) (Figure 1.5); however, this is not very efficient in terms of diacetyl production as it mainly results in the production of acetic acid and acetoin (Platteeuw et al. 1995). Another approach is to control the intracellular redox balance, that is, the NADH/NAD$^+$ ratio. In homofermentative LAB, NADH is oxidized through the reduction of pyruvate to lactate, leading to the regeneration of NAD$^+$ and, thus, to the continuation of the glycolysis (Figure 1.5). This approach involves the oxidation of NADH by a NADH-oxidase, which can be overexpressed in *Lactococcus lactis* using the NICE expression system (López de Felipe et al. 1998). Combining the second strategy with a mutant LAB strain lacking the enzyme α-acetolactate decarboxylase (Figure 1.5) resulted in the conversion of more than 50% of the sugar to diacetyl (Hugenholtz 2008). Another metabolic engineering approach, which was recently used in *Lactobacillus casei*, involved mutations at the LDH gene and the pdhC gene of the pyruvate dehydrogenase and resulted in the production of significant amounts of diacetyl/acetoin (~1.4 g/l) (Nadal et al. 2009).

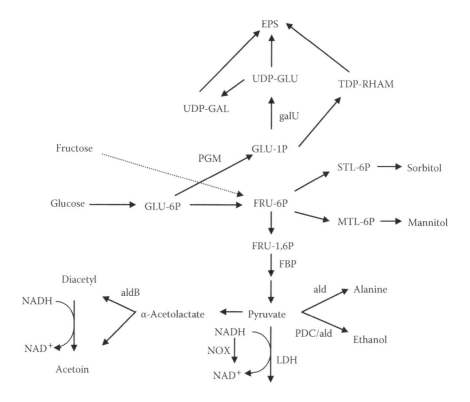

FIGURE 1.5 Overview of metabolic pathways in homofermentative LAB leading to the synthesis of diacetyl, EPS, and low-calorie sugars. (Adapted from Hugenholtz, J. and Kleerebezem, M., *Current Opinion* in Biotechnology, 10:492–497, 1999; Ladero, V. et al., *Applied and Environmental Microbiology*, 73:1864–1872, 2007). Abbreviations: ald, L-alanine dehydrogenase; aldb, α-acetolactate decarboxylase; EPS, exopolysaccharide; FBP, fructose bisphosphatase; FRU, fructose; galU, UDP-glucose pyrophosphorylase; GLU, glucose; LDH, lactate dehydrogenase; NAD+/NADH, nicotinamide adenine dinucleotide; NOX, NADH oxidase; PGM, phosphoglucomutase; MTL, annitol; PDC, pyruvate decarboxylase; Pi, phosphate; RHAM, rhamnose; STL, sorbitol; TDP, thymidine diphosphate; UDP, uridine diphosphate.

1.4.2 EPS

Exopolysaccharides (EPS) are long-chain, high-molecular-mass polymers produced mainly from LAB. The common EPS-producing bacterial species used in food applications include *Streptococci* (Qin et al. 2011; Sawen et al. 2010), *Lactobacilli* (Rodriguez-Carvajal et al. 2008; Wang et al. 2010), *Lactococci* (Costa et al. 2010), and *Bifidobacteria* (Prasanna et al. 2012). EPS produced by various species of bacteria have been shown to have higher viscosifying, thickening, stabilizing, gelling, and emulsifying activities over some other commercially used polymers, such as guar gum, locust bean gum, and gum arabic (Kanmani et al. 2011; Wang et al. 2008). Furthermore, religious and vegetarian lifestyle choices restrict some consumers from eating foods (yogurt, ice cream, and whipped desserts) containing animal-based hydrocolloids, for example, gelatin. In addition, there is an increase in demand for smooth and creamy yogurt products, commonly achieved by increasing the content of fat, sugars, proteins, or stabilizers (pectin, starch, or gelatin). However, consumer demand for products with low fat or sugar content and low levels of additives, as well as cost factors, make EPS a viable alternative (Jolly et al. 2002b). Also, as it has no taste, EPS could be used to develop food products without hindering their original distinctive flavors (Duboc and Mollet 2001). Therefore, EPS produced by starter cultures in fermented dairy products in situ are potential sources of food hydrocolloids, which could be used to overcome the shortcomings related to plant and animal hydrocolloids.

In LAB, EPS are synthesized through the intracellular assembly of sugar nucleotide building blocks, such as UDP-glucose, UDP-galactose, and TDP-rhamnose, from the sugar nucleotide precursor glucose-1-phosphate, which is formed from the glycolytic intermediate glucose-6-phosphate. These reactions are catalyzed by glycosyltransferases and result in the synthesis of heteropolysaccharides, consisting of more than one type of monosaccharide (Gänzle 2009). Considering the EPS biosynthesis reactions, the controlling points for metabolic engineering are the conversion of glucose-6-phosphate to glucose-1-phosphate, catalyzed by the enzyme phosphoglucomutase (PGM); the synthesis of the sugar nucleotides, for example, UDP-glucose from glucose-1-phosphate; and the EPS gene cluster, which expresses the glucosyltransferases involved in EPS synthesis. Experiments with *Lactococcus lactis* have shown that overexpression of the PGM gene or the galU gene results in the increased accumulation of UDP-glucose and UDP-galactose (Kleerebezem et al. 1999) and, after further optimization of the system, in increased EPS production (Looijesteijn et al. 1999). Positive results following a similar approach have also been obtained for *Lactobacillus casei* (Rodriguez-Diaz and Yebra 2011; Sanfelix-Haywood et al. 2011). A considerable amount of research has also been carried out to understand the function of the EPS gene cluster and the role of specific genes toward EPS in the case of *Lactococci* (Dabour and LaPointe 2005; Groot and Kleerebezem 2007), *Streptococci* (Tyvaert et al. 2006; De Vuyst et al. 2011), and *Lactobacilli* (Lebeer et al. 2009; Jolly et al. 2002a; Lamothe et al. 2002). However, further experimental work is needed to evaluate the possibility of over-expressing the EPS gene, aiming to increase EPS production in industrial food strains and, if successful, combine this approach with one that aims to increase the intracellular levels of the sugar nucleotide building blocks.

1.4.3 Low-Calorie Sweeteners

Another example of metabolic pathway engineering for LAB is the conversion of sugar substrate (e.g., glucose, lactose, sucrose, etc.) into sugar alcohols (polyols), such as mannitol and sorbitol. These are low-calorie sugars that are not metabolized by the human host and have similar taste and sweetness to glucose and sucrose (Ladero et al. 2007). With obesity being a growing problem in the Western world, the food industry is interested in developing products that contain low-calorie sweeteners. To this end, the conversion, through LAB fermentation, of high-calorie sugars to low-calorie sugars is very promising and, as a result, has attracted a lot of attention. Mannitol and sorbitol are often produced by LAB that are deficient in LDH activity; this has been shown for *Lactococci* (Gaspar et al. 2004, 2011) and *Lactobacilli*

(Ladero et al. 2007; De Boeck et al. 2010). In the absence of LDH, intermediate hexoses from the glycolysis pathway play the role of electron acceptor instead of pyruvate, resulting in the synthesis of mannitol or sorbitol with concomitant production of NAD^+. More specifically, these sugar alcohols are formed from the reduction of fructose-1-phosphate through the action of mannitol-1P dehydrogenase and sorbitol-6P dehydrogenase, respectively (Figure 1.5). The metabolic engineering approach employed by researchers is to over-express these dehydrogenases. Following this kind of approach, high mannitol production was achieved in *Lactococcus lactis* with the conversion yield of glucose to mannitol ranging between 20% and 50% (Wisselink et al. 2004, 2005; Gaspar et al. 2011). Moreover, high sorbitol production in *Lactobacillus plantarum* was obtained with a conversion yield of glucose to mannitol of approximately 40% (Ladero et al. 2007). The theoretical maximum yield based on the stoichiometric equations is 67% in both cases. Further work in this field should focus on the application of engineered strains in real food systems, targeting the development of low-calorie fermented foods.

1.5 SUMMARY

Metabolic engineering is a very promising approach for the enhanced synthesis of food ingredients from microbial cell factories as well as the production of fermented foods with specific characteristics, for example, organoleptic, textural, and functional characteristics. In this respect, lactic acid bacteria (LAB) have received considerable interest as they are used extensively for the production of fermented foods from various raw materials, including milk, meat, cereals, and vegetables. The knowledge of the complete genome sequencing of various LAB strains and the advances in functional genomics and genetic engineering have been key in the application of metabolic engineering for LAB. Several successful works have been published in which the metabolism was directed to the synthesis of specific compounds, such as diacetyl, exopolysaccharides, and low-calorie sweeteners. However, further work is necessary, ideally with industrial strains, to overcome several bottlenecks, focusing mainly on increasing the process yields.

REFERENCES

Adams, M.R. and Moss, M.O. (2008) *Food Microbiology*, Cambridge: The Royal Society of Chemistry.

Ammor, M.S. and Mayo, B. (2007) Selection criteria for lactic acid bacteria to be used as functional starter cultures in dry sausage production: An update. *Meat Science*, 76:138–146.

Axelsson, L. (1998) Lactic acid bacteria: Classification and physiology. In S. Salminen, A. von Wright (ed.) *Lactic Acid Bacteria: Microbiology and Functional Aspects*, New York: Marcel Dekker, pp. 1–72.

Boylston, T.D. (2006) Dairy products. In Y.H. Hui (ed.) *Food Biochemistry and Food Processing*, Oxford: Blackwell Publishing, pp. 595–614.

Buckenhuskes, H.J. (1993) Selection criteria for lactic acid bacteria to be used as starter cultures for various food commodities. *FEMS Microbiology Reviews*, 12:253–272.

Condon, S. (1987) Responses of lactic acid bacteria to oxygen. *FEMS Microbiology Reviews*, 46:269–280.

Costa, N.E., Hannon, J.A., Guinee, T.P., Auty, M.A.E., McSweeney, P.L.H. and Beresford, T.P. (2010) Effect of exopolysaccharide produced by isogenic strains of *Lactococcus lactis* on half-fat Cheddar cheese. *Journal of Dairy Science*, 93:3469–3486.

Dabour, N. and LaPointe, G. (2005) Identification and molecular characterization of the chromosomal exopolysaccharide biosynthesis gene cluster from *Lactococcus lactis* subsp. *cremoris* SMQ-461. *Applied and Environmental Microbiology*, 71:7414–7425.

Daeschel, M.A., Anderson, R.E. and Fleming, H.P. (1987) Microbial ecology of fermenting plant materials. *FEMS Microbiology Reviews*, 46:357–367.

De Boeck, R., Adriana Sarmiento-Rubiano, L., Nadal, I., Monedero, V., Perez-Martinez, G. and Yebra, M.J. (2010) Sorbitol production from lactose by engineered *Lactobacillus casei* deficient in sorbitol transport system and mannitol-1-phosphate dehydrogenase. *Applied Microbiology and Biotechnology*, 85:1915–1922.

De Vuyst, L., Weckx, S., Ravyts, F., Herman, L. and Leroy, F. (2011) New insights into the exopolysaccharide production of *Streptococcus thermophiles*. *International Dairy Journal*, 21:586–591.

Driessen, F.M., Kingma, F. and Stadhouders, J. (1982) Evidence that *Lactobacillus bulgaricus* in yogurt is stimulated by carbon dioxide produced by *Streptococcus thermophiles*. *Netherlands Milk and Dairy Journal*, 36:135–144.

Duboc, P. and Mollet, B. (2001) Applications of exopolysaccharides in the dairy industry. *International Dairy Journal*, 11:759–768.

FAO/WHO (2001) Joint report of expert consultation: Health and nutritional properties of probiotics in food including powder milk with live lactic acid bacteria.

Fleming, H.P., McFeeters, R.F. and Daeschel, M.A. (1985) The *Lactobacilli*, *Pediococci* and *Leuconostocs*: Vegetable products. In S.E. Gilliland (ed.) *Bacterial Starter Cultures for Food*, Boca Raton: CRC Press, pp. 97–118.

Fox, P.F., McSweeney, P., Cogan, T.M. and Guinee, T.P. (2000) *Fundamentals of Cheese Science*, Gaithersburg: Aspen Publishers.

Gänzle, M.G. (2009) From gene to function: Metabolic traits of starter cultures for improved quality of cereal foods. *International Journal of Food Microbiology*, 134:29–36.

Gaspar, P., Neves, A.R., Gasson, M.J., Shearman, C.A. and Santos, H. (2011) High yields of 2,3-butanediol and mannitol in *Lactococcus lactis* through engineering of NAD(+) cofactor recycling. *Applied and Environmental Microbiology*, 77:6826–6835.

Gaspar, P., Neves, A.R., Ramos, A., Gasson, M.J., Shearman, C.A. and Santos, H. (2004) Engineering *Lactococcus lactis* for production of mannitol: High yields from food-grade strains deficient in lactate dehydrogenase and the mannitol transport system. *Applied and Environmental Microbiology*, 70:1466–1474.

Goh, K.K.T., Haisman D.R. and Singh H. (2005) Development of an improved procedure for isolation and purification of exopolysaccharides produced by *Lactobacillus delbrueckii* subsp. *bulgaricus* NCFB 2483. *Applied Microbiology and Biotechnology*, 67:202–208.

Groot, M.N. and Kleerebezem, M. (2007) Mutational analysis of the *Lactococcus lactis* NIZO B40 exopolysaccharide (EPS) gene cluster: EPS biosynthesis correlates with unphosphorylated EpsB. *Journal of Applied Microbiology*, 103:2645–2656.

Hammes, W.P., Bantleon, A. and Min, S. (1990) Lactic acid bacteria in meat fermentation. *FEMS Microbiology Reviews*, 87:165–173.

Hammes, W.P. and Hertel, C. (1998) New developments in meat starter cultures. *Meat Science*, 49:S125–S138.

Hammes, W.P. and Vogel, R.F. (1995) The genus *Lactobacillus*. In B.J.B. Wood, W.H. Holzapfel (ed.) *The Genera of Lactic Acid Bacteria*, London: Chapman and Hall, pp. 19–54.

Hungenholtz, J. (1993) Citrate metabolism in lactic acid bacteria. *FEMS Microbiology Reviews*, 12:165–178.

Hugenholtz, J. (2008) The lactic acid bacterium as a cell factory for food ingredient production. *International Dairy Journal*, 5:466–475.

Hugenholtz, J. and Kleerebezem, M. (1999) Metabolic engineering of lactic acid bacteria: Overview of the approaches and results of pathway rerouting involved in food fermentations. *Current Opinion in Biotechnology*, 10:492–497.

Jolly, L., Newell, J., Porcelli, I., Vincent, S.J.F. and Stingele, F. (2002a) *Lactobacillus helveticus* glycosyltransferases: From genes to carbohydrate synthesis. *Glycobiology*, 12:319–327.

Jolly, L., Vincent, S.J.F., Duboc, P. and Neeser, J.R. (2002b) Exploiting exopolysaccharides from lactic acid bacteria. *Antonie van Leeuwenhoek*, 82:367–374.

Joshi, N.S., Godbole, S.H. and Kanekar, P. (1989) Microbial and biochemical changes during Dhokla fermentation with special reference to flavour compounds. *Journal of Food Science and Technology*, 26:113–115.

Kandler, O. (1983) Carbohydrate metabolism in lactic acid bacteria. *Antonie van Leeuwenhoek*, 49:209–224.

Kanmani, P., Satish Kumar, R., Yuvaraj, N., Paari, K.A., Pattukumar, V., and Arul, V. (2011) Production and purification of a novel exopolysaccharide from lactic acid bacterium *Streptococcus phocae* PI80 and its functional characteristics activity *in vitro*. *Bioresource Technology*, 102:4827–4833.

Kleerebezem, M., Hols, P. and Hugenholtz, J. (2000) Lactic acid bacteria as a cell factory: rerouting of carbon metabolism in Lactococcus lactis by metabolic engineering. *Enzyme and Microbial Technology*, 26:840–848.

Kleerebezem, M. and Hugenholtz, J. (2003) Metabolic pathway engineering in lactic acid bacteria. *Current Opinion in Biotechnology*, 14:232–237.

Kleerebezem, M., van Kranenburg, R., Tuinier, R., Boels, I.C., Zoon, P., Looijesteijn, E., Hugenholtz, J. and de Vos, W.M. (1999) Exopolysaccharides produced by *Lactococcus lactis*: From genetic engineering to improved rheological properties. *Antonie Van Leeuwenhoek*, 76:357–365.

Ladero, V., Ramos, A., Wiersma, A., Goffin, P., Schanck, A., Kleerebezem, M., Hugenholtz, J., Smid, E.J. and Hols, P. (2007) High-level production of the low-calorie sugar sorbitol by *Lactobacillus plantarum* through metabolic engineering. *Applied and Environmental Microbiology*, 73:1864–1872.

Lamothe, G.T., Jolly, L., Mollet, B. and Stingele, F. (2002) Genetic and biochemical characterization of exopolysaccharide biosynthesis by *Lactobacillus delbrueckii* subsp. *Bulgaricus*. *Archives of Microbiology*, 178:218–228.

Lebeer, S., Verhoeven, T.L.A., Francius, G., Schoofs, G., Lambrichts, I., Dufrene, Y., Vanderleyden, J. and De Keersmaecker, S.C.J. (2009) Identification of a gene cluster for the biosynthesis of a long, galactose-rich exopolysaccharide in *Lactobacillus rhamnosus* GG and functional analysis of the priming glycosyltransferase. *Applied and Environmental Microbiology*, 75:3554–3563.

Lee, B.H. (1996) *Fundamentals of Food Biotechnology*, Cambridge: VCH Publishers.

Leroy, F. and De Vuyst, L. (2004) Lactic acid bacteria as functional starter cultures for the food fermentation industry. *Trends in Food Science and Technology*, 15:67–78.

Leuchtenberger, W., Huthmacher, K. and Drauz, K. (2005) Biotechnological production of amino acids and derivatives: Current status and prospects. *Applied Microbiology and Biotechnology*, 69:1–8.

Lomax, A.R. and Calder, P.C. (2009) Prebiotics, immune function, infection and inflammation: A review of the evidence. *British Journal of Nutrition*, 101:633–658.

Looijesteijn, P.J., Boels, I.C., Kleerebezem, M. and Hugenholtz, J. (1999) Regulation of exopolysaccharide production by *Lactococcus lactis* subsp. *cremoris* by the sugar source. *Applied and Environmental Microbiology*, 65:5003–5008.

López de Felipe, F.L., Kleerebezem, M., de Vos, W.M. and Hugenholtz, J. (1998) Cofactor engineering: A novel approach to metabolic engineering in *Lactococcus lactis* by controlled expression of NADH oxidase. *Journal of Bacteriology*, 180:3804–3808.

Marshall, V.M. (1987) Lactic acid bacteria: Starters for flavor. *FEMS Microbiology Reviews*, 46:327–336.

Mikelsaar, M., Mandar, R. and Sepp, E. (1998) Lactic acid microflora in the human microbial ecosystem and its development. In S. Salminen, A. von Wright (ed.) *Lactic Acid Bacteria: Microbiology and Functional Aspects*, New York: Marcel Dekker, pp. 279–342.

Mills, S., O'Sullivan, O., Hill, C., Fitzgerald, G. and Ross, R.P. (2010) The changing face of dairy starter culture research: From genomics to economics. *International Journal of Dairy Technology*, 63:149–170.

Nadal, I., Rico, J., Perez-Martinez, G., Yebra, M.J. and Monedero, V. (2009) Diacetyl and acetoin production from whey permeate using engineered *Lactobacillus casei*. *Journal of Industrial Microbiology and Biotechnology*, 36:1233–1237.

Oberman, H. and Libudzisz, Z. (1998) Fermented milks. In B.J.B. Wood (ed.) *Microbiology of Fermented Foods*, London: Blackie Academic and Professional, pp. 308–349.

Palles, T., Beresford, T., Condon, S. and Cogan, T.M. (1998) Citrate metabolism in *Lactobacillus casei* and *Lactobacillus plantarum*. *Journal of Applied Bacteriology*, 85:147–154.

Parkes, G.C., Sanderson, J.D. and Whelan, K. (2009) The mechanisms and efficacy of probiotics in the prevention of *Clostridium difficile*-associated diarrhea. *Lancet Infectious Diseases*, 9:237–244.

Pfeiler, E.A. and Klaenhammer, T.R. (2007) The genomics of lactic acid bacteria. *Trends in Microbiology*, 15:546–553.

Platteeuw, C., Hugenholtz, J., Starrenburg, M., Vanalenboerrigter, I. and Devos, W.M. (1995) Metabolic engineering of *Lactococcus lactis*: Influence of the overproduction of alpha-acetolactate synthase in strains deficient in lactate-dehydrogenase as a function of culture conditions. *Applied and Environmental Microbiology*, 61:3967–3971.

Prasanna, P.H.P., Grandison, A.S. and Charalampopoulos, D. (2012) Screening human intestinal *Bifidobacterium* strains for growth, acidification, EPS production and viscosity potential in low fat milk. *International Dairy Journal*, 23:36–44.

Qin, Q.Q., Xia, B.S., Xiong, Y., Zhang, S.X., Luo, Y.B. and Hao, Y.L. (2011) Structural characterization of the exopolysaccharide produced by *Streptococcus thermophilus* 05-34 and its in situ application in yogurt. *Journal of Food Science*, 76:C1226–C1230.

Rodrigues, D., Rocha-Santos, T.A.P., Gomes, A.M., Goodfellow, B.J. and Freitas, A.C. (2012) Lipolysis in probiotic and synbiotic cheese: The influence of probiotic bacteria, prebiotic compounds and ripening time on free fatty acid profiles. *Food Chemistry*, 131:1414–1421.

Rodríguez-Carvajal, M.A., Ignacio Sánchez, J., Campelo, A.B., Martínez, B., Rodríguez, A. and Gil-Serrano, A.M. (2008) Structure of the high-molecular weight exopolysaccharide isolated from *Lactobacillus pentosus* LPS26. *Carbohydrate Research*, 343:3066–3070.

Rodriguez-Diaz, J. and Yebra, M.J. (2011) Enhanced UDP-glucose and UDP-galactose by homologous overexpression of UDP-glucose pyrophosphorylase in *Lactobacillus casei*. *Journal of Biotechnology*, 154:212–215.

Ruemmele, F.M., Bier, D., Marteau, P., Rechkemmer, G., Bourdet-Sicard, R., Walker, W.A. and Goulet, O. (2009) Clinical evidence for immunomodulatory effects of probiotic bacteria. *Journal of Pediatric Gastroenterology and Nutrition*, 48:126–141.

Salovaara, H. (1998) Lactic acid bacteria in cereal-based products. In S. Salminen, A. von Wright (ed.) *Lactic Acid Bacteria: Microbiology and Functional Aspects*, New York: Marcel Dekker, pp. 115–137.

Sanfelix-Haywood, N., Coll-Marques, J.M. and Yebra, M.J. (2011) Role of alpha-phosphoglucomutase and phosphoglucose isomerase activities at the branching point between sugar catabolism and anabolism in *Lactobacillus casei*. *Journal of Applied Microbiology*, 111:433–442.

Sawen, E., Huttunen, E., Zhang, X., Yang, Z. and Widmalm, G. (2010) Structural analysis of the exopolysaccharide produced by *Streptococcus thermophilus* ST1 solely by NMR spectroscopy. *Journal of Biomolecular NMR*, 47:125–134.

Settanni, L. and Moschetti G. (2010) Non-starter lactic acid bacteria used to improve cheese quality and provide health benefits. *Food Microbiology*, 27:691–697.

Siezen, R.J., van Enckevort, F.H., Kleerebezem, M. and Teusink, B. (2004) Genome data mining of lactic acid bacteria: The impact of bioinformatics. *Current Opinion in Biotechnology*, 15:105–115.

Smit, G., Smit, B.A. and Engels, W.J.M. (2005) Flavour formation by lactic acid bacteria and biochemical flavour profiling of cheese products. *FEMS Microbiology Reviews*, 29:591–610.

Sybesma, W., Hugenholtz, J., de Vos, W.M. and Smid, E.J. (2006) Safe use of genetically modified lactic acid bacteria in food: Bridging the gap between consumers, green groups, and industry. *Electronic Journal of Biotechnology*, 9:424–448.

Tamime, A.Y. (2005) *Fermented Milks*, Oxford: Blackwell.

Teusink, B. and Smid, E.J. (2006) Modelling strategies for the industrial exploitation of lactic acid bacteria. *Nature Reviews in Microbiology*, 4:46–56.

Thoreux, K. and Schmucker, D.L. (2001) Kefir milk enhances intestinal immunity in young but not old rats. *Journal of Nutrition*, 131:807–812.

Tseng, C.P. and Montville, T.J. (1993) Metabolic regulation and end product distribution in *Lactobacilli*: Causes and consequences. *Biotechnology Progress*, 9:113–121.

Tyvaert, G., Morel, C., Joly, J.P., Decaris, B. and Charron-Bourgoin, F. (2006) The EPS locus of *Streptococcus thermophilus* IP6756 is not involved in exopolysaccharide production. *International Dairy Journal*, 16:467–473.

Uteng, M., Hauge, H.H., Brondz, I., Nissen-Meyer, J. and Fimland, G. (2002) Rapid two-step procedure for large-scale purification of pediocin-like bacteriocins and other cationic antimicrobial peptides from complex culture medium. *Applied and Environmental Microbiology*, 68:952–956.

Vinderola, C.G., Duarte, J., Thangavel, D., Perdigon, G., Farnworth, E. and Matar, C. (2005) Immunomodulating capacity of kefir. *Journal of Dairy Research*, 72:195–202.

Wang, Y., Ahmed, Z., Feng, W., Li, C. and Song, S. (2008) Physicochemical properties of exopolysaccharide produced by *Lactobacillus kefiranofaciens* ZW3 isolated from Tibet kefir. *International Journal of Biological Macromolecules*, 43:283–288.

Wang, Y., Li, C., Liu, P., Ahmed, Z., Xiao, P. and Bai, X. (2010) Physical characterization of exopolysaccharide produced by *Lactobacillus plantarum* KF5 isolated from Tibet kefir. *Carbohydrate Polymers*, 82:895–903.

Wisselink, H.W., Mars, A.E., van der Meer, P., Eggink, G. and Hugenholtz, J. (2004) Metabolic engineering of mannitol production in *Lactococcus lactis*: Influence of overexpression of mannitol 1-phosphate dehydrogenase in different genetic backgrounds. *Applied and Environmental Microbiology*, 70:4286–4292.

Wisselink, H.W., Moers, A., Mars, A.E., Hoefnagel, M.H.N., de Vos, W.M. and Hugenholtz, J. (2005) Overproduction of heterologous mannitol 1-phosphatase: A key factor for engineering mannitol production by *Lactococcus lactis Applied and Environmental Microbiology*, 71:1507–1514.

2 Isolation, Improvement, and Preservation of Microbial Cultures

Syed G. Dastager

CONTENTS

2.1 INTRODUCTION

During past decades, the industrial application of biotechnologies was characterized by rapid development. Microbiologists, biotechnologists, ecologists, engineers,
and specialists in different areas have experimented and developed new methods
for efficient usage of the products produced from microorganisms present in the
given environment (Labeda and Shearer 1990; Swings 1992; Sharafi et al. 2010).
The availability of microorganisms with highly specific activity, that is, strains pos-
sessing useful characteristics, is the basic element for the development of applied
and fundamental biotechnology products (Adachi et al. 2003). Observing different
fermentation processes in nature, man has chosen different microorganisms for producing bakery products, cheese, wine, and beer (Beishir 1991). Modern microbiology operates actively with strains of actinomycetes, bacteria, yeasts, fungi, viruses,
and other microorganisms for which morphological, taxonomical, and biochemical
characteristics are well studied and documented. A large number of these microorganisms are already in use in the industry. Genetic engineers have performed many
experiments for the selection of productive strains and construction of microorganisms with new characteristics. The process of selection of a single industrial strain
is a result of the efforts of tens of researchers or, sometimes, several generations of
scientists (Sambrook et al. 1989). The development of the modern biotechnology in
a given country can be estimated through the quality and quantity of the applied
strains in the economy, industrial development, ecological culture, and human
health. As a result of this purposeful research, a number of immunological, diagnostic, prophylactic, curative, hormonal, and stimulating bioproducts were prepared
and applied in the fields of human and veterinary medicine (Hosoya et al. 1998).
Mankind successfully fights a lot of infectious diseases by means of bacterial and
viral vaccines nowadays (Hu and Ochi 2001). Microbiological science is also important for agriculture because of its application in forage and silage preparation. With
the help of microorganisms, soil fertility can be improved and the processes of biosynthesis and plant bioconservation can be regulated and directed. Microorganisms
are widely applied in the food-processing industry for preparation of new foodstuffs;
for improvement and preservation of food flavor, taste, and digestibility; and for protein and carbohydrate synthesis (Vinci and Byng 1999). The process of preservation
of a variety of foods is performed by microorganisms. Microorganisms are also used
in the chemical and pharmacological industry for preparation of different organic
compounds, such as organic acids, enzymes, vitamins, antibiotics, alcohols, vinegar,

etc. (Donev 001). The biotechnology has found its place in the oil- and ore-extracting industry for obtaining oil and metals from rich deposits. The waste-free technologies and the biological-type purification plants using microbiological methods for environmental protection are of significance for the industry (Claus 1989). It is important to emphasize that the microorganisms were an essential part of the natural cycles, and their preservation is as important for any other biological process. The main work for any microbiologist consists of isolation, selection, investigation, improvement, and genetic modification of different natural strains for different purposes. The conservation of the microbial vitality, specificity, activity, and immunogenicity, that is, the preservation of the scientific achievements in the area of the applied microbiology, is of prime significance. The conservation and preservation of microbiological samples is an important task. At present, it is related to the processes of standardization and preparation of products of high quality, but it also involves foresight into the future development of the biotechnology (Boeswinkel 1976; Vinci and Byng 1999).

2.2 ISOLATION OF MICROORGANISMS

The first stage in the screening for microorganisms of potential industrial application is their isolation. Isolation involves obtaining either pure or mixed cultures followed by their assessment to determine the desired reaction or process for the desired product. The isolate must eventually carry out the process economically. Therefore, the selection of the culture to be used is a compromise between the productivity of the organism and the economic constraints of the process criteria as being important in the choice of organism (Anon. 1979, 1984):

- The nutritional characteristics of the organism: using a very cheap medium or a predetermined one
- The optimum temperature of the organism
- The reaction of the organism with the equipment to be employed and the suitability of the organism to the type of process to be used
- The stability of the organism and its amenability to genetic manipulation
- The productivity of the organism, measured in its ability to convert substrate into product and to give a high yield of product per unit time
- The ease of product recovery from the culture

2.2.1 ISOLATION METHODS

Microorganisms are thought ubiquitous in their occurrence; common sources for their isolation are soils, lakes, and river muds. The technique of isolating microorganisms also varies according to the nature and physiological properties of the microbe to be isolated. For example, isolation of fungi from a mixture of fungi and bacteria can be achieved easily by incorporating antibiotics to which the bacteria are sensitive in the growth medium. Similarly, from a mixture of sporulating (spore forming) and nonsporulating bacteria, spore-forming bacteria can be isolated by heating the mixture to 70°C–80°C for 5–10 min, leading to the death of all vegetative forms so that the remaining spores will belong only to the sporulating bacteria (Dang 1971; Gerhardt et al. 1981). Small samples of microbial cultures can also be obtained from permanent cultures, which are maintained by the government as well as private agencies in most countries (Grigomva and Nonis 1990). A number of methods are available for the isolation of bacteria, molds, yeasts, and algae. The following are a few important methods.

2.2.1.1 Surface Plating

Surface plating is also known as the streak culture method. The specimen to be cultured is taken in a platinum loop. One loopful of the specimen is transferred onto the well-dried surface of a convenient medium. It is then spread over a small area at the periphery. The inoculum is then distributed thinly over the plate. This is done by streaking with the loop in a series of parallel lines in different segments of the plate. The loop should be flamed and cooled between different sets of streaks. After incubating the plate, it can be seen that growth is confluent at the original site of inoculation, but it becomes progressively thinner. In the final series of streaks, well-separated individual colonies of bacteria can be obtained.

2.2.1.2 Enrichment and Selective Media

Bacteria can be isolated by growing it in enriched or selective media. Enriched media is a preparation to which additional growth factors have been added. These may be added individually or in complex mixtures. Enriched medium therefore may be chemically defined or chemically undefined, simple or complex. An example of enriched medium employing the latter would be blood agar, which is made from complex ingredients to which blood is added separately.

2.2.1.3 Aerobic and Anaerobic Conditions

Oxygen is a universal component of cells and is always provided in large amounts by H_2O. However, prokaryotes display a wide range of responses to molecular oxygen, O_2 (Table 2.1). Obligate aerobes require O_2 for growth; they use O_2 as a final electron

TABLE 2.1
Terms Used to Describe O_2 Relations of Microorganisms

	Environment		
Group	**Aerobic**	**Anaerobic**	**O_2 Effect**
Obligate aerobe	Growth	No growth	Required (utilized for aerobic respiration)
Microaerophilic	Growth if level is not too high	No growth	Required but at levels below 0.2 atm
Obligate anaerobe	No growth	Growth toxic	
Facultative anaerobe (facultative aerobe)	Growth	Growth	Not required for growth but utilized when available
Aero-tolerant anaerobe	Growth	Growth	Not required and not utilized

acceptor in aerobic respiration. Obligate anaerobes (occasionally called aerophobes) do not need or use O_2 as a nutrient. In fact, to them, O_2 is a toxic substance, which either kills or inhibits their growth. Obligate anaerobic prokaryotes may live by fermentation, anaerobic respiration, photosynthetic bacterium, or the novel process of methanogenesis (Hu et al. 2002; Frigaard and Bryant 2004;). Facultative anaerobes (or facultative aerobes) are organisms that can switch between aerobic and anaerobic types of metabolism. Under anaerobic conditions (no O_2), they grow by fermentation or anaerobic respiration, but in the presence of O_2, they switch to aerobic respiration. Aero-tolerant anaerobes are bacteria with an exclusively anaerobic (fermentative) type of metabolism, but they are insensitive to the presence of O_2. They live by fermentation alone whether or not O_2 is present in their environment (Table 2.1).

2.2.1.4   Effect of pH on Growth

The pH, or hydrogen ion concentration (H^+), of natural environments varies from about 0.5 in the most acidic soils to about 10.5 in the most alkaline lakes. Appreciating that pH is measured on a logarithmic scale, the H^+ of natural environments varies by more than a billion-fold, and some microorganisms are living at the extremes as well as every point between the extremes. Most free-living prokaryotes can grow over a range of 3 pH units, about a thousand-fold changes in H^+. The range of pH over which an organism grows is defined by three cardinal points: the minimum pH, below which the organism cannot grow; the maximum pH, above which the organism cannot grow; and the optimum pH, at which the organism grows best. For most of the microorganisms, there is an orderly increase in growth rate between the minimum and the optimum pH and a corresponding orderly decrease in growth rate between the optimum and the maximum pH, reflecting the general effect of changing H^+ on the rates of enzymatic reaction (Figure 2.1).

Microorganisms that grow at an optimum pH well below neutrality (7) are called acidophiles. Those that grow best at neutral pH are called neutrophiles, and those that grow best under alkaline conditions are called alkaliphiles. Obligate acidophiles, such as some *Thiobacillus* species, actually require a low pH for growth because their

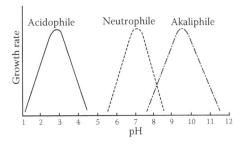

FIGURE 2.1   Growth rate versus pH for three environmental classes of prokaryotes. Most free-living bacteria grow over a pH range of about three units. Note symmetry of curves below and above optimum pH for growth. (From Membré, J.M. et al., *Appl. Environ. Microbiol.*, 65(11):4921–4925, 1999. With permission.)

TABLE 2.2

Minimum, Maximum, and Optimum pH for Growth of Certain Prokaryotes

Organism	Minimum pH	Optimum pH	Maximum pH
Thiobacillus thiooxidans	0.5	2.0–2.8	4.0–6.0
Sulfolobus acidocaldarius	1.0	2.0–3.0	5.0
Bacillus acidocaldarius	2.0	4.0	6.0
Zymomonas lindneri	3.5	5.5–6.0	7.5
Lactobacillus acidophilus	4.0–4.6	5.8–6.6	6.8
Staphylococcus aureus	4.2	7.0–7.5	9.3
Escherichia coli	4.4	6.0–7.0	9.0
Clostridium sporogenes	5.0–5.8	6.0–7.6	8.5–9.0
Erwinia carotovora	5.6	7.1	9.3
Pseudomonas aeruginosa	5.6	6.6–7.0	8.0
Thiobacillus novellus	5.7	7.0	9.0
Streptococcus pneumoniae	6.5	7.8	8.3
Nitrobacter sp.	6.6	7.6–8.6	10.0

Source: Todar, K., Todar's Online Textbook of Bacteriology. www.textbookofbacteriology.net, 2008–2012. With permission.

membranes dissolve and the cells lyse at neutrality. Several genera of archaea, including *Sulfolobus* and *Thermoplasma*, are obligate acidophiles. Among eukaryotes, many fungi are acidophiles, but the champion of growth at low pH is the eukaryotic alga *Cyanidium*, which can grow at a pH of 0. In the construction and use of culture media, one must always consider the optimum pH for growth of a desired organism and incorporate buffers in order to maintain the pH of the medium in the changing milieu of bacterial waste products that accumulate during growth. Many pathogenic bacteria exhibit a relatively narrow range of pH over which they will grow. Most diagnostic media for the growth and identification of human pathogens have a pH near 7 (Table 2.2).

2.2.1.5 Isolation by Difference in Temperature

Microorganisms have been found growing in virtually all environments where there is liquid water, regardless of its temperature. Prokaryotes have been detected growing around black smokers and hydrothermal vents in the deep sea at temperatures at least as high as 120°C. Microorganisms have been found growing at very low temperatures as well. A particular microorganism will exhibit a range of temperature over which it can grow, defined by three cardinal points in the same manner as pH. Considering the total span of temperature where liquid water exists, the prokaryotes may be subdivided into several subclasses on the basis of one or another of their cardinal points for growth. For example, organisms with an optimum temperature near 37°C (the body temperature of warm-blooded animals) are called mesophiles. Organisms with an optimum temperature between approximately 45°C and 70°C are thermophiles. Some archaea with an optimum temperature of 80°C or higher and a maximum temperature as high as 115°C are now referred to as extreme thermophiles or hyperthermophiles. The cold-loving organisms are psychrophiles, defined by their ability to grow at 0°C. A variant of a psychrophile (which usually has an optimum temperature of 10–15°C) is a psychrotroph, which grows at 0°C but displays an optimum temperature in the mesophile range, nearer to room temperature (Tables 2.3 and 2.4). Psychrotrophs are the scourge of food storage in refrigerators because they are invariably brought in from their mesophilic habitats and continue to grow in the refrigerated environment where they spoil the food. Of course, they grow

TABLE 2.3

Terms Used to Describe Microorganisms in Relation to Temperature Requirements for Growth

Group	Minimum °C	Optimum °C	Maximum °C	Comments
Psychrophile	Below 0	10–15	Below 20	Grow best at relatively low temperatures
Psychrotroph	0	15–30	Above 25	Able to grow at low temperatures but prefer moderate temperatures
Mesophile	10–15	30–40	Below 45	Most bacteria, especially those living in association with warm-blooded animals
Thermophile	45	50–85	Above 100 (boiling)	Among all thermophiles there is a wide variation in optimum and maximum temperatures

TABLE 2.4

Minimum, Maximum, and Optimum Temperatures for Growth of Certain Bacteria and Archaea

Bacterium	Minimum (°C)	Optimum (°C)	Maximum (°C)
Listeria monocytogenes	1	30–37	45
Vibrio marinus	4	15	30
Pseudomonas maltophilia	4	35	41
Thiobacillus novellus	5	25–30	42
Staphylococcus aureus	10	30–37	45
Escherichia coli	10	37	45
Clostridium kluyveri	19	35	37
Streptococcus pyogenes	20	37	40
Streptococcus pneumoniae	25	37	42
Bacillus flavothermus	30	60	72
Thermus aquaticus	40	70–72	79
Methanococcus jannaschii	60	85	90
Sulfolobus acidocaldarius	70	75–85	90
Pyrobacterium brockii	80	102–105	115

Source: Todar, K., Todar's Online Textbook of Bacteriology. www.textbookofbacteriology.net, 2008–2012. With permission.

more slowly at 2°C than at 25°C. Think about how quickly milk spoils on the countertop versus in the refrigerator. Psychrophilic bacteria have adapted to their cool environment by having largely unsaturated fatty acids in their plasma membranes. Some psychrophiles, particularly those from the Antarctic have been found to contain polyunsaturated fatty acids, which generally do not occur in prokaryotes. The degree of unsaturation of a fatty acid correlates with its solidification temperature or thermal transition stage (i.e., the temperature at which the lipid melts or solidifies). Unsaturated fatty acids remain liquid at low temperatures but are also denatured at moderate temperatures; saturated fatty acids, as in the membranes of thermophilic bacteria, are stable at high temperatures, but they also solidify at relatively high temperatures. Thus, saturated fatty acids (like butter) are solid at room temperature while unsaturated fatty acids (like safflower oil) remain liquid in the refrigerator. Whether fatty acids in a membrane are in a liquid or a solid phase affects the fluidity of the membrane, which directly affects its ability to function. Psychrophiles also have enzymes that continue to function, although at a reduced rate, at temperatures at or near 0°C. Usually, the psychrophile's proteins and/or membranes, which adapt them to low temperatures, do not function at the body temperatures of warmblooded animals (37°C), so they are unable to grow at even moderate temperatures (Figure 2.2). Thermophiles are adapted to temperatures above 60°C in a variety of ways (Todar 2008–2012). Often thermophiles have a high G + C content in their DNA such that the melting point of the DNA (the temperature at which the strands of the double helix separate) is at least as high as the organism's maximum temperature for growth. But this is not always the case, and the correlation is far from perfect, so thermophiles' DNA must be stabilized in these cells by other means. The membrane fatty acids of thermophilic bacteria are highly saturated, allowing their membranes to remain stable and functional at high temperatures. The membranes of hyperthermophiles isolated from superheated environments, virtually all of which are archaea, are not composed of fatty acids but of repeating subunits of the C5 compound phytane, a branched, saturated, isoprenoid substance, which contributes heavily to the ability of these bacteria to adapt to the environment. The structural proteins (e.g., ribosomal proteins and transport proteins or permeases) and enzymes

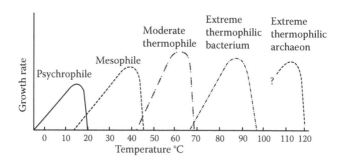

FIGURE 2.2 Growth rate versus temperature for five environmental classes of prokaryotes (From Todar, K., Todar's Online Textbook of Bacteriology. www.textbookofbacteriology.net, 2008–2012. With permission.)

of thermophiles and hyperthermophiles are very heat stable compared with their mesophilic counterparts. The proteins are modified in a number of ways, including dehydration and through slight changes in their primary structure, which accounts for their thermal stability (Hewlett 1918).

Thermophilic bacteria grow at 60°C. Some bacteria, such as *N. magnitudes*, grow at 22°C. By incubation at different temperatures, bacteria can be selectively isolated.

2.2.1.6   Separation of Vegetative and Spore-Forming Bacteria
Vegetative bacteria are killed at 80°C. But spore-forming bacteria, such as tetanus bacilli, survive at this temperature. So by heating to 80°C, vegetative bacteria can be eliminated, and spore-forming bacteria can be isolated.

2.2.1.7   Separation of Motile and Nonmotile Bacteria
This can be achieved by using a Craig tube or a U tube. In the U tube, the organisms are introduced in one limb and the motile organism can be isolated at the other limb (Ashabil and Burhan 2007).

2.2.1.8   Animal Inoculation
Pathogenic bacteria can be isolated by inoculation into appropriate animals; for example, *Anthrax bacilli* can be isolated by inoculation into mice or guinea pigs.

2.2.1.9   Filtration
Bacteria of different sizes may be separated by using selective filters. For example, the ISO-GRID (made by the Neogen Corporation, Lansing, MI) is a dual-filtration method developed for food products (Taitt et al. 2004). FTA filters (from Whatman, Springfield, KY) have been developed for rapid isolation of nucleic acids from environmental, clinical, or food samples (Lampel et al. 2004).

2.2.1.10   Micromanipulation
By means of micromanipulation, single bacterium can be separated and cultured. Fastidious soil bacteria are mixed microbial communities. Isolation establishes microcolonies from these membranes by using fluorescence viability staining and micromanipulation. This approach facilitates the recovery of diverse, novel isolates, including the recalcitrant bacterium and some host-plant pathogens, which have never been isolated outside the host (Ferrari and Gillings 2009).

2.3   STRAIN IMPROVEMENT

Fermentation has been used for preserving food for hundreds of years, and virtually every culture has, as part of its diet, a variety of fermented milk, meat, vegetable, fruit, or cereal products. Microorganisms, including bacteria, yeasts, and mold, produce a wide range of metabolic end products that function as preservatives, texturizers, stabilizers, and flavoring and coloring agents. Several traditional and nontraditional methods have been used to improve metabolic properties of food fermentation microorganisms (Hesketh and Ochi 1997; Hosoya et al. 1998). These include mutation and selection techniques; the use of natural gene transfer methods,

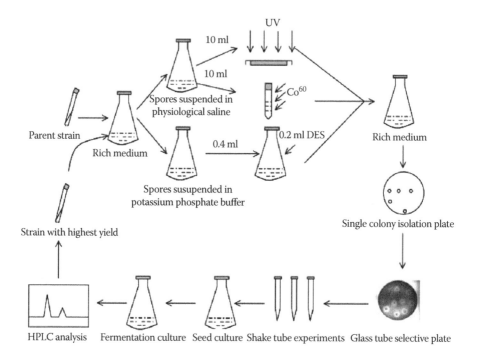

FIGURE 2.3 Stepwise strain improvement process. (From Parekh, S. et al., *Appl. Microbiol. Biotechnol.*, 54(3):287–301, 2000. With permission.)

such as transduction, conjugation, and transformation; and, more recently, genetic engineering. These techniques will be briefly reviewed with emphasis on the advantages and disadvantages of each method for genetic improvement of microorganisms used in food fermentation (Figure 2.3).

2.3.1 TRADITIONAL GENETIC IMPROVEMENT STRATEGIES

2.3.1.1 Mutation and Selection

In nature, mutations (changes in the chromosome of an organism) occur spontaneously at very low rates (one mutational event in every 10^6 to 10^7 cells per generation. These mutations occur at random throughout the chromosome, and a spontaneous mutation in a metabolic pathway of interest for food fermentation would be an extremely rare event. The mutation rate can be dramatically increased by exposing the microorganisms to mutagenic agents, such as ultraviolet light or various chemicals, which induce changes in the deoxyribonucleic acid (DNA) of host cells. Mutation rates can be increased to one mutational event in every 10^1 or 10^2 cells per generation for auxotrophic mutants (the inability of an organism to synthesize a particular organic compound required for its growth) and to one in 10^3 to 10^5 for the isolation of improved secondary metabolite producers. A method of selection is critical for effective screening of mutants as several thousand individual isolates may need to be evaluated to find one strain with improved activity in the property of interest (Hu and Ochi 2001). Mutation and selection techniques have been used to improve the metabolic properties of microbial starter cultures used for food fermentation; however, there are severe limitations with this method. Mutagenic agents cause random mutations; thus specificity and precision are not possible. Potentially deleterious undetected mutations can occur because selection systems may be geared for only the mutation of interest. Additionally, traditional mutation procedures are extremely costly and time-consuming, and there is no opportunity to expand the gene pool (Lee and Rho 1999). In spite of these limitations, mutation and selection techniques have been used extensively to improve industrially important microorganisms, and in some cases, yields of greater than 100 times the normal production level of bacterial secondary metabolites have been achieved (Jin and Gross 1988).

2.3.1.2 Natural Gene Transfer Methods

The discovery of natural gene transfer systems in bacteria has greatly facilitated the understanding of the genetics of microbial starter cultures and, in some cases, has been used for strain improvement. Genetic exchange in bacteria can occur naturally by three different mechanisms: transduction, conjugation, and transformation (Hu et al. 2002).

2.3.1.3 Transduction

Transduction involves genetic exchange mediated by a bacterial virus (bacteriophage). The bacteriophage acquires a portion of the chromosome or plasmid from the host strains and transfers it to a recipient during subsequent viral infection. Although transduction has been exploited for the development of a highly efficient gene transfer system in the gram-negative organism *Escherichia coli*, it has not been used extensively for improving microorganisms used in food fermentation. In general, transduction efficiencies are low, and gene transfer is not always possible between unrelated strains, limiting the usefulness of the technique for strain improvement. In addition, bacteriophages have not been isolated and are not well characterized for most strains (Brüssow et al. 2004).

2.3.1.4 Conjugation

Conjugation, or bacterial mating, is a natural gene transfer system that requires close physical contact between donors and recipients and is responsible for the dissemination of plasmids in nature. Numerous genera of bacteria harbor plasmid DNA. In most cases, these plasmids are cryptic (the functions encoded are not known), but in some cases, important metabolic traits are encoded by plasmid DNA. If these plasmids are also self-transmissible or mobilizable, they can be transferred to recipient strains. Once introduced into a new strain, the properties encoded by the plasmid can be expressed in the recipient. The lactic acid bacteria naturally contain between one and more than 10 distinct plasmids, and metabolically important traits, including lactose-fermenting ability, bacteriophage resistance, and bacteriocin production, have been linked to plasmid DNA. Conjugation has been used to transfer these plasmids into recipient strains for the construction of genetically improved commercial dairy starter cultures.

There are some limitations in the application of conjugation for strain improvement. To exploit the use of conjugative improvement requires an understanding of plasmid biology and, in many cases, few conjugative plasmids encoding genes of interest have been identified or sufficiently characterized. Conjugation efficiencies vary widely, and not all strains are able to serve as recipients for conjugation. Moreover, there is no opportunity to expand the gene pool beyond those plasmids already present in the species (Griffiths et al. 2000).

2.3.1.5   Transformation

Certain microorganisms are able to take up naked DNA present in the surrounding medium. This process is called transformation, and this gene transfer process is limited to strains that are naturally competent. Competence-dependent transformation is limited to a few, primarily pathogenic, genera and has not been used extensively for genetic improvement of microbial starter cultures. For many species of bacteria, the thick peptidoglycan layer present in Gram-positive cell walls is considered a potential barrier to DNA uptake. Methods have been developed for enzymatic removal of the cell wall to create protoplasts. In the presence of polyethylene glycol, DNA uptake by protoplasts is facilitated. If maintained under osmotically stabilized conditions, transformed protoplasts regenerate cell walls and express the transformed DNA. Protoplast transformation procedures have been developed for some of the lactic acid bacteria; however, the procedures are tedious and time-consuming, and frequently, parameters must be optimized for each strain. Transformation efficiencies are often low and highly variable, limiting the application of the technique for strain improvement (Lal et al. 1996).

2.3.1.6   Electroporation

The previously mentioned gene transfer systems have become less popular since the advent of electroporation, a technique involving the application of high-voltage electric pulses of short duration to induce the formation of transient pores in cell walls and membranes. Under appropriate conditions, DNA present in the surrounding medium may enter through the pores. Electroporation is the method of choice for strains that are recalcitrant to other gene transfer techniques although optimization of several parameters (e.g., cell preparation conditions, voltage and duration of the pulse, regeneration conditions, etc.) is still required.

2.3.2   Genetic Engineering

Genetic engineering provides an alternative method for improving microbial starter cultures. This rapidly expanding area of technology provides methods for the isolation and transfer of single genes in a precise, controllable, and expedient manner. Genes that code for specific desirable traits can be derived from virtually any living organism (plant, animal, microbe, or virus). Genetic engineering is revolutionizing the science of strain improvement and is destined to have a major impact on the food fermentation industry (Zhang et al. 2002).

Although much of the microbial genetic engineering research since the advent of recombinant DNA technology in the early 1970s has focused on the gram-negative

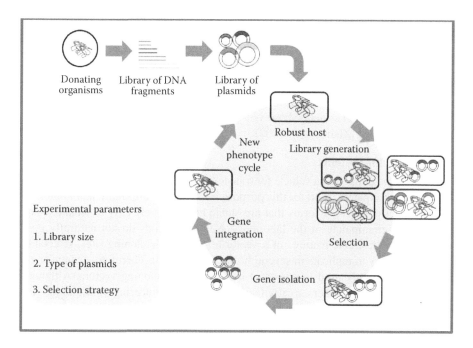

FIGURE 2.4 Gene transfer systems.

bacterium *Escherichia coli*, significant progress has been made with lactic acid bacteria and yeast. Appropriate hosts have been identified; multifunctional cloning vectors have been constructed; and reliable, high-efficiency gene transfer procedures have been developed. Further, the structural and functional properties, as well as expression in the host strains, of several important genes have been reported. Engineered bacteria, yeasts, and molds could also be used for the production of other products, including food additives and ingredients; processing aids, such as enzymes; and pharmaceuticals (Figure 2.4).

2.3.2.1 Metabolism and Biochemistry of the Host

A necessary prerequisite for the application of genetic engineering to any microorganism is a fundamental understanding of the metabolism and biochemistry of the strain of interest. Although for hundreds of years the metabolic potential of microbial starter cultures has been exploited, in many cases, little is known about specific metabolic pathways, the regulation of metabolism, or the structural and functional relationships of critical genes involved in metabolism. This information is essential for the design of genetic improvement strategies as it provides the rationale for selection of desirable gene(s) and assures that once inserted into a new host, the gene(s) will be appropriately expressed and regulated as predicted (Vinci and Byng 1999).

2.3.2.2 Transformable Hosts

Plasmid-free, genetically characterized, and highly transformable hosts, coupled with multifunctional expression vectors, provide the necessary tools for transfer, maintenance, and optimal expression of cloned DNA in microbial starter cultures. Many microbial starter cultures harbor plasmid DNA, and although most plasmids remain cryptic, resident plasmids interfere with identification of plasmid-containing transformants. Use of plasmid-free hosts also eliminates plasmid incompatibility problems and the possibilities of co-integrate formation between transforming and endogenous plasmids. It is important to note that plasmid-free strains are used for the development of model systems; however, ultimately it will be necessary to engineer commercial strains (Zhang et al. 2002).

2.3.2.3 Vector Systems

A vector can be defined as a vehicle for transferring DNA from one strain to another. Plasmids are frequently used for this purpose because they are small, autonomously replicating, circular DNA forms that are stable and relatively easy to isolate, characterize, and manipulate in the laboratory. Native plasmids do not naturally possess all of the desirable features of a vector (e.g., multiple cloning sites, selectable marker(s), ability to replicate in several hosts, and so forth). Therefore, genetic engineering is frequently used to construct multifunctional cloning vectors. Although antibiotic resistance markers greatly facilitate genetic engineering in microbial systems, vectors derived solely from food-grade organisms may be critical in obtaining regulatory approval for use of the organisms as antibiotic resistance determinants may not be acceptable in food systems. An alternative vector strategy involves the development of linear fragments of DNA that are capable of integrating into the host chromosome via homologous recombination. Although transformation frequencies are very low, the advantage of the integrative vector is that transformed genetic information is targeted to the chromosome where it will be more stably maintained. Insertion sequences (IS elements) naturally present in the chromosome that can transpose chromosomal DNA to plasmids could be used as an alternative strategy for developing integrative vectors for some strains of lactic acid bacteria (Lee and Stanton 2008).

2.3.2.4 Efficient Gene Transfer Systems

Once gene(s) have been identified and cloned into the appropriate vector in the test tube, they must be introduced into a viable host. Because the recombinant DNA is a naked DNA molecule, gene transfer systems based on protoplast transformation and electroporation are most applicable in genetic engineering experiments. High transformation efficiencies (greater than 104 to 105 transformants per kilogram of DNA) greatly facilitate screening and identification of appropriate transformants. Electroporation is the transformation procedure of choice for most microbial strains (Enríquez et al. 2006; Oyane et al. 2010).

2.3.2.5 Expression Systems

Transfer of structural genes to a new host using genetic engineering does not guarantee that the genes will be expressed. To optimize expression of cloned genes, efficient promoters, ribosome-binding sites, and terminators must be isolated, characterized, and cloned along with the gene(s) of interest. Identification of signal sequences essential for secretion of proteins outside the cell may be useful for situations where microbial starter cultures are used to produce high-value food ingredients and processing aids. Secretion into the medium greatly facilitates purification of such substances (Penn 1991).

2.3.3 PROPERTIES OF INTEREST

Several properties could be enhanced using genetic engineering. For example, bacteriocins are natural proteins produced by certain bacteria that inhibit the growth of other, often closely related, bacteria. In some cases, these antimicrobial agents are antagonistic to pathogens and spoilage organisms commonly found as contaminants in fermented foods. Transfer of bacteriocin production to microbial starter cultures could improve the safety of fermented products. Acid production is one of the primary functions of *Lactobacilli* during fermentation. Increasing the number of copies of the genes that code for the enzymes involved in acid production might increase the rate of acid production, ensuring that the starter will dominate the fermentation and rapidly destroy less-acidic competitors.

Certain enzymes are critical for proper development of flavor and texture of fermented foods. For example, *Lactococcal* proteases slowly released within the curd are responsible for the tart flavor and crumbly texture of aged Cheddar cheese. Cloning of additional copies of specific proteases involved in ripening could greatly accelerate the process. An engineered *Saccharomyces cerevisiae* (baker's yeast), which is more efficient in leavening of bread, has been approved for use in the United Kingdom and is the first strain to attain regulatory approval. This strain produces elevated levels of two enzymes, maltose permease and maltase, involved in starch degradation (Donev 2001).

2.3.4 LIMITATIONS

There are a number of issues that must be resolved before genetically engineered starter cultures could be used in food. Engineered strains will need to be approved for use by appropriate regulatory agencies. To date, no engineered organisms have been approved in the United States, and specific criteria for approval have not been established by the Food and Drug Administration. The public must be assured that the products of biotechnology are safe for consumption. If consumers have the perception that the products are not safe, the technology will not be utilized. Although genetic engineering is probably safer and more precise than strain-improvement methods used in the past, most US consumers are not aware of the role of bacteria in fermented foods and do not have a fundamental understanding of recombinant DNA technology, and they may be unwilling to accept the technology. This may be less of a problem in developing countries where improved microbial starter cultures could provide significantly safer and more nutritious foods with longer shelf life and higher quality. Another limitation is that genetic improvement of microbial starter cultures requires sophisticated equipment and expensive biological materials that may not be available in developing countries. Where equipment and materials are available in industrialized countries, there may be little incentive for researchers to improve strains that would probably not be used in their own countries. Genetic improvement of microbial starter cultures is most appropriate for those fermentations that rely solely or primarily on one microorganism. In many cases, our knowledge about the fermentation is limited, making selection of the target strain very difficult. Because many food fermentation processes are complex and involve several microorganisms, genetic improvement of just one of the organisms may not improve the overall product (Meier et al. 1994).

2.4   MICROBIAL PRESERVATION

The majority of the biomass and biodiversity of life on the Earth is accounted for by microbes. They play pivotal roles in biogeochemical cycles and harbour novel metabolites that have industrial uses. For these reasons the conservation of microbial ecosystems, communities and even specific taxa should be a high priority. The conservation results could be affected by lots of factors during the preliminary culture preparation, by the choice of protectants, preservation and regeneration methods with minimum consequences for the strains. With the development of microbiology the requirements for the culture preservation increase. It is not enough to perform a successful conservation; Microorganisms are rarely considered by conservation biologists and yet they form the base of most food chains and accomplish biogeochemical transformations of critical importance to the biosphere (Hawksworth, 1997). For example, each year c. 120 million tonnes of nitrogen gas are removed from the atmosphere by microbial nitrogen fixation and made available to the rest of the biosphere (Freiberg et al. 1997). More than 40% of the carbon dioxide drawn down from the atmosphere is accomplished by microorganisms in the marine environment, giving them an important role in climate regulation (Behrenfeld et al. 2006). Microorganisms take part in a range of processes that are essential for nutrient flux in the biosphere, including rock weathering, organic decomposition and reductive and oxidative transformations of a range of essential elements, changing their mobility and biological availability. Marine microorganisms are responsible for producing and degrading dimethyl sulphide (DMS), a major climate-cooling trace gas. The health of marine microbial ecosystems is therefore directly linked to the sulphur cycle and the climate system of the Earth (Gondwe et al. 2003). From a biodiversity perspective, microorganisms dominate life on Earth and it is estimated that, 10% of the Earth's microbial diversity has been characterized (Budhiraja et al. 2002). In the deep subsurface, where multicellular organisms cannot persist, microorganisms are the only biological entities. Here, we consider microorganisms to cover fungi, bacteria, archaea and protists (unicellular eukaryotes) and viruses, although the examples we discuss are necessarily limited to just some of these groups. The impact of human activity and pollution on microbial ecosystems is not well-known because it has not received the attention afforded to animals and plants. Weinbauer and Rassoulzadegan (2007) discussed concerns about microbial extinction and suggested that habitat fragmentation, a pervasive problem for animals and plants, is unlikely to cause microbial extinctions. They highlighted microbial communities that may be at risk from extinction, such as those closely associated with host animals and plants that themselves are at risk. A possible example is the high microbial diversity associated with marine sponges or microorganisms endemic to particular environments such as endemic Sulfolobus species, heat-loving microorganisms that inhabit hot springs (Whitaker et al. 2003). However, as with animals and plants, extinction is not necessarily required to justify conservation efforts. Regional extinctions, or even merely a reduction in microbial community diversity, can be sufficient to have important knock-on effects on other organisms warranting conservation. Despite overwhelming facts about the role of microorganisms in the biosphere and examples of where they might be threatened, formulating a consistent environmental ethic for the conservation of microorganisms is notoriously difficult (Cockell 2008). On what basis should they be protected? it is also necessary to keep the strain for a long–term period. The long-term preservation of bacteria, accompanied by the stable maintenance of their properties, is important for any laboratory. To this end, the lyophilization (freeze-drying) method has been widely used since Flosdorf and Mudd (1938) established basic experimental conditions. However, due to the cost of the necessary instruments and the complexity of the lyophilization procedures, this method cannot always be used in small laboratories. In the microbiological practice, priority is given to the cryogenic and lyophilic methods for conservation and preservation of microorganisms. They are considered to be the best ones for strain conservation. These methods guarantee reproducibility and maintenance of standards—necessary conditions for the high quality, efficiency, and competitive power of the biotechnological productions. Therefore, banks for the conservation and preservation of microorganisms are found in many countries. These institutions are characterized by modern equipment and suitable conditions for the application of a lot of methodologies for microbial conservation (Flosdorf and Mudd 1935, 1938). They act as sorts of scientific insurance centers. The problem for the foundation of microbial banks is of special importance for biotechnological departments and research institutes (Howard 1959). In many cases, the results of long-term research work are lost as a result of unsuitable conditions for preservation and a lack of knowledge about the process of conservation. The establishment of laboratories for microbial conservation is a necessity for every microbiological department. It introduces an element of insurance in their work. The production of standardized industrial products is impossible without the availability of tested-in-advance starter cultures, conserved in suitable packaging and size. The long-term preservation of microbial samples in a condition close to their initial state is the main purpose of the conservation process. Therefore, not only the preservation of the vitality and the number of the microbial populations is required. Undesired mutations, causing significant genotype and phenotype changes, should be avoided, too. The efficient reception of mutants is helped by the short generation time of the microorganisms (Ellis 1979; Feniss 1984). At the same time, the preservation of the achieved results in the selection of useful intracellular (metabolic) properties, taxonomic features, and other valuable characteristics for the practice causes problems. Reversible or irreversible changes in the structure, shape, and size of the cells and the colonies; the pigmentation; the physiological, biochemical characteristics; the productivity, etc., can be observed. They depend on the nature of the microorganisms, the conservation period, and the applied method. Therefore, knowledge and practical skills are very important for the preservation of culture identity. The research of the microbial conservation is directly related to the reversibility of the transition from the active, vital state (biosis) into hypobiosis or anabiosis or vice versa. Most of the conservation research work is empirical and is a result of the necessity for decision of concrete problems for a given strain or a group of micro-organisms. The large variety of microbiological objects and the differences in their resistance to the different methods for conservation and reactivation result in a lack of generally accepted "recipes for efficient preservation." The experimental work is a necessity, but the individual approach can ensure the desired results. At present, a lot of methods and modifications and combinations of methods for microbial conservation are known (Yukie et al. 2000).

Culture collections are considered to be the best ones for conservation of strains. They act as sorts of scientific insurance centers. Subcultivation or periodical reculti-vation on nutrient medium occurs

- Under layers of mineral oils
- In sterile salt solutions with tap water
- With drying (thermal-conductional or convectional, spray-drying, vacuum L-drying)
- With drying on different inert carriers
- With cooling and storage at positive temperatures
- With freezing at different temperatures (–20°C to –196°C)
- With vacuum-sublimation drying (lyophilization)

Experimental data show that, practically, the most efficient methods include transition into the anabiotic state of the cell. The methods causing slowing down of the metabolism (i.e., hypobiosis) have more limited applications. Based on the experimental results from research work all over the world, it is concluded that there are no good or bad methods for microbial conservation. Each one of the previously mentioned methods is important and used in research and applied microbiology and biotechnology. The method chosen should correspond to the specific characteristics of the object (the strain), the duration, and the purposes of its storage and future application.

2.4.1 Conservation of Microorganisms under Laboratory Conditions

One of the basic objectives of fundamental and industrial microbiology is the pres-ervation of microorganisms under laboratory conditions. The conservation of their vitality and number is not the only important task; the microbial identity and phe-notype properties should be preserved, too. Methods for conservation that prevent accidental selection and spontaneous mutations, causing changes in the microbial phenotype and genotype, should be chosen. At present, many methods for conserva-tion and their modifications are known. Described here, are some of the methods for conservation, applied in the microbiological practice.

2.4.1.1 Subcultivation

The method of subcultivation is the oldest and is still performed. The basic require-ment for performance of this technique is to take cells for transfer from a large number of colonies. This decreases the possibility for selection of spontaneous mutants. The composition of the solid agar media and the content of mineral salts in suitable concentrations are very important for the correct application of the method. The choice of suitable nutrient media aims at the maximal preservation of the taxonomical and biochemical properties of the microorganisms for about 10 years. Adding some specific microelements and vitamins is necessary in many cases. Some authors recommend periodical replacement of the nutrient media with other suitable media during the process of cultivation. In many cases, this has a positive effect on the vitality and activity of the culture. Recently, there have been reported applications of synthetic media and successive usage of liquid and solid media according to a given scheme. Cultures are usually immobilized at temperatures of 4°C–8°C in order to prolong the interval for culture transfer. Test tubes with screw or rubber caps, sealed with paraffin, are used to prevent the drying of the agar medium. Many microbial species are conserved by the method of subcultivation in collections all over the world. With a connection to the variety of the microbial cultures, thousands of recipes are worked out in the American-type cultural collection (ATCC). They ensure the conservation of those groups of microorganisms that are processed through the method of subcultivation. Some authors believe that better results can be obtained after a precise choice of media and longer intervals of time between the culture transfers. For example, one of the longest periods for transfer that has been reported is for the microscopic fungi from the genus *Aspergillus*—more than 5 years. Some species of actinomycetes and bacteria require transfer after 8 to 12 months. According to information from the Holland collection of microorganisms, CBS (one of the oldest in the world), a large number of taxonomical groups of microscopic fungi can be conserved through subcultivation on agar media. Using this approach, the oldest strain there has been preserved since 1895. This method is also described as being suitable for conservation of thermophilic actinomycetes in the form of spores on agar media. Besides the reports in favor of this method for conservation, there are many reports concerning its disadvantages, which are supported by experimental results. During the frequent and elaborate manipulations, there is a danger of contamination with side microflora and technical mistakes. Subcultivation can cause the selection of undesirable mutants or a gradual decrease in culture vitality and activity. For example, morphological changes in some yeasts and microscopic fungi are observed over the course of their storage. Investigations on 580 strains of yeasts from 250 genera from the English-type collection (NCTC) revealed significant changes, damages, and even losses for 40 years of preservation using the previously mentioned method. The unsuitable choice of nutrient media resulted in the loss of valuable properties related to the production of antibiotics in some strains of actinomycetes. The replacement of highly active, fast-growing variants with inactive ones of the same strain was observed, too. Biochemical changes and a gradual exhausting of the dextran-producing strain *Leuconostoc mesenterioides*, preserved on sucrose medium through frequent culture transfer, were detected. The highly productive industrial microorganisms and mutants, received after many years of selection, were especially susceptible to reversion changes. This can be explained by the decreased biological potential and the property of microorganisms to preserve their initial stable state, grounded by nature (Reasoner and Geldreich 1985).

2.4.1.2 Under Mineral Oils

Another method for microbial preservation is under mineral oils. It was proposed by Lumier in 1914 for a gonorrhea-causing agent. In 1921, Michel applied the method for some pathogenic microorganisms, such as *Gonococci*, *Meningococci*, and *Pneumococci*, which were difficult to preserve under laboratory conditions. The method consists of the application of sterile mineral oil (1 cm layer) on test tubes with well-developed microbial cultures on liquid or solid media. This limits the access of oxygen and diminishes the culture growth and metabolism, that is, hypobiosis is achieved. The oils used should be highly viscous with a relative weight equal to 0.8–0.9 at 20°C. They should not contain toxic or oxidized products. Moisture should be avoided during their sterilization. Paraffin and Vaseline are recommended for microbial conservation. There are two procedures for sterilization of mineral oils through autoclaving at 1 atm for 30 min followed by drying at 150°C and through heating at 170°C for 1 h. However, the thermal processing of oils results in the formation of side toxic products. The requirements for the media used for the microbial development in this method are the same as those for the method of subcultivation. In the specialized literature, there are different data about the conservation temperature. The view that in the interval between 4°C and 20°C there are no significant changes in the microbial properties is predominant. All the disadvantages of the method of the subcultivation technique are characteristic for this method, too. Its advantage is the prolongation of the intervals between the culture transfers. For bacteria, the period is 1 to 7 years; for microscopic fungi, it is 1 to 5 years; for yeasts, it is up to 7 years. This information is based on the observations of separate representatives from the mentioned groups (da Silva et al. 1994).

2.4.1.3 Water and Water–Salt Solutions

The preservation of microbial cultures in water and water–salt solutions is a method for transition into hypobiosis. The cells in this state are also called resting cells. The first experiments were carried out with some species of phytopathogenic bacteria and microscopic fungi in a physiological solution and distilled water. The samples were stored for one or two years. Bacteria from the genus *Acinetobacter* and the yeast species *Saccharomyces cerevisiae* were reported to have been successfully preserved. This method is applied for almost all the microbiological objects. It is suitable for a short-term conservation at 4°C–8°C for 1 to 4 weeks. This way, cultures are better preserved than at temperatures below zero (–4°C to –8°C). Experiments for conservation of yeasts in a nonfrozen state at –20°C by means of kerosene-paraffin cryo-protective medium were successful, too. This medium is characterized by a low crystallization point (Ellis 1979).

2.4.1.4 Preservation through Drying

The method for conservation through drying has been used since the end of the eighteenth century. It is also based on the ability of the microorganisms to undergo anabiosis. Actually, it reproduces natural conditions. Suspensions are usually dropped on carriers: sand, mud, soil, talcum, active carbon, grains of wheat, shavings, glass beads, gelatinous or synthetical granules or tablets, filter paper, etc. The carriers are characterized by large surfaces and can absorb part of the moisture. The drying process at room temperature or upon heating at 36°C–40°C is performed. Hydroscopic chemical compounds, such as gypsum, silica gel, and phosphorus pentoxide, are used to absorb the vapor. Convectional drying with warm air is applied for samples of bigger volumes. Becker et al. (1981) performed various experiments for drying and conserving yeasts from the *Saccharomyces* genus. These scientists performed the widest industrial application of the method. In 1954, Annear used a vacuum to rate up the drying process of the microbial suspensions. This modification is called L-drying. It is applied for the conservation of a lot of microorganisms under laboratory conditions, for example, representatives of the genera *Leptospira*, *Spirochaeta*, *Salmonella*, and some yeast strains. Despite the percentage of viability variations within broad limits, the method has industrial applications. After approximately 20 years of observations, some researchers found that the method was applicable for the conservation of actinomycetes, bacteria, and yeasts, but it was not accepted as a basic conservation method. The information on microbial conservation through L-drying concerns mainly its survival rate. The data regarding the preservation of the biological activity and the physiological, morphological, and biochemical properties are insufficient (Lange and Boyd 1968).

2.4.1.5 Vacuum Drying

Sealing off under vacuum is another method for conservation of dense suspensions combined with protectors. In 1950, Rodes reported the successful preservation of bacterial strains for 12 years. Combined methods for conservation are also applied, for example, partial dehydration by, or on a carrier, followed by preservation in a frozen or lyophilic state. The most widely applied methods for conservation of microorganisms in an anabiotic state are the ones for freezing and preservation at low and very low temperatures and at the lyophilized state (Nuzum 1989). They are described in the following sections.

2.4.2 CRYOBIOLOGICAL METHODS FOR CONSERVATION AND PRESERVATION OF MICROORGANISMS

The processes of cooling and freezing of microbial cultures are regular ones, but at the same time, they are characterized by many specific features. According to the requirements of the microorganisms, the content of the suspension medium can vary in its composition and concentration of its components. On the other hand, each strain variant has its own cryogenic characteristics: osmolarity, supercooling and crystallization points, eutectic temperature, point of solidification, etc. The large variety of recipes for cryoprotective media, which influence the crystallization process inside and outside the cells, should also be mentioned. After including the kind of packing (plastic tube, straw, glass ampoule or bank, metallic box, a container, other, etc.), its physical parameters (thermal conductivity, shape, contact surface, etc.), the amount of the suspension for freezing (for collection work, 0.1–1 ml; for vaccine production, 0.1–250 ml; for preparation of starter cultures, up to 2 l), the number of influencing factors increases a great deal. The type of the freezing system (by convection or conduction) is also important. Microorganisms are characterized by different resistance

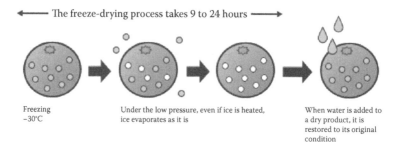

FIGURE 2.5   Freeze-drying process. (From Yukie, M.S. et al., *Cryobiology*, 41(3):251–255, 2000. With permission.)

to the rate of cooling and freezing. The basic purpose of the cryogenic conservation and the vacuum-sublimation drying is the maintenance of the treated microorganisms in an anabiotic state and preservation of their vitality in the utmost degree (Figure 2.5). The achievement of this purpose requires not only theoretical knowledge, but much experimental work, too. There is a lot of information concerning the problems of the freezing of biological objects and the mechanisms of cryo-damage and cryo-protection in the specialized literature. Medical investigations on the cryo-conservation of organs for transplantation, bone marrow cells, blood plasma and cells, bone fragments, skin, embryos, sperm, etc., are predominant. Cryobanks are found in many countries. Most of the animal and plant cell lines are preserved by freezing them in liquid nitrogen (Moore and Carlson 1975). A lot of experiments are carried out in the areas of human medicine and animal breeding. As a result, theoretical and applied cryobiology has achieved considerable success (Maussori 1975; MacKenzie 1975).

As a method for conservation, lyophilization established itself in the 1940s. However, the results from the long-term storage of lyophilized cultures reveal that changes in their initial properties may occur. Thus the low-temperature conservation is determined to be the much more reliable method. That is why, at present, microbial banks store a large percentage of their strains in liquid nitrogen (Moore and Carlson 1975). A limited number of investigations on the optimization of the conditions and parameters of vacuum-sublimation drying and freezing have been performed. One of the reasons is that the Budapest International Treaty from 1997 requires only the presence of viable microorganisms to be proven. Another reason is the large number and the variety of the preserved microorganisms. Standard, frozen or lyophilized, starter cultures of high quality with a high microbial number, high biological activity, and strict specificity are very important for biotechnological production. The correct choice of the conditions for freezing determines the achievement of the desired results in the cryogenic and lyophilic conservation of strains (Nei 1973). The microbial suspensions represent the liquid dispersion medium (i.e., aqueous solution of nutrient medium, buffering salts, or cryoprotectors) and dispersed cells. For a deeper understanding of the processes, which occur in the course of freezing of such a multicomponent system, the phenomenon should be observed at first on the level of chemical solutions (Yukie et al. 2000).

ACKNOWLEDGMENTS

The author thanks Dr. S.R. Shetye, director of the National Institute of Oceanography, for facilities and encouragement, and acknowledges the Council of Scientific and Industrial Research (CSIR), New Delhi, India.

REFERENCES

Adachi, O., Moonmangmee, D., Toyama, H., Yamada, M., Shinagawa, E., and Matsushita, K. (2003) New developments in oxidative fermentation. *Appl. Microbiol. Biotechnol.*, 60: 643–653.

Anon. (1979) The Oxoid Manual, 4th edn. London: Oxoid Ltd. New edition in preparation.

Anon. (1984) Difco Manual, 10th edn. Detroit: Difco Laboratories.

Ashabil, A., and Burhan, A. (2007) An overview on bacterial motility detection. *Int. J. Agri. Biol.*, 1:193–196.

Becker, M.E., Danberg, B.E., and Rapoport, A.I. (1981) *Anabiosis of Microorganisms*. Riga: Zinatne, 253 (in Russian).

Behrenfeld, M.J., O'Malley, R.T., Siegel, D.A., Mcclain, C.R., Sarmiento, J.L., Feldman, G.C. et al. (2006) Climate-driven trends in contemporary ocean productivity. *Nature*, 444, 752–755.

Beishir, L. (1991) *Microbiology in Practice: A Self-Instructional Laboratory Course*, 5th edn. New York: Harper Collins.

Boeswinkel, H.J. (1976) Storage of fungal cultures in water. *Transactions of the British Mycological Society*, 66:183–185.

Brüssow, H., Canchaya, C., and Hardt, W.D. (2004). Phages and the evolution of bacterial pathogens: From genomic rearrangements to lysogenic conversion. *Microbiol. Mol. Biol. Rev.*, 68(3):560–602.

Budhiraja, R., Basu, A. and Jain, R. (2002) Microbial diversity: significance, conservation and application. *National Academy Science Letters* (India), 25, 189–201.

Claus, G.W. (1989) *Understanding Microbes: A Laboratory Textbook for Microbiology*. New York: W.H. Freeman and Company.

Cockell, C.S. (2008) Environmental ethics and size. *Ethics and the Environment*, 13, 23–40.

da Silva, A.M., Borba, C.M., and de Oliveira, P.C. (1994) Viability and morphological alterations of *Paracoccidioides brasiliensis* strains preserved under mineral oil for long periods of time. *Mycoses*, 37(5–6):165–169.

Dang, D.M. (1971) Techniques for microscopic preparation. In *Methods in Microbiology*, edited by C. Booth, 4, 95–111. London: Academic Press.

Donev, T. (2001) *Methods for Conservation of Industrial Microorganisms*. Sofia: National Bank for Industrial Microorganisms and Cell Cultures.

Ellis, J.J. (1979) Preserving fungus strains in sterile water. *Mycologia*, 71:11072–11075.

Enríquez, L.L., Mendes, M.V., Antón, N., Tunca, S., Guerra, S.M., Martín, J.F., and Aparicio, J.F. (2006) An efficient gene transfer system for the pimaricin producer *Streptomyces natalensis*. *FEMS. Microbiol. Lett.*, 257(2):312–318.

Feniss, R.S. (1984) Effects of microwave oven treatment on microorganisms in soil. *Phytopathology*, 74(12):1–126.

Ferrari, B.C., and Gillings, M.R. (2009) Cultivation of fastidious bacteria by viability staining and micromanipulation in a soil substrate membrane system. *Appl. Environ. Microbiol.*, 75(10):3352–3354.

Flosdorf, E.W., and Mudd, S. (1935) Procedure and apparatus for preservation in "lyophile" form of serum and other biological substances. *J. Immunol.*, 29:389–425.

Flosdorf, E.W., and Mudd, S. (1938) An improved procedure and apparatus for preservation of sera, microorganisms and other substances: The Cryochem-Process. *J. Immunol.*, 34:469–490.

Freiberg, C., Fellay, R., Bairoch, A., Broughton, W.J., Rosenthal, A. and Perret, X. (1997) Molecular basis of symbiosis between Rhizobium and legumes. *Nature*, 387, 394–401.

Frigaard, N.U., and Bryant, D. (2004) Seeing green bacteria in a new light: Genomics-enabled studies of the photosynthetic apparatus in green sulfur bacteria and filamentous anoxygenic phototrophic bacteria. *Arch. Micro.*, 182:265–276.

Gerhardt, P. et al. eds. (1981) *Manual of methods for general bacteriology.* Washington DC: American Society for Microbiology. [An authoritative classic, but many of the culture media are for medical use.]

Gondwe, M., Krol, M., Gi Eskes, W., Klaassen, W. and De Baar, H. (2003). The contribution of ocean-leaving DMS to the global atmospheric burdens of DMA, MSA, SO2, and NSS SO42. *Global Biogeochemical Cycles*, 17, 1056.

Griffiths, A.J.F., Miller, J.H., Suzuki, D.T. et al. (2000) *An Introduction to Genetic Analysis*, 7th edn. New York: W.H. Freeman.

Grigomva, R., and Nonis, J.R. eds. (1990) Techniques in microbial ecology. In *Methods in Microbiology.* London: Academic Press. [Includes several chapters on observation and enumeration of bacteria.]

Grivell, A., and Jackson, J. (1969) Microbial culture preservation with silica gel. *J. Gen. Microbiol.*, 58:423–435.

Hawksworth, D.L. (1997) The Biodiversity of Microorganisms and Invertebrates. Oxford University Press, Oxford, UK.

Heckley, R. (1961) Preservation of bacteria by lyophilization. *Adv. Appl. Microbiol.*, 3:1.

Hesketh, A., and Ochi, K. (1997) A novel method for improving *Streptomyces coelicolor* A3(2) for production of actinorhodin by introduction of *rpsL* (encoding ribosomal protein S12) mutations conferring resistance to streptomycin. *J. Antibiot.*, 50:532–535.

Hewlett, R.W. (1918) *Manual of Bacteriology*, 6th edn. 535 pp.

Hosoya, Y., Okamoto, S., Muramatsu, H., and Ochi, K. (1998) Acquisition of certain streptomycin-resistant (*str*) mutations enhances antibiotic production in bacteria. *Antimicrob. Agents Chemother.*, 42:2041–2047.

Howard, D.H. (1959) The preservation of bacteria by freezing in glycerol broth. *J. Bacteriol.*, 71:625.

Hu, H., and Ochi, K. (2001) Novel approach for improving the productivity of antibiotic-producing strains by inducing combined resistant mutations. *Appl. Environ. Microbiol.*, 67:1885–1892.

Hu, H., Zhang, Q., and Ochi, K. (2002) Activation of antibiotic biosynthesis by specified mutations in the *rpoB* gene (encoding the RNA polymerase subunit) of *Streptomyces lividans. J. Bacteriol.*, 184:3984–3991.

Jin, D.J., and Gross, C.A. (1988) Mapping and sequencing of mutations in the *Escherichia coli rpoB* gene that lead to rifampicin resistance. *J. Mol. Biol.*, 202:45–58.

Labeda, P.D., and Shearer, M.C. (1990) *Isolation of Actinomycetes for Biotechnological Applications.* McGraw-Hill Publishing Company.

Lal, R., Khanna, R., Kaur, H., Khanna, M., Dhingra, N., Lal, S., Gartemann, K.H., Eichenlaub, R., and Ghosh, P.K. (1996) Engineering antibiotic producers to overcome the limitations of classical strain improvement programs. *Crit. Rev. Microbiol.*, 22:201–255.

Lampel, K., Dyer, D., Kornegay, L., and Orlandi, P. (2004) Detection of *Bacillus* spores using PCR and FTA filters. *J. Food. Prot.*, 67:1036–1038.

Lange, J., and Boyd, W. (1968) Preservation of fungal spores by drying on porcelain beads. *Phytopathology*, 58:1711–1712.

Lee, S.H., and Rho, Y.T. (1999) Improvement of tylosin fermentation by mutation and medium optimization. *Lett. Appl. Microbiol.*, 28:142–144.

Lee, L.Y., and Stanton, B.G. (2008) T-DNA binary vectors and systems. *Plant. Physiology*, 146(2):325–332.

MacKenzie, P. (1975) Collapse during freeze-drying: Qualitative and quantitative aspects. In *Freeze-Drying and Advanced Food Technology*. London, New York, San Francisco: Academic Press, 277–307.

Maussori, G.A. (1975) Kinetics of H_2O loss from cell at subzero centigrade temperatures. *Cryobiology*, 12:34–46.

Meier, A., Kirschner, P., Bange, F.C., Vogel, U., and Bottger, E.C. (1994) Genetic alterations in streptomycin-resistant *Mycobacterium tuberculosis*: Mapping of mutations conferring resistance. *Antimicrob. Agents Chemother.*, 38:228–233.

Membré, J.M., Kubaczka, M., and Chéné, C. (1999) Combined effects of pH and sugar on growth rate of *Zygosaccharomyces rouxii*, a bakery product spoilage yeast. *Appl. Environ. Microbiol.*, 65(11):4921–4925.

Moore, L.W., and Carlson, R.V. (1975) Liquid nitrogen storage of phytopathogenic bacteria. *Phytopathology*, 65:246–250.

Nei, T. (1973) Growth of ice crystals in frozen specimen. *J. of Microscopy*, 99(2):227–233.

Nuzum, C. (1989) A simple method for the preservation of some non-sporing fungi. *Australasian Plant Pathology*, 18(4):104–105.

Oyane, A., Tsurushima, H., and Ito, A. (2010) Highly efficient gene transfer system using a laminin-DNA-apatite composite layer. *J. Gene. Med.*, 12(2):194–206.

Parekh, S., Vinci, V.A., and Strobel, R.J. (2000) Improvement of microbial strains and fermentation processes. *Appl. Microbiol. Biotechnol.*, 54(3):287–301.

Penn, C. (1991) *Handling Laboratory Microorganisms*. Milton Keynes: Open University Press.

Reasoner, D.J., and Geldreich, E.E. (1985) A new medium for the enumeration and subculture of bacteria from potable water. *Appl. Environ. Microbiol.*, 49(1):1–7.

Sambrook, J., Fritsch, E.F., and Maniatis, T. (1989) *Molecular Cloning, A Laboratory Manual (M)*. 2nd edn. New York: Cold Spring Harbor Laboratory Press, 1.21–1.52, 2.60–2.80, 7.3–7.35, 9.14–9.22.

Sharafi, S.M., Rasooli, I., and Beheshti-Maal, K. (2010) Isolation, characterization and optimization of indigenous acetic-acid bacteria and evaluation of their preservation methods. *Iran. J. Microbiol.*, 2(1):41–48.

Swings, J. (1992) The genera *Acetobacter* and *Gluconobacter*: 15. In *The Prokaryotes: A Handbook on the Biology of Bacteria: Ecophysiology, Isolation, Identification, Applications*, edited by A. Balows, H.G. Truper, M. Dworkin, W. Harder, and K.H. Schleifer. New York: Springer-Verlag, vol. III, pp. 2268–2286.

Taitt, C.R., Shubin, Y.S., Angel, R., and Ligler, F.S. (2004) Detection of Salmonella enterica serovar Typhimurium by using a rapid, array-based immunosensor. *Appl. Environ. Microbiol.*, 70:152–158.

Todar, K. (2008–2012) Todar's Online Textbook of Bacteriology. www.textbookofbacteriology .net.

Vinci, V.A., and Byng, G. (1999) *Manual of Industrial Microbiology and Biotechnology*. Washington, DC: ASM Press.

Weinbauer, M.G. and Rassoulzadegan, F. (2007) Extinction of microbes: evidence and potential consequences. *Endangered Species Research*, 3, 205–215.

Whitaker, R.J., Grogan, D.W. and Taylor, J.W. (2003) Geographic barriers isolate endemic populations of hyperthermophilic archaea. *Science*, 301, 976–978.

Yukie, M.S., Takashi, I., Junji, S., Yukie, M., Sugio, K., and Yasuhiko, K. (2000) Survival rate of microbes after freeze-drying and long-term storage. *Cryobiology*, 41(3):251–255.

Zhang, Y.X., Perry, K., Vinci, V.A., Powell, K., Stemmer, W.P., and del Cardayre, S.B. (2002) Genome shuffling leads to rapid phenotypic improvement in bacteria. *Nature*, 415:644–646.

3 Physical and Chemical Factors Affecting Fermentation in Food Processing

Antonella Amore and Vincenza Faraco

CONTENTS

3.1   INTRODUCTION

Microbial fermentation is diffusely exploited in the food industry, allowing the accomplishment of several different goals, such as food preservation, an increase in food nutritional quality by increasing digestibility, and delivery of functional characteristics to foods.

Food preservation is achieved through the control of growth of food spoilage microorganisms by the end products of fermentation—generally acids, alcohol, and carbon dioxide—and it is mostly dependent on fermentation by lactic acid bacteria (LAB). Besides preservation, enhancement of the flavor, aroma, and texture of foods is provided by the end products of LAB metabolism. The delivery of functional characteristics to a food by fermentation includes both traditional activities, such as the release of probiotic bacteria in products such as fermented milks, and, more recently, the introduced activities of functional components, such as vitamins, antioxidants, and other compounds in a variety of different fermented foods.

Fermentation processes are strongly affected by physical and chemical factors, such as temperature, pH, aeration, and medium composition. In the following sections, the effects of physical and chemical factors on fermentations in the dairy, bakery, and alcoholic beverage industries are described.

3.2   FERMENTATION IN THE DAIRY INDUSTRY

Fermented milks are obtained from milk through fermentative processes performed by specific microorganisms that are able to produce acids, thus favoring pH reduction and coagulation. Dairy products are produced all over the world with enormous variety in their properties and characteristics, strictly connected to the traits of the countries where they are manufactured (Table 3.1). Figure 3.1 represents a "family tree" where fermented milks are reported and divided considering, above all, the nature of the microorganisms (mesophilic or thermophilic) involved in their fermentation (Khurana and Kanawjia 2007).

Global sales of dairy products are increasing quickly, and the manufacture of cultured dairy products represents the second most important fermentation industry after the production of alcoholic drinks. In this contest, the increasing demand for dairy products with "functional" properties is a key factor driving growth of value sales in developed markets. The Foreign Agricultural Service of the US Department of Agriculture has recently analyzed dairy market trends by collecting statistical data on milk, cheese, and other dairy product manufacturing all over the world (USDA 2011). The dairy production yield for each country is strongly affected by milk production, imports, and exports. As reported in *The FAO World Milk Report* (FAO 2012), world milk production in 2012 is forecast to grow by 3 % to 760 million tons, with Asia, particularly India, expected to account for most of this increase.

TABLE 3.1
Broad Classification of Fermented Milks

Name	Country of Origin	Microflora
Acidophilus milk	Australia	*L. acidophilus*
Yogurt (bio-ghurt)	Middle Asia, Balkans	*S. salivarius* ssp. *thermophilus, L. delbreukii* ssp. *bulgaricus, Micrococcus* and other lactic acid, cocci, yeasts, molds
Kefir	Caucasus	*L. lactis* ssp. *lactis, Leuconostoc* spp. *L. delbreukii* ssp. *caucasiucu, Saccharomyces kefir, Torula kefir, Micrococci,* spore-forming bacilli
Kumiss	Asiatic steppes	*L. delbreukii* ssp. *bulgaricus, L. acidophilus, Torula kumiss, Saccharomyces lactis, Micrococci,* spore-forming bacilli
Dahi (dadhi)	India, Persia	*L. lactis* ssp. *lactis, S. salivarius* ssp. *thermophilus, L. delbreukii* ssp. *bulgaricus, plantarum,* lactose-fermenting yeasts, mixed culture (not defined)
Shrikhand (chakka)	India	*S. salivarius* ssp. *thermophilus, L. delbreukii* ssp. *bulgaricus*
Lassi	India	*L. lactis* ssp. *lactis, S. salivarius* ssp. *thermophilus, L. delbreukii* ssp. *bulgaricus*
Cultured buttermilk	Scandinavian and European countries	*L. lactis* ssp. *lactis, L. lactis* ssp. *diacetylactis, Lueconostoc dextranicum* ssp. *citrovorum*
Leben, Labneh	Lebanon, Arab countries	*L. lactis* ssp. *lactis, S. salivarius* ssp. *thermophilus, L. delbreukii* ssp. *bulgaricus, plantarum,* lactose-fermenting yeasts

Source: Adapted from Batish, V.K. et al., *Biotechnology: Food Fermentation, Vol. II,* Kerela, India: Educational Publishers and Distributors Press, 1999. With permission.

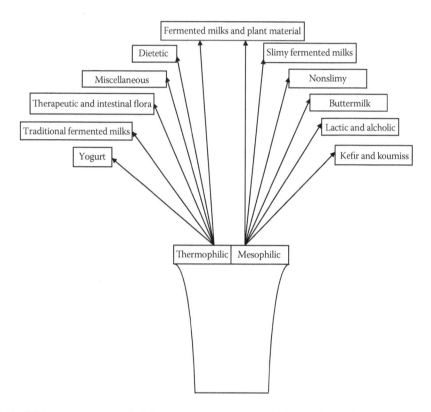

FIGURE 3.1 "Family tree" of fermented milk products. (Adapted from Khurana, H.K. and Kanawjia, S.K., *Current Nutrition and Food Science*, 3:91–108, 2007. With permission.)

EU milk production is forecast to rise by only 1.5% to 157.9 million tons in 2012; milk production in the United States is forecast to rise to 90.3 million tons, an increase of almost 2%; output in Canada is set to remain stable at 8.3 million tones. South American milk production rate of increase is foreseen by over 5%, similarly to 2011, Argentina, Ecuador and Uruguay being the major producers, followed by Brazil and Chile. In Africa, a small increase in milk output is anticipated for 2012, even though the adverse weather conditions negatively affect dairy production (http://www.fao.org/).

Enormous work has been carried out to improve the technological aspects of fermented milks, focusing attention on the physiology of starter cultures and their role in fermentation, the heart of dairy product manufacturing.

Physical and chemical parameters affecting milk fermentation are discussed in Section 3.2.1, and the manufacturing processes of cheese and yogurt are reported as case studies. Starting from a short description of the main stages of their production, the principal factors affecting fermentative processes are described.

3.2.1 MILK FERMENTATION STARTER CULTURES

Organisms responsible for milk fermentation generally can be divided into *Streptococci/Lactobacilli*, *Propionibacteria*, yeasts, and coliform bacteria.

LAB are the main bacteria involved in milk fermentation, and they are responsible for acid production obtained by the anaerobic breakdown of milk carbohydrates, that is, lactose, into lactic acid and other organic acids. *Lactococcus lactis* is an example of LAB involved in milk fermentation, and its metabolic pathways are well known. LAB metabolism varies from strain to strain, but a common mark is the involvement of permease enzymes, acting as carriers for lactose transport in the cells, and β-galactosidases, which are responsible for lactose hydrolysis into glucose and galactose.

Depending on their fermentative metabolism, LAB are classified into two groups: homofermentative or homolactic, which produce more than 85% of lactic acid by fermenting glucose in the Embden–Meyerhof pathway, and heterofermentative, which produce 50% of lactic acid and considerable amounts of ethanol, acetic acid, and carbon dioxide in the pentose phosphate pathway.

Examples of homolactic LAB genera include *Lactococcus*, *Enterococcus*, *Streptococcus*, *Pediococcus*, and some species of *Lactobacillus*. Obligate heterofermentative LAB include *Leuconostoc*, *Oenococcus*, *Weissella*, and some species of *Lactobacillus* (Toro 2005).

The primary function of almost all starter cultures is to develop acid in the products, and the secondary effects of acid production include coagulation, expulsion of moisture, texture formation, and initiation of flavor production. In addition to these, the starters also help in imparting a pleasant acidic taste, confirming protection against potential pathogens and microorganisms that can cause spoilage, and providing a longer shelf life for the product.

Generally, three different types of starter cultures are used in the dairy industry for the manufacture of a variety of fermented products: single-strain starters (pure culture of LAB such as *L. lactis* subsp. *lactis or L. lactis* subsp. *cremoris*), mixed-strain starters (combinations of *L. lactis* subsp. *lactis* or *L. lactis* subsp. *cremoris* and the gas- and aroma-producing mesophilic LAB *L. lactis* subsp. *diacetilactis* and *Leuconostoc* spp.), and multiple-strain starters (mixtures of known compatible, nonphage-related, carefully selected strains).

Thermophilic LAB (37°C–45°C), including *Streptococcus thermophilus* and *Lactobacillus species*, are also used in the dairy industry for manufacturing some fermented products, such as yogurt, acidophilus milk, and high-temperature-scalded cheese (e.g., Swiss cheese).

3.2.2 Factors Affecting Fermentation Process of Starter Cultures

The factors affecting microbial growth also affect milk fermentation. Fermentation rates are generally associated with the microbial growth curve, up to the stationary phase. The fermentation processes of LAB starters can be influenced by several factors, which are both intrinsic when inherent to the milk properties and extrinsic— related to the storage environment. As far as intrinsic factors are concerned, nutrient content, moisture content, pH, available oxygen, biological structures, and antimicrobial constituents can be considered. Temperature relative humidity, and the gases surrounding the food are among the main extrinsic factors influencing milk fermentation.

3.2.2.1 Milk Composition

Milk and dairy products are generally very rich in nutrients, thus providing an ideal growth environment for many microorganisms, especially LAB.

Cow's milk, which is mainly used for dairy manufacturing, has the following composition—which varies depending on the species (cow, goat, sheep), the location, and the animal feed: 87% water, 4.8% carbohydrates, 4% fats, 3.5% proteins, and 0.7% vitamins and minerals. The main carbohydrate is lactose, which is found in two anomeric forms, α-anomer and β-anomer, whose solubility depends on temperature.

Milk fats contain approximately 65% saturated, 30% monounsaturated, and 5% polyunsaturated fatty acids, and as a consequence of their variability, melting points can vary.

There are two main classes of milk proteins: the casein family, consisting of several types of caseins (α-s1, α-s2, ß, and 6), and serum (whey) proteins, such as ß-lactoglobulin, α-lactalbumin, blood serum albumin, immunoglobulins, lactoferrin, transferrin, and many minor proteins and enzymes.

Milk is a good source of both water-soluble vitamins, such as thiamin (vitamin B1), riboflavin (vitamin B2), and cobalamin (vitamin B12), and water-insoluble vitamins, such as vitamins A, D, E, and K. It also contains traces of niacin (vitamin B3), pantothenic acid (vitamin B5), pyridoxine (vitamin B6), vitamin C, and folate.

Calcium, magnesium, phosphorus, potassium, selenium, and zinc are the main minerals that can be found in milk, most of them in the form of associated salts (e.g., calcium phosphate). Traces of manganese, copper, iron, and sodium can also be found, depending on the source of the milk.

3.2.2.2 Nutrient Requirements

As for most of the bacteria, LAB involved in milk fermentation require the following basic nutrients: water, an energy source, a carbon/nitrogen source, vitamins, and minerals.

LAB tolerate high salt concentrations, and among them, *Leuconostoc* possesses the highest salt tolerance; thus it is involved in the initiation of the majority of lactic acid fermentations.

Nitrogen is mainly required for the synthesis of both the enzymes involved in sugar metabolism and those responsible for transportation of nutrients into the cells. The presence of free amino acids leads to an increase in the LAB growth rate. LAB have a limited capacity to produce amino acids; thus proteins and peptides are used as a nitrogen source. As matter of fact, LAB possess a full armory of proteases and peptidases, and most of them are intracellular. These proteases are useful for hydrolysis of milk proteins into peptides to be used as a nitrogen source, confirming that milk composition is ideal for LAB growth.

Vitamins are fundamental for LAB fermentation, and their absence results in very slow growth. Vitamin B, folic acid, biotin, riboflavin, pantothenate, and nicotinic acid are known to positively regulate the growth of *Lactobacilli*, favoring their growth and being, and in some cases, essential for LAB development.

Few studies have been reported concerning the effect of metal ions on LAB growth. Calcium, magnesium, manganese, iron, potassium, and sodium are the main minerals to be analyzed. They generally have a positive effect on the LAB growth rate, which is different from heavy metals, for example, copper, zinc, and cadmium, which have inhibitory effects or can even kill the bacteria if they exist in an excessive dosage. However, there is a strong variability among *Lactobacillus* species in response to inorganic substances.

Calcium strongly affects LAB growth. For instance, it can have both inhibitory and activating roles in bacterial growth, and it can also influence the requirement of other components in the growth medium. Manganese and magnesium are generally required in LAB media preparation, especially because of their action in protecting cells against zinc and oxidative stress. Iron supplementation to LAB media is generally not necessary, particularly when manganese is present in the environment. For instance, Imbert and Blondeau (1998) have monitored the growth ability of several *Lactobacillus* species in the presence of iron, showing that supplementation of chelated iron does not affect the growth of species in the presence of manganese. Elli et al. (2000) demonstrated that LAB iron sensitivity can depend on the nucleotides present in the medium, suggesting a role for iron in the purine or pyrimidine metabolism of *Lactobacilli*. Copper is an important cofactor that is biologically relevant as a redox factor. However, based on the few studies carried out on LAB copper homeostasis, none of them strictly need it for growing. However, Kaneko et al. (1990) observed increased growth of *L. lactis* subsp. *lactis* 3022 when it was grown aerobically in the presence of copper. *Enterococcus hirae* is the only LAB for which response to copper has been widely studied; thus it is a model for LAB general response to metals (Solioz and Stoyanov 2003). Zinc has the role of cofactor for many enzymes, but if concentrations are too high, it can have very damaging effects on cells. Thus, zinc concentration in the medium must be strictly controlled. Few manuscripts have so far reported about LAB zinc homeostasis; thus efforts need to better characterize the response to zinc, which is, instead, known for other gram-positive bacteria. Bolotin et al. (2001) studied a zinc transporter system in a strain of *Lactobacillus lactis*, suggesting the strong influence of zinc concentration on its regulation. In *S. thermophilus*, the chromosomal *cadC* and *cadA* genes were shown to be involved in a cadmium–zinc resistance system (Schirawski et al. 2002).

3.2.2.3 Water Activity (a_w)

Bacteria dominates in foods with high water activity, a_w (with a minimum value of approximately 0.90 a_w), and yeasts and molds, which require less moisture, and dominates in low a_w foods (minimum 0.70 a_w). The water activity of fluid milk is approximately 0.98 a_w, which is the amount of water generally required by LAB.

3.2.2.4 Antimicrobial Activity

Milk contains several nonimmunological proteins, which inhibit the growth and metabolism of many microorganisms. Lactoperoxidase, lactoferrin, lysozyme, and xanthine are the most common antimicrobial constituents of protein nature. Moreover, LAB growth is strictly affected by the presence of residual antibiotics and sanitizers in milk, mainly as a result of their indiscriminate use in the treatment of mastitis or udder diseases. Hence, milk must be deeply analyzed for the presence of residual antibiotics before the addition of starter cultures. The methods based on immunological reactions as well as isotopic tracer dilution procedures (a Charm test) are very effective and commonly used.

It is also worth noting that antibiotic-like substances (bacteriocins), which are produced by certain wild strains of *L. lactis* subsp. *lactis* and other lactic cultures, are present in raw milk.

3.2.2.5 Bacteriophages

The slow acid production by LAB is mainly attributed to the presence of bacteriophages in the fermentation environment. The use of phage-resistant LAB as an alternative to phage-sensitive LAB is the most promising solution to overcome this problem strongly affecting the yield of dairy fermentative processes (Sanders 1988). The use of defined single strains and their phage-resistant mutants is becoming increasingly popular throughout the world for the manufacture of various fermented foods (Moineau 1999). For instance, Hill (2006) has deeply reviewed the nature and variety of resistance systems used by LAB in phage-attack conditions, underlining the recent progress in the construction of novel resistance mechanisms.

3.2.2.6 Available Oxygen

Most LAB appear to be indifferent to aeration of the medium, slightly preferring a reduced oxygen tension. Acid production is faster under controlled oxygen tension conditions. For this reason, LAB are generally defined as aerotolerant anaerobes. However, oxygen can play an important role, especially in the metabolism of homofermentative strains, where it acts as an electron acceptor. In this case, LAB are protected from oxygen byproducts (e.g., H_2O_2) by the action of their peroxidases. Thus, oxygen substantially influences sugar metabolism, energy and biomass yields, and the end products of sugar metabolism (Condon 1987). On the contrary, heterofermentative LAB use, anaerobically, electron acceptors other than oxygen, such as fructose, citrate, and glycerol.

Van de Guchte et al. (2002) have largely reviewed oxygen stress in LAB. Interest in understanding how LAB metabolism is affected by the presence of oxygen has increased in recent years as a consequence of LAB increased use in the food industry. As reported by Miyoshi et al. (2003), new stress-resistant strains have been produced by both site-directed and random mutagenesis in order to understand how genes are implicated in oxygen stress response, but efforts are still needed to deepen the understanding of the problem and construct stress-resistant mutant strains more suitable for food industry needs.

3.2.2.7 pH

The control of pH is a crucial issue in milk fermentation because pH influences not only the growth of LAB, but also their metabolic activity. Even if milk generally has a pH of 6.6, which is ideal for the growth of many microorganisms, the fermentative process by LAB causes a decrease of the pH in the milk medium as a result of the fact that they produce lactic acid with a general rate of 10% of their weight per minute. Thus, it is important to maintain pH between 5.5 and 6.5 (Peeva and Peev 1997) in order to avoid both metabolic repression and loss of viability of the bacteria as a consequence of the extreme acidic pH.

Currently, there are several methods to control pH, such as the use of media with a fine chemical composition to control pH internally; the removal of end products by means of different techniques, including dialysis, adsorption, extraction, or membrane separation; and the addition of alkalis, for example, ammonium, calcium, sodium, or potassium hydroxide and sodium carbonate (Joglekar et al. 2006).

To accomplish perfect control of pH medium in the fermentative process of dairy products, it might be attractive to take advantage of the genetic manipulation of LAB in order to improve their metabolic pathway and better regulate the genes involved in acid production. For instance, John et al. (2008) developed a low pH-tolerant, fast-growing mutant of *Lactobacillus delbrueckii* combining nitrous acid-based mutagenesis of the wild-type strain with genome shuffling between the obtained mutant and an amylase-producing nonfastidious *Bacillus amyloliquefaciens*.

3.2.2.8 Temperature

Temperature is one of the operative parameters mainly affecting milk fermentation because of its direct influence on LAB growth (Tamine 2002; Haque et al. 2001). The majority of lactic starters grow optimally in the range of temperatures between 27°C and 32°C, but some thermophilic LAB are currently used in milk fermentative processes; for example, *S. thermophilus* has an optimal temperature for growth in the temperature range 37°C to 42°C. Interest in psychrotrophic LAB strains able to grow at temperatures less than 10°C (Kato et al. 2000) is growing because of their potential use in preventing the spoilage of refrigerated dairy products.

Thus, temperature is a very important physical factor affecting milk fermentation, and it must be taken under control, particularly in the case of mixed and multiple starters that are composed of different strains diverging for their optimal temperature requirement.

Moreover, temperature modulation is required to control the properties of the final cheese products because temperatures higher than 40°C can affect milk properties, thus influencing flavor, aroma, and whey expulsion.

3.2.2.9 Incubation Period

The period of incubation is another important factor influencing LAB growth, and generally, it strongly changes depending on the dairy product and the manufacturer's empirically fixed parameters. However, a 16 to 24 h incubation period is generally adequate for the maximal growth of these organisms at their optimal temperatures. As evidence of the variability of the incubation period in manufacturing procedures of milk fermentation, it is worth noting that production of yogurt—using *S. thermophilus* and *L. bulgarian* as starters—at 37°C to 50°C takes 2 to 4 h. Acidophilus milk—using *L. acidophilus*—at 38°C requires 18 to 20 h. Mozzarella cheese with *Lactococcus lactis* and *Lactococcus cremoris* and butter with *Lactococcus lactis*, *Lactococcus cremoris*, *Lactococcus diacetylactis*, and *Leuconostoc cremoris* takes 5 to 15 h at 20°C to 30°C.

3.2.2.10 End Product Inhibition

It is well known that the most abundant end product of lactic fermentation, that is, lactic acid, has inhibitory effects on LAB growth as largely reviewed in past years (Luedeking and Piret 1959; Friedman and Gaden 1970; Hanson and Tsao 1972; Rogers et al. 1978; Loubier et al. 1997). To overcome this problem, many efforts have been made by researching new strategies for removal of end products from the fermentative medium, and the integration of fermentation and separation of fermentative products is the main route. Several separation methods can be applied; the use of a membrane for dialysis is mainly considered. Recently, the potential of LAB and yeast immobilization for food fermentation, including milk for cheese production, has been discussed by Bekatorou and Kanellaki (2010).

3.2.2.11 Effect of Carbon Dioxide

A minimum concentration of carbon dioxide is essential for the initiation of bacterial growth because the complete removal of carbon dioxide from the medium results in an extended lag phase. For most of the LAB, the optimum initial concentration of CO_2 varies from 0.2% to 2% by volume.

3.2.3 CASE 1: CHEESE

Most cheeses are generally made from heat-treated or pasteurized cow's milk. When the milk is not heat-treated, the cheese must be ripened for at least 60 days at a maximum temperature of 4°C to avoid growth of pathogenic organisms.

The main steps commonly involved in manufacturing most types of cheeses are described below along with the key factors affecting them.

3.2.3.1 Milk Preparation

The milk usually undergoes pasteurization, a heat treatment designed to create the optimum conditions for milk conversion, especially into fresh cheese.

Pasteurization must be sufficient to kill bacteria capable of affecting the quality of the cheese, for example, coliforms, which can cause early "blowing" and a disagreeable taste. Thus, pasteurization at 72°C to 73°C for 15 to 20 s is most commonly applied.

However, pasteurization does not kill spore-forming microorganisms, which, surviving in the spore state, can negatively influence the safety and properties of the final product. Several chemicals can be added to the milk before starting fermentation, for example, sodium nitrate ($NaNO_3$) and hydrogen peroxide (H_2O_2), in order to avoid the effects arising from the presence of heat-resistant spore-forming bacteria (principally *Clostridium tyrobutyricum*). However, the use of chemicals has been widely criticized as being another external factor to be controlled in the fermentation stage, which also affects LAB activities and growth.

3.2.3.2 Starter Cultures

The starter culture is a very important factor in cheese-making. The ability to produce lactic acid in the curd, to break down the proteins, and to produce carbon dioxide are among the main characteristics to be possessed by a starter culture.

The most frequently used cultures in cheese fermentation are mixed-strain cultures, in which two or more strains of both mesophilic (20°C–40°C) and thermophilic (40°C–50°C) bacteria exist in symbiosis, achieving their mutual benefit. These cultures not only produce lactic acid, but also aroma components and CO_2.

Development of acids lowers the pH, which is important in assisting syneresis (i.e., contraction of the coagulum accompanied by elimination of whey). Furthermore, salts of calcium and phosphorus are released, influencing the consistency of the cheese and helping to increase the firmness of the curd.

Moreover, the acid produced by LAB is useful to attack any microorganism that survived after pasteurization or bacteria of contamination, which needs lactose quickly converted by LAB. Production of lactic acid stops when all the lactose has been fermented, which takes very little time.

Carbon dioxide produced by CO_2-forming bacteria is eventually present in the starter culture and is useful for the production of cheeses that are characterized by cavities, such as Emmental cheese. In these cases, it is necessary to control their metabolism, thus assuring the accumulation of gas, which is then released, causing the formation of the typical cavity. Bacteria in the culture also offers the complete pattern of proteolytic enzymes that are involved in the ripening process together with rennet enzymes.

3.2.3.3   Curdling

In this stage of cheese manufacturing, there is a coagulation of milk into the so-called curd, which is then cut and prestirred. To regulate curd size, a new heating step is required at a temperature of approximately 40°C, where the activity of the mesophilic LAB is retarded and only the most heat-resistant lactic acid-producing bacteria survive, including *Propionibacterium freudenreichii* ssp. *shermanii*. The whey is thus finally removed, and the curd is handled by pressing.

At the final crucial step, salting of the cheese is performed. Salts are added approximately 6 h after the beginning of fermentation; thus salt concentration, which is generally estimated to be 1% w/w of the final cheese product, must be controlled to avoid inhibition of starter culture metabolism.

3.2.4   CASE 2: YOGURT

Yogurt is a semisolid milk product whose manufacturing is very simple compared to that of other dairy products, such as cheese. It is obtained through milk fermentation by mixed cultures of LAB added at an initial bacteria concentration in the range of 10^{-6} to 10^{-9} cfu/m. Several types of yogurt are available, depending on their fat content: very low fat, nonfat or skim yogurt (0.2%–0.5% fat), half-fat or semiskim yogurt (0.5%–2.0% fat), whole milk or full-fat yogurt (more than 2.0% fat), and high-fat or cream yogurt with fat content as high as 10% (Early 1998).

The main steps of yogurt preparation and the key operative factors are reported below.

3.2.4.1   Milk Preparation

Milk preparation is a key step in yogurt preparation because its physical and chemical properties strongly influence both the fermentative process and the yogurt properties.

As in the case of cheese production, in yogurt manufacturing, pasteurization is necessary in order to produce a relatively sterile milieu for starter cultures, by killing undesired microorganisms; to denature proteins that would affect the viscosity and texture of the final product; and to affect other additives to ensure the required physical and chemical conditions that lead to the formation of gels. Before being heat treated, milk has to be clarified and mixed with the other dairy ingredients, for example, stabilizers and gums, which are chosen on a case-by-case basis, considering the desired final product. Gelatin, pectin, agar, and starch are commonly added as stabilizers in quantities that vary in the range from 0.1% to 0.5%. Besides the addition of stabilizers, which also help to avoid the problem of syneresis, it is important to guarantee the absence of any substances that could negatively interfere with the growth of the starter cultures, for example, antibiotics, preservatives, disinfectants, or bacteriophages.

Before starting the fermentative process, the mixture is generally concentrated by evaporation or the addition of milk concentrate or whey powder, in order to modulate both the water content available for further fermentation, and the dry-matter content of the final product, which strongly affects yogurt's physical properties.

Furthermore, depending on the type of yogurt being produced, the addition of sugars or sweeteners to the fermentative milk is also needed. In this case, it is important not to overcome a sugar concentration of 10% v/v because of the negative influence on the starter culture's growth, for example, changes in osmotic pressure. For this reason, in the case of a high sugar concentration requirement, sugars are added immediately after the fermentation process.

3.2.4.2   Homogenization and Heating

Homogenization of the mixture is required for the maintenance of the physical quality of the final product, preventing the formation of cream and/or whey.

There is also a heating step in the mixture, which is a fundamental action, considering that the best temperature for starter culture growth must be assured.

3.2.4.3   Starter Culture Concentration, Temperature, pH, Time of Incubation, and Storage

Streptococcus thermophilus and *Lactobacillus delbrueckii* subsp. *bulgaricus* are among the LAB mainly used in yogurt manufacturing; they are generally added to the fermentative milk medium in a ratio of 1:1.

The starting concentration is a key factor to be assessed because if cell concentrations are too low, it could result in the slow growth of the strain of interest and/or the fast growth rate of microorganisms other than LAB still present in the fermentative milk.

Considering that *Streptococcus thermophilus* and *Lactobacillus delbrueckii* subsp. *bulgaricus* are thermophilic and mesophilic strains, respectively, the choice of temperature is crucial in this fermentative process, and it is generally defined around the value of 45°C. Both of them produce acids and carbon dioxide and, thus, are responsible for a pH drop to approximately 4.5 to 5, a pH value that strongly affects yogurt's properties and requires an efficient adjustment to avoid negative influences on the LAB metabolism.

The time of incubation is another parameter strongly affecting milk fermentation and, as a consequence, yogurt characteristics, since flavor changes according to the amount of chemicals produced by the LAB. A short incubation time can result in low chemical production, but too much incubation time, on the contrary, causes a strong pH drop, which negatively affects LAB metabolism.

3.3 FERMENTATIONS IN BAKERY INDUSTRY

Sourdough, that is, a spontaneously fermenting mixture of ground cereals and water, has been used since ancient times to improve the properties of the bread dough, enhance bread texture and flavor, and delay bread spoilage and staling (Rehman et al. 2006; De Vuyst et al. 2009).

According to the flavor preferences of different countries or regions, a broad variety of sourdough products were developed in different countries. Table 3.2 reports a list of different sourdough products and dough acidifiers that are currently available on the market.

Sourdough fermentations require the symbiosis of certain mostly heterofermentative LAB, commonly including *Lactobacillus sanfranciscensis*, *L. brevis*, and *L. plantarum*, with certain yeasts, among which *Saccharomyces cerevisiae* is the

---

TABLE 3.2
Overview of Sourdough Products and Dough Acidifiers (Not Exhaustive)

Products	Raw Materials	Characteristics	TTA (Total Titratable Acid)[a]	Dosage (on Flour) (%)
Dried sourdoughs	Rye, wheat, durum, oat, rice, spelt, buckwheat flours, cereal germ flours, malted flours, etc.	Concentrated flavor, low dosage	40–220	1–6
Pasty sourdoughs	Rye, wheat, spelt flours, cereal germ flours, malted flours, grains, sprouts, hydrocolloids (guar gum, xanthane, etc.)	Volatile flavors of sourdough, high amount of fermented and swelled flour in the resulting bread dough	25–100	3–30
Liquid sourdoughs	Rye, wheat, durum, spelt, oat flours, cereal germ flours, malted flours, salt, etc.	Pumpable, volatile flavors	30–150	1–10
Dough acidifier (powder)	Pregelatinized flours, organic acids, sometimes dried sourdoughs are added	Concentrated acid, low dosage	Up to 1000	0.5–1.5
Dough acidifier (liquid/pasty)	Rye or wheat flour, organic acids, hydrocolloids, sometimes sourdough	Concentrated acid, low dosage, pumpable (liquid)	Up to 400	0.5–3

Source: Retrieved from Brandt, M.J., *Food Microbiology*, 24:161–164, 2007. With permission.

[a] TTA: ml 0.1M NaOH per 10 g product.

most frequently present (Gobbetti 1998; De Vuyst and Neysens 2005). Properties of these microbial strains, including their highly adapted metabolism and growth requirements, the production of antimicrobial compounds by LAB, and their stress responses, represent the most important key parameters for stable sourdough fermentation (De Vuyst and Neysens 2005). Besides the prevailing previously mentioned microbial species, secondary microflora, including LAB species such as *L. alimentarius, L. acidophilus, L. fructivorans, L. fermentum, L. reuteri,* and *L. pontis* (Gobbetti et al. 1994; Vogel et al. 1996; Corsetti et al. 2001) and yeast species such as *S. exiguus, Candida krusei,* and *C. milleri* (Rossi 1996), are present in spontaneous sourdough fermentations, and they can have an effect on the ecosystem, either directly or indirectly by affecting the dominant microbiota.

3.3.1   CLASSIFICATION OF SOURDOUGH PRODUCTION PROCESSES AND EFFECTS OF OPERATION CONDITIONS ON DIFFERENT TYPES OF SOURDOUGH PRODUCTION PROCESSES

Three different types of sourdoughs can be distinguished, depending on the technology adopted for their production (Böcker et al. 1995): type I sourdoughs (traditional sourdoughs), type II sourdoughs (accelerated sourdoughs), and type III sourdoughs (dried sourdoughs). The characteristics of the sourdough production processes of different types are reported below.

3.3.1.1   Characteristics of Type I Sourdough Production Processes

Sourdoughs of type I are produced by traditional procedures involving frequent refreshments in order to maintain the microorganism activity, thus allowing high gas production. Examples of bakery products of this type include panettone and other brioches, Pugliese, San Francisco sourdough French bread, pane Toscano, and Altamura bread.

Fermentation is performed at 20°C to 30°C and a pH of 4. Starter cultures for type I sourdoughs can be pure cultures from natural sourdough preparations (type Ia), mixed cultures prepared from wheat and rye or their mixtures through multiple-stage fermentation processes (type Ib), and sourdoughs from fermentations at high temperatures in tropical regions (type Ic) (Stolz 1999).

Type Ia culture sourdoughs contain a microflora characterized by a stable composition, a high souring activity, and resistance against microbial contamination. A starter preparation containing *L. sanfranciscensis* for the production of San Francisco French bread represents an example of type Ia sourdough. *L. sanfranciscensis* is a heterofermentative LAB having souring activity resulting from the large amounts of lactic acid and acetic acid produced from maltose and contributing dough leavening by gas production (Gobbetti et al. 1996; Gobbetti and Corsetti 1997). Maltose-negative strains of the yeast *S. exiguus* are mainly responsible for the leavening function in this bread (Sugihara et al. 1971).

Type Ib microflora are composed mainly of obligate heterofermentative *L. sanfranciscensis,* and other species (i.e., obligate heterofermentative *L. brevis* and related *Lactobacillus* spp., *L. buchneri, L. fermentum, L. fructivorans, L. pontis,*

L. reuteri, and *W. cibaria*; facultative heterofermentative *L. alimentarius*, *L. casei*, *L. paralimentarius*, and *L. plantarum*; and obligate homofermentative *L. acidophilus*, *L. delbrueckii*, *L. farciminis*, and *L. mindensis*) can be present depending on environmental conditions (Hammes and Gänzle 1998; Vogel et al. 1999).

Fermentation for manufacturing wheat and rye sourdoughs takes between 3 and 48 h (Stephan and Neumann 1999a,b). Type Ib rye sourdough usually derives from three fermentation steps. The starter serves as the inoculum for each batch of bread dough, and a new batch is consecutively reinoculated from the previous one (backslopping). Besides contributing to the acidification and leavening of the dough, the microflora plays an important role in aroma formation. Gas production is required for leavening of the dough unless baker's yeast is added.

Examples of Type Ic sourdoughs are African sorghum sourdoughs that are produced at temperatures higher than 35°C. This microflora consists of the obligate heterofermentative *L. fermentum*, *Lactobacillus* spp. related to *L. pontis*, and *L. reuteri* species as well as the obligate homofermentative *L. amylovorus* (Hamad et al. 1992, 1997). The yeast most often associated with this type of sourdough is *I. orientalis*.

3.3.1.2   Characteristics of Type II Sourdough Production Processes

Type II sourdoughs are semifluid preparations developed as a consequence of the industrialization of rye bread production asking for faster and more efficient fermentations. These sourdoughs are mainly used as dough acidifiers. Continuous sourdough fermentation plants have being installed in industrial bakeries (Stolz and Böcker 1996).

Type II fermentations typically take 2 to 5 days and are often performed at temperatures higher than 30°C to speed up the process (Böcker et al. 1995; Hammes and Gänzle 1998). Those sourdoughs show a high acid content with a pH of 3.5 after 24 h of fermentation.

In type II sourdoughs, *L. sanfranciscensis* is not competitive enough to dominate the fermentation. The obligate homofermentatives *L. acidophilus*, *L. delbrueckii*, *L. amylovorus* (rye), *L. farciminis*, and *L. johnsonii* and obligate heterofermentatives *L. brevis*, *L. fermentum*, *L. frumenti*, *L. pontis*, *L. panis*, and *L. reuteri* as well as *Weissella* (*W. confusa*) species are found (Vogel et al. 1999; Müller et al. 2001).

3.3.1.3   Characteristics of Type III Sourdough Production Processes

Type III sourdoughs are dried dough developed by defined starter cultures and used in baking for acidification and aroma production (Stolz and Böcker 1996). They are composed by LAB resistant to drying and are able to survive in that form, for example, heterofermentative *L. brevis*, facultative heterofermentative *P. pentosaceus*, and *L. plantarum* strains. The drying process leads to an increased shelf life of the sourdough providing a stock product until further use.

3.3.2   Effect of Carbohydrates

Rye and especially wheat flours contain low amounts of soluble carbohydrates with a total concentration of maltose, sucrose, glucose, and fructose from 1.55% to 1.85% (Martinez-Anaya 1996). The use of soluble carbohydrates by LAB is influenced by the associated yeasts. There is no competition for the carbon sources in LAB–yeast associations in food fermentations. However, LAB grow and ferment sugars more slowly in mixtures with yeasts than in pure cultures (Collar 1996, Merseburger et al. 1995). Investigations of coculture model systems, including the prevailing sourdough LAB (*Lb. sanfranciscensis* and *Lb. plantarum*) and yeast (*S. cerevisiae*), showed that bacterial growth and fermentation decreased as a result of the faster consumption of maltose and, especially, glucose by *S. cerevisiae* (Gobbetti et al. 1994a).

Most of the yeasts assume hexose and maltose by high affinity transport systems while the uptake of disaccharide is less efficient in LAB (*Lb. brevis*) (Collar 1996). In wheat sourdough fermentation, the discrepancy between the high consumption of sugars by yeast and the low starch hydrolysis by flour enzymes provides a rapid depletion of soluble carbohydrates, which decreases LAB acidification (Rouzaud and Martinez-Anaya 1993; Gobbetti et al. 1994b). On the other hand, the greater flour enzyme activity in rye dough fermentation increases the availability of soluble carbohydrates (Röcken and Voysey 1993).

Cofermentations can be adopted to enable LAB to use nonfermentable substrates. For instance, cofermentation of fructose and maltose or glucose in a fructose-negative strain of *Lb. sanfranciscensis* allows the use of fructose as an additional electron acceptor through its reduction to mannitol (Gobbetti et al. 1995; Stolz et al. 1995).

Lb. sanfranciscensis strains are very efficient in metabolizing maltose and are not susceptible to glucose repression. Hydrolysis of maltose by *Lb. sanfranciscensis* produces glucose (Gobbetti et al. 1994c; Stolz et al. 1993) that is not metabolized, but it is excreted and may be used by maltose-negative yeasts, such as *S. exiguous*, or may prevent competitors from using maltose by glucose repression, thereby giving an advantage to *Lb. sanfranciscensis*.

In the presence of sucrose as a carbon source, cell yield and lactic acid production by *Lb. plantarum* increase when it is associated with yeasts (*S. cerevisiae* or *S. exiguus*) as a result of the hydrolysis of sucrose by yeasts into glucose and fructose, which are more rapidly metabolized than the sucrose by LAB (Aksu and Kutsal 1986; Gobbetti et al. 1994a).

S. cerevisiae frequently disappears from the microbial population of sourdough during consecutive fermentations because of the repression of genes involved in maltose fermentation (Nout and Creemers-Molenaar 1987) and its susceptibility to the acetic acid produced by LAB. To balance the poor survival of yeasts in consecutive sourdough fermentations, large amounts of baker's yeast are used (Suihko and Makinen 1984).

3.3.3 Effect of Nitrogen Compounds

Peptides and amino acids have a key role as flavor precursors of baked sourdough products (Spicher and Nierle 1984) and interfere with the physical properties of the dough (Mascaros et al. 1994; Collar 1996). Amino acids accumulate during wheat sourdough fermentation (Collar et al. 1991), resulting from proteolysis by flour enzymes (Spicher and Nierle 1988) and various LAB strains (Spicher and Nierle 1988; Mascaros et al. 1994) or deriving from the cell mass of microorganisms (Rothenbuehler et al. 1982; Gobbetti et al. 1994d) and, to a lesser extent, from other metabolisms (Okuhara and Harada 1971). Lysis of microbial cells, induced by the addition of mixing ingredients into the dough, liberates specific amino acids—γ-aminobutyric acid, proline, valine, isoleucine, glycine, and alanine—and peptides are mainly released by *S. cerevisiae* (Collar 1996) while glycine and alanine are released by LAB.

In coculture model systems, it was demonstrated that growth of *Lb. sanfranciscensis* and *Lb. plantarum* is stimulated by *S. cerevisiae* and *S. exiguous* because of a lack of competition for the nitrogen source (in a mixture of NH_4Cl and amino acids, yeasts preferentially use NH_4^+) and the yeast's excretion of specific amino acids and small peptides (Thorne 1957; Suomalainen and Oura 1971; Lyons and Rose 1977).

It is worth noting that bacteria prefer peptides rather than free amino acids because direct transport of peptides into the cell reduces the metabolic energy used for amino acid uptake. The use of *Lb. sanfranciscensis* and *Lb. plantarum* during sourdough fermentation causes a considerable increase in the total concentration of free amino acids because LAB proteolysis increases the concentration of aliphatic, dicarboxylic, and hydroxy amino acid groups, which, for the most part, are stimulatory for bacterial growth and are used by yeasts (Gobetti et al. 1994a). Among sourdough LAB, *Lb. sanfranciscensis* strains show the highest aminopeptidase, dipeptidase, tripeptidase, and iminopeptidase activities (Gobetti et al. 1996b).

3.4   FERMENTATIONS IN ALCOHOLIC BEVERAGE MANUFACTURING

Wine and beer are the main alcoholic beverages produced in the world, and their market is still growing. The European Union (EU) is the world's largest wine producer, consumer, exporter, and importer, and its market is expected to grow further as a result of the increasing demand from both industrialized (United States, Canada, Japan) and developing (Brazil, Russia, India, and China) countries. The United States remain the leading export market (24.6% of the total volume and 30.7% of the total value) for the EU-27. As far as the beer market is concerned, it is growing all over the world, too, rapidly increasing, particularly in emerging countries, for example, China, Africa, and South America (EUROSTAT 2011).

Alcoholic fermentation is the basis of wine fermentation, and it is performed by both wild yeast that are naturally present in the grape must and malt wort and commercial yeast starter cultures. For instance, *S. cerevisiae* is the most commonly encountered species in wine and beer fermentations, having good response to the stress occurring in alcoholic fermentation.

Parameters affecting the fermentative process of alcoholic beverage manufacturing are discussed below.

3.4.1   Factors Affecting Alcoholic Fermentation Production

3.4.1.1   Ethanol Concentration

Ethanol is the primary end product of yeast fermentation of sugars, and its inhibitory effects on yeast cell growth have been largely reviewed in recent years. Ethanol has been shown to negatively affect both growth and fermentation rate as its concentration increases in the medium; thus its concentration must be controlled during the fermentation process of alcoholic beverages. As deeply reviewed by D'Amore (1992), it has been demonstrated that ethanol can modify the cell membrane by altering both its structural organization and permeability, resulting in a lack of cofactors and coenzymes. Moreover, ethanol can cause the denaturation of many intracellular enzymes, including glycolytic enzymes; inhibition of transport systems of sugars and amino acids; and enhancement of thermal death.

3.4.1.2   pH

Acidity is one of the most important organoleptic parameters in alcoholic beverages, for example, wine, and it is mainly a result of the presence of weak organic acids. Most of them are already present in the grape, but small amounts of them, for example, succinic and acetic acids, are also produced by the yeast during fermentation. Thus, wine acid composition depends on both the quality of the grapes and the microbial activity. The pH of the must can vary in the range between 2 and 4, depending on both the climate conditions and the strength of pressing the grape in the first steps of wine manufacturing when large amounts of potassium can be extracted from the raw material, having a key role in the control of pH in the milieu. The effects of potassium concentration on yeast fermentation were largely studied by Kudo et al. (1998), who showed that potassium is an ion exchanger involved in the balance of cytoplasmic acidification, besides favoring glucose uptake by the cell. *Saccharomyces* strains are known to grow in acidic media at pH values not lower than 2.5. Very high concentrations of potassium cause a strong pH reduction, which results in inhibition of the fermentative process.

3.4.1.3   Temperature

Yeast growth and metabolism are severely affected by extreme temperatures, particularly as a result of the physical effects on membrane structure composition. For instance, high temperatures change the fatty acid composition, resulting in the increase of saturated fatty acids esterified into the membrane lipids. Moreover, an increase in fermentation temperature causes an increase in ethanol inhibitory effects and reduction of the ethanol production rate (D'Amore 1992).

Non-*Saccharomyces* strains have an important contribution to beginning alcoholic fermentation. These strains, for example, *Candida* and *Kloeckera* genera, start fermentation having an optimum growth temperature lower than 20°C, where *Saccharomyces* strains are not able to grow. Depending on the strains, the end products of non-*Saccharomyces* yeast fermentation contributes to defining wine composition and quality. For instance, at temperatures lower than 20°C an increase of ethyl acetate and acetaldehyde production can occur as a consequence of *Kloeckera* genera fermentation, and higher temperatures stimulate production of higher alcohols, especially 3-methylbutanol attributed to *Saccharomyces* strains, as studied by Erten (2002). Results achieved by Erten (2002) agree with those achieved in the past by Rankine (1967), who found an increase in n-propanol, i-butanol, and 2-methylbutanol production at high temperatures, and Killian and Ough (1979) and Cottrel and McLellan (1986), who demonstrated the increase of ethyl acetate at low temperatures.

Growth of non-*Saccharomyces* strains does not negatively affect *Saccharomyces* growth and metabolism. Nevertheless, it is worth noting that *Saccharomyces* strains are very sensitive to the quick switches of temperature, which can occur, for example, in the budding stage of wine manufacturing. Thus, temperature must be finely tuned.

Glycerol is a byproduct of ethanol fermentation by *S. cerevisiae*, and it is known to have a beneficial effect on wine quality by providing sweetness, fullness, and smoothness (Remize et al. 2000). Therefore, many studies have been performed to increase its production during alcoholic fermentation, and it has been largely demonstrated that many factors, such as temperature, pH, substrate type and initial concentration, inoculation ratio, and nutriment type and concentration (nitrogen, sulfur dioxide) affect glycerol production. High amounts of glycerol are generally produced at temperatures of around 30°C, and an alkaline milieu is known to positively affect its production, too (Yalcin-Karasu and Ozbas 2008).

3.4.1.4   Nitrogen Concentration

Low initial nitrogen levels in grape must be perhaps the most studied cause of slow fermentation (Bell et al. 1979; Ingledew and Kunkee 1985; Cramer et al. 2001). Nitrogen is mainly used for protein synthesis, but many authors have remarked on its important role in affecting yeast fermentation kinetics. Nitrogen concentration strongly affects the pH in the medium because its consumption favors a great acidification. As a matter of fact, ammonia and amino acids are mainly used as nitrogen sources. Ammonia is more easily assimilated, but its consumption produces less acidification in comparison to amino acids. Glutamine and glutamate are the mainly used amino acid, followed by asparagines, aspartate, serine, and alanine (Monteiro and Bisson 1991).

Nitrogen requirements change according to the strains. For instance, enological yeast strains have been studied by Manginot et al. (1998), demonstrating that the nitrogen requirements of industrial strains are very different from those of yeast already present in the grape. Raw materials, both grape and malt, have a nitrogen composition variation depending on the fertilization, maturation, water content, and soil type. The most variability is found in grape composition, and different authors have found different ranges of nitrogen concentrations in grape must, generally ranging between 60 and 2400 mg/L (Alexandre et al. 1994; Sabbayrolles et al. 1996).

3.4.1.5   Sulfur Dioxide

Sulfur dioxide (SO_2) is used in alcohol fermentation, particularly in wine manufacturing because of its antiseptic and antioxidant properties. Although SO_2 is highly toxic to microorganisms, commercial wine yeast strains show a high tolerance to SO_2, which is thus added to the must before fermentation. In this way, its concentration must be efficiently controlled to avoid inhibiting the growth of the starter culture. Stratford and Rose (1986) described the uptake of SO_2 by *S. cerevisiae*, showing that simple diffusion is the main method of SO_2 uptake, which is favored by the low intracellular pH that characterizes *S. cerevisiae* cells.

3.4.1.6   Vitamins and Minerals

Vitamins are needed in alcoholic fermentation, mainly acting as cofactors for enzymatic conversions. Biotin, nicotinic acid, vitamin B (thiamine), pantothenic acid, and vitamin C are the major vitamins required for yeast growth and fermentation. Biotin is the most important vitamin because it is involved in many enzymatic reactions, including synthesis of proteins, DNA, carbohydrates, and fatty acids. The effect of vitamins on the production of higher alcohols, which mainly contribute to the flavor and aroma of alcoholic beverages, has been largely demonstrated. Thiamine is essential for pyruvate decarboxylase activity and pantothenic acid production (Dixon and Webb 1964); pantothenic acid is necessary for coenzyme A and acetyl-CoA synthesis (Nordstrom 1963) and, therefore, for isoamyl alcohol production (Webb and Ingraham 1963). Aspartic acid is necessary for n-propyl alcohol formation (Webb and Ingraham 1963).

Potassium, calcium, magnesium, copper, and zinc are the main minerals involved in yeast metabolism and other processes, such as flocculation and cell division. The rate of uptake and utilization of these ions strongly depends on their concentration in the medium, which is determined by their solubility and the eventual presence of chelators. As previously reported, potassium strongly affects pH; thus its concentration must be opportunely monitored. Magnesium is the most important mineral required by the yeast, particularly because of its influence on yeast growth as a critical component in ATP production. Normally, grapes and malt contain a sufficient mineral content to ensure a good fermentation; thus mineral addition is not needed. However, mineral concentrations must be controlled to avoid their deficiency in the medium, which negatively impacts cell growth and biomass yield as well as fermentation rate.

It is worth noting that yeasts, when used many times in an anaerobic process, can vary their metabolic performances as a result of the effects of dynamic metal ion accumulation and release (Aleksander et al. 2009).

3.4.1.7 Sugar Concentration

Sugar content in grapes and malt affects yeast fermentation yields. *S. cerevisiae* is able to ferment only hexoses, and it is well known that the non-*Saccharomyces* strains are also able to ferment pentose. Sugar concentration in the raw material varies, depending on the grape or malt variety and their geographical localization. High sugar concentrations, such as those registered, for example, in well-matured grapes, negatively affect yeast fermentation because of the high production of ethanol, having toxic effects on the cells and causing hyperosmotic stress. Accumulation of sugars can arise also from inactivation of sugar transport systems because of the loss of activity of involved proteins whose synthesis can be arrested as a consequence of nitrogen exhaustion (Bustuna and Lagunas 1986; Salmon 1996).

3.4.1.8 Available Oxygen

For both wine and beer manufacturing, oxygen is a fundamental parameter in the yeast fermentative process because it strongly affects the growth rate through its involvement in sterol and fatty acid biosynthesis. Insufficient oxygenation results in inadequate yeast growth, which means inconsistent or long fermentation while overoxygenation is not a problem because yeasts are generally able to use all the available oxygen within the 4 to 10 h of fermentation. For instance, the amount of oxygen generally required for efficient yeast growth is in the range of 5 to 10 mg/L (Sablayrolles et al. 1996). It is certain that the oxygen requirement is variable, depending on the yeast strain and the quantity of dissolved oxygen present in the grape and brew must.

Grapes must contain a sufficient amount of dissolved oxygen, depending on its source and treatment, but the addition of the yeast is always preceded by the introduction of soluble oxygen through an injection of sterile air. Mechanisms by which oxygen is added (sprays, whipping, splashing, shaking, air pumping, or injection of pure oxygen through a sintered stone) also influence the yeast strain's growth.

3.4.1.9   Water Activity and Osmotic Pressure

High concentrations of solutes can reduce water availability and inhibit yeast growth, which generally requires an optimal water activity between 0.9 and 1. For instance, low water activity affects both structure and function of cell components, particularly enzymes and membranes. Osmotic stress occurs as a consequence of a decrease of water activity in the environment. For example, 0.94 represents the minimum possible value in the case of brewing (Jones and Greenfield 1986).

Many studies have demonstrated that viability and growth of yeast is strongly affected by osmotic pressure as most yeasts are sensitive to osmotic stresses. An increase in osmotic pressure results in cell growth slowdown or even cell death. Marechal and Gervais (1994) showed that a slow increase in osmotic pressure can result in high cell resistance to extremely high levels of osmotic pressure while osmotic shock is lethal for cells. However, resistance to osmotic pressure stress can be coupled with temperature effect, and it has been shown that the viability of *S. cerevisiae* can be preserved at very high hyperosmotic shocks at low temperatures of growth (Beney et al. 2001).

Hallsworth (1998) deeply studied the effect of ethanol on water activity, showing that the increase of ethanol concentration in the milieu causes a strong reduction of water availability: an ethanol concentration of 20% (w/v) reduces the water activity to 0.895, a considerably lower value than that recommended for yeast growth and fermentation.

D'Amore et al. (1988) studied the correlation between osmotic stress and ethanol production, which showed that an increase in ethanol concentration and osmotic pressure have inhibitory effects on both growth and fermentation rates. The same negative effect of osmotic stress on yeast fermentation occurs in conditions of nutrient limitation or toxic end products (octanoic and decanoic acids) accumulation.

3.4.1.10   Pitching Rate

The pitching rate, that is, the inoculum size, influences both yeast growth and physiology besides affecting the final quality of alcoholic beverages as a consequence. The increase of cell concentration in the reactor is one of the strategies used to improve the fermentation rate in beer and wine manufacturing. The pitching rate considerably affects the production of diacetyl and acetoin as studied by Kringstad and Rasch (1966), who demonstrated that yeast strains respond differently in this respect.

It is generally known that a higher pitching rate favors the increase of the volumetric productivity of fermentation processes. The pitching rate is strain-specific, and it is linked to oxygen availability at the time of inoculum as reported by Verbelen et al. (2009).

Aeration occurring in the pitching step strongly influences yeast growth and metabolism, influencing the final content of esters in alcoholic beverages (Ahvenainen and Maekinen 1981; Ahvenainen 1982; Cheong et al. 2007). In the case of brewing, for example, the amount of yeast needed for fermentation depends on the original gravity of beer and the fermentation temperature. The amount of yeast (measured as millions of yeast cells per milliliter of wort) needed for good fermentation increases with the increase of the original gravity and the decrease in temperature of fermentation.

Increasing the pitching rate has been shown to be a beneficial strategy for high-gravity brewing, even more advantageous than supplementation of nutrients to the medium. For instance, Nguyen and Viet Man (2009) studied the effect of a high pitching rate on high-gravity beer manufacturing, demonstrating that the increase in pitching rate improves several aspects, such as the fermentation time, the sugar uptake, the ethanol production rates, and the ethanol concentration. In addition, they demonstrated that high pitching rates even decrease the dependence of yeast growth on nutrients.

REFERENCES

Ahvenainen, J. (1982) Lipid composition of aerobically and anaerobically propagated brewers bottom yeast. *J. Inst. Brew.,* 88:367–370.

Ahvenainen, J. and Maekinen, V. (1981) The effect of pitching yeast aeration on fermentation and beer flavor. *EBC Congress.*

Aksu, Z. and Kutsal, T. (1986) Lactic acid production from molasses utilizing *Lactobacillus delbrueckii* and invertase together. *Biotechnol. Lett.,* 8:157–160.

Aleksander, P., Piotr, A., Tadeusz, T. and Makarewicz, M. (2009) Accumulation and release of metal ions by brewer's yeast during successive fermentations. *J. Inst. Brew.,* 115(1):78–83.

Alexandre, H., Rousseaux, I. and Charpentier, C. (1994) Relationship between ethanol tolerance, lipid composition and plasma membrane fluidity in *Saccharomyces cerevisiae* and *Kloeckera apiculata*. *FEMS Microbiol. Lett.,* 124:7–22.

Batish, V.K., Grover, S., Pattnaik, P. and Ahmed, N. (1999) Fermented milk products. In V.K. Joshi, A. Pandey (ed.) *Biotechnology: Food Fermentation, Vol. II,* Kerela, India: Educational Publishers and Distributors Press.

Bekatorou, A. and Kanellaki, M. (2010) Fermentation technology: Immobilized cells and low temperature processes. Available at http://www.scitopics.com/Fermentation_technology_Immobilized_cells_low_temperature_processes.html (accessed December 1, 2011).

Bell, A.A., Ough, C.S. and Kliewer, W.M. (1979) Effects of must and wine composition, rates of fermentation, and wine quality of nitrogen fertilization of *Vitis vinifera* var. *Thompson seedless* grapevine. *Am. J. Enol. Vitic.,* 30:124–129.

Beney, L., Marechal, P.A. and Gervais, P. (2001) Coupling effects of osmotic pressure and temperature on the viability of *Saccharomyces cerevisiae*. *Appl. Microbiol. Biotechnol.,* 56:513–516.

Böcker, G., Stolz, P. and Hammes, W.P. (1995) Neue erkenntnisse zum ö̈kosystem sauerteig und zur physiologie des sauerteigtypischen sta̋mme *Lactobacillus sanfrancisco* und *Lactobacillus pontis*. *Getreide Mehl und Brot,* 49:370–374.

Bolotin, A., Wincker, P., Mauger, S., Jaillon, O., Malarme, K., Weissenbach, J., Ehrlich, S.D. and Sorokin, A. (2001) The complete genome sequence of the lactic acid bacterium *Lactococcus lactis* ssp. *lactis* IL1403. *Genome Res.,* 11:731–753.

Brandt, M.J. (2007) Sourdough products for convenient use in baking. *Food Microbiology*, 24:161–164.

Bustuna, A. and Lagunas, R. (1986) Catabolite inactivation of glucose transport system in *Saccharomyces cerevisiae*. *J. Gen. Microbiol.*, 132:379–385.

Caplice, E. and Fitzgerald, G.F. (1999) Food fermentations: Role of microorganisms in food production and preservation, *Int. J. of Food Microbiology*, 50:131–149.

Cheong, C., Wackerbauer, K. and Kang S.A. (2007) Influence of aeration during propagation of pitching yeast on fermentation and beer flavor. *J. Microbiol. Biotechnol.*, 17(2):297–304.

Collar, C. (1996) Biochemical and technological assessment of the metabolism of pure and mixed cultures of yeast and lactic acid bacteria in breadmaking applications. *Food Sci. Technol. Intern.*, 2:349–367.

Collar, C., Mascaros, A.F., Prieto, J.A. and Benedito de Barber, C. (1991) Changes in free amino acids during fermentation of wheat doughs started with pure culture of lactic acid bacteria. *Cereal Chem.*, 68:66–72.

Condon, S. (1987) Responses of lactic acid bacteria to oxygen. *FEMS Microb. Lett.*, 46(3):269–280.

Corsetti, A., Lavermicocca, P., Morea, M., Baruzzi, F., Tosti, N. and Gobbetti, M. (2001) Phenotypic and molecular identification and clustering of lactic acid bacteria and yeasts from wheat (species *Triticum durum* and *Triticum aestivum*) sourdoughs of southern Italy. *Int. J. Food Microbiol.*, 64:95–104.

Corsetti, A. and Settanni, L. (2007) *Lactobacilli* in sourdough fermentation. *Food Res. Int.*, 40:539–558.

Cottrell, T.H.E. and McLellan, M.R. (1986) The effect of fermentation temperature on chemical and sensory characteristics of wines from seven white grape cultivars grown in New York State. *Am. J. of Enol. Vitic.*, 37:190–194.

Cramer, A.C., Vlassides, S. and Block, D.E. (2001). Kinetic model for nitrogen-limited wine fermentations. *Biotech. Bioeng.*, 77:49–60.

D'Amore, T. (1992) Improving yeast fermentation performance. *J. Inst. Brew.*, 98:375–382.

D'Amore, T., Panchal, C.J. and Stewart, G.C. (1988) Intracellular ethanol accumulation in *Saccharomyces cerevisiae* during fermentation. *Appl. Environm. Microbiol.*, 54(1):110–114.

De Vuyst, L. and Neysens, P. (2005) The sourdough microflora: Biodiversity and metabolic interactions. *Trends Food Sci. Technol.*, 16:43–56.

De Vuyst, L., Vrancken, G., Ravyts, F., Rimaux, T. and Weckx, S. (2009) Biodiversity, ecological determinants, and metabolic exploitation of sourdough microbiota. *Food Microbiol.*, 26:666–675.

Dixon, M. and Webb, E.G. (1964) *Enzymes*, 2nd edn. New York: Academic Press.

Early, R. (1998) *Technology of Dairy Products*, 2nd edn. Blackie Academic and Professionals.

Elli, M., Zink, R., Rytz, A., Reniero, R. and Morelli, L. (2000) Iron requirement of *Lactobacillus* spp. in completely chemically defined growth media. *J. Appl. Microb.*, 88:695–703.

Erten, H. (2002) Relations between elevated temperatures and fermentation behavior of *Kloeckera apiculata* and *Saccharomyces cerevisiae* associated with winemaking in mixed cultures. *World J. of Microb. Biotech.*, 18:373–378.

EUROSTAT. (2011) Available at http://epp.eurostat.ec.europa.eu/statistics_explained/index .php/Beverages_production_statistics (accessed December 10, 2011).

FAO. (2012) The FAO World Milk Report. www.fao.org.

Friedman, M.R. and Gaden, E.L. (1970) Growth and acid production of *Lactobacillus delbrueckii* in a dialysis culture system. *Biotech. Bioeng.*, 12:961–974.

Gobbetti, M. (1998) The sourdough microflora: Interactions of lactic acid bacteria and yeasts. *Trends Food Sci. Technol.*, 9:267–274.

Gobbetti, M. and Corsetti, A. (1997) *Lactobacillus sanfrancisco* a key sourdough lactic acid bacterium: A review. *Food Microbiol.*, 14:175–187.

Gobbetti, M., Corsetti, A. and Rossi, J. (1994a) The sourdough microflora: Interactions between lactic acid bacteria and yeasts: Metabolism of carbohydrates. *Appl. Microbiol. Biotechnol.*, 41:456–460.

Gobbetti, M., Corsetti, A. and Rossi, J. (1994b) The sourdough microflora: Evolution of soluble carbohydrates during the sourdough fermentation. *Microbiol. Aliments Nutr.*, 12:9–15.

Gobbetti, M., Corsetti, A., Rossi, J., La Rosa, F. and De Vincenzi, M. (1994c) Identification and clustering of lactic acid bacteria and yeasts from wheat sourdoughs of central Italy. *Ital. J. Food Sci.,* 1:85–93.

Gobbetti, M., Simonetti, M.S., Rossi, J., Cossignani, L., Corsetti, A. and Damiani, P. (1994d) Free D- and L-amino acid evolution during sourdough fermentation and baking. *J. Food Sci.*, 59:881–884.

Gobbetti, M., Corsetti, A. and Rossi, J. (1995) Maltose-fructose co-fermentation by *Lactobacillus brevis* subsp. *lindneri* CB1 fructose-negative strain. *Appl. Microbiol. Biotechnol.*, 42:939–944.

Gobbetti, M., Corsetti, A. and Rossi, J. (1996a) *Lactobacillus sanfrancisco*, a key sourdough lactic acid bacterium: physiology, genetic and biotechnology. *Advances in Food Science*, 18:167–175.

Gobbetti, M., Smacchi, E. and Corsetti, A. (1996b) The proteolytic system of *Lactobacillus sanfrancisco* CB1: Purification and characterization of a proteinase, dipeptidase and aminopeptidase. *Appl. Environm. Microbiol.*, 62:3220–3226.

Hallsworth, J.E. (1998) Ethanol-induced water stress in yeast. *J. Ferm. Bioeng.*, 85(2):125–137.

Hamad, S.H., Böcker, G., Vogel, R.F. and Hammes, W.P. (1992) Microbiological and chemical analysis of fermented sorghum dough for Kisra production. *Appl. Microbiol. and Biotechnol.*, 37:728–731.

Hamad, S.H., Dieng, M.C., Ehrmann, M.A. and Vogel, R.F. (1997) Characterization of the bacterial flora of Sudanese sorghum flour and sorghum sourdough. *J. of Appl. Microbiol.*, 83:764–770.

Hammes, W.P. and Gänzle, M.G. (1998) Sourdough breads and related products. In B.J.B. Woods (ed.) *Microbiology of Fermented Foods*. London: Blackie Academic/Professional Press.

Hanson, T.P. and Tsao, G.T. (1972) Kinetic studies of the lactic acid fermentation in batch and continuous cultures. *Biotech. Bioeng.*, 14:233–252.

Haque, A., Richardson, R.K. and Morris, E.R. (2001) Effect of fermentation temperature on rheology of set and stirred yogurt. *Food Hydrocolloids*, 15:593–602.

Hill, C. (2006) Bacteriophage and bacteriophage resistance in lactic acid bacteria. *FEMS Microb. Rev.*, 12:87–108.

Imbert, M. and Blondeau, R. (1998) On the iron requirement of *Lactobacilli* grown in chemically defined medium. *Curr. Microb.*, 37:64–66.

Ingledew, W.M. and Kunkee, R. (1985) Factors influencing sluggish fermentations of grape juice. *Am. J. Enol. Vitic.*, 36:65–76.

Joglekar, H.G., Rahman, I., Babu, S., Kulkarni, B.D., and Joshi, A. (2006) Comparative assessment of downstream processing options for lactic acid. *Sep. Purif. Technol.*, 52(1):1–17.

John, R.P., Gangadharan, D. and Nampoothiri, K.M. (2008) Genome shuffling of *Lactobacillus delbrueckii* mutant and Bacillus amyloliquefaciens through protoplasmic fusion for l-lactic acid production from starchy wastes. *Biores. Tech.*, 99(17):8008–8015.

Jones, R.P. and Greenfield, P.F. (1986) Role of water activity in ethanol fermentations. *Biotechnol. Bioeng.*, 28:29–40.

Kaneko, T., Takahashi, M. and Suzuki, H. (1990) Acetoin fermentation by citrate-positive *Lactococcus lactis* subsp. *lactis* 3022 grown aerobically in the presence of hemin or Cu. *Appl. Environ. Microb.*, 56:2644–2649.

Kato, Y., Sakal, R.M., Hayashidani, H., Kiuchi, A., Kaneuki, C. and Ogawa, M. (2000) *Lactobacillus algidus* sp. Nov., a psychrophilic lactic acid bacterium isolated from vacuum-packaged refrigerated beef. *Int. J. Syst. Evol. Microbiol.*, 50:1143–1149.

Khurana, H.K. and Kanawjia, S.K. (2007) Recent trends in development of fermented milks. *Current Nutrition and Food Science*, 3:91–108.

Killian, E. and Ough, C.S. (1979) Fermentation esters–Formation and retention as affected by fermentation temperature. *Am. J. Enol. Vitic.*, 30:301–305.

Kringstad, H. and Rasch, S. (1966) The influence of the method of preparation of pitching yeast on its production of diacetyl and acetoin during fermentation. *J. Inst. Brew.*, 72:56–61.

Kudo, M., Vagnoli, P. and Bisson, L.F. (1998) Imbalance of potassium and hydrogen ion concentrations as a cause of stuck enological fermentations. *Am. J. Enol. Vitic.*, 49:295–301.

Loubier, P., Cocaign-Bousquet, M., Matos, J., Goma, G. and Lindley, N.D. (1997) Influence of end-products inhibition and nutrient limitations on the growth of *Lactococcus lactis* subsp. *Lactis. J. Appl. Microbiol.*, 82:95–100.

Luedeking, R. and Piret, E.L. (1959) A kinetic study of the lactic acid fermentation. *J. Biochem. Microbiol. Tech. Eng.*, 1:393–412.

Lyons, T.P. and Rose, A.H. (1977) Whiskey. In A.H. Rose (ed.) *Alcoholic Beverages*. London: Academic Press.

Manginot, C., Roustan, J.L. and Sablayrolles, J.M. (1998). Nitrogen demand of different yeast strains during alcoholic fermentation: Importance of the stationary phase. *Enz. Micro. Techn.*, 23:511–517.

Marechal, P.A. and Gervais, P. (1994) Yeast viability related to water potential variation: Influence of the transient phase. *Appl. Microbiol. Biotechnol.*, 42:617–622.

Martinez-Anaya, M.A. (1996) Enzymes and bread flavour. *J. Agric. Food Chem.*, 44:2469–2480.

Mascaros, A.F., Martinez, C.S. and Collar, C. (1994) Metabolism of yeasts and lactic acid bacteria during dough fermentation relating functional characteristics of fermented doughs. *Rev. Esp. Cienc. Tecnol. Alimen.*, 34:623–642.

Merseburger, T., Ehret, A., Geiges, O., Baumann, B. and Schmidt-Lorenz, W. (1995) Microbiology of dough preparation VII: Production and use of preferments to produce wheat-bread. *Mitt. Gebiete Lebensm. Hyg.*, 86:304–324.

Miyoshi, A., Rochat, T., Gratadoux, J.J., Le Loir, Y., Costa Oliveira, S., Langella, P. and Azevedo, V. (2003) Oxidative stress in *Lactococcus lactis*. *Genet. Mol. Res.*, 2(4):348–359.

Moineau, S. (1999) Application of phage resistance in lactic acid bacteria. *Antoine van Leeuwenhoek*, 76:377–382.

Monteiro, F.F. and Bisson, L.F. (1991) Biological assay of nitrogen content of grape juice and prediction of sluggish fermentations. *Am. J. Enol. Vitic.*, 42:47–57.

Muller, M.R.A., Wolfrum, G., Stolz, P., Ehrmann, M.A. and Vogel, R.F. (2001) Monitoring the growth of *Lactobacillus* species during a rye flour fermentation. *Food Microbiology*, 18:217–227.

Nguyen, T.H. and Viet Man, L.V. (2009) Using high pitching rate for improvement of yeast fermentation performance in high gravity brewing. *Food. Res. Int.*, 16:547–554.

Nordstrom, K. (1963) Formation of ethyl acetate in fermentation with brewer's yeast. IV. Metabolism of acetyl-coenzyme a. *J. Inst. Brew.*, 69:142–153.

Nout, M.J.R. and Creemers-Molenaar, T. (1987) Microbiological properties of some wheat-meal sourdough starters. *Chem. Mikrobiol. Technol. Lebensm.* 10:162–167.

Okuhara, M. and Harada, T. (1971) Formation of N-acetyl-L-alanine and N-acetylglycine from glucose by *Candida tropicalis* OH23. *Biochim. Biophys. Acta*, 244:16–22.

Peeva, L. and Peev, G. (1997) A new method for pH stabilization of the lactoacidic fermentation. *Enz. Microb. Technol.*, 21:176–181.

Rankine, B. (1967) Formation of higher alcohols by wine yeasts and relationship to taste thresholds. *J. Sci. Food Agric.*, 18:584–589.

Rehman, S.U., Paterson, A. and Piggott, J.R. (2006) Flavour in sourdough breads: A review. *Trends Food Sci. Technol.*, 17:557–566.

Remize, F., Sabbayrolles, J.M. and Dequin, S. (2000) Re-assessment of the influence of yeast strain and environmental factors on glycerol production in wine. *J. Appl. Microbiol.*, 88:371–378.

Röcken, W. and Voysey, P.A. (1993) Sourdough fermentation in bread making. *J. Appl. Bacteriol.* (Supplement), 79:38S–39S.

Rogers, P.L., Bramall, L. and MacDonald, I.J. (1978) Kinetic analysis of batch and continuous culture of *Streptococcus cremoris*. *Can. J. Microbiol.*, 24:372–380.

Rossi, J. (1996) The yeasts in sourdough. *Adv. Food Sci. (CMTL)*, 18:201–211.

Rothenbuehler, E., Amado, R. and Solms, J. (1982) Isolation and identification of amino acid derivates from yeast. *J. Agric. Food Chem.*, 30:439–442.

Rouzaud, O. and Martinez-Anaya, M.A. (1993) Effect of processing conditions on oligosaccharide profile of wheat sourdoughs. *Z. Lebensm. Unters. Forsch.*, 197:434–439.

Sablayrolles, J.M., Dubois, C., Manginot, C., Roustan, J.L. and Barre, P. (1996) Effectiveness of combined ammoniacal nitrogen and oxygen additions for completion of sluggish and stuck fermentations. *J. Ferm. Bioeng.*, 82:377–381.

Salmon, J.M. (1996) Sluggish and stuck fermentations: Some actual trends on their physiological basis. *Vitic. Enol. Sci.*, 51:137–140.

Sanders, M.E. (1988) Phage resistance in lactic acid bacteria. *Biochimie*, 70(3):411–422.

Schirawski, J., Hagens, W., Fitzgerald, G.F. and Van Sinderen, D. (2002) Molecular characterization of cadmium resistance in *Streptococcus thermophilus* strain 4134: An example of lateral gene transfer. *Appl. Environ. Microbiol.*, 68:5508–5516.

Solioz, M. and Stoyanov, J.V. (2003) Copper homeostasis in *Enterococcus hirae*. *FEMS Microb. Rev.*, 27:183–195.

Spicher, G. and Nierle, W. (1984) The microflora of sourdough. XVIII. Communication: The protein degrading capabilities of the lactic acid bacteria of sourdough. *Z. Lebensm. Unters. Forsch.*, 178:389–392.

Spicher, G. and Nierle, W. (1988) Proteolytic activity of sourdough bacteria. *Appl. Microbiol. Biotechnol.*, 28:487–492.

Stephan, H. and Neumann, H. (1999a) Technik der roggen-sauerteigführung. In G. Spicher and H. Stephan (ed.) *Handbuch Sauerteig: Biologie, Biochemie, Technologie*. Hamburg: Behr's Verlag Press.

Stephan, H. and Neumann, H. (1999b) Technik der weizenvorteigund weizensauerteigführung. In G. Spicher and H. Stephan (ed.) *Handbuch Sauerteig: Biologie, Biochemie, Technologie*. Hamburg: Behr's Verlag Press.

Stolz, P. (1999) Mikrobiologie des sauerteiges. In G. Spicher and H. Stephan (ed.) *Handbuch Sauerteig: Biologie, Biochemie, Technologie*. Hamburg: Behr's Verlag Press.

Stolz, P. and Böcker, G. (1996) Technology, properties and applications of sourdough products. *Advances in Food Science*, 18:234–236.

Stolz, P., Böcker, G., Hammes, W.P. and Vogel, R.F. (1995) Utilization of electron acceptors by *Lactobacilli* isolated from sourdough. *Z. Lebensm. Unters. Forsch.* 201:91–96.

Stolz, P., Böcker, G., Vogel, R.F. and Hammes, W.P. (1993) Utilization of maltose and glucose by *Lactobacilli* isolated from sourdough. *FEMS Microbiol. Letters*, 109:237–242.

Stratford, M. and Rose, H.A. (1986) Transport of sulphide dioxide by *Saccharomyces cerevisiae*. *J. Gen. Microbiol.*, 133:2173–2179.

Sugihara, T.F., Kline, L. and Miller, M.W. (1971) Microorganisms of the San Francisco sourdough bread process. I. Yeasts responsible for the leavening action. *Appl. Microbiol.*, 21:456–458.

Suihko, M.L. and Makinen, V. (1984) Tolerance of acetate, propionate and sorbate by *Saccharomyces cerevisiae* and *Torulopsis holmii*. *Food Microbiol.*, 1:105–110.

Suomalainen, H. and Oura, E. (1971) Yeast nutrition and solute uptake. In A.H. Rose and J.S. Harrison (ed.) *Yeasts*. London: Academic Press.

Tamine, A.Y. (2002) Fermented milks: A historical food with modern applications—A review. *Eur. J. Clin. Nutr.*, 56(4):2–15.

Thorne, R.S. (1957) Brewer's yeast. In W. Roman (ed.) *Yeasts*. London: Academic Press.

Toro, C.R. (2005) Uso de bactérias lácticas probióticas na alimentação de camarões *Litopenaeus vannamei* como inibidoras de microrganismos patogênicos e estimulantes do sistema imune. Originally presented as doctorate thesis, Federal University of Paraná. Curitiba.

USDA (United States Department of Agriculture) (2011) Dairy: World markets and trade. Available at http://www.sabmiller.com/index.asp?pageid=39 (accessed December 9, 2011).

Van de Guchte, M., Serror, P., Chervaux, C., Smokvina, T., Ehrlich, S.D. and Maguin, E. (2002) Stress responses in lactic acid bacteria. *Antonie van Leeuwenhoek*, 82:187–216.

Verbelen, P.J., Dekoninck, T.M.L., Saerens, S.M.G., Van Mulders, S.E., Thevelein, J.M. and Delvaux, F.R. (2009) Impact of pitching rate on yeast fermentation performance and beer flavor. *Appl. Microbiol. Biotechnol.*, 82:155–167.

Vogel, R.F., Knorr, R., Müller, M.R.A., Steudel, U., Ganzle, M.G. and Ehrmann, M.A. (1999) Non-dairy lactic fermentations: The cereal world. *Antonie van Leeuwenhoek*, 76: 403–411.

Vogel, R.F., Mueller, M., Stolz, P. and Ehrmann, M. (1996) Ecology in sourdoughs produced by traditional and modern technologies. *Adv. Food Sci. (CMTL)*, 18:152–159.

Webb, A.D. and Ingraham, J.L. (1963) Fusel oil. *Adv. Appl. Microbiol.*, 5:317–353.

Yalcin-Karasu, S. and Ozbas, Z.Y. (2008) Effects of pH and temperature on growth and glycerol production kinetics of two indigenous wine strains of *Saccharomyces cerevisiae* from Turkey. *Braz. J. Microbiol.*, 39:325–332.

4 Upstream Operations of Fermentation Processes

Parameswaran Binod, Raveendran Sindhu,
and Ashok Pandey

CONTENTS

4.1 INTRODUCTION

The rapid development of biotechnology has made a tremendous impact on various sectors of the economy over the last several years. Industries such as agriculture, food, chemicals, and pharmaceuticals are major players that are showing a rapid economic growth. Advances in fermentation techniques have made the scale-up of products easier and cheaper. Breakthroughs in the field of microbiology have also contributed a great deal to the development of the fermentation industry. Even though fermentation is an old process, it still attracts lots of industries for its ability to develop newer products in economical ways. The art of trial and error was the best technique to be handed on to the next generation so as to achieve the best end result. Even today, some producers of fermented products, such as beer brewers, rely very much on art and received wisdom.

4.2   BASICS OF FERMENTATION

"Fermentation" is the term used to describe any process in which production is done by means of the mass culture of a microorganism. Simply put, it is a chemical change brought on by the action of microorganisms. The product can be the cell itself (referred to as biomass production) or the microorganism's own metabolite (referred to as product). The two key components in the fermentation process are the microorganism and the substrate. The organisms employed in fermentation are diverse, and it should maintain GRAS (generally regarded as safe) status, especially if it is used in the food industry. In most fermentation processes, the final end product depends on the substrate used.

The process of fermentation can be carried out by submerged or solid-state cultivation. Optimization of the nutritional requirements of organisms and investigations into their responses to environmental variables are indispensable for enhanced product yield. The general objectives of fermentation medium optimization are to maximize productivity, minimize byproducts and costs, and ensure product quality.

4.3   SOLID-STATE FERMENTATION

Solid-state fermentation (SSF) is a low-cost fermentation process, particularly suitable to the needs of developing countries. One of the major advantages of solid-state fermentation is that it is usually carried out using naturally occurring agricultural byproducts, such as straw, bran, etc. (Pandey 1992; Lonsane et al. 1992). SSF has several advantages compared to submerged fermentation (SmF), such as resulting in a more concentrated extract and increasing the substrate amount, etc. However, this system can be susceptible to water content, pH, oxygen gradients, and accumulation of metabolic heat, which make scaling up difficult.

In SSF, the moisture necessary for microbial growth exists either in an absorbed state or complexed within the solid matrix. Although most workers consider solid-state and solid-substrate fermentation to be essentially one and the same, Pandey et al. (2001) distinguished these two as separate processes. According to them, solid-substrate fermentation includes those processes in which the substrate itself acts as the carbon source and occurs in the absence or near-absence of free water, whereas solid-state fermentation is defined as any fermentation process occurring in the absence or near-absence of free water and employing a natural substrate or an inert substrate as solid support. The lower moisture level at which SSF can occur is approximately 12% because, below this level, all biological activities cease.

Two types of SSF systems have been distinguished, depending upon the nature of the solid phase used; the most commonly used system involves cultivation of a natural material, and the less frequently used one involves cultivation on an inert support impregnated with a liquid medium (Ooijkaas et al. 2000). Based on the type of microorganisms involved, SSF processes can be classified into two major categories: natural or indigenous SSF and pure-culture SSF, where individual strains or a mixed culture is used (Pandey et al. 2000). Pure cultures are generally used in industrial SSF processes as they help in optimum substrate utilization for the targeted product, whereas mixed cultures are used for bioconversion of agro-industrial residues (Krishna 2005).

4.4 SUBMERGED FERMENTATION

Most industrial fermentations are carried out by submerged fermentation (SmF). The medium in the SmF is liquid, which remains in contact with the microorganism. A supply of oxygen is essential in SmF. There are four main ways of growing microorganisms in SmF. They are batch culture, fed-batch culture, perfusion-batch culture, and continuous culture. In batch culture, microorganisms are inoculated in a fixed volume of the medium. In the case of fed-batch culture, concentrated components of the nutrient are gradually added to the batch culture. In perfusion-batch culture, the addition of the culture and withdrawal of an equal volume of used cell-free medium is performed. In continuous culture, fresh medium is added into the batch system at the exponential phase of microbial growth with a corresponding withdrawal of medium containing microorganisms. Continuous cultivation gives a near-balanced growth with little fluctuation of nutrients, metabolites, cell numbers, or biomass.

4.5 UPSTREAM OPERATIONS

The fermentation process involves two broad operations: upstream and downstream. All the operations before starting the fermenter are collectively called upstream operations, and they include sterilization of the reactor, preparation and sterilization of the culture media, preparation and growth of suitable inoculums of microbial strains, etc. The major steps in the upstream process are shown in Figure 4.1. All the other operations after the fermentation are known as the downstream process. The upstream operations are discussed in detail in the following sections.

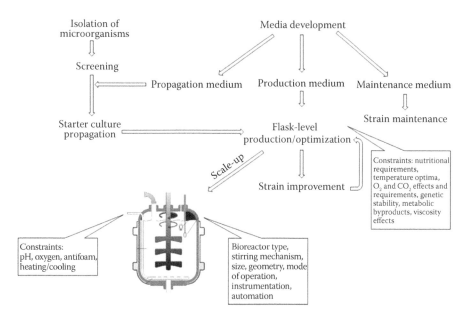

FIGURE 4.1 Upstream operations in fermentation.

4.5.1 SELECTION AND TYPE OF BIOPRODUCTS

There are five main groups of commercially important fermentations. These include the production of microbial cells, enzymes, metabolites, and recombinant products as well as biotransformation. Many products of primary metabolism are economically important and are produced by fermentation. These include citric acid, glutamic acid, vitamins, lysine, polysaccharides, nucleotides, and phenylalanine. Secondary metabolites include those compounds that are synthesized during the stationary phase, and they do not have any role in cell metabolism. Many secondary metabolites have antimicrobial activity; some are growth promoters, some act as enzyme inhibitors, and some have pharmacologically important properties. Wild-type microorganisms produce primary and secondary metabolites in low concentrations.

4.5.2 ISOLATION AND SELECTION OF ORGANISM

Identification and isolation of the required microorganism is critical for any microbiological process. Isolation of microorganisms also helps to screen them as to whether they can be used in industrial processes. The microorganisms should satisfy specific criteria, which include utilization of a cheap media for growth that should be safe, stable, and allow genetic manipulation and should convert the substrate into product rapidly, and the product should be easily recovered from the culture medium.

Several techniques were employed for isolating microorganisms. This includes the liquid-culture method and the solid-culture method. The liquid-culture method is carried out in shaker flasks containing the liquid-culture medium, and the solid-culture method is carried out in a solid-culture medium containing a substrate.

Screening involves the use of highly selective procedures for the detection and isolation of microorganisms of interest from among a large microbial population. Screening eliminates valueless microorganisms. The screening approach has been extensively employed in the search for microorganisms capable of producing industrially important enzymes, amino acids, vitamins, antibiotics, organic acids, etc. Primary screening allows the detection and isolation of microorganisms that possess potentially interesting industrial applications. This screening is followed by a secondary screening to further test the capabilities and gain information about these organisms (Casida 1993).

Microorganisms producing organic acids can be detected either by incorporating a pH-indicating dye, such as neutral red/bromothymol blue or calcium carbonate. The production of these compounds is indicated by a change in color of the indicating dye or the production of a cleared zone of dissolved calcium carbonate around the colony. The simplest screening technique for antibiotic producers is the crowded-plate technique. Colonies producing antibiotic activity are indicated by an area of agar around the colony that is free of growth of other colonies. Microorganisms capable of synthesizing extracellular enzymes, amino acids, vitamins, and other secondary metabolites can be detected using a similar screening approach. Secondary screening allows further sorting of microorganisms to evaluate the true potential of a microorganism for industrial usage. The secondary screening can be either quantitative or qualitative. The qualitative approach indicates which microbes produce the compound of interest. The quantitative approach indicates the yield of the compound that has been expected when the microorganism is grown in various differing media.

4.5.3 STRAIN IMPROVEMENT FOR DEVELOPING HYPER-PRODUCERS

The product yield will be low when naturally available microorganisms are used for fermentation. Overproduction of primary or secondary metabolites is a complex process, and successful development of improved strains requires knowledge of physiology, pathway regulation and control, and the design of creative screening procedures. To increase productivity, it is essential to modify the genetic structure because it determines the productivity of the organism. The genetic makeup of the microbes can be altered by mutation, sexual recombination, transduction, transformation, phage conversion, etc. Rational metabolic and cellular engineering approaches have been successfully used for strain improvement. These approaches were limited to the manipulation of genes encoding enzymes and regulatory proteins.

A continuous program of mutation and selection is required during process development. Before initiating a mutation and selection program, it is essential to select a high-yielding strain and use best possible production medium. Inducing mutations in the microorganisms is the most important strain improvement technique. Mutation and selection programs employ ionizing radiation, ultraviolet light, alkylating agents, nitrous acid, purine and pyramidine analogues, or other mutagenic agents to adjust or change the hereditary material of the organism (Stanbury et al. 1994). The probability of mutation occurring may be increased by exposing the culture to a mutagenic agent. Most of the improved strains currently available are produced by induced mutations. Survivors of the mutagen exposure contain hyper-producers and also mutants, which produce lower levels of the desired product. Hence, a screening procedure is carried out for the selection of the desired mutants (hyper-producers) from the inferior ones. The screening approach is easier for strains producing primary metabolites than those producing secondary metabolites. Feedback control is involved in the inhibition of enzyme activity and the repression of enzyme synthesis by the end product when it is present in the cell at a sufficient concentration. A hyper-producing mutant lacks feedback control systems, which results in the overproduction of the end product. Mutants of *Corynebacterium glutamicum* capable of producing lysine lack control systems. In *C. glutamicum*, the first enzyme in the lysine synthesis pathway, aspartokinase, is inhibited only when lysine and threonine are synthesized above a threshold level. Nakayama et al. (1961) used this technique to isolate a homoserine auxotroph of *C. glutamicum* that can produce 44 g/l of lysine.

Another method for strain improvement is protoplast fusion. This method has greatly increased the prospects of combining characteristics found in different strains. Cell fusions followed by nuclear fusion occur between protoplasts of strains. This has been achieved in filamentous fungi, bacteria, *Streptomyces*, and yeasts. Tosaka et al. (1982) produced an improved strain, producing lysine by protoplast fusion of a lysine producer *Brevibacterium flavum* with a nonlysine producer of *B. flavum*. Chang et al. (1982) used this technique to combine the desirable properties of two strains of *Penicillium crysogenum* producing penicillin to one producer strain.

The problems associated with gene transmission are circumvented by recombinant DNA technology. A recombinant DNA molecule that contains a gene from one organism joined with regulatory sequences from other organisms may be produced. The capacity to produce recombinant DNA molecules has created the power and opportunity to produce novel gene combinations to suit specific needs. The recombinant DNA molecules are produced to obtain a large number of copies of specific DNA fragments, to recover large quantities of proteins produced by the concerned gene, or to integrate the gene into the chromosome of the target organism where it expresses itself. Debabof (1982) reported an improvement in threonine production by *Escherichia coli* by incorporating the entire threonine operon of a threonine analogue-resistant mutant into a plasmid and transforming it into the bacterium. The activity of threonine operon enzymes was increased 40 to 50 times. The recombinant strain produced 30 g/l of threonine when compared to 2–3 g/l produced by the wild type.

The starting point of any fermentation involves inoculation with a pure-culture, single-cell isolate. It is essential to prepare a stock culture of the pure culture. The working cultures are generated in a consistent manner from the stock culture. One of the most common methods used for preservation of microbes is lyophilization or freeze-drying. In this process, water and other solvents are removed from a frozen aqueous sample by sublimation. The application of lyophilization has been limited primarily to preservation of sporulating cultures that have the potential to survive the freezing and drying process. Bacteria are prepared as glycerol stocks in 5%–10% v/v glycerol or 5%–10% DMSO. Log-phase cells were used for preparation of the stock culture. The cryovials containing the stock culture were stored at −80°C in a freezer. To ensure long-term storage with genetic and phenotypic stability, all the bacterial stock cultures should be stored in the vapor phase of liquid nitrogen. For preservation of actinomycetes, the culture was harvested after 24–48 h of incubation and resuspended in a medium containing 10%–20% v/v glycerol as a cryoprotectant. Alternatively, the cultures can be frozen in the vapor phase of liquid nitrogen or in a mechanical freezer at −80°C. Freezing and storage in liquid nitrogen is the preferred method of culture preservation in filamentous fungi. The advantage of cryopreservation is that it has been successfully used for preservation of both sporulating and nonsporulating cultures (Dietz and Currie 1996).

4.5.4 Media Selection, Preparation, and Sterilization

Selection of a good medium is important to the success of an industrial fermentation. Media requirements depend on the type of microorganism used in the fermentation process. The medium should contain a source for energy, water, carbon sources, nitrogen sources, vitamins, minerals, buffers, chelating factors, air, and antifoaming agents. The culture medium should produce the desired product at a faster rate and lower yield than undesired products. Thus, the type and amount of nutrient components of a medium are critical. Natural media ingredients have high batch variation, so it is desirable to use defined or formulated media that have very little batch variation. A good medium would contain the minimum components needed to produce the maximum product yield.

Carbon serves as a major energy source for organisms. The product formation depends on the rate at which the carbon source is metabolized, and the main product of fermentation depends on the type of carbon source used. Carbon enters the pathway of the energy-yielding respiratory mechanism. The carbon sources for fermentation can be simple or complex carbohydrates, organic acids, proteins, peptides, amino acids, oils, fats, and hydrocarbons. Many microorganisms can use a single organic compound to supply both carbon and energy needs. The cost of carbon substrate considerably affects the final prices of any fermentative product (Sindhu et al. 2011). Hence, the use of industrial byproducts and biomass-derived carbon sources in fermentation are more cost-effective. There are several reports are available on the use of biodiesel industry-generated glycerol as a good carbon source for various microbial fermentation processes (Deepthi et al. 2011; Yoshikazu and Sei-ichi 2010; Mothes et al. 2007).

Following carbon, nitrogen is the next most plentiful substance used in fermentation media. Few microbes can utilize nitrogen as an energy source. It occurs in the organic compounds of the cell and in a reduced form in amino acids. The commonly used nitrogen sources in fermentation media are ammonia, ammonium salts, and urea. Other nitrogen sources include amino acids, proteins, sulfite waste liquor, corn steep liquor, and molasses. Nitrogen is used for the anabolic synthesis of nitrogen-containing cellular substances, such as amino acids, purines, DNA, and RNA.

Minerals supply the essential elements required for cells during their cultivation. The essential minerals for all media include calcium, chlorine, magnesium, phosphorous, potassium, and sulfur. Other minerals, such as copper, cobalt, iron, manganese, molybdenum, and zinc, are required in trace amounts. The trace elements contribute to primary and secondary metabolite production. The specific concentration of the different minerals depends upon the type of microorganism being used. The functions of trace elements include coenzyme functions to catalyze many reactions, vitamin synthesis, and cell wall support (Vogel and Todaro 1996). Primary metabolite function is not sensitive to trace element composition; secondary metabolite production is sensitive to trace element concentration.

Vitamins, amino acids, and fatty acids are used as growth factors in the fermentation process. Vitamins function as coenzymes to catalyze many reactions. The most frequently required vitamins are thiamine and biotin. Zabriskie and Zabriskie (1980) demonstrated the importance of biotin in the production of glutamic acid by *C. glutamicum*. A fivefold improvement in penicillin production was observed when corn steep liquor was added to the fermentation medium. The corn steep liquor contained phenylalanine and phenylethylamine, which are precursors for penicillin G (Vogel and Todaro 1996).

Chelating agents prevent formation of insoluble metal precipitates. One of the commonly used chelating agents is EDTA. It forms complexes with metal ions present in the medium and can be utilized by the microorganisms.

Fermentation media should contain buffers to retard changes in pH during microbial growth. Decarboxylation of organic acids; deamination of organic amines; and utilization of peptides, amino acids, or proteins can alter the medium pH value. Commonly used buffers are calcium carbonate, ammonia, and sodium hydroxide. The medium pH affects the ionic states of the medium components and the cell surface. Shifts in pH affect growth by affecting enzymes in the cytoplasmic membrane or enzymes associated with the cell wall.

Fermentation produces large amounts of foam in the fermenter. This is a result of microbial proteins or other media components. Foam will reduce the working volume of the fermenter vessel, decrease the rate of heat transfer, and deposit cells on the top of the fermenter. The air filter becomes wet, allowing growth of contaminating organisms. The usual procedure for controlling foam is to add an antifoaming agent. It lowers surface tension, which, in turn, decreases the stability of the foam bubbles so that they burst. Commonly used antifoaming agents are cottonseed oil, olive oil, castor oil, sulfonates, cod liver oil, silicones, linseed oil, and stearyl alcohol.

The factors that must be considered in developing a medium for large-scale fermentations include nutrients required by the selected organism, composition of the nutrients, and cost of the ingredients. For the development of media for an industrial process, the stability of the nutrient components is very important. Product concentration, yield, and productivity are the important process variables in determining conversion costs. Hence, a medium should be formulated in such a way that the process is economically feasible. Performing multivariable experiments with interaction of process variables, such as nutrients, aeration, temperature, incubation time, inoculum concentration, and media pH, allows the determination of the optimum level of a process. This can be done using a statistical approach. The various media constituents must be carefully sterilized by autoclaving or by filtration. Heat-stable constituents are autoclaved at 121°C for 15 min, and heat-labile components, such as vitamins, growth factors, proteins, serum, etc., must be sterilized through membrane filters.

4.5.5 STERILIZATION OF FERMENTER

Sterilization is one of the most important processes in industrial fermentation. It is usually carried out using a pure culture in which only selected strains are allowed to grow. If any contaminant exists in the medium or on any part of the equipment, the production organisms have to compete with the contaminants for limited nutrients. Hence, before starting fermentation, the medium and the fermenter have to be sterilized. It is normally achieved by heating jackets or coils of the fermenter with steam and sparging steam into the vessel through all entries. Steam pressure is held in the vessel for 20 min at 15 psi. It is essential to sparge sterile air into the fermenter after the cycle is complete and to maintain a positive pressure; otherwise, a vacuum may develop and draw unsterile air into the vessel (Stanbury et al. 1994). Sterilization of the fermenter or the media can be accomplished by destroying all living organisms by means of moist or dry heat; radiation, such as x-rays or ultraviolet light; or by mechanical means using ultrasonic vibrations, filtration, or high-speed centrifugation. Moist heat is the most widely used and effective method of sterilization of both liquid medium and heatable solid objects. Autoclaves are commonly used for sterilization.

4.5.6 PREPARATION OF INOCULUMS AND THE INOCULATION PROCESS

Inoculum media differ in composition from production media. The media is formulated in such a way as to quickly yield a large number of microbial cells in their proper physiological and morphological states while maintaining the genetic stability of the cells. The inoculum media usually contain a lower level of the main nutritive carbon source. The quality and reproducibility of the inoculum are critical factors that determine the reproducibility of product yields from one production run to another.

Substantial losses occur in the fermentation industry because of the variability in yields and productivity (Webb and Kamat 1993). In all fermentation processes, inoculum consistency in terms of size and quality is very important. The reproducibility of inocula was found to govern the reproducibility of the culture. During loop transfer, inoculum size varies depending upon the handling technique. Replacing the traditional loop transfer technique with a liquid-based method reduces variability in the inoculum from agar slant to the shaken flask stage. Liquid-based techniques can be automated to improve the industrial fermentation process.

In most SSF processes involving fungi, spore inocula are commonly used. A spore inoculum does not always give the best results. It allows greater flexibility in coordination of the inoculum preparation with the cultivation process. The spores retain viability for longer periods than the fungal mycelia and, therefore, can be stored and used when required. A disadvantage of the spores is that they are metabolically dormant. One of the most important variables of fungal growth is spore germination time because it determines the duration of the lag phase and, consequently, the duration of the fungal fermentation. A prolonged lag phase in fungal fermentation leads to an increased cost of production. Hence, it is essential to select a fungal strain with a short lag phase (Gaitis and Marakis 1994). The sporulation of the fungi is generally better on solid media than in liquid media. The even distribution and the density of the inoculum are also important. Several factors influence spore germination. These include initial humidity, initial pH, inoculum size, temperature, and nutrients. Spore germination and respiratory activity depend on the strain and culture conditions. The liquid inocula are easier to distribute evenly than the solid inocula. The inoculum density is also an important factor because overcrowding of the spores can inhibit the germination and development. However, the optimum density differs for each application; hence, no generalization can be drawn (Binod and Pandey 2009). Most industrial fermentation is done as batch operations. Seed cultures maintained under carefully controlled conditions are used for inoculation of petri plates or in liquid media in shake flasks. After sufficient growth, a preculture is used to inoculate the seed fermenter.

4.6 TYPES OF FERMENTATION AND FERMENTATION CONDITIONS

Fermentation may be carried out in batch, fed-batch, or continuous processes. Batch culture is a closed system, which contains an initial limited amount of nutrients. Constant conditions cannot be maintained because biomass, substrate, and substrate concentration vary with time. Growth at a constant rate is possible in the exponential phase of the culture when there are sufficient nutrients and growth is not inhibited by the accumulation of metabolites. Growth can be altered by changing the type of substrate or the temperature. Batch fermentations may be used to produce biomass, primary metabolites, and secondary metabolites. For the production of biomass, conditions that favor the fastest growth rate and maximum cell population would be used. In the production of primary metabolites, the conditions are set to extend the exponential phase accompanied by product formation, and for the production of a secondary metabolite, conditions giving a short exponential phase and an extended production phase are to be used (Stanbury et al. 1994).

Continuous culture starts as a batch culture first, and after establishment of growth, it is switched to continuous operation. There is no accumulation or depletion of nutrients; hence, the concentration remains constant as long as the dilution rate is kept constant. One of the major risks associated with continuous culture is the risk of contamination because of the flowing streams. Two-stage continuous cultures are used for the production of secondary metabolite. There is cell growth in the first reactor, and product formation occurs in the second reactor. Chemostat and auxostat systems are the major continuous systems used for physiological studies.

4.6.1 Monitoring the Fermentation Process

The success of fermentation depends on the defined conditions for biomass and product formation. So it is essential to monitor and control the process in order to obtain optimal operating conditions. Parameters, such as temperature, pH, degree of agitation, and oxygen concentration, may be monitored throughout the fermentation process. This can be achieved by careful monitoring of the fermentation so that any deviation from the optimum conditions can be corrected by a control system.

Shear is critical for the scale-up of the microbial culture process, irrespective of reactor configuration. It may influence the culture, causing damage, which may result in cell death or metabolic changes. Mijnbeek (1991) reported that damage to cells in airlift and stirred fermenters was a result of sparging and break-up of air bubbles on the medium surface. This type of damage causing cell death can be reduced by increasing the height-to-diameter ratio in the vessel, increasing the bubble size, decreasing the gas flow rate, and adding protective agents.

Temperature is one of the most important parameters to monitor and control in any process. Metal-resistant thermometers and thermistors are used in most fermentation processes. Thermistors are relatively cheap and very stable and give reproducible readings. Temperature control is monitored by probes positioned at suitable points in the vessel. Water jackets or pipe coils were used in the fermenter as a means of temperature control. In large fermenters, regulatory valves at the cooling water inlet may be sufficient to control the temperature.

The pH can be controlled by the addition of acid and base. Rapid changes in pH can be reduced by properly designing the media, especially the carbon and nitrogen sources. For maintaining pH, the controller is set to a predetermined pH value. When a signal is received, the valves open and pump acid or alkali into the fermenter for a short time. The addition is followed by mixing, and at the end of the mixing cycle, a reading is taken to check whether adequate correction for pH drift was made.

For filamentous fungi, when grown in submerged culture, the type of growth varies from pellet form to filamentous form. Mycelia morphology may be influenced by both concentration of the spores in a spore inoculum and the inoculum-development medium. Foster (1949) reported that a high spore inoculum will tend to produce a dispersed form of growth, and low concentrations favor pellet formation. In the commercial production of fungal products, it is necessary to grow the organism in the desired form. Whitaker (1992) observed that actinomycetes are capable of producing different morphological types. Mycelial forms of *Streptomyces griseus* and *S. hygroscopicus* are desirable for the production of streptomycin and turimycin, and the pelleted form is desirable for the production of glucose isomerase by *S. nigrificans*.

Dissolved oxygen is an important variable in the fermentation process. The dissolved oxygen level can be controlled by the speed of the agitator or the volume of the oxygen in the gas sparge. An alternative approach is to increase the ratio of oxygen to nitrogen in the input gas while maintaining a constant flow rate (Stanbury et al. 1994). In aerobic fermentations, it is essential to maintain the dissolved oxygen concentration above the specified minimal level. Steam sterilizable oxygen electrodes are used for monitoring.

Flow measurement of both gases and liquids is important in process management. Oxygen and carbon dioxide gas analyzers are used for measurement of gas analysis. Pressure measurement is crucial in many processes. It is essential for several reasons, and the most important is in terms of safety. Maintenance of pressure is important for media sterilization. It will influence the solubility of gases and contribute to the maintenance of sterility when a positive pressure is present. It is essential to monitor and record atmospheric pressure if oxygen concentrations in inlet and exit gases are to be determined by a gas analyzer. The pressure changes should be monitored continuously to make appropriate corrections.

Foam formation during fermentation causes serious problems if it is not controlled. Foaming is caused by several factors, such as salts, temperature, pH, media components, agitation, and airflow. The common practice is to add an antifoaming agent. The type of antifoam to be used needs to be tested to determine whether it produce any change in physiological behavior or protein quality and to check its effect on downstream processing. Foam depletes the dissolved oxygen level by causing an increase in the residence time of air bubbles in the reactor. It is usually measured using a conductance or capacitance probe. The presence of foam is detected when it touches the tip of the probe, thus completing a circuit within the vessel, which, in turn, actuates a pump to add a chemical antifoaming agent (McNeil and Harvey 2008). Mechanical antifoam systems, such as spinning discs and cones, are feasible in larger fermenters.

4.6.2 STIRRING MECHANISM

Mixing is usually carried out in a stirred tank. Cylindrical tanks are commonly used, and the base of the tank is rounded at the edges, which prevents sharp corners and pockets into which fluid currents may not penetrate and discourages formation of stagnant regions (Doran 2005). Many impeller designs are available for fermentation applications. The choice of an impeller depends on several factors, including viscosity of the liquid to be mixed and sensitivity of the system to mechanical shear. For low- to medium-viscosity liquids, propellers and flat-blade turbines are recommended. The most commonly used impeller in the fermentation industry is the Rushton turbine. Recently, a number of agitators have been developed to overcome problems associated with high-viscosity fermentations. The Scaba 6SRGT, dual impeller, intermig agitator, and multirod mixing agitator are a few among them. The Scaba 6SRGT is a radial-flow agitator, which can overcome problems associated with efficient bulk blending of high-viscosity fermentations. Good mixing and aeration in high-viscosity broths is achieved using a dual impeller, where the lower impeller acts as the gas disperser and the upper impeller acts as a device for circulating the vessel's contents. Intermig agitators are more complex in design; a large-diameter air sparger is used for optimization of air dispersion. The loss of power is lowered when compared to a Rushton turbine.

4.6.3 VISCOSITY EFFECTS

Viscosity is one of the most important factors that affect the flow behavior of a fluid. It has a marked effect on pumping, mixing, mass transfer, heat transfer, and aeration of fluids (Doran 2005). Viscosity of fermentation fluids is affected by the presence of cells, substrates, products, and air. Cell morphology has a profound influence on broth rheology. Filamentous growth produces structure in the broth, resulting in pseudo-plasticity, and broths containing pelleted cells tend to be more Newtonian. The pelleted form gives rise to much lower broth viscosity, resulting in lower power consumption. A medium with high viscosity needs a higher power input for effective stirring. Modern agitators like the Scaba 6SRGT, the Prochem Maxflo T, the lightning A315, and the Ekato intermig, which are derived from open turbines, can be used for mixing high-viscosity broths.

4.7 QUALITY CONTROL DURING UPSTREAM PROCESSING OF FERMENTATION

The aim of quality control in fermentation is to ensure that each batch of the fermented product has a satisfactory and uniform quality. Quality-control procedures are essential for the production of safe products and contribute to the success of the food-processing business. Inadequate quality control can have an adverse effect on local demand for the product. Hence, quality measures should be carefully monitored during the upstream stages of fermentation. Appropriate quality control procedures need to be developed and implemented. Careful monitoring is needed for selecting good-quality substrate and fermentation conditions, and high standards of personal hygiene should be ensured by the food processors.

4.8 CONCLUSION

Upstream processing in fermentation technology encompasses any technology that leads to the synthesis of a product. It includes screening and selection of a suitable organism for a particular product, development of a suitable medium, and finally the mass culturing of the organism for the product. The upstream processing costs about

20%–50%, whereas the downstream processing costs about 50%–80%. Upstream processing is crucial for getting the desired end product. The techniques of fermentation and fermentation parameters will affect the product yield, and proper quality-control measures should be taken during upstream operations for getting the desired product quality.

REFERENCES

Binod, P. and Pandey, A. (2009) Bioprocess technology and product development, in *Bioprocess and bioproducts—Technology, trends and opportunities*, S. Biswas, N. Kaushik and A. Pandey (eds), Asiatech Publishers, Inc., New Delhi, pp. 10–27.

Casida, L.E. (1993) *Industrial microbiology*, Wiley Eastern Limited, New Delhi, pp. 55–63.

Chang, L.T., Terasaka, D.T. and Elander, R.P. (1982) Protoplast fusion in industrial fungi, *Dev. Ind. Microbiol.*, 23, 21–29.

Debabof, V.G. (1982) *Overproduction of microbial products*, V. Krumphanzl, B. Sikyta and Z. Vanek (eds), Academic Press, London, p 345.

Deepthi, S.K., Binod, P., Sindhu, R. and Pandey, A. (2011) Media engineering for the production of poly-β-hydroxybutyrate production by *Bacillus firmus* NII 0830, *J. Sci. Ind. Res.*, 70, 968–975.

Dietz, A. and Currie, S. (1996) Actinomycetes, in *Maintaining cultures for biotechnology and industry*, J.C. Hunter-Cevera and A. Belt (eds), Academic Press, Inc., San Diego, CA, pp. 85–99.

Doran, P.M. (2005) *Bioprocess engineering principles*, Academic Press, California, pp. 129–163.

Foster, H.W. (1949) *Chemical activities of the fungi*, Academic Press, New York.

Gaitis, F. and Marakis, S. (1994) Tannin acid effects on spore germination time and mycelial morphology of *Aspergillus carbonarius, Micol. Neotrop. Appl.*, 7, 5–16.

Krishna, C. (2005) Solid-state fermentation systems: an overview, *Crit. Rev. Biotechnol.*, 25, 1–30.

Lonsane, B.K., Castenada, G.S., Raimbault, M., Roussos, S., Gonzalez, G.V., Ghildyal, N.P., Ramakrishna, M. and Krishnaiah, M.M. (1992) Scale-up strategies for solid-state fermentations, *Process Biochem.*, 27, 259–273.

McNeil, B. and Harvey, L.M. (2008) *Practical fermentation technology*, John Wiley and Sons, England, pp 271–288.

Mijnbeek, G. (1991) Shear stress effects on cultured animal cells, *Bioteknowledge (Applicon)*, 1, 3–7.

Mothes, G., Schnorpfeil, C. and Ackermann, J.U. (2007) Production of PHB from crude glycerol, *Eng. Life Sci.*, 7, 5, 475–479.

Nakayama, K., Kituda, S. and Kinoshita, S. (1961) Studies on lysine fermentation 1: The control mechanism of lysine accumulation by homoserine and threonine, *J. Gen. Appl. Microbiol.*, 7, 145–154.

Ooijkaas, L.P., Weber, F.J., Buitelaar, R., Tramper, J. and Rinzema, A. (2000) Defined media and inert supports: Their potential as solid state fermentation production systems, *Tibtech.*, 18, 356–360.

Pandey, A. (1992) Recent process developments in solid state fermentation, *Process Biochem.*, 27, 109–117.

Pandey, A., Soccol, C.R., Rodriguez-Leon, J.A. and Nigam, P. (2001) Solid state fermentation, in *Biotechnology fundamentals and application*, Asiatech Publisher, New Delhi, pp 17.

Pandey, A., Soccol, C.R., Nigam, P., Brand, D., Mohan, R. and Roussos, S. (2000) Biotechnological potential of coffee pulp and coffee husk for bioprocesses, *Biochem. Eng. J.*, 6, 153–162.

Sindhu, R., Ammu, B., Binod, P., Deepthi, S.K., Ramachandran, K.B., Soccol, C.R. and Pandey, A. (2011) Production and characterization of poly-3-hydroxybutyrate from biodiesel industry generated glycerol by *Bacillus sphaericus* NII 0838, *Braz. Arch. Biol. Technol.*, 54, 4, 783–794.

Stanbury, P., Whitaker, A. and Hall, S.J. (1994) Principles of fermentation technology (2nd edition), Elsevier, Burlington, MA, pp. 1–367.

Tosaka, O., Karasawa, M., Ikeda, S. and Yoshii, H. (1982) Production of lysine by fermentation, in Proceedings of the fourth International Symposium on Genetics and Industrial Microorganisms, Kodansha, Tokyo, p. 61.

Vogel, H.C. and Todaro, C.L. (1996) *Fermentation and biochemical engineering handbook*, second edition, Noyes Publications, New Jersey, pp. 122–160.

Webb, C. and Kamat, S.B. (1993) Improving fermentation consistency through better inoculum preparation, *World J. Microbiol. Biotechnol.*, 9, 308–312.

Whitaker, A. (1992) Actinomycetes in submerged culture, *Appl. Biochem. Biotechnol.*, 32, 23–35.

Yoshikazu, K. and Sei-ichi, A. (2010) Poly(3-hydroxybutyrate) production by isolated Halomonas sp. KM-1 using waste glycerol, *Biosci. Biotechnol. Biochem.*, 74, 175–177.

Zabriskie, D.W. and Zabriskie, D.H. (1980) *Trader's Guide to Fermentation Media Formulation*, Trader's Oil Mill Co., Ft. Worth, TX.

5 Theoretical Tools to Predict Physicochemical Properties of Industrial Foods and Cultivation Media

André Lebert

CONTENTS

5.1   INTRODUCTION

A major goal for food researchers and engineers in the food and biotechnology industry is to find effective simulation tools permitting the increase of nutritional and
safety qualities in transformed products as well as to assure a better homogeneity
of production and a cost allowing an increased competitiveness for the companies.

To achieve this ambitious goal, the project is voluntarily located in the field of
optimization and creation of products and processes. It requires the use of numeric
simulation. This is a method that consists of analyzing a phenomenon, a process, or
the behavior of a system through a mathematical model that correctly describes the
reactions of the studied object. However, to describe the transformation of an agricultural product in a food, it is necessary to describe it according to time and space:

- The heat and mass transfer that occur between the product and its environment or inside the product. These transfers result in the apparition of
  temperature, water content, and solute gradients, therefore modifying the
  chemical composition at every point of the product. They are characterized
  by exchange coefficients that depend on the environmental conditions (room
  temperature, relative humidity) and by diffusion coefficients that depend on
  the physicochemical properties of the product (temperature, water activity,
  pH, redox potential, etc.).
- The chemical, biochemical, or microbiological kinetics that have a place on
  the surface or in the product. These kinetics also lead local modifications of
  the chemical composition.
- The local modifications of chemical compositions that are led. These result in
  changes in the chemical potentials of the different present chemical species
  and, therefore, of the activities (in the thermodynamic sense) of these as the
  pH, the water activity (a_w), or the redox potential. But these activities depend
  notably on the rate constant of the kinetics or the availability of the solutes.
- The evolution of the mechanics linked to the structural, nutritional, microbial, and sensory properties of the product according to the transfers and
  the kinetics. These properties are the result of the evolutions described
  previously.

In a general way, the simulation tools, thus completed, will permit the description
of the spatial and time evolution of the physicochemical properties (temperature,
water content, activities of the media compounds) in a given product. The knowledge
of these properties is indispensable to describing and understanding the kinetic (bio)
chemical or microbiological changes occurring in a food or on its surface. These
generic tools will act as the basis of a tool for the conception of a new food or the
optimization of an existing food, according to sensory, nutritional, or microbiological criteria.

Water activity (a_w) and pH, with temperature, are the major physicochemical properties in biotechnology and the food industry: They are used to characterize the microbial behavior as well as the evolution of chemical reactions in the aqueous liquid solution. In the aqueous solutions used in pharmaceuticals, biology, or microbiology, many solutes can be added to the mixtures, and their concentrations may vary from one composition to another. These mixtures, in which water is the major component, called the solvent, can contain many different solutes in different concentrations. These solutes can be neutral molecules, such as sugars or alcohol, or electrolytes, such as acids, bases, or amino acids; all of them are used as a depressor of the water activity or used to modify the pH. The aim of this chapter is to show how it is possible to represent these properties (pH and a_w) in multicomponent solutions. Such a model would make it possible to respond to the expectations of biotechnology and food engineers for the design of equipment and processes. Indeed, a large amount of reliable data on the equilibrium properties of materials is necessary to describe the transformation of raw materials into finished foods or to formulate new foods with defined characteristics. However, the limited availability of experimental data can hardly satisfy such an enormous demand.

5.2   FROM EMPIRICISM TO KNOWLEDGE

5.2.1   Definitions

In thermodynamics, the activity of a component i is linked to the component concentration by the following relationship:

$$a_i = \gamma_i c_i$$

where γ_i is the activity coefficient (1 mol^{-1}), a_i is the solute activity, and c_i is the concentration (mol l^{-1}).

The activity coefficient describes deviation from the ideal solution behavior, that is, the difference that exists between the component activity and its concentration. Three types of solutions can be distinguished:

1. A *pure* solvent is an ideal solution in the sense of Raoult's law with $\gamma_i = 1$.
2. An *ideal* dilute solution: The solvent is present in excess compared to the solute. Henry's law can be applied to the solute, and γ_i approaches unity.
3. A *real* solution: γ_i is different from unity.

The deviation from ideal solution behavior has two consequences:

1. On the method to measure pH: pH is defined based on the activity of the proton, but it is usually believed that pH = $-\log(c_i)$, where c_i is the proton concentration. In the case of a strong acid, this expression is only valid when the concentration of the mixture is low.
2. On the estimation of pH and a_w: As soon as a salt is added to a medium, even at a very low concentration, the salt can greatly modify the chemical equilibrium as indicated by Gibbs–Duhem's law (Prausnitz et al. 1999). In

media containing nonelectrolyte species, deviation from the ideal increases when the concentration of a component increases or when the number of the components, even at a low concentration, increases.

Water activity is also defined as the ratio of the vapor pressure of water in a system to the vapor pressure of pure water at the same temperature or the equilibrium relative humidity (ERH) of the air surrounding the system at the same temperature. Thus, for a system at a uniform temperature θ, water activity can be expressed as follows:

$$a_w = \frac{p_w}{P_w^S} = ERH$$

where a_w is the water activity, p_w is the vapor pressure of water in the system, P_w^S is the vapor pressure of pure water at saturation, and ERH is the equilibrium relative humidity of air.

5.2.2 Empirical and Semiempirical Models

Many empirical and semiempirical models were developed during the last 50 years for predicting water activity in liquid solutions, and very few were developed for pH. For example, the buffering theory model (Wilson et al. 2000) is limited to the prediction of pH in a medium containing low acid concentrations and cannot predict the a_w during drying as a function of product composition and airflow properties. However, these models—water activity or pH—suffer numerous limitations, the slightest of which is the absence of generalization. Indeed, there are one or several parameters adjusted on experimental data. In consequence, only the principal empirical and semiempirical models of water activity will be evocated.

5.2.2.1 Empirical Models

Norrish (1966) derived an equation to predict the water activity of nonelectrolyte solutions on thermodynamic grounds. The equation is as follows:

$$a_w = m_w \exp\left(km_s^2\right)$$

where m_w is the mole fraction of water, m_s is the mole fraction of the solute, and k is an empirical constant.

For electrolytes, a deviation from linearity is observed and can be eliminated by the introduction of an intercept:

$$\ln\left(\frac{a_w}{m_w}\right) = km_s^2 + b$$

Lewicki (2008) summarizes the values of k constant for some biological molecules of interest: sugars, amino acids, or organic acids. Rahman and Perera (1997) modified the Norrish equation when the molecular weight of the solute is not known:

$$a_w = \frac{X_w}{X_w + EX_s} \left[\exp\left[k\left(1 - \frac{X_w}{X_w + EX_s} \right)^2 \right] \right]$$

where $E = M_w/M_s$ is the ratio of the molecular weight of water to the molecular weight of the solute, M_w is the mass of water, and M_s is the mass of solute.

5.2.2.2 Semiempirical Models Based on Raoult's Law

These models are based on Raoult's law for an ideal solution:

$$a_w = \frac{n_w}{n_w + n_s}$$

where n_w is the number of moles of water, and n_s is the number of moles of solute.

At solvation equilibrium, an average hydration number denotes the average number of molecules of bound solvent per solute molecule. For real solutions, the hydration number is given by the following equation:

$$h = \frac{55.51}{m} - \frac{a_w}{1 - a_w}$$

where a_w is the water activity, and m is the molality.

This approach was used by several authors (Stokes and Robinson 1966; Baranowska et al. 2001) to develop equations predicting activity of solution of several sugar solutions.

Raoult's law was modified by Palnitkar and Heldman (1971) by introducing the effective molecular weight of the solute:

$$a_w = \frac{(x_w/M_w)}{(x_w/M_w) + (x_s/M_{sw})}$$

where x_w is the mass fraction of water (kg water/kg solution), x_s is the mass fraction of the solute (kg solute/kg solution), M_w is the molecular weight of water (kg/mol), and M_s is the molecular weight of the solute (kg/mol).

Chen and Karmas (1980) used the above equation to calculate activity coefficients for sugars, polyols, amines and amino acids, organic acids, inorganic salts, and isolated soy proteins.

In order to take into account the interactions between water and solutes, indeed the nonideality of the solutions, Caurie (1983) derived an equation based on Raoult's law:

$$a_w = \frac{55.5}{m_s + 55.5} - \frac{m_s}{m_s + 55.5}(1 - a_w)$$

where m_s is the molal concentration of the solute (number of moles of solute/kg solvent).

5.2.2.3 Semiempirical Models Based on the Gibbs–Duhem Equation

A change in the activity of the solution resulting from changes in its composition in a rigorous form is given by the Gibbs–Duhem equation:

$$m_w dln\,(a_w) + m_1 dln\,(a_1) + m_2 dln\,(a_2) + m_3 dln\,(a_3) + \cdots = 0$$

where n_w is the mole fraction of water, a_w is the water activity, m_i is the mole fraction of solute i, and a_i is the activity of solute i.

From the Gibbs–Duhem equation, Ross (1975) deduced a simple expression of the water activity. There are two main steps in Ross's development:

1. Developing the ln function to its series through its Taylor development to the first order.
2. Neglecting the interactions between diluted components: The method consists of the substitution of the relative activities of each component in the specific solution by its standard activity in a binary water solution. In practice, the interactions between the different components are neglected in the water activity.

Thus, in the only case of multicomponent dilute solutions, the water activity a_w can be approximated by the following equation:

$$a_w = a_1 a_2 a_3 \cdots$$

where a_i is the standard binary activity of each component.

The errors in water activities calculated by the above equation are relatively small and do not exceed 1% in the concentration ranges that are usual for food products.

Caurie (1985) criticized the fact that interactions between components were neglected and modified in consequence of the equation. But Kitic and Chirife (1988) and Chen (1990) showed that the modified equation had so many deficiencies that the equation was useless in predicting a_w of simple and multicomponent mixtures.

5.2.3 Toward Knowledge Models

All the previous models give satisfactory predictions in the restricted domain of water activity and composition of media. If the composition of the biological medium or of the food is changed, the parameters have to be reidentified with new experimental data. It is clear that this is not the best way to solve the problem. A new approach must be developed with a single goal: The model has to be as predictable as possible. This constraint explains why the different developments in chemical engineering during the 1970s and later are of use.

5.2.3.1   A Brief History

Since the early years of physical chemistry, thousands of articles have been written in an effort to understand the behavior of mixed fluids. While there is not a general theory of liquid mixtures, there is, instead, a variety of restricted theories and models, each useful for a particular type of mixture (Prausnitz et al. 1999). All theories are based on excess partial molar free energy (g^E) and activity coefficient (γ) estimation, parameters that allowed the calculation of physicochemical properties (Figure 5.1). To construct a theory of liquid mixture, two kinds of information are required: the structure of the liquids (the way the molecules in a liquid are arranged in space) and the intermolecular forces between like and unlike molecules. Unfortunately, information of either kind is inadequate and, as a result, all theories must make simplifying assumptions to overcome this disadvantage (Prausnitz et al. 1999). More theoretical work has been concerned with mixtures of liquids whose molecules are nonpolar and spherical: For example, the regular solution theory of Scatchard and Hildebrand (Prausnitz et al. 1999) frequently provides a good approximation for mixtures of hydrocarbons. All theories are then extended, often semiempirically, to more complicated molecules. Among these theories, predictive activity coefficient methods were initially based on other models but are now mostly based on group-contribution models: analytical solutions of groups (ASOG) (Wilson and Deal 1962) and the universal functional activity coefficient (UNIFAC) (Fredenslund et al. 1975). These methods are used daily in the chemical industry, and new developments

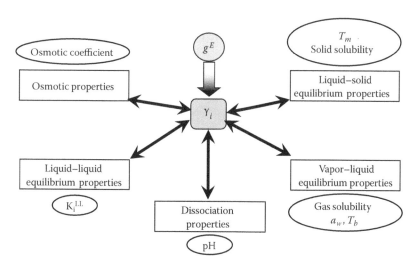

FIGURE 5.1   Excess Gibbs energy and physicochemical properties.

continue to be made in the cases of UNIFAC (Weidlich and Gmehling 1987; Larsen et al. 1987; Gmehling 1998) and ASOG (Kojima and Tochigi 1979; Tochigi et al. 1990) improving the range and accuracy of the methods. New methods, such as COSMOS-RS (Eckert and Klamt 2002; Klamt 1995), the group contribution solvation model (Lin and Sandler 1999), and the segment contribution solvation models (Lin and Sandler 2002), are being developed and improved.

Models for electrolyte activity coefficients result largely from that of Debye–Hückel for the long-range ion–ion interaction contribution like that of Pitzer (1973). Chen et al. (1982) extended the nonrandom two-liquid (NRTL) local composition model (Renon and Prausnitz 1968) to electrolyte solutions by considering two critical characteristics of electrolyte solutions (local electroneutrality and like-ion repulsion). Kikic et al. (1991), in order to take into account salt effects on vapor-liquid equilibria, added a Debye–Hückel term to the UNIFAC equation. Nevertheless, in all cases, Robinson and Stokes (1959), and then Achard et al. (1994), clearly identified the need to define the ionic entity in terms of its degree of hydration.

Finally, in the food industry, Lebert and Richon (1984), Le Maguer (1992), Achard et al. (1992), and Catté et al. (1994, 1995) did pioneering work on aqueous carbohydrate systems. Peres and Macedo (1997), and then Spiliotis and Tassios (2000), introduced new main groups to describe nonaqueous sugar solutions and mixed-solvent mixtures, such as ethanol–water or n-hexane–water. A UNIFAC equation is also used to predict the retention of aroma compounds by food (Sancho et al. 1997) or the a_w and pH of bacterial growth media (Coroller et al. 2001; Lebert et al. 2005).

5.2.3.2 Group-Contribution Models

Given the thousands of biochemical compounds of interest in food or biological processes, and the lack of thermodynamic data for many of them, it was decided to focus on predictive group-contribution methods. Indeed, in any group-contribution method, the basic idea is that, whereas there are thousands of chemical compounds of interest, the number of functional groups that constitute these compounds is smaller (Figure 5.2). Extending the group-contribution idea to mixtures is attractive because, although the number of pure compounds is very large, the number of

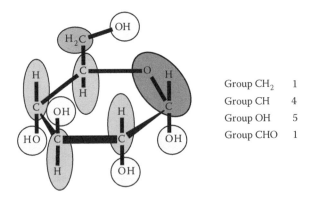

FIGURE 5.2 Group decomposition of glucose.

different mixtures is larger by many orders of magnitude. Millions of multicomponent mixtures of interest in the food industry can be constituted from perhaps 30, 50, or, at most, 100 functional groups.

The fundamental assumption of a group-contribution method is additivity: The contribution made by one group within a molecule is assumed to be independent of that made by any other group in that molecule. This assumption is valid only when the influence of any group in a molecule is not affected by the nature of other groups within that molecule. In consequence, any group-contribution method is necessarily approximate because the contribution of a given group in one molecule is not necessarily the same as that in another molecule.

5.2.3.3   Selection of Thermodynamic Model

Foods are complex media, which can be liquid or solid. Food constituents are diverse in size and molecular behavior: water, organic acids, bases, electrolytes, amino acids, nucleic acids, and metabolites. By characterizing the types of interactions occurring between these constituents and by considering the dissociation reactions in complex aqueous mixtures and the thermodynamic theories of solutions, researchers have developed models that predict the physicochemical properties of such mixtures (pH, a_w, concentration of the constituents) (Pinho et al. 1994, Achard et al. 1994; Sereno et al. 2001). These models determine the activity coefficient of each constituent present in the solution and calculate the activity:

$$a_i = \gamma_i x_i$$

where γ_i is the activity coefficient, a_i is solute activity, and x_i is the molar fraction of the component i in the medium determined from the medium composition.

When the activity coefficients of all the components in the solution are known, the activity coefficients of water and hydrogen ions can be calculated and used to determine two major parameters affecting bacterial growth, the pH (Equation 4) and the a_w (Equation 5):

$$\text{pH} = -\log_{10}\left(a_{H^+}^m\right) = -\log_{10}\left(\gamma_{H^+}^m m_{H^+}\right)$$

where $\gamma_{H^+}^m$ is the activity coefficient of H^+ defined on the molality scale, $a_{H^+}^m$ is the activity of H^+ defined on the molality scale, and m_{H^+} is the molality of H^+ (mol H^+/kg water) defined on the molar fraction scale with standard pure water as a reference state.

$$a_w = \gamma_{H_2O} x_{H_2O}$$

where γ_{H_2O} is the activity coefficient of water, a_w is the activity of water, and x_{H_2O} is the molar fraction of water.

The model by Achard et al. (1994) is a good example because it can take the equilibrium properties of water, from which pH and a_w, are derived into account as well as all the equilibrium properties of all the constituents of the liquid solutions.

5.2.3.4 Structure of Achard's Thermodynamic Model

The thermodynamic properties of a mixture depend on the forces that exist between the species in the mixture. When electrolytes are considered, the system is characterized by the presence of both molecular species and ionic species, resulting in three different types of interactions: ion–ion, molecule–molecule, and ion–molecule. Ion–ion interactions are governed by electrostatic forces between ions that have a much longer range than other intermolecular forces. Molecule–molecule and ion–molecule interaction forces are known to be short-range in nature. The excess Gibbs energy of systems containing electrolytes can be considered as the sum of two terms, one related to long-range forces between ions, and the other related to short-range forces between all the species.

The solution model developed by Achard et al. (1994) was used to estimate the order of magnitude of the deviation for all of the species, charged or not, in relation to an ideal infinitely diluted solution (Figure 5.3). It is formed by the juxtaposition of three contributions that take account of factors related to the size of the molecules present, the electrostatic effects for the charged species, and the solvation of some ionic forms, respectively.

1. The UNIFAC model modified by Larsen et al. (1987) was used to predict the activities of nonionic molecules in mixtures such as sugars, alcohols, etc. This model is based on the concept of group contributions: Each molecule is decomposed into functional groups. The activity coefficient is the sum of the two terms: The first takes account of the size and surface areas of individual functional groups using structural parameters and the second of the energetic interactions between groups.
2. The model describing long-term interactions between ionic species is based on the Pitzer–Debye–Hückel theory (Pitzer 1973). Ions are considered to be UNIFAC independent groups. The crystal ionic radii of the elements necessary to calculate the group volume and surface area parameters were those of Weast (1972).

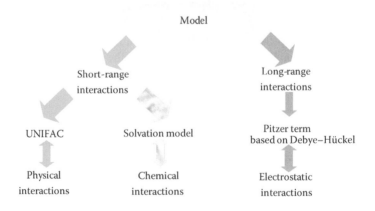

FIGURE 5.3 Structure of thermodynamic model.

3. Solvation of charged species giving clusters was taken into account by means of a hydration number for each ion at infinite dilution. A mixture containing water, one anion, and one cation was then described using six group interaction parameters and two hydration numbers; this number was reduced, based on the following three assumptions:

    i. Repulsive forces between ions of like charge are extremely large.

    ii. Interactions between ions of opposite charge are dominated by electrostatic forces and are accounted for by the PDH term.

    iii. The energy of interaction between two molecules of water can be obtained from the difference of internal energy between the gaseous and liquid states.

With this approach, two interaction parameters (water–anion and water–cation interactions) were sufficient to characterize a water–salt system. These interaction parameters and the hydration number for each ion were evaluated for 43 anions and cations from a database of 110 water–salt systems (Achard et al. 1994). Table 5.1 summarizes the different groups and subgroups taken into account in the modified UNIFAC model by Larsen et al. (1987) and the ionic groups introduced by Achard et al. (1994). Ben Gaïda et al. (2006, 2010) modified the solvation term by introducing the concept of variable solvation.

The combination of these three contributions is the basis of the solution model. It makes it possible to calculate the activities of different species. These calculations are combined with thermodynamic equilibrium equations that use the equilibrium constants (pK) defined at infinite dilution and generally provided in the reference tables. The detailed composition of the solution (concentrations of different charged species) and the activities of the different solutes are determined by solving the equation system. At least, activity coefficients give access to properties, such as water activity, osmotic pressure, freezing-point depression, boiling-temperature increase, pH, and acidity.

The software manages several types of data (Figure 5.4) that allow the predictions of pH and a_w of aqueous mixtures:

- The list of functional groups and subgroups (Table 5.1)
- The physical parameters (R_k surface parameter and Q_k volume parameter) of each functional group
- The energetic interaction parameters (a_{mn} and a_{nm}) between the m and n functional groups
- The list of each chemical compound described in the software
- The characteristic data of each compound (Figure 5.5):
  - Neutral compounds are described by one file with the molecular weight and the decomposition in functional groups.
  - Ionic compounds are described (salts, acids, bases, amino acids) and represented by several files: one for the neutral form and one for each ionized form.

When the next compound is integrated in the software, these data must be delivered.

TABLE 5.1
List of Groups and Subgroups Used in Thermodynamic Model

(a) Nonelectrolyte Compounds

Group Number	Group Formula	Subgroup Number	Subgroup Formula	Group Number	Group Formula	Subgroup Number	Subgroup Formula
1	CH_2	1	CH_3	8	CHO	17	CHO
		2	CH_2	9	CCOO	18	CH_3COO
		3	CH			19	CH_2COO
		4	C	10	CH_2O	20	CH_3O
2	C=C	5	$CH_2=CH$			21	CH_2O
		6	CH=CH			22	CHO
		7	$CH_2=C$			23	FCH_2O
		8	CH=C	11	NH_2	24	NH_2
		9	C=C	12	CH_2NH	25	CH_3NH
3	ACH	10	ACH			26	CH_2NH
		11	AC			27	CHNH
4	OH	12	OH	13	CH_2N	28	CH_3N
5	CH_3OH	13	CH_3OH			29	CH_2N
6	H_2O	14	H_2O	14	ANH_2	30	ANH_2
7	CH_2CO	15	$CH_3-C=O$				
		16	$CH_2-C=O$				

Group Number	Group Formula	Subgroup Number	Subgroup Formula
15	Pyridine	31	C_5H_5N
		32	C_5H_4N
		33	C_5H_3N
16	CH_2CN	34	CH_3CN
		35	CH_2CN
17	COOH	36	COOH
18	CCl	37	CH_2Cl
		38	CHCl
		39	CCl
19	CCl_2	40	CH_2Cl_2
		41	$CHCl_2$
20	CCl_3	42	CCl_2
		43	$CHCl_3$
		44	CCl_3

(b) Electrolyte Compounds

No	Species	No	Species	No	Species	No	Species	No	Species	No	Species
22	H^+	46	H^+	40	Fe	68	Fe^{2+}	59	ClO_4^-	89	ClO_4^-
23	K^+	47	K^+			69	Fe^{3+}	60	BrO_3^-	90	BrO_3^-
24	Na^+	48	Na^+	41	Sr^{2+}	70	Sr^{2+}	61	CNS^-	91	CNS^-
		49	NH_4^+	42	Ce^{3+}	71	Ce^{3+}	62	ClO_3^-	92	ClO_3^-
		50	NH_3^+	43	Nd^{3+}	72	Nd^{3+}	63	CrO_4^{2-}	93	CrO_4^{2-}
25	NH_x	51	NH_2^+	44	La^{3+}	73	La^{3+}	64	SO_x	94	SO_4^{2-}
		52	NH^+	45	Cr^{3+}	74	Cr^{3+}			95	SO_3^{2-}
		53	N	46	Pr^{3+}	75	Pr^{3+}			96	HSO_3^-
26	Ag^+	54	Ag^+	47	Sm^{3+}	76	Sm^{3+}			97	HSO_4^-
27	Cs^+	55	Cs^+	48	Al^{3+}	77	Al^{3+}			98	$S_2O_3^{2-}$
28	Li^+	56	Li^+	49	Sc^{3+}	78	Sc^{3+}			99	$S_2O_4^{2-}$
29	Ca^{2+}	57	Ca^{2+}	50	Y^{3+}	79	Y^{3+}	65	Misc	100	HS^-
30	Mg^{2+}	58	Mg^{2+}	51	Eu^{3+}	80	Eu^{3+}			101	S^{2-}
31	Mn^{2+}	59	Mn^{2+}	52	Rb^+	81	Rb^+			102	HCO_3^-
32	Ba^{2+}	60	Ba^{2+}	53	OH^-	82	OH^-			103	CO_3^{2-}
33	Cd^{2+}	61	Cd^{2+}	54	Cl^-	83	Cl^-			104	CN^-
34	Cu^{2+}	62	Cu^{2+}	55	Br^-	84	Br^-			105	$H_2PO_4^-$
35	Zn^{2+}	63	Zn^{2+}	56	F^-	85	F^-			106	HPO_4^{2-}
36	Be^{2+}	64	Be^{2+}	57	I^-	86	I^-			107	PO_4^{3-}
37	Ni^{2+}	65	Ni^{2+}	58	NO_x	87	NO_3^-			108	$-COO^-$
38	Co^{2+}	66	Co^{2+}			88	NO_2^-			109	$HCOO^-$
39	Pb^{2+}	67	Pb^{2+}								

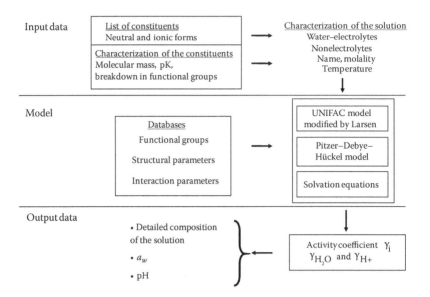

FIGURE 5.4   Data structuration.

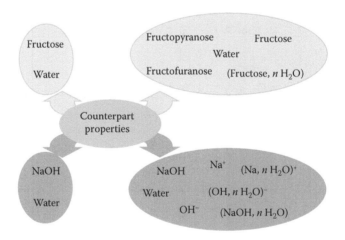

FIGURE 5.5   Management of chemical species.

5.2.3.5   Extension to Complex Media

As stated previously, the thermodynamic model was developed for aqueous mixtures. But gelatin or beef are complex media with a very large number of compounds that are not necessarily identified or well characterized. To extend the thermodynamic model to such complex media, three possible ways can be followed:

1. An exhaustive description of the media, which is impossible in the case of gelatin or beef.

2. A simplified composition having the same behavior as that of the complex media. A major obstacle is the choice of the molecules of interest that can be used in the simplified mixture: Are they majority molecules? Are they only molecules with well-known physicochemical properties (pKas, etc.)?
3. A virtual molecule having the same behavior as that of the complex media: It is necessary to determine the equivalent molecular weight, the functional groups, and the pKas.

5.3 SOME APPLICATIONS OF THERMODYNAMIC MODEL

In this section, some examples of prediction of pH and a_w will be given for more and more complex media: from binary mixtures of water with one solute (salt, organic acid, amino acid, or sugar) to meat products through microbiological broth media.

5.3.1 BIOLOGICAL MOLECULES

5.3.1.1 Sugars and Salts

The thermodynamic model can predict the water activity of pH of binary and ternary aqueous solutions made of water and one or two solutes:

- Salts: NaCl, $NaNO_2$, $NaNO_3$, and KCl
- Polyols: glycerol, glucose, sucrose, maltose, and fructose

Water activity predictions are accurate in a large domain of molality for salts and sugars (Figure 5.6): Indeed the absolute average relative errors are less than 0.1% when molalities are less than 1.3. For higher molalities, up to saturation in the case of sucrose and glucose, the absolute average relative errors are less than 1%.

Predictions of pH obtained for mixtures containing increasing concentrations of NaCl, $NaNO_2$, and $NaNO_3$ up to 3 mol solute/kg water and buffered at pH 7 or 5.8 with a KH_2PO_4/K_2HPO_4 buffer system are accurate (Figure 5.7).

Water activity predictions (Table 5.2) are accurate in a large domain of molality: Indeed the absolute average relative errors are less than 0.1% when molalities are less than 1.3. For higher molalities, up to saturation in the case of sucrose and glucose, the absolute average relative errors are less than 1.2%.

5.3.1.2 Amino Acids

Eighteen amino acids can be used in the thermodynamic model: Indeed, for cysteine and methionine, two amino acids containing groups with sulfur, it is not possible to decompose them into functional groups in the UNIFAC model modified by Larsen because the groups containing sulfur are not available.

Titration curves were realized with solutions of NaOH 0.1 M and HCl 0.1 M for the majority of the acids and solutions at 0.01 M for L-tyrosine. With the exception of glutamic acid, the amino acids tested were in the L series. The pH values

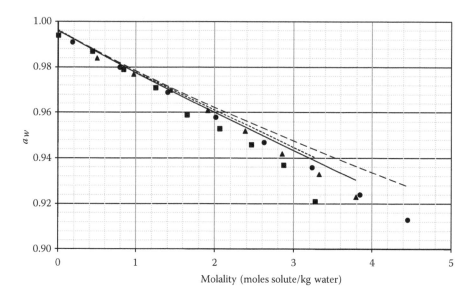

FIGURE 5.6 Comparison of water activity of three (water + sugar) mixtures at 20°C (symbols) and predicted using thermodynamic model (lines) in KH_2PO_4/K_2HPO_4 buffer at pH 5.6 or 7.0: glycerol (●, – – –), glucose (▲, ——), and saccharose (■, ·········).

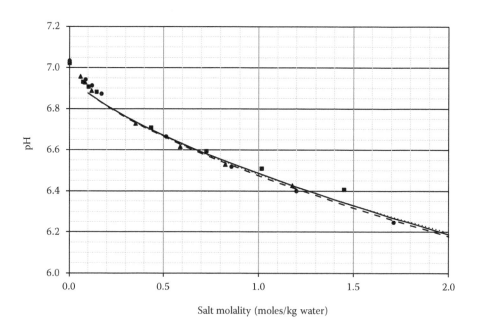

FIGURE 5.7 Comparison of water activity of three (water + salt) mixtures at 20°C (symbols) and predicted using thermodynamic model (lines): NaCl (●, ——), $NaNO_3$ (▲, – – –), and $NaNO_2$ (■, ·········).

TABLE 5.2

a_w Absolute Average Relative Error for Binary Aqueous Solutions

Solute	Molality Range (mol/kg of water)	a_w Average Relative Error (%)		Molality Range (mol/kg of water)	a_w Average Relative Error (%)	
Glucose	0.0–1.1	0.04	(13)	0.0–5.5	0.68	(9)
Fructose	0.0–1.7	0.07	(15)	0.0–11.1	0.99	(11)
Maltose	0.0–1.1	0.09	(16)	0.0–2.8	0.19	(9)
Sucrose	0.0–1.2	0.12	(16)	0.0–5.8	1.16	(11)
Glycerol	0.0–1.1	0.06	(13)	0.0–5.4	0.32	(9)
NaCl	0.0–1.0	0.03	(13)	0.0–3.4	0.14	(9)
KCl	0.0–1.3	0.04	(15)	0.0–3.4	0.42	(10)

were accurately predicted by the Achard model for amino acids, such as glycine, L-aspartic acid, and L-glutamic acid (Figure 5.8).

The predictions for the other amino acids are also correct. Small differences between measurements and predictions were observed for some of the amino acids. It was possible to correct these differences by estimating the pK values a second time. This was justified by the fact that the values proposed in the tables were generally obtained by extrapolation at the infinite dilution from measurements in diluted solutions that were almost ideal, whereas the estimates really do take activity corrections into account. These values are given in Table 5.3.

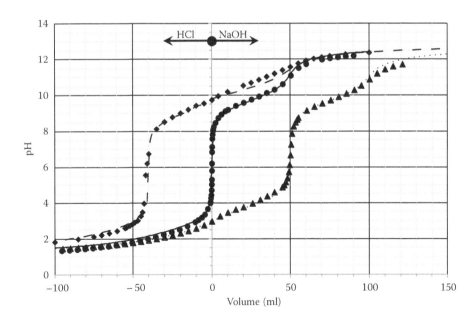

FIGURE 5.8 Comparison between experimental (symbols) and predicted (line) titration curves: glycine (●, ——), L-glutamic acid (▲, ⋯⋯⋯), and L-alanine (◆, – – –).

TABLE 5.3
pKa Estimation for Amino Acids

	pKa	pKb	pKc
L-alanine	2.35[a]	9.87[a]	
L-arginine			
L-asparagine	2.16[b]	8.73[b]	
L-aspartic acid	1.99[a]	10.00[a]	3.90[a]
L-glutamic acid	2.16[a]	[9.90]	4.27[a]
L-glutamine	2.18[b]	9.00[b] [9.25]	
Glycine	2.35[a]	9.78[a]	
L-histidine			
L-isoleucine	2.26[b]	9.60[b] [9.85]	
L-leucine	2.33[a]	9.74[a]	
L-lysine			
L-phenylalanine	1.85[a] [2.18]	9.31[a]	
L-proline	1.95[a]	10.64[a] [10.80]	
L-serine	2.19[a]	9.21[a] [9.30]	
L-threonine	2.09[a]	9.10[a]	
L-tryptophane	2.38[b]	9.34[b]	
L-tyrosine	2.20[a]	10.54[a] [9.04]	
L-valine	2.29[a]	9.72[a] [9.90]	

Sources: [a]Ould-Moulaye, C.-B., Calcul des propriétés de formation en solution aqueuse des composés impliqués dans les procédés microbiologiques et alimentaires: Prédiction et réconciliation de données, Modélisation des équilibres chimiques et des équilibres entre phases. PhD thesis, Université Blaise Pascal de Clermont-Ferrand, France, 1998; [b]Lide, D.R., Section 7: Biochemistry. In *Handbook of chemistry and physics*, 80th edition, CRC Press, Boca Raton, 7.1–7.11, 1999–2000.

Note: Brackets indicate data estimated a second time.

In the case of glutamic acid, two titration curves were performed, one for the L-glutamic acid and the other for the DL-glutamic acid. The L-glutamic acid curve is accurately predicted by the model, whereas a shift was observed for DL-amino acid. The presence of the two forms of amino acid in the solution leads to a different conformation of the molecules because of a dimerization or a complexation phenomenon that is not taken into account by the model. The Achard model can be used for the prediction of pH for mixtures with two ternary or three amino acids (Figure 5.9).

5.3.2 Broth Media and Gelatin Gel

5.3.2.1 Composition of Broth Media and Gelatin Gel

Bacteria are grown in complex bacterial media, which were broth or gelatin gel. Their constituents are diverse in size and molecular behavior: water, organic acids,

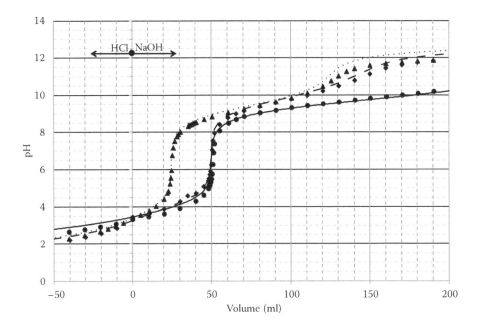

FIGURE 5.9 Experimental titration curve for mixture (a) (●), mixture (b) (▲), and mixture (c) (◆). Comparison with predicted curves, respectively, for mixture (a) (——), mixture (b) (·········), and mixture (c) (– – –). Mixture (a): L-alanine (0.10 mol l⁻¹)–L-aspartic acid (0.10 mol l⁻¹). Mixture (b): L-aspartic acid (0.05 mol l⁻¹)–L-alanine (0.05 mol l⁻¹)–L-serine (0.10 mol l⁻¹). Mixture (c): L-aspartic acid (0.05 mol l⁻¹)–L-glutamic acid (0.05 mol l⁻¹)–L-proline (0.05 mol l⁻¹)–L-serine (0.05 mol l⁻¹).

bases, electrolytes, amino acids, nucleic acids, and metabolites. In this study, broths were made of water, glucose, a phosphate buffer, and three peptones. The peptones (meat extract, proteose peptone, and tryptone) are composed of a digest of protein-aceous material. The gelatin gel is made of water, glucose, the three previous peptones, and gelatin (Lebert et al. 2004). Gelatin is composed of polymers of amino acids obtained after the hydrolysis of collagen.

The tryptic meat broth (TMB), sterilized by filtration, contained meat extract (10 g l⁻¹), proteose peptone (10 g l⁻¹), tryptone (5 g l⁻¹), and glucose (5 g l⁻¹). The medium was buffered with a K_2HPO_4–KH_2PO_4 0.1 M solution (Table 5.5). NaCl, KCl, D(+)-anhydrous glucose, D(+)-sucrose, and glycerol were added to TMB to obtain a_w ranging from 0.80 to 0.99. These media are, respectively, referred to as broth + NaCl, broth + KCl, broth + glucose, broth + sucrose, and broth + glycerol.

A pure gelatin gel was prepared by melting gelatin granules in pure distilled water at 58°C for 45 min with slight agitation. The water content was adjusted to 3 kg of water for 1 kg of anhydrous gelatin (moisture content of 75% w/w). This pure gelatin preparation was gelatinized at 4°C for at least 24 h.

A gel (TMB_g) was derived from the TMB and did not contain the phosphate buffer. It was made of 120 g gelatin powder, 2.5 g proteose peptone, 1.25 g tryptone, 2.5 g meat extract, and 1.25 g glucose.

TABLE 5.4

Decomposition of Peptones into Functional Groups

	Mean Apparent Mass	Decomposition	pKa
Proteose peptone	214.3	3 CH$_3$, 4 CH$_2$, 4 CH, 1 NH$_3^+$, 1 COO–	4.0–9.5
Tryptone	227.3	3 CH$_3$, 4 CH$_2$, 5 CH, 1 NH$_3^+$, 1 COO–	4.5–9.7
Meat extract	172.2	2 CH$_3$, 3 CH$_2$, 3 CH, 1 NH$_3^+$, 1 COO–	4.2–9.9
Gelatin	1763	12 CH$_3$, 14 CH$_2$, 3 CH, 24 NH$_2$, 1 NH$_3^+$, 11 COOH, 1 COO$^-$, 7 CH$_2$CO, 7 CH$_2$NH	1.2–2.2–3.9–8.4–10.9–12.9

5.3.2.2 Titration and Sorption Curves of Broth Media

Table 5.11 gives an estimation of the mean apparent molar mass and the two pK for each peptone that can be found in broth. They were then broken down into functional groups. From these values, it was possible to propose a decomposition of the peptones and gelatin into functional groups (Table 5.4).

Accurate predictions of pH were obtained in broth containing increasing concentrations of glycerol up to 4.45 mol solute/kg water (Figure 5.10). In the presence of NaCl, good predictions were obtained: A slight deviation between the experimental and predicted values was observed in broth at pH 5.8 and reached 0.2 pH unit for

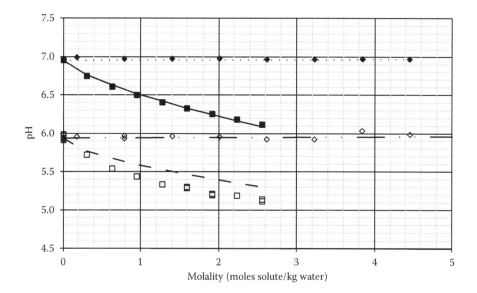

FIGURE 5.10 Comparison between experimental (symbols) and predicted (lines) pH in TMB buffered at pH 7 or 5.8 with KH$_2$PO$_4$/K$_2$HPO$_4$, 0.1 M with increasing concentrations of NaCl or glycerol: pH 7.0 + NaCl (■, ——), pH 5.8 + NaCl (□, – – –), pH 7.0 + glycerol (◆, ·········), and pH 5.8 + glycerol (◇, – ···).

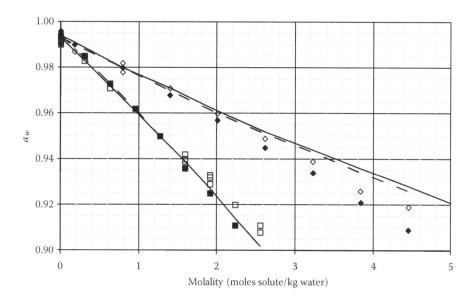

FIGURE 5.11 Comparison between experimental (symbols) and predicted (lines) a_w in TMB buffered at pH 7 or 5.8 with KH_2PO_4/K_2HPO_4, 0.1 M with increasing concentrations of NaCl or glycerol: pH 7.0 + NaCl (■, ——), pH 5.8 + NaCl (□, – – –), pH 7.0 + glycerol (◆, ·········), and pH 5.8 + glycerol (◊, – ···).

concentrations of 2.5 mol NaCl/kg water. Accurate predictions of a_w were obtained in broth containing increasing concentrations of NaCl up to 2.5 mol solute/kg water regardless of the pH tested (Figure 5.11). The differences between experimental and predicted values were less than 0.005 a_w unit for molality below 2 mol NaCl/kg water. In the presence of glycerol, deviations were observed but were less than 0.006 a_w unit for molality below 3 mol glycerol/kg water. Similar results were obtained for broth containing KCl, glucose, or saccharose.

5.3.2.3   Titration and Sorption Curves of Gelatin

In the case of gelatin, for the first time, it was necessary to measure the volume variations that one can observe when water and different chemical compounds are mixed to take account of them in the model: Thus, a variation of 15% of the volume was observed for miscellaneous water–gelatin–NaCl mixtures. To take account of the peptidic bonds, CH_2CO and CH_2NH groups were introduced because the group CO-NH does not exist in the thermodynamic databases. A chemical compound—equivalent molecular mass of 1763 and 6 pKas—was found to have, in many conditions, the same behavior as gelatin. The decomposition of this compound into functional groups is given in Table 5.4.

The model predicts the pH and the a_w of solid media made of water, gelatin, and salts. Good predictions for the sorption curve of pure gelatin are shown in Figure 5.12. The deviations between experimental and predicted values are less than 0.01 a_w unit for gel at a_w above 0.90. Deviation was higher (between 0.01 and 0.02 a_w unit) for a gel having a measured a_w between 0.70 and 0.90. Water activity is also correctly predicted when increasing the sodium chloride content up to 42% (w/w) (Figure 5.12). The addition of different salts

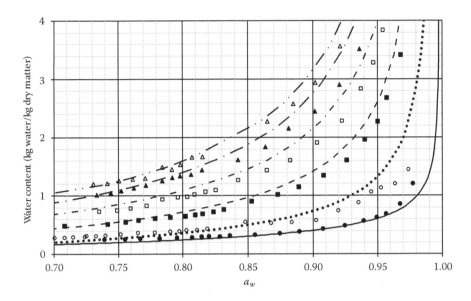

FIGURE 5.12 Incidence of sodium chloride content on gelatin sorption curves measured at 20°C (symbols) and predicted using the thermodynamic model (lines): pure gelatin (●, ——), gelatin + 5.8% of NaCl (O, ·········), gelatin + 15.0% of NaCl (■, – – –), gelatin + 25.9% of NaCl (□, – · – ·), gelatin + 35.8% of NaCl (▲, – · · –), and gelatin + 42.2% of NaCl (△, – · · · –) in mass on anhydrous gel basis.

(sodium acetate, sodium lactate, potassium lactate, and potassium sorbate) to gelatin is, at least, predicted (Figure 5.13) with good precision (mean error less than 0.02 a_w unit).

Titration curves are also correctly predicted with an average error of 0.2 pH unit (the maximum error is less than 0.5 pH unit) when gelatin is mixed with lactic acid (Figure 5.14). It must be noted that the noneffect of the NaCl addition is well predicted (Figure 5.14). Similar results were obtained with the addition of acetic, lactic, sorbic, and citric acid.

5.3.3 FOOD PRODUCTS

5.3.3.1 Sugar Mixtures

Ben Gaïda et al. (2006) applied the thermodynamic to multicomponent mixtures more representative of food products. For the first time, they showed that a thermodynamic approach can give a phase diagram for glucose and fructose, which show the existence of hydrated forms (glucose, 1 H_2O and fructose, 2 H_2O) and anhydrous forms. The model is also able to represent the metastable line observed experimentally, far below the eutectic point.

They also studied the ability of the model to predict water activity of concentrated apple juice and honey. The prediction of water activity in Greek honeys of different geographical and botanical origins is satisfactory in a range of 0.5–0.8.

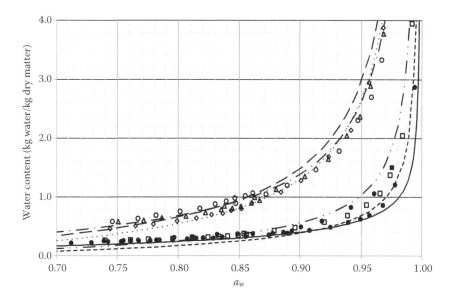

FIGURE 5.13 Incidence of different salt content on gelatin or TMBg sorption curves measured at 20°C (symbols) and predicted using thermodynamic model (lines): pure gelatin (●, ——), TMBg (■, ⋯⋯), gelatin + 11% of sodium lactate (○, – – –), gelatin + 10% of potassium lactate (◇, ⋯⋯), gelatin + 5.3% of sodium acetate (△, – · –), and TMBg + 2.0% of potassium sorbate (□, – · · –) in mass on anhydrous gel basis.

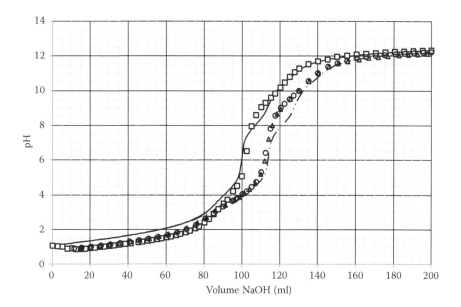

FIGURE 5.14 Incidence of different NaCl and lactic acid content on gelatin titration curves measured (symbols) and predicted (lines) at 25°C: gelatin (□, ——), gelatin + 5% lactic acid (△, ⋯⋯), and gelatin + 5% lactic acid + 5% NaCl (○, – – –).

5.3.3.2 Beef

A chemical compound—equivalent molecular mass of 2229 and 6 pKas—was found to have, under many conditions, the same behavior as beef. The model predicts the pH and the a_w of beef with or without the addition of different additives (salts or organic acids or salts of organic acids). Excellent predictions are obtained for the titration curve of beef (Figure 5.15): The deviations between experimental and predicted values are less than 0.2 pH unit in the pH domain from 2.0 up to 10.0. Good results are obtained when organic acids are added: The average error is equal to 0.3 pH unit, and the maximum error is less than 0.6 pH unit.

Sorption curves are also correctly predicted for beef when sodium chloride is added (Figure 5.16). The deviations between experimental and predicted values are less than 0.01 a_w unit for gel at a_w below 0.95. Deviation was higher (between 0.01 and 0.02 a_w unit) for a gel having a measured a_w above 0.95. The addition of different salts (sodium acetate, sodium lactate, potassium lactate, and potassium sorbate) to gelatin is, at least, predicted with a good precision (mean error less than 0.02 a_w unit).

5.3.4 APPLICATION TO PREDICTIVE MICROBIOLOGY

The quality and safety of food products depend on the microorganisms, the food characteristics, and the process. Food products may become unsafe because of the growth of pathogenic or spoilage bacteria if they are not properly stored, processed, packaged, or distributed. The concept of predictive microbiology was developed to evaluate the effect

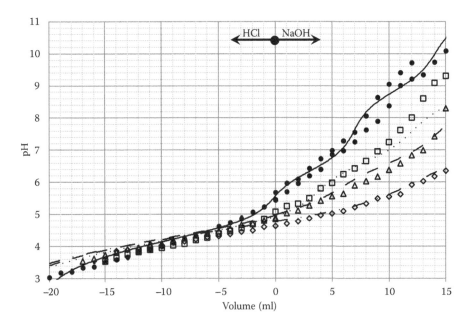

FIGURE 5.15 Incidence of different NaCl and lactic acid content on gelatin titration curves measured (symbols) and predicted (lines) at 25°C: gelatin (□, ——), gelatin + 5% lactic acid (△, ·········), and gelatin + 5% lactic acid + 5% NaCl (●, – – –).

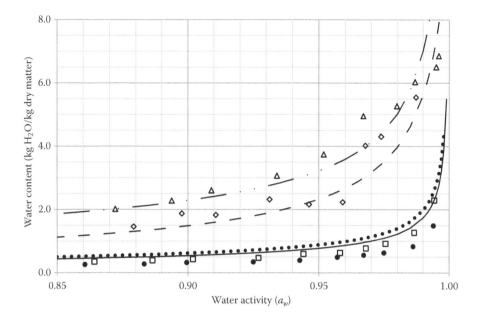

FIGURE 5.16 Incidence of different NaCl content on beef meat sorption curves measured (symbols) and predicted (lines) at 25°C: beef meat (●, ———), beef meat + 0.5% NaCl (□, ·········), beef meat + 5% NaCl (◇, – – –), and beef meat + 10% NaCl (△, – · · –).

of processing, distribution, and storage operations on the microbiological safety of foods (McMeekin et al. 2002), and is based on the use of mathematical models that correlate bacterial growth or death response to the environmental conditions of the food.

During any process, environmental conditions may vary as a function of time or space in the food. In many preservation techniques, temperature varies (cooling, pasteurization), pH decreases (acidification resulting from bacterial metabolism, fermentation), or the water activity (a_w) decreases (drying process, addition of solutes). As a consequence, phenomena of diffusion (water or mass transfer), as well as heat transfer, occur in the product, indicating that in a food, gradients of temperature, pH, or a_w occur. The prediction of conditions that prevent growth in such complex situations is not easy. Additional information is needed on how the variables that control the process and the food affect the main environmental parameters (temperature, pH, and a_w) used in predictive microbiology. Such information cannot be obtained by the predictive microbial models alone. A combined modeling approach (Figure 5.17) to predict the bacterial growth at the surface of a solid medium subjected to air-drying.

Lebert et al. (2005) previously showed how the variables that control the evolution of the water activity at the surface of a product (air velocity, relative humidity, thickness, diffusion properties) during the drying process can be determined to predict bacterial growth. In general, most variables (Figure 5.17) can be measured, and in the best cases, models exist to predict them. For example, many predictive bacterial models exist that predict growth parameters or bacterial kinetics (Ratkowsky et al. 1983; Lebert et al. 1998; Augustin 1999; Augustin and Carlier 2000a,b).

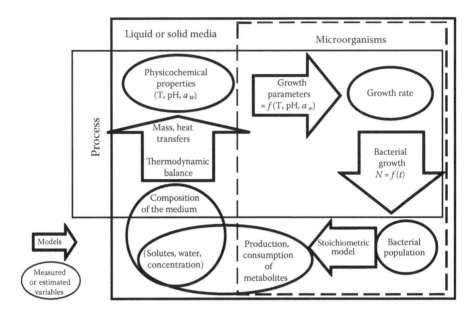

FIGURE 5.17 Variables and models involved in global modeling approach to predict bacterial growth.

In order to predict the growth of *Listeria innocua* on the surface of a gelatin gel submitted to a drying process (changes in relative humidity and air velocity), Lebert et al. (2005) developed a combined modeling approach by integrating three models:

1. A *bacterial* model that predicts the bacterial growth from the physicochemical properties of the media.
2. A *water transfer* model that predicts the effects of drying-process variables on medium characteristics. The model—based on Fick's law with a mass diffusivity depending on the water content—takes into account the shrinkage of the gelatin gel.
3. A *thermodynamic* model (see below).

The three models—bacterial, water transfer, and thermodynamic, separately validated—were combined according to an integrated modeling strategy (Figure 5.18). The growth of *L. innocua* was performed on the surface of gelatin gel placed in a wind tunnel under three conditions of relative humidity and velocity (Robles-Olvera et al. 1999; Desnier-Lebert 2004). The initial water contents of the gelatin gel in the three experiments were equal to 2 kg water/kg dry matter. The inoculum level was between 10^3 and 10^4 CFU/cm^2 (colony forming unit). When the relative humidity decreased from 95.5% to 92% at the same velocity, the growth was slowed down after 48 h of incubation, and the maximal population decreased from 8.9 to 7.3 $\log_{10}$ (CFU/cm^2). When the velocity was twice as much at a similar relative humidity of about 92%, a decrease in growth was observed as well. Growth was stopped after 24 h of incubation at a relative humidity of 92.5% and a velocity of 4.6 m/s, and the

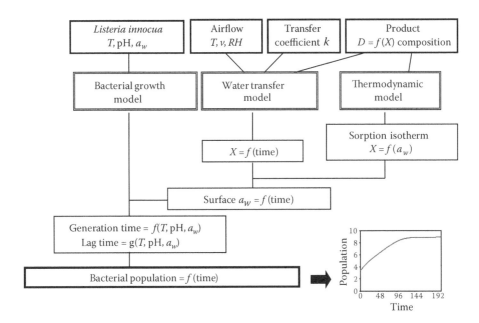

FIGURE 5.18 Global modeling strategy a_w, water activity; T, temperature; v, velocity of air; RH, relative humidity of air; k, mass transfer coefficient at the interface air/product; D, diffusivity; X, water content; GT, generation time; L, lag time.

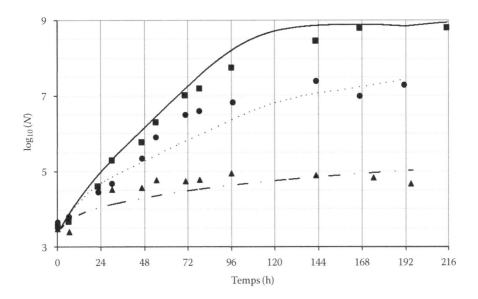

FIGURE 5.19 Measured (symbols) and predicted (lines) growth of *L. innocua* at 18°C during process drying as a function of relative humidity (RH) and velocity (v) of air at $RH = 95.5\%$, $v = 2.3$ m/s (■, ———); at $RH = 92\%$, $v = 2.3$ m/s (●, ·········); at $RH = 92.5\%$, $v = 4.6$ m/s (▲, – · –).

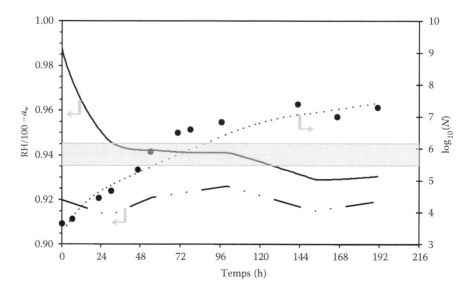

FIGURE 5.20 a_w calculated (———) by combination of water transfer model and thermo-dynamic model, on the surface of gelatin gel at 18°C, $RH = 92\%$ (– · · –), $v = 2.3$ m/s and measured (●), and predicted (··········) growth of *L. innocua*.

maximum population did not exceed 5 $\log_{10}$ (CFU/cm²). The different behavior of *L. innocua* under the three drying conditions was accurately predicted by the combined model (Figure 5.19).

The water transfer model coupled with the thermodynamic model predicted a_w on the gel surface. The predicted surface a_w explained why growth inhibition was observed. Indeed, growth stopped at a predicted surface $a_w < 0.94$, corresponding to the *L. innocua* minimum a_w during the drying process. The global model satisfactorily predicted *L. innocua* growth on the surface of the gel (Figure 5.20). This study proved the validity of the approach and showed that the combination of the water transfer and thermodynamic models compensates for the lack of a_w measurement techniques.

5.4 CONCLUSION

Contrary to the received ideas, some tools now exist to calculate *a priori* the physico-chemical properties of complex liquid solutions, understanding numerous molecules of biological interest (salts, organic acids, amino acids, peptides, sugars) with a satisfactory precision. Indeed, they take into account the deviation from nonideal values of the solutions. They permit a better understanding of experimental data: They predict the evolution of the physicochemical properties in combination with models of heat and mass transfer as these properties are difficult to measure during a process. When the media or the food is complex, it is necessary to find equivalent molecules having the same behavior. It requires a database of sorption and titration curves.

Currently, the research on the thermodynamic approach of the physicochemical properties of food or media used in biotechnology follows several ways in order to extend its application:

- Incorporating new functional groups, such as sulfur groups in the Larsen–UNIFAC model.
- Describing new foods (pork, milk, and cheese) or new culture broths
- Extending to the prediction of redox potential (Eh), which is linked to the activity of electrons
- Developing a simplified thermodynamic model (neural networks) in order to allow integration in process simulators

REFERENCES

Achard, C., Dussap, C.G., and Gros, J.-B. (1992). Prédiction de l'activité de l'eau, des températures d'ébullition et de congélation de solutions aqueuses de sucres par un modèle UNIFAC. *I.A.A.*, 109, 93–101.

Achard, C., Dussap, C.G., and Gros, J.-B. (1994). Prediction of pH in complex aqueous mixtures using a group-contribution method. *AIChE Journal*, 40, 1210–1222.

Augustin, J.-C. (1999). Modélisation de la dynamique de croissance des populations de Listeria monocytogenes dans les aliments. 153 p., PhD thesis, Université Claude Bernard—Lyon 1, France.

Augustin, J.-C., and Carlier, V. (2000a). Mathematical modelling the growth rate and lag time for *Listeria monocytogenes*. *International Journal of Food Microbiology*, 56, 29–51.

Augustin, J.-C., and Carlier, V. (2000b). Modelling the growth rate of *Listeria monocytogenes* with a multiplicative type model including interactions between environmental factors. *International Journal of Food Microbiology*, 56, 53–70.

Baranowska, H.M., Klimek-Poliszko, D., and Poliszko, S. (2001). *Effect of hydration water on the properties of saccharide solutions in properties of water in foods*, Lewicki, P.P. (Ed.), Warsaw Agricultural University Press, Warsaw, pp. 12–20.

Baucour, P., and Daudin, J.D. (2000). Development of a new method for fast measurement of water sorption isotherms in the high humidity range validation on gelatine gel. *Journal of Food Engineering*, 44, 97–107.

Ben Gaïda, L., Dussap, C.G., and Gros, J.B. (2006). Variable hydration of small carbohydrates for predicting equilibrium properties in diluted and concentrated solutions. *Food Chemistry*, 96, 387–401.

Ben Gaïda, L., Dussap, C.G., and Gros, J.B. (2010). Activity coefficients of concentrated strong and weak electrolytes by a hydration equilibrium and group contribution model. *Fluid Phase Equilibria*, 289, 40–48.

Catté, M., Dussap, C.G., and Gros, J.-B. (1994). Excess properties and solid–liquid equilibria for aqueous solutions of sugars using a UNIQUAC model. *Fluid Phase Equilibria*, 96, 33–50.

Catté, M., Dussap, C.G., and Gros, J.-B. (1995). A physical chemical UNIFAC model for aqueous solutions of sugars. *Fluid Phase Equilibria*, 105, 1–25.

Caurie, M. (1983). A research note: Raoult's law, water activity and moisture availability in solutions. *Journal of Food Science*, 48, 648–649.

Caurie, M. (1985). A corrected Ross equation. *Journal of Food Science*, 50, 1445–1447.

Chen, C.S. (1986). Effective molecular weight of aqueous solutions and liquid foods calculated from the freezing point depression. *Journal of Food Science*, 51, 6, 1537–1539.

Chen, C.S. (1990). Predicting water activity in solutions of mixed solutes. *Journal of Food Science*, 55, 494–497, 515.

Chen, C.-C., Britt, H.I., Boston, J.F., and Evans, L.B. (1982). Local composition model Gibbs energy of electrolyte systems. Part 1. Single solvent, single completely dissociated electrolyte system. *AIChE Journal*, 28, 588–596.

Chen, A.C.C., and Karmas, E. (1980). Solute activity effect on water activity. *Lebensmittel Wissenschaft und Technologie*, 13, 101–104.

Coroller, L., Leguérinel, Y., and Mafart, P. (2001). Effect of water activities of heating and recovery media on apparent heat resistance of Bacillus cereus spores. *Applied and Environmental Microbiology*, 67, 1, 317–322.

Desnier-Lebert, I. (2004). Prédiction de la croissance de Listeria innocua par une approche phénoménologique: Modélisations complémentaires des propriétés du milieu, des transferts d'eau et des cinétiques. PhD thesis, Université Blaise Pascal de Clermont-Ferrand, France.

Eckert, F., and Klamt, A. (2002). Fast solvent screening via quantum chemistry: COSMO-RS approach. *AIChE Journal*, 48, 369–385.

Edgar, T.F., and Himmelblau, D.M. (1988). Optimization of chemical processes, McGraw Hill, New York.

Fredenslund, A., Jones, R.L., and Prausnitz, J.M. (1975). Group contribution estimation of activity coefficients in nonideal liquid mixtures. *AIChE Journal*, 21, 1086–1099.

Gmehling, J. (1998). Present status of group-contribution methods for synthesis and design of chemical processes. *Fluid Phase Equilibria*, 144, 37–47.

Kikic, I., Fermeglia, M., and Rasmussen, P. (1991). UNIFAC prediction of vapour–liquid equilibria in mixed solvent-salt systems. *Chemical Engineering Science*, 46, 11, 2775–2780.

Kitic, D., and Chirife, J. (1988). Technical note: Criticism of a method for predicting the water activity of simple and multicomponent mixtures of solubles and non-solutes. *International Journal of Food Science and Technology*, 23, 199–201.

Klamt, A. (1995). Conductor-like screening models for real solutions: A new approach to the quantitative calculation of solvation phenomena. *Journal of Physical Chemistry*, 99, 2224.

Kojima, K., and Tochigi, K. (1979). Prediction of vapor-liquid equilibria by the ASOG method. Kodansha-Elsevier, New York.

Larsen, B.L., Rasmussen, P., and Fredenslund, A. (1987). A modified UNIFAC group-contribution model for prediction of phase equilibria and heats of mixing. *Industrial Engineering Chemistry Research*, 26, 2274–2286.

Lebert, I., Bégot, C., and Lebert, A. (1998). Development of two *Listeria monocytogenes* growth models in a meat broth and their application to beef meat. *Food Microbiology*, 15, 499–509.

Lebert, I., Dussap, C.G., and Lebert, A. (2004). Effect of a_w, controlled by the addition of solutes or by water content, on the growth of *Listeria innocua* in broth and in a gelatine model. *International Journal of Food Microbiology*, 94, 67–78.

Lebert, I., Dussap, C.G., and Lebert A. (2005). Combined physico-chemical and water transfer modelling to predict bacterial growth during food processes. *International Journal of Food Microbiology*, 102, 305–322.

Lebert, A., and Richon, D. (1984). Infinite dilution activity coefficients of n-alcohols as a function of dextrin concentration in water-dextrin systems. *Journal of Agricultural and Food Chemistry*, 32, 1156–1161.

Le Maguer, M. (1992). Thermodynamics and vapor–liquid equilibria. In Schwartzberg, H.G. and Martel, R.W. (Eds.), *Physical chemistry of foods*, Marcel Dekker, New York.

Lewicki, P.P. (2008). Data and models of water activity: I solutions and liquid foods. In Rahman, M.S. (Ed.), *Food properties handbook*, CRC Press, pp. 33–66.

Lide, D.R. (1999–2000). Section 7: Biochemistry. In *Handbook of chemistry and physics*, 80th edition, CRC Press, Boca Raton, 7.1–7.11.

Lin, S.-T., and Sandler, S.I. (1999). Infinite dilution activity coefficients from Ab initio solvation calculations. *AIChE Journal*, 45, 2606–2618.

Lin, S.-T., and Sandler, S.I. (2002). A priori phase equilibrium prediction from a segment contribution solvation model. *Ind. Eng. Chem. Res.*, 41, 899–913.

Norrish, R.S. (1966). An equation for the activity coefficients and equilibrium relative humidities of water in confectionery syrups. *Journal of Food Technology*, 1, 25–39.

McMeekin, T.A., Olley, J., Ratkowsky, D.A., and Ross, T. (2002). Predictive microbiology: Towards the interface and beyond. *International Journal of Food Microbiology*, 73, 395–407.

Ould-Moulaye, C.-B. (1998). Calcul des propriétés de formation en solution aqueuse des composés impliqués dans les procédés microbiologiques et alimentaires: Prédiction et réconciliation de données, Modélisation des équilibres chimiques et des équilibres entre phases. PhD thesis, Université Blaise Pascal de Clermont-Ferrand, France.

Palnitkar, M.P., and Heldman, D.R. (1971). Equilibrium moisture characteristics of freeze-dried beef components. *Journal of Food Science*, 36, 1015–1018.

Peres, A., and Macedo, E.A. (1997). A modified UNIFAC model for the calculation of thermodynamic properties of aqueous and nonaqueous solutions containing sugars. *Fluid Phase Equilibria*, 139, 47–74.

Pinho, S., Silva, C.M., and Macedo, E.A. (1994). Solubility of amino acids: A group-contribution model involving phase and chemical equilibria. *Industrial and Engineering Chemistry Research*, 33, 1341–1347.

Pitzer, K.S. (1973). Thermodynamics of electrolytes. 1. Theoretical basis and general equation. *Journal of Physical Chemistry*, 77, 268–277.

Prausnitz, J.M., Lichtenthaler, R.N., and de Azevedo, E.M. (1999). *Molecular thermodynamics of fluid-phase equilibria*, third edition, Prentice Hall, Inc., Upper Saddle River, New Jersey.

Rahman, S.M., and Perera, C.O. (1997). Evaluation of the GAB and Norrish models to predict the water sorption isotherms in foods. In Jowitt, R. (Ed.), *Engineering and Food at ICEF 7*, Sheffield Academic Press, Sheffield, U.K., pp. A101–A104.

Ratkowsky, D.A., Lowry, R.K., McMeekin, T.A., Stockes, A.N., and Chandler, R.E. (1983). Model for bacterial culture growth rate throughout the entire biokinetic temperature range. *Journal of Bacteriology*, 154, 1222–1226.

Renon, H., and Prausnitz, J.M. (1968). Local compositions in thermodynamic excess functions for liquid-mixtures. *A.I.Ch.E.*, 14, 135–144.

Robinson, R.A., and Stokes, R.H. (1959). *Electrolytes solutions*, second edition, Butterworths, London.

Robles-Olvera, V., Bégot, C., Lebert, I., and Lebert, A. (1999). An original device to measure bacterial growth on the surface of meat at relative air humidity close than 100%. *Journal of Food Engineering*, 38, 425–437.

Ross, K.D. (1975). Estimation of water activity in intermediate moisture foods. *Food Technology*, 29, 3, 26–34.

Sancho, M.F., Rao, M.A., and Downing, D.L. (1997). Infinite dilution activity coefficients of apple juice aroma compounds. *Journal of Food Engineering*, 34, 145–158.

Sereno, A.M., Hubinger, M.D., Comesaña, J.F., and Correa, A. (2001). Prediction of water activity of osmotic solutions. *Journal of Food Engineering*, 49, 103–114.

Spiliotis, N., and Tassios, D. (2000). A UNIFAC model for phase equilibrium calculations in aqueous and non aqueous sugar solutions. *Fluid Phase Equilibria*, 173, 39–55.

Stokes, R.H., and Robinson, R.A. (1966). Interactions in aqueous nonelectrolyte solutions. I. Solut-solvent equilibria. *The Journal of Physical Chemistry*, 70, 2126–2130.

Tochigi, K., Tiegs, D., Gmehling, J., and Kojima, K. (1990). Determination of new ASOG parameters. *Journal of Chemical Engineering of Japan*, 23, 453–463.

Weast, R.C. (1972). *Handbook of chemistry and physics*, 53rd edition, The Chemical Rubber Company, Cleveland, OH.

Weidlich, U., and Gmehling, J. (1987). A modified UNIFAC model. I—Prediction of VLE, h^E and γ^α. *Ind. Eng. Chem. Res.*, 26, 1372–1381.

Wilson, G.M., and Deal, C.H. (1962). Activity coefficients and molecular structure. *Ind. Eng. Chem. Fund.*, 1, 20–23.

Wilson, P.D.G., Wilson, D.R., and Waspe, C.R. (2000). Weak acids: Dissociation in complex buffering systems and partitioning into oils. *Journal of the Science of Food and Agriculture*, 80, 471–476.

6 Characterization of Bioreactors Using Computational Fluid Dynamics

Christophe Vial and Youssef Stiriba

CONTENTS

6.1   INTRODUCTION

Computational fluid dynamics (CFD) is usually defined as a set of numerical modeling tools able to simulate and predict the behavior of fluid flows. CFD is also described as one of the three dimensions of fluid mechanics, complementary to analytical methods and experimental techniques (Anderson 1995). It corresponds to the most recent of these fields, as CFD appeared in the 1960s, mainly for aeronautic applications. CFD corresponds, therefore, to a transdisciplinary field between fluid mechanics and applied mathematics. As a result, the development of CFD is closely linked to that of computers as flow predictions are always highly intensive in terms of CPU space and memory usage. A brief history of CFD is presented in Figure 6.1, but it must be kept in mind that CFD is still a rapidly evolving area of research, not only in terms of flow modeling, but also in terms of the fastest solution algorithms and simulation strategies.

In chemical engineering, applications of CFD emerged in the early 1990s and have only been mature enough since the late 1990s, when CFD described chemical reactions in multicomponent fluid flows, and, more recently in the 2000s, when they became able to cope with mass transfer–chemical reaction coupling in multiphase flows (Ranade 2002). Fluid flows of the process industries, including the chemical, pharmaceutical, cosmetics, food, and fuel and oil industries, actually correspond to a high level of complexity and high computational requirements. Since the late 1980s, CFD has, therefore, gained an increasing interest for design and scale-up purposes, first in pumping, mixing, and heat transfer or power generation, and, more recently, in separation unit operations and (bio)chemical reaction engineering. Another reason for which CFD has spread in chemical and biochemical design and optimization methodology is that, at the same time, commercial CFD software packages became available with the pros and cons of "ready-to-wear" versus "custommade". This has reduced the need for time-consuming and fastidious programming tasks, especially for complex situations, such as multiphase flows, that require high levels of computer-programming skills. Although the first commercially available software tool was Phoenics from CHAM Ltd. in 1978, only the engineer-friendly

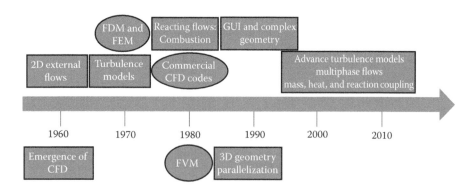

FIGURE 6.1   Brief history of CFD (FDM, FEM, and FVM are finite difference, finite element, and finite volume methods, respectively; GUI means graphical user interface).

user interfaces that were accessible at the end of the 1990s popularized CFD in the process industries. Finally, the last reason of CFD development in R&D methodology was the rising speed and falling cost of computer processing in the last decades, coupled with the progress of algorithm development, which made simple CFD simulations feasible, even using laptops.

As a result, this chapter will focus on the effective use of computational methods for bioreactors, mainly using commercial solutions. These are, now, considered to be standard numerical tools, widely used within industry although they are yet to be fully integrated into the design methodology. Conversely, the details of CFD-algorithm development remain beyond the scope of the discussion. Readers interested in these topics can find complementary information in Ferziger and Perić (2001). Consequently, the objectives will be to provide an overview of CFD methodology for bioreactors and an understanding of the basic methods and terminology and also to develop engineering judgment of CFD results and limitations. Consequently, the first section of this chapter will briefly describe what can be expected from the characterization of bioreactors using CFD. The second section will depict the main steps of CFD analysis. The third and fourth sections will address the physical modeling and numerical issues of CFD, focusing on the specificity of bioreactors. The last section will analyze the recent progresses and limitations of CFD, using the example of bubble column and airlift reactors so as to illustrate the steps and issues developed previously on bioreactors.

6.2 WHAT CAN CFD DO FOR BIOREACTORS?

Bioreactors encompass a wide variety of equipment and applications for which the reader will find information in the specialized literature of biochemical engineering and bioreactors (Scragg 1991; Van't Riet and Tramper 1991; Cabral et al. 2001). Most of them involve large amounts of fluids and are multiphase, usually gas–liquid but also liquid–solid and sometimes gas–liquid–solid. For example, the volume of activated sludge reactors for sewage and industrial wastewater treatment can reach $10,000 \text{ m}^3$. Similarly, bubble columns, that is, pneumatically agitated bioreactors that can be used for protein production, can reach 2000 m^3, and the industrial production of antibiotics usually involves bioreactors between 6 and 100 m^3 in stirred tanks. The behavior of these reactors is, therefore, highly sensitive to mixing conditions: As a result, scale-up from lab scale (and sometimes from shake flasks) to industrial scale may be tricky when final reactor volume is higher than 10 m^3. The standard tools of bioreactor scale-up methodology have been recently reviewed by Marques et al. (2010); these do not include CFD and are mainly based on the assumption of ideal reactors. However, the ideal reactor models, such as perfectly mixed conditions that can be validated in lab-scale and sometimes in pilot-scale stirred tanks, usually fall short of reality in industrial reactors, especially when highly viscous broth, froth formation, or stress-sensitive microorganisms are involved. As CFD is the numerical simulation of local fluid motion (Marshall and Bakker 1995), it obviously constitutes an adequate alternative tool for predicting local mixing conditions and also mass transfer rates in industrial chemical and biochemical reactors. Another major problem may be heat transfer, for example, when oxygen consumption is high as aerobic fermentations are exothermic and generate about 14.4 MJ/kg O_2. CFD can also be helpful to circumvent heat-transfer limitations.

Generally speaking, the main advantage of CFD is to predict local flow accurately when analytical tools are unusable and when conventional, one-dimensional (1D) reactor models of chemical engineering combining plug flow (or axially dispersed plug flow), perfectly mixed (or a cascade of perfectly mixed) tank reactors, and even advanced network-of-zone models (Zahradník et al. 2001) lack accuracy. Examples of these conventional models are presented in Figure 6.2a for a bubble column, a pneumatically agitated gas–liquid bioreactor, and Figure 6.2b illustrates the more realistic two-dimensional (2D) models developed for the same type of reactor. In Figure 6.2a, both phases are described using either a dispersed plug flow model or a cascade of perfectly mixed tanks-in-series with back mixing. These popular models that account for incomplete mixing in large industrial bioreactors are deduced from tracer studies and residence time distribution (RTD) measurements between the inlet and the outlet of the reactor. As a result, they do not represent the local hydrodynamics and are not unique, that is, several reactor models are able to fit the RTD data and, inversely, different reactors may have the same tracer response. While the semiempirical analytical 2D models of Figure 6.2b should have been more robust, these have never gained general acceptance and suffer from the same level of empiricism. Conversely, CFD data gives access to local concentrations and temperatures, which should be specific to reactor geometry and be unique as CFD is based on first principles. The only condition should be that key biological, chemical, and physical phenomena are appropriately accounted for in CFD modeling although this constitutes an oversimplified view of CFD issues that we will discuss later.

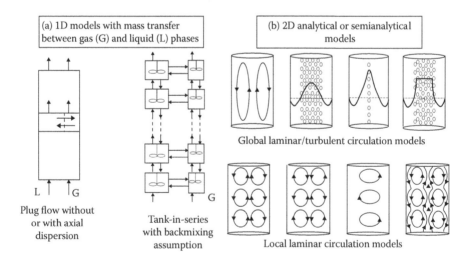

FIGURE 6.2 Conventional gas–liquid mixing models of bubble column bioreactors (Vial 2000); (a) One-dimensional models based on plug flow or stirred tank-in-series; and (b) two classes of 2D circulation models, based either on local or global circulation patterns.

As a result, CFD has found applications not only for the design of new biosystems, but also for analyzing, diagnosing, and optimizing the operation of existing systems or for reactor troubleshooting purposes. In detail, CFD is able to do the following:

- Provide a much better and deeper understanding of local hydrodynamics, including mixing and the spatial distribution of phases, reactants, or temperature as a function of operating conditions.
- Simulate various alternative designs and compare their performance under various operating conditions, reducing process development time and equipment or manpower costs associated with lab-scale or pilot testing.
- Simulate conditions for which measurements or even experiments are difficult or not possible, such as a dangerous environment.
- Improve the robustness of a reactor model, which should, in turn, reduce scale-up problems as equipment at its full scale can be analyzed by simulation.

Some examples of CFD use for the three main steps of chemical or biochemical reactor design are reported in Table 6.1. A few practical examples can be found in Sharma et al. (2011). For instance, a robust CFD model should be able to optimize stirring speed and aeration rate in a mechanically stirred aerobic bioreactor or to accurately predict how higher pressure or pure or enriched O_2 atmospheres can affect yield. Now, CFD seems to be able to determine whether the oxygen-transfer rate, maximum shear, and local energy dissipation rate are acceptable for microorganisms. It seems to be able to identify flow regimes, determine the minimum mixing time to achieve homogeneity, detect dead volume, and numerically estimate the previously mentioned RTD. For mechanically stirred tanks, it is able to optimize impeller location and multiple impeller design and to minimize power requirements with constraints on aeration, mixing homogeneity, and particle settling. CFD is, therefore, a valuable tool for equipment manufacturers, but also for design, production, and process engineers and for R&D staff.

TABLE 6.1
Example of CFD Use in Design Methodology of Chemical and Biochemical Reactors

Traditional Design Method	Information	Use of CFD
Cold flow and tracer studies (e.g., with the air/water system)	Mixing, flow regimes, local hydrodynamics, phase maldistribution, mass transfer	Cold flow simulations, numerical tracer study
Hot flow study (with reaction)	Local heat transfer	Avoid hot spots, improve local heat transfer
Preliminary design	Reactor modeling, reactor sizing, selection of operating conditions	Improved reactor model, study of alternative configurations and operating conditions
Final design/optimization	Design validation, refinements in design	Model validation, cost or waste reduction, performance optimization

6.3   HOW DOES CFD WORK IN PRACTICE?

In the field of biochemical engineering, flows are complex as they usually involve reacting multiphase turbulent systems with several reacted and unreacted species in each phase and mass transfer between phases, possibly coupled with heat generation and transfer resulting from a chemical reaction. This explains why the use of CFD is mainly based on commercial CFD software packages. These are, now, robust and easy to use but may be specific, meaning they usually include the same standard model capabilities for common situations, such as single-phase flows, but sometimes implement different models for specific applications, such as multiphase flows. For example, the multiple size group (MUSIG) model that accounts for a bubble-size distribution and bubble coalescence and breakup was implemented in the CFX code only in the late 1990s (Olmos et al. 2001). As already explained, these general codes avoid the need for complex programming but are usually not optimized for the particular applications of biochemical engineers. To circumvent this problem, customized models can be implemented in most commercial software packages through user-defined subroutines or functions in Fortran or C language in general. This is typically the case for basic Monod or Herbert kinetic models of biochemical reactions that are nonstandard in CFD codes. Again, restrictions specific to each package may exist. In this case, the use of free open codes constitutes an alternative solution but remains a challenging task. Table 6.2 summarizes the main commercial and free CFD software packages. Most of them (except some free ones) are GUI-based and can be supported by many platforms, such as Windows or Linux workstations or clusters using Intel CPUs under 32- or 64-bit operating systems. Most of them can also be used on high-performance computing servers, for example, from IBM (USA), HP (USA), SGI (USA), or Cray, Inc. (USA). The most common commercial distribution solution is annual retail software licensing. Typically, if commercial packages may be limiting in terms of physical models on the one hand, they give more time to

TABLE 6.2
Most Common Commercial and Free CFD Software Packages

Software Package	Company	Licensing	Platform	Multiphase
CFD-ACE+	ESI Group (France)	Retail	Linux, Windows	No
CFX	Ansys, Inc. (USA)	Retail	Linux, Windows	Yes
Fluent	Ansys, Inc. (USA)	Retail	Linux, Windows	Yes
Star-CD/Star-CCM+	CD-Adapco (USA)	Retail	Linux, Windows	Yes
OpenFOAM	OpenFOAM Foundation	Open source	Linux	Yes
Phoenics	CHAM Ltd. (UK)	Retail	Linux, Windows	Yes
NUMECA	Numeca Int. (India)	Retail	Linux, Windows	No
Flow-3D	Flow Science, Inc. (USA)	Retail	Linux, Windows	Limited
Comsol Multiphysics	Comsol Group (Sweden)	Retail	Linux, Windows	To define
FLoEFD	Mentor Graphics (USA)	Retail	Linux, Windows	No
CFDesign	Autodesk, Inc. (USA)	Retail	Linux, Windows	No
Code_Saturne	EDF (France)	GNU GPL	Linux	Fluid–solid

analyze the physics of the problem and to develop smart numerical strategies on the other hand. CFD software packages have intrinsic limitations, as does any experimental tool, but experienced users are able to use them at their maximum potential.

In practice, a CFD software package is a set of two to four executable programs able to carry out the three main steps of CFD methodology: namely, preprocessing, processing, and postprocessing. These are detailed in Table 6.3 in which a preliminary step has been added. CFD can, indeed, improve the understanding and the quality of the predictions of reactor behavior in comparison to conventional chemical engineering models, but always at the expense of higher CPU times. A compromise must, therefore, be found between the available CPU power, the time allowed for process development, and the information needed. Reasonable objectives must be defined as modeling complexity is expensive. Roughly, CFD should be able to draw a correct picture of what happens within a flow and where it happens as shown, for example, for oxygen-mass transfer in a wastewater treatment process by Cockx et al. (2001). It should also be able to reasonably predict the differences between several reactor designs or operating conditions. Conversely, it is not expected to exactly predict the yield and productivity of industrial bioreactors as this should require

TABLE 6.3
Main Steps of Treatment of CFD Problem

1. Preliminary steps	• Define the objectives (is CFD the most appropriate tool?). • Define whether the problem is transient or steady or if transient calculations can help more efficiently obtain a steady solution. • Analyze the physical phenomena involved. • Define the geometry and the domain that should be modeled with the possible simplifications (symmetry, etc.). • Define the time allowed for calculations (check whether CFD is the most appropriate tool), computing requirements, and also the accuracy required.
2. Preprocessing	• A solid modeler software must be used to create the geometry with its physical bounds (boundary conditions). This geometry must be accurate enough to grasp the required information, and unnecessary details must be omitted. • Mesh must be defined on the domain with a compromise between accuracy (fine grid) and computing time (coarse grid).
3. Processing	• Define the numerical models that account for the physical phenomena among the available models of the CFD solver or use user-defined functions. • Define material properties. • Define boundary conditions. • Run the calculations to simulate the behavior of the system using an adequate solver and optimize the solving procedure.
4. Postprocessing	• Extract and visualize data from the flow solver using a viewer. • Criticize the results from physical and numerical points of view. • Compare with global experimental data (and with local data if available). • Adapt or redesign the mesh to improve simulation quality except if objectives are achieved.

calculations on such a large range of time and space scales that simulations would require years even today. Further details on the three main steps of CFD methodology are developed in the following paragraphs.

6.3.1 PREPROCESSING

Basically, a CFD software package includes a solid modeler (i.e., a computer aided design or CAD software), a mesh-generation code, a solver, and a graphical post-processing viewer for visualizing and analyzing solutions (Table 6.3). Major commercial CFD software packages comprise, indeed, their own GUI-based CAD software and mesh generator that can sometimes be embedded in one unique executable program. These are also compatible with most commercial external CAD and mesh-generation software and with those of other commercial CFD codes. The situation may, however, be different for free CFD codes. While the need for a solid modeler to create the geometry is intuitive, meshing remains an unintuitive but essential task for beginners to CFD. In practice, the numerical solution of flow equations can be obtained only by splitting the computational domain (i.e., the regions of the geometry on which flow, heat, or concentrations have to be estimated) into small subdomains (denoted as control cells) on which the partial differential equations resulting from the first principles (mass, momentum, energy balances, etc.) can be approximated by algebraic equations. These must be solved numerically in each cell simultaneously. The physical limits of the domain, called the boundaries, deserve a special treatment. This step, denoted as discretization, requires the definition of the mesh (or the grid) that is the collection of all these subdomains (see, e.g., Figure 6.3). The grid may be structured when all cells are quadrilaterals in 2D and hexahedra in three-dimensional (3D) domains or unstructured otherwise; the most common unstructured mesh corresponds to triangles and tetrahedra in 2D and 3D domains, respectively. Hybrid meshes combine structured and unstructured regions.

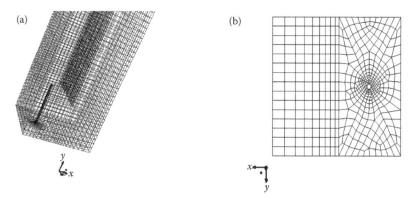

FIGURE 6.3 (a) Geometry and grid of the bottom of split-rectangular airlift bioreactor with internal baffle and a single-orifice nozzle for air injection; (b) two-dimensional projection of bottom-surface grid of the same bioreactor, showing structured, nonuniform mesh in down-comer section and unstructured mesh in riser section resulting from gas inlet tube around which mesh is refined.

Structured meshes may be uniform when all cells are of a similar size and are non-uniform otherwise. Fine-mesh grid and mesh refinement involve the reduction of cell sizes, that is, an increase in their number. This usually leads to higher computing requirements but also to improved accuracy for numerical solutions.

Meshing constitutes a critical step because it may strongly affect the accuracy of the numerical solution. A rule of thumb is that the quality of the mesh determines the quality of the solution provided adequate physical modeling has been retained. It is, however, possible to start the calculations on coarse grids so as to rapidly obtain a rough estimation of the solution and then to refine the mesh where necessary as a function of the coarse solution. This strategy is known as mesh adaptation. This implies that the last three steps of Table 6.3 may be embedded in an iterative loop until the desired accuracy is achieved. This also highlights the need for preliminarily defining the desired accuracy. In practice, the number of cells lies commonly between 20,000 and 2,000,000 for three to 20 unknown parameters per cell (pressure, velocity, temperature, and concentrations, sometimes in several phases). This results in up to 20,000,000 equations to solve simultaneously. In addition, boundary conditions that describe the flow behavior at the frontiers of the computational domain must be assigned. This explains why CFD simulations are so highly time-consuming. Unstructured meshes are more memory- and time-consuming than structured meshes but can handle more complex geometries as in Figure 6.3; they also allow more versatile mesh adaptation. The quality of the mesh depends on the flow field, but geometrical parameters can be used for a rough estimation of mesh quality. Typically, quadrilateral cells with a too-high length-to-width ratio (denoted aspect ratio) have to be avoided. Many shape parameters defined to analyze the quality of structured, unstructured, and hybrid meshes are reported in the literature and can be estimated by mesh generation software; they can be grouped into two classes: ratios that range between 1 (the best) and infinity and skewness, which varies between 0 (the best) and 1 (Ansys, Inc. 2009).

Finally, it can be concluded that meshing requires expertise and remains partly an art. First, it requires a good understanding of the physics of the problem as local grid refinement must be applied a priori where high gradients are supposed to be resolved together with a high skill level in grid generation. Mesh adaptation may be automated (using an auto-adaptative mesh) or manually driven as the best mesh can only be defined if the exact flow field is known. The use of automated mesh refinement procedures is, however, limited in bioreactors that often involve multiphase flows because calculations on fine meshes are too CPU-intensive.

6.3.2 Processing

Once the mesh has been generated, it can be imported into the solver code. The solver constitutes the cornerstone of a CFD software package (Table 6.3). Free CFD codes sometimes reduce to a solver that only accepts meshed geometries from external software. Basically, the solver discretizes the flow equations on each cell of the grid for a predefined flow model and is able to numerically solve the resulting set of algebraic equations. In the 1990s, it was usual to have different solvers for structured and unstructured mesh problems or for single-phase and multiphase flows. In recent GUI-based CFD software packages (Figure 6.4), users are able to select the

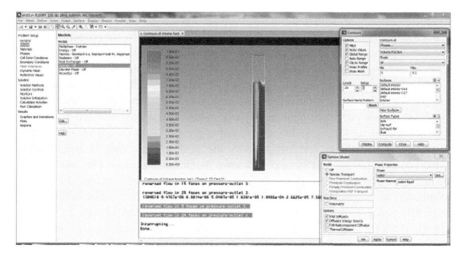

FIGURE 6.4   Example of GUI-based interface of Fluent in Ansys 12.

physical models, which directly modifies the discretized equations, of course, only for the available physical models. This means that CFD does not cover all the flow situations: for example, commercial solvers able to deal with the multiphase flows of aerobic fermenters emerged in the early 1990s. They were able to cope with coupled mass transfer and reaction in the late 1990s, and to couple heat transfer in such complex flows in the early 2000s. Physical modeling also includes the definition of material properties (viscosity or rheological models for complex fluids, etc., but also thermodynamic properties, such as heat capacity, thermal conductivity, heat of reaction, and diffusion coefficients, etc., when species and energy have to be accounted for). In recent packages, material property libraries are available, but diffusion coefficients must often be manually defined.

The selection of physical models in the solver constitutes the second major problem of CFD applied to bioreactors. This corresponds to two typical situations:

- The physical model selected in the preliminary step is not available in the solver. In a limited way, the implementation of additional user-defined models is allowed, usually for source/sink terms, such as reaction kinetics or mass transfer in bioreactors. Otherwise, another model must be retained in the solver and must be as close as possible to the physics of the problem.
- Several mathematical models are sometimes available for the same phenomena. They may be based on different physical assumptions or only differ by their numerical accuracy and CPU time. The typical example is turbulence for which further discussion is available in Section 6.4. In this case, the selection of the most adequate modeling strategy is not always straightforward, and it must be remembered that CFD requires experienced users because adequate choice simultaneously involves physical and numerical aspects.

Together with poor mesh quality, inadequate physical modeling is the second main reason responsible for discrepancies between experiments and CFD simulations.

Once solver models have been selected, the resulting set of numerical equations can be solved using the solver code until convergence is achieved. A key point that must be emphasized is that, although only commercial codes are addressed, these do not eliminate all the numerical issues the users have to face. For example, as solvers are iterative, an initialization step is necessary in which constant pressure and zero velocity are usually assumed in the domain (except for boundary conditions) when no information is available. CFD uses, indeed, iterative methods to solve the set of equations: Convergence measures the amount by which these equations are not satisfied. Similarly, convergence criteria are fixed by CFD users. When they are reached, this implicitly means that the numerical solution has been approached enough and will not change significantly for the level of accuracy required if further calculations are conducted. This choice also needs expertise as required values of convergence criteria may drastically vary with the types of flows and of unknown parameters (pressure, velocity, temperature, concentrations, etc.). Further details on convergence will be reported in Section 6.5. A key point is that nonconverged results have no physical meaning and are not a true estimated solution of the problem!

6.3.3 Postprocessing

Once a numerical solution has been obtained, postprocessing can start. It does not only consist of the visualization of flow but should also include confrontation to the literature data or to an experimental data bank, coupled with a physical analysis of the solutions (Table 6.3). Postprocessing may be carried out with viewers included in CFD software packages or external viewers. In Figure 6.4, the same execution contains solver and postprocessing utilities. Visualization tools have rapidly evolved with GUI over the last decades. The classical XY, 2D, and 3D contour, streamlines, isosurface, and vector plots that can be displayed for all unknown parameters can, now, be completed by animations involving simulations obtained for transient flows at several time steps. Typical examples are provided in Figure 6.5. Numerical data, including volume-average and surface-average quantities can also be estimated. Consequently, volume-average gas fraction in aerobic fermenters and volume-average molar productivity in all-type bioreactors, and also surface-average velocity, heat transfer requirements, or the molar flow rate of any reaction product, can be deduced from local data. Similarly, torque and mechanical power requirements for impellers in stirred-tank bioreactors can be numerically estimated (see, e.g., Ahmed et al. 2010).

Even though CFD should be able to provide an accurate idea of reactor behavior representative of physical phenomena, a key point is that the solution is always a numerical approximation that should be mesh-independent. Testing mesh dependency is one of the basics, often forgotten, of CFD. This can be done a priori by parallel computations on different mesh or by using a posteriori mesh adaptation on a coarse solution. Basically, mesh refinement should be applied in high-gradient regions while coarsening is possible for over-resolved low-gradient regions. Auto-adaptive

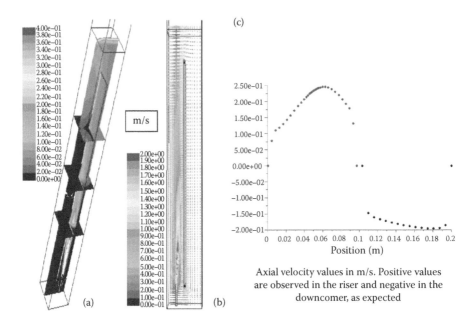

FIGURE 6.5   Examples of projected contours of local gas fraction on selected planes of 3D airlift bioreactor; (a) of water phase velocity vector field in the median plane; (b) and of the axial water velocity profile at mid-height of this plane; (c) in the same device.

mesh refinement procedures during processing are now available in recent CFD software packages but do not seem mature enough for reacting and multiphase flows. When both convergence and accuracy are in agreement with the objectives and the physical analysis of the problem, results may be used for engineering purposes. Nonconvergence (i.e., convergence criteria cannot be reached) may also be instructive as it may stem first from nonadequate physical models or from a too-coarse grid. Other causes may be the inadequate definition of the domain on which calculations are carried out, or of the boundary conditions at the frontiers of this domain.

6.4   FUNDAMENTALS OF CFD MODELING FOR BIOREACTORS: THE PHYSICAL ISSUE

It has been emphasized that CFD requires experts who do not necessarily have to be biochemical engineers: For a same commercial code, the quality of the simulations widely depends on the numerical strategy. This point will be addressed in Section 6.5. On the contrary, CFD experts, biochemists, or microbiologists are not always aware of the physics of biochemical reactors. This section is dedicated to biochemical engineers who need the basics to dialog with CFD experts. The guidelines of the physical modeling of flows in bioreactors will be established. We will assume all the necessary data on the biochemical system (in particular, kinetic data) are available.

6.4.1 BASICS OF CFD MODELING: NAVIER–STOKES EQUATIONS AND CONSERVATIVE SCALARS

Basically, CFD is aimed at giving access to the numerical solution of fluid flow and its primary quantities: namely, the local pressure and velocity fields. In addition, local temperature and species concentrations can be estimated, leading to chemical reaction parameters, such as local reaction rates, and by spatial integration, reaction yield, productivity, selectivity, etc. In practice, CFD methods are based on first principles of mass, momentum, and energy conservation, and they involve the local solution of conservation equations for mass, momentum, and any additional conservative scalar quantity (such as energy or species) over a small control volume, that is, each cell of the grid. These correspond first to the continuity and Navier–Stokes equations in laminar flow conditions for single-phase flows expressed as a function of time t in a 3D space (x,y,z). For Cartesian coordinates, transient Navier–Stokes equations for velocity v, pressure P, and density ρ can be written as follows (Ferziger and Perić 2001):

$$\frac{\partial \rho}{\partial t} + \sum_{j=x,y,z} \frac{\partial \rho v_j}{\partial x_j} = 0 \qquad (6.1)$$

$$\rho \frac{\partial v_i}{\partial t} + \rho \sum_{j=x,y,z} v_j \frac{\partial v_i}{\partial x_j} = -\frac{\partial}{\partial x_i}\left[P + \underbrace{\frac{2}{3}\mu \sum_{j=x,y,z} \frac{\partial v_j}{\partial x_j}}_{\text{dilatation}} \right]$$

$$\underbrace{-\rho g_i}_{\text{gravity}} + \underbrace{\frac{\partial}{\partial x_i}\left[\sum_{j=x,y,z} \mu \left(\frac{\partial v_i}{\partial x_j} + \frac{\partial v_j}{\partial x_i} \right) \right]}_{\text{shear stress}} + \underbrace{S_V}_{\text{source/sink}} \qquad (6.2)$$

This set of four equations, as $i = x,y,z$ in Equation 6.2, can sometimes be reduced to three because of symmetry (2D flow). For incompressible flows, the dilatation term of the right-hand-side term of Equation 6.2 vanishes, and Equations 6.1 and 6.2 become the following:

$$\sum_{j=x,y,z} \frac{\partial v_j}{\partial x_j} = 0 \qquad \rho \underbrace{\frac{\partial v_i}{\partial t}}_{\text{transient}} + \rho \underbrace{\sum_{j=x,y,z} v_j \frac{\partial v_i}{\partial x_j}}_{\text{convention}} = -\underbrace{\frac{\partial P}{\partial x_i}}_{\text{pressure}} - \underbrace{\rho g_i}_{\text{gravity}}$$

$$+ \underbrace{\frac{\partial}{\partial x_i}\left[\sum_{j=x,y,z} \mu \left(\frac{\partial v_i}{\partial x_j} + \frac{\partial v_j}{\partial x_i} \right) \right]}_{\text{shear stress}} + \underbrace{S_V}_{\text{source/sink}} \qquad (6.3)$$

As a result, the incompressible Navier–Stokes equations link the velocity vector $v = (v_j)_{j=x,y,z}$ to pressure P forces in a defined gravity field ($g_i = 0$, except in the vertical direction). A source/sink term S_V has been added in Equation 6.2 to account for possible additional external forces, such as centrifugal force. This formulation can be applied to Newtonian fluids for which molecular viscosity μ depends only on temperature. It also applies to non-Newtonian viscous fluids in which μ is an apparent viscosity that can be expressed as a function of velocity derivatives (see, e.g., Wu 2010). More complex descriptions of the shear stress term in Equation 6.3 are required only when viscoelastic fluids are involved, but standard CFD solvers are usually not able to account for viscoelasticity effects up to now. Even in this simple framework, too simple for bioreactor modeling, the numerical solution of Navier–Stokes equations is tricky and requires special treatments because

- They are partial differential equations and depend simultaneously on x,y,z, and time.
- They are highly coupled as a result of the convection term and exhibit non-linearity in the convection and sometimes in the source/sink terms.
- The four unknown parameters are the three velocity components and pressure, but no explicit equation is available for pressure.
- The final set of equations at each grid cell must be accompanied by the corresponding boundary conditions at the frontiers of the computational domain on which equations are solved.
- The transport of any conservative scalar quantity ϕ, including concentration or energy, can be expressed using the generic convection/diffusion transport equation in unsteady flows, which increases the number of equations to solve per grid cell:

$$\underbrace{\rho \frac{\partial \phi}{\partial t}}_{\text{transient}} + \underbrace{\sum_{j=x,y,z} \rho v_j \frac{\partial \phi}{\partial x_j}}_{\text{convection}} = \underbrace{\sum_{j=x,y,z} \frac{\partial}{\partial x_j}\left(D_\phi \cdot \frac{\partial \phi}{\partial x_j}\right)}_{\text{diffusion}} + \underbrace{S_\phi}_{\text{source/sink}} \tag{6.4}$$

In Equation 6.4, D_ϕ accounts for the diffusive properties of ϕ. An advantage, however, is that the continuity and the Navier–Stokes equations also follow this generic form when $\phi = 1$ and $\phi = v_i$, respectively, which means that similar numerical tools can be applied.

Modified versions of these equations can be used when rotating components are present in the computational domain, such as an impeller in stirred-tank bioreactors. These are based on the rotating reference frame approach that accounts for centrifugal effects and for which several levels of complexity are available. Basically, it assumes that the region near the impeller rotates at the same speed as the impeller, and the regions near static walls are stationary. A typical example is illustrated by Kerdouss et al. (2008). This approach does, however, not modify the general pattern.

For biochemical engineering cases, the nature of the coupling between equations rapidly increases as the conservation of other scalar quantities must be studied: Each scalar quantity implies one additional equation per scalar. For example, for an incompressible liquid, energy balance leads to

$$\rho C_p \left(\frac{\partial T}{\partial t} + \sum_{j=x,y,z} v_j \frac{\partial T}{\partial x_j} \right) = \sum_{j=x,y,z} \frac{\partial}{\partial x_j} \left(\lambda \frac{\partial T}{\partial x_j} \right) + S_T \qquad (6.5)$$

in which temperature T is the scalar quantity, λ is thermal conductivity, C_p is heat capacity, and S_T accounts for source/sink terms, such as heat of reaction. As reaction kinetics and, more generally, all physicochemical parameters depend on temperature, this induces strong coupling and high nonlinearity in the governing flow equations. Obviously, this also implies that the temperature dependence of all the parameters must be known. Similarly, a local mass balance on the concentration of species m, noted C_m, corresponds to an additional equation in which S_m is the source/sink term resulting from biochemical reaction and D_m to the diffusion coefficient of species m:

$$\frac{\partial C_m}{\partial t} + \sum_{j=x,y,z} v_j \frac{\partial C_m}{\partial x_j} = \sum_{j=x,y,z} \frac{\partial}{\partial x_j} \left(D_m \frac{\partial C_m}{\partial x_j} \right) + S_m \qquad (6.6)$$

In Equation 6.6, coupling and nonlinearity mainly result from S_m. In addition, the number of concentration scalars may be high. For bioreactors, the main consequences are that the number of concentrations to monitor using CFD must be reduced to the minimum value (only key compounds must be analyzed, such as oxygen, substrate, and the main product) and temperature-dependence must be investigated only when necessary as coupling and nonlinearity increase CPU requirements. For the same reasons, an adequate mesh must be defined: Too-fine mesh grids are unrealistic because of unaffordable memory and CPU requirements, and too-coarse grids provide inaccurate simulations. An alternative strategy is to reduce the size of the computational domain, simulating only the most interesting portion of a bioreactor. In both cases, a physical analysis is compulsory as mesh refinement supposes a guess on where high- and low-gradient zones can be found, and selecting a portion of a reactor geometry assumes that the coherent flow structures in this zone do not depend heavily on the other regions of the reactor.

Another key issue for all these equations is the definition of their boundary conditions at each boundary of the computational domain. This is, however, less difficult for chemical reaction engineering applications than in other fields of CFD as the boundaries usually correspond to the physical frontiers of the reactor: inlet and outlet flow streams, reactor walls, etc. Thus, boundary conditions can be deduced from physical considerations. The situation may be more complex when simulations involve only a portion of the reactor, which reinforces the need for careful selection of the region to simulate.

6.4.2 The Issue of Turbulence Modeling in Single-Phase Bioreactors

In biochemical reactors, the situation is more complex than described in Section 6.4.1 as flow is usually turbulent. This implies velocity and pressure fluctuations versus time on a broad range of length and times scales and the presence of coherent structures that are mainly responsible for mixing, mass transfer, and heat transfer. These cannot be ignored. A physical picture of turbulence for chemical engineers can be found in Ranade (2002). For engineering calculations, the most common approach consists in solving the Reynolds-averaged Navier–Stokes (RANS) equations. This approach assumes a mean flow field on which fluctuations are superimposed. Equations 6.1 through 6.6 are time-averaged over an appropriate time interval so as to obtain the RANS transport equations of the mean flow field. More complex alternatives, such as large eddy simulation (LES), are too computer-intensive to be used in the field of biochemical engineering except for in fundamental research (see, e.g., Zadghaffari et al. 2009); in addition, the modeling of multiphase flows using LES remains in the future. In practice, the RANS method gives access to a new set of partial differential equations that are similar to Equations 6.1 through 6.6 for the mean flow field but include extra terms resulting from turbulence as velocity and pressure, but also temperature and concentrations, are affected by turbulence. Turbulent stress, eddy-induced heat transport, and turbulent diffusion have to be accounted for, but turbulence also may modify the source/sink terms related to the chemical reaction. The modeling of these turbulent extra terms is known as the turbulence closure problem and remains one of the main cornerstones of CFD as it still combines complex theories and empiricism.

The RANS approach covers a wide range of models in which one finds the historical one-equation turbulent viscosity model of Boussinesq or the mixing-length model of Prandtl (Hinze 1975). This simple framework has, however, been superseded in the 1970s by two-equation models. These present the major advantage of removing the assumptions of a spatially constant turbulent viscosity or mixing length and account for the transport of turbulence kinetic energy by convection and diffusion from the regions where it is generated to the regions where dissipation prevails. Among the two-equation models, the semiempirical standard k-ε model of Launder and Spalding (1974) remains, even today, the cornerstone of engineering calculations for turbulent flows. This assumes that turbulence is isotropic. As a result, the intensity of velocity fluctuations can be estimated using only the turbulent kinetic energy k while ε is the turbulent energy dissipation rate resulting from molecular viscosity. Another advantage is that k and ε emerge as conservative scalars that can be described using the generic equation of scalar transport (Equation 6.4) and are related to turbulent viscosity μ_t using an empirical coefficient C_μ:

$$\mu_t = C_\mu \frac{\rho k^2}{\varepsilon} \tag{6.7}$$

High μ_t values imply low velocity gradients in the mean flow field and good mixing conditions but also high turbulent stress that can damage microorganisms, such as mammalian cells. It is commonly admitted that damages occur when the size of stress-sensitive microorganisms is larger than Kolmogorov's length scale l_K, which is governed by ε and depends on fluid density ρ and molecular viscosity μ:

$$l_K = \left(\frac{\mu^3}{\rho^3 \varepsilon} \right)^{\frac{1}{4}} \tag{6.8}$$

Conversely, mixing and interfacial mass transfer are enhanced by high ε values. Therefore, ε is, a major parameter for bioreactor design. Contrary to Boussinesq's approach, turbulent viscosity, k, and ε are not spatially constant and can be predicted locally in the k–ε model. Thus, CFD constitutes a versatile tool for impeller and gas sparger design because it gives access to a local parameter able to optimize mixing and mass transfer and simultaneously prevent microorganism disruption.

From a numerical point of view, turbulence modeling using the standard k–ε model corresponds to a conventional set of Navier–Stokes equations (Equation 6.3) in which molecular viscosity is replaced by turbulent viscosity μ_t in the stress–tensor term (except in the wall region) with two additional conservative scalars to solve: k and ε. Similarly, molecular diffusivity and thermal conductivity in Equations 6.5 and 6.6 must be replaced by turbulent diffusion/conduction coefficients that are simply assumed to be proportional to μ_t. The main drawbacks of this approach are the use of five empirical coefficients for closure and the assumption of isotropic turbulence, valid only far from the walls.

Many improvements on the standard version have been proposed, in particular the RNG (renormalization group) k–ε model that theoretically predicts some of the empirical coefficients or the realizable k–ε model that better accounts for low Reynolds regions and should better predict turbulent viscosity in rotating flows. A key problem is that these models have been far less validated than the standard k–ε model in engineering calculations. Alternatives also include other various, less-popular, two-equation models, but also the more complex Reynolds stress models (RSMs) that remove the assumption of turbulence isotropy. RSMs constitute the most advanced tools in RANS methodology and also cover a wide variety of models that include the transport of up to seven additional scalars to solve. Although RSM should drastically improve turbulence modeling in the rotating flows of stirred tanks, the increased complexity of these models does not avoid the empiricism of closure assumptions. Another problem is that the RSM approach has not yet been validated, especially in multiphase flows because it still involves too computationally intensive calculations. Typical comparisons of turbulence modeling in multiphase-flow bioreactors can be found in Laborde-Boutet et al. (2009) for bubble columns and Wu (2010) for aerobic digesters. No definitive answer can be found in the literature as the limitations of these models vary with flow conditions.

6.4.3 Specific Issues of Multiphase-Flow Bioreactors

In biochemical engineering, another level of complexity concerns multiphase flows in which fluid–fluid and fluid–solid systems can be distinguished. While fluid–solid systems usually correspond to dispersed solid particles of biomass in a fluid phase, fluid–fluid gas–liquid systems (in which we can include gas–liquid–solid systems) cover a wide variety of situations in bioreactors as various interface topologies are possible, such as dispersed and stratified flows (Figure 6.6). Dispersed flows are common in bioreactors as biomass is often a solid-dispersed phase, and a dispersed-gas phase in the form of bubbles is commonly used for aerobic fermentation. However, stratified flows correspond to surface-aeration bioreactors (see, e.g., Huang et al. 2009). Various dispersed flow patterns also exist in stirred tanks as a function of impeller speed (Azzopardi et al. 2011).

First, multiphase flow models are always CPU- and time-consuming. So they must be used only when necessary. For liquid–solid systems, pseudo-homogeneous single-phase simulations can be carried out when the solid phase is diluted, has a density close to that of the liquid, and does not significantly affect the flow field of the liquid phase. The local solid content can therefore be treated as a species concentration. This approach only needs to modify density and viscosity as a function of solid content. Conversely, it fails for gas–liquid systems as the slip velocity between both fluid phases is not negligible. In practice, four strategies are available in CFD to analyze these situations. Three of them are compared in Figure 6.7. They can be summarized as follows:

1. The volume of fluid (VOF) approach is limited to fluid–fluid systems and consists in interface tracking techniques. It is devoted to solving free-surface problems and stratified flows (Huang et al. 2009). Another possibility is to analyze bubble shape, bubble breakup, and bubble coalescence on a limited number of particles (Ma et al. 2011), but this remains of limited use in aerated bioreactors in which the amount of gas is too high.

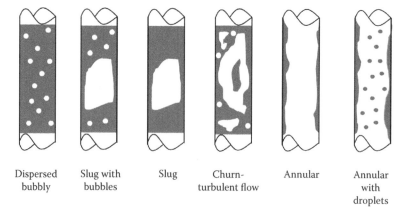

| Dispersed bubbly | Slug with bubbles | Slug | Churn-turbulent flow | Annular | Annular with droplets |

FIGURE 6.6   Different topologies of gas–liquid flows.

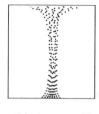
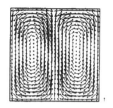

VOF Grid cell is smaller than bubble size	Euler–Lagrange model Grid cell is larger than bubbles that are followed individually	Euler–Euler model Grid cell is larger than bubbles for which only local gas fraction and velocity field are known

FIGURE 6.7   Comparison between main multiphase flow CFD models.

2. The Euler–Lagrange approach applies to systems involving a dispersed fluid or solid phase and can handle easily dispersed phases with a size distribution. The continuous phase is treated as a continuum by solving RANS equations as in single-phase flows. The dispersed phase is formed by large numbers of particles, the trajectories of which are followed individually in the flow field of the continuous phase. Momentum transfer between both phases implies that the velocity field of each phase can affect the other one. This corresponds to a physical approach that is limited, numerically, by the number of individual particles that can be monitored. The consequence is that the Euler–Lagrange approach is limited to dilute systems in which the volume fraction of the dispersed phase is lower than 10% to 15%. A typical example in a bubble column can be found in Laín et al. (2002). Another typical application of Euler–Lagrange models in single-phase or multiphase flows is the estimation of numerical RTD so as to analyze mixing properties of chemical reactors in place of conventional experimental tracing experiments. In this case, an additional dispersed phase of particles corresponding to passive tracers is injected, meaning that they do not affect the flow field of other phases. Typical applications of this technique in airlift reactors are available in Talvy et al. (2007a) and Le Moullec et al. (2008).

3. The mixture model is an improvement over the pseudo-homogeneous single-phase model in which phases can have slip velocity and exchange momentum. These quantities are usually estimated using a simple algebraic model. The mixture model solves RANS equations for the mixture and the conservation of the local volume fraction for the dispersed phase. If the presence of a particle size distribution for the dispersed phase is accounted for, an interfacial area transport equation may be added. This method can replace the Euler–Lagrange approach when the volume fraction of the dispersed phase is above 10% although the dispersed phase is treated as a continuum in the mixture model. It can also deal with several dispersed phases. This model usually fails, however, when density differences between phases are too high, for example, for gas–liquid bubbly flows.

4. The Euler–Euler approach is a general methodology that treats all the phases as interpenetrating continuous media. The probability of the presence of each of them at any location of a reactor over time corresponds to its local volume fraction. In this case, each phase is described by the RANS methodology, and the set of equations is similar for all phases even when dispersed flows are considered. This constitutes the only way for modeling concentrated dispersed flows with strong coupling, for example, in bubble column reactors or stirred tank bioreactors. For dispersed flows, a key disadvantage is that the standard treatment assumes only a unique particle size. Solutions to circumvent this limitation are, however, available now.

The RANS equations of the continuous phase in the Euler–Lagrange approach and of all phases in the Euler–Euler frame of reference can be expressed as follows for any phase n:

$$\frac{\partial \rho_n \alpha_n}{\partial t} + \sum_{j=x,y,z} \frac{\partial \rho_n \alpha_n v_{jn}}{\partial x_j} = \sum_{m \neq k} \dot{m}_{kn} \tag{6.9}$$

$$\rho_n \frac{\partial v_{in} \alpha_n}{\partial t} + \rho_n \sum_{j=x,y,z} \alpha_n v_{jn} \frac{\partial v_{in}}{\partial x_j} = -\frac{\partial \alpha_n P}{\partial x_i} - \rho \alpha_n g_i$$

$$+ \frac{\partial}{\partial x_i} \left[\sum_{j=x,y,z} \mu_n \alpha_n \cdot \left(\frac{\partial v_{in}}{\partial x_j} = +\frac{\partial v_{jn}}{\partial x_i} \right) \right] + S_{Vn} \tag{6.10}$$

Strong coupling between the phase equations results not only from the local volume phase fraction α_n, $\sum_n \alpha_n = 1$, but also from the source terms that account for mass transfer between the phase n and other phases in Equation 6.9 and from momentum interfacial transfer in SV_n in Equation 6.10. Multiphase closure involves not only turbulence modeling, but also the description of interfacial momentum transfer. The most common interfacial force terms that must be added in Equation 6.10 are summarized in Table 6.4. Additional memory and wall effects have been sometimes accounted for, but their use remains uncommon.

Typically, drag force takes into account the interfacial momentum transfer in the flow direction between the mean flow velocity values. It determines the slip velocity between phases and affects, therefore, the residence time and the volume fraction of the dispersed phase. Conversely, lift force is a radial effect, and in a bubble swarm, the turbulent dispersion and added-mass force of Table 6.4 mainly play a role on the radial dispersion of the dispersed phase. However, except for drag force, no model has received a general acceptance in the literature, especially for the lift force. A comparison between possible formulations of the source term in Equation 6.10 can be found in Tabib et al. (2008) and in Selma et al. (2010a) for bubble column bioreactors. The same stands for the conventional part of turbulence modeling, already

TABLE 6.4

Summary of the Most Common Interfacial Momentum Transfer Sink/Source Terms in Turbulent Multiphase Flows

Interfacial Force	Direction	Type	Exists?
Buoyancy	Vertical	Volume	Always
Drag	Direction opposite to velocity	Surface	Flow
Lift	Orthogonal to velocity vector	Surface	Flow
Turbulent dispersion	Toward regions of low gas fraction	From RANS	Turbulent flow
Added mass	Opposite to inertia forces	Volume	Unsteady flow

described in Section 6.4.2 (see, e.g., Laborde-Boutet et al. 2009; Selma et al. 2010a; and Tabib et al. 2008 although opposite and contradictory results can be found in the literature).

Drag, lift, and added-mass forces are nonlinear effects that exist even in the laminar flow pattern. The consequence from RANS methodology is that turbulent components of these forces may appear in the source term of Equation 6.10. In addition, a turbulent contribution proportional to the gradient of local volume fractions may also emerge from RANS methodology. These effects are commonly denoted turbulent dispersion forces and correspond to various models in the literature. Contributions to the modeling of the turbulent drag, lift, and added-mass forces in gas–liquid bubbly flows can be found in Talvy et al. (2007a), Selma et al. (2010a), and Ali et al. (2011) although this remains a field of active research in CFD. In practice, only the RANS term based on the local volume fraction is sometimes accounted for.

As a result, multiphase flows remain one of the most challenging and evolving fields of CFD because of the variety of flow topologies and to the rapid evolution of models. The CFD analysis of bioreactors strongly contributes to this evolution. Recent advances for modeling of dispersed phase flows involve the population balance model (PBM), which describes the local evolution of the size distribution of the dispersed phase, taking into account nucleation, growth, and coalescence/breakup for bubbles and droplets or aggregation/agglomeration/breakup for solid particles. As a result, a better description of local gas fraction, interfacial area, and mass transfer rates should be achieved. The presence of local heterogeneities resulting from the gas phase could be detected and superimposed on the mixing properties of the liquid phase. PBM may be coupled to the Euler–Euler approach and deems to be able to better account for gas–liquid mass transfer in aerated bioreactors and also for oxygen mass transfer to microorganisms. Examples of application to bioreactors can be found in Sanyal et al. (2005) and Bannari et al. (2008). The two main approaches under development use additional scalar transport equations denoted population-balance equations (PBE); they correspond to the method of class (MOC) and the method of moments (MOM), respectively. A comparison between MOC and MOM can be found in Selma et al. (2010b). A summary of all the issues of turbulent multiphase-flow modeling encountered in bioreactors is provided in Table 6.5.

TABLE 6.5
Summary of the Most Common Simulation Strategies to Deal with Turbulent Multiphase Flows

Turbulence	Turbulence is shared by all phases.	k–ε and more complex models can be used as in single-phase flows.
	Turbulence differs in each phase.	One set of k–ε RANS equations must be used per phase. k and ε transfer between phases must be accounted for.
	Turbulence in the continuous phase is partially generated by a dispersed phase.	k–ε RANS equations apply only to the continuous phase. Closure laws describe turbulence induced by the dispersed phase (limited to dispersed flows).
Drag force	Several models available for fluid–fluid and fluid–solid systems.	Key problems when particle morphology deviates from sphericity and when interactions in a particle swarm must be accounted for (Leon-Becerril et al. 2002; Talvy et al. (2007a); Simmonet et al. 2008 for bubbles in gas–liquid flows)
Other forces	Lift force, added mass force, and turbulent components	These forces are sometimes accounted for in the literature, depending on flow topology (Selma et al. 2010a).
Population balance equations (dispersed flow)	Method of classes (MOC)	Size distribution is divided into classes characterized by a particle number/m³. One scalar transport equation per class is necessary with closure for nucleation, growth, breakup, and coalescence/aggregation.
	Method of moments (MOM)	Involves only three to four additional scalar transport equations corresponding to some statistical moments of the size distribution. Similar requirements as MOC for closure. Reconstruction of size distribution from moments is not straightforward.

6.5 FUNDAMENTALS OF CFD SOLVER FOR BIOREACTORS: THE NUMERICAL ISSUE

The third pillar that governs the quality of CFD simulations, after mesh quality and physical modeling, consists of the numerical strategy. Poor numerical strategies can, indeed, lead to poor solution quality and even divergence, although proper physical models have been selected. This section will, however, be limited to the necessary guidelines to understand the terminology of the numerical aspects of CFD-solving strategies and to apprehend the numerical tricks that may impair simulation accuracy. It will be assumed that geometry, mesh, and boundary conditions have already been defined.

6.5.1 DISCRETIZATION METHODOLOGY

As already mentioned, discretization is the necessary step to transform partial differential equations into ordinary differential equations and finally algebraic equations that form a large system of nonlinear equations resulting from convective and source terms (Equations 6.3 and 6.4). This must be applied on each cell for each unknown parameter. This means that for 10^6 cells, a 2D laminar case combines at least two 10^6 equations for velocity components to 10^6 equations for pressure to which boundary conditions have to be added. Discretization must be consistent; that is, the solution of the algebraic system must tend toward the solution of the partial differential equations when cell size tends to zero. It is governed both by mathematical concepts and numerical knowledge on error truncation. Another requirement of discretization is that algebraic equations must be linearized as only linear solvers can be used when such a large number of unknowns have to be estimated simultaneously. The art of discretization consists in transforming each generic convection/diffusion transport equation (Equation 6.4) for the ϕ_P value in a particular P cell of the domain into $d\phi_P/dt = \sum_{i,j,k} a_{ijk}\phi_{ijk} + b_P$ after spatial discretization for transient flows or into $a_P\phi_P = \sum_{i,j,k\neq P} a_{ijk}\phi_{ijk} + b_P$ for steady flows. For transient flows, time discretization is necessary to convert $d\phi_P/dt = \sum_{i,j,k} a_{ijk}\phi_{ijk} + b_P$ into $a_P\phi_P = \sum_{i,j,k\neq P} a_{ijk}\phi_{ijk} + b_P$. In these expressions, ϕ_{ijk} corresponds to the unknown ϕ values in the neighboring cells close to the P cell in the three directions of space. Also, a_P and a_{ijk} are coefficients that can be estimated locally at time t. The main advantage of CFD software packages is that this complex task is fully automated and does not need programming. However, the analysis of simulations requires knowing which discretization methodology has been used by the CFD software package as this may affect CPU requirements and postprocessing. Three main numerical techniques can be used for discretization:

- The finite difference method (FDM) consists of the mathematical approximation of derivatives using truncated Taylor series expansions. The consequence is that the solution is known only at the nodes of the mesh (i.e., the "corners" of the cells). This technique is mainly restricted to structured meshes and is no longer used in commercial software codes. It is, however, the easier to implement and can be found in industrial codes (see, e.g., Ali et al. 2011).

- The finite element method (FEM) assumes a predefined evolution of each parameter in each cell, using a piecewise linear or higher-order polynomial approximation. It has been developed specifically to handle complex geometries and unstructured grids and is widely used except for CFD, for which calculations are too slow and costly: In particular, it is poorly efficient for turbulence modeling. Consequently, the key applications of FEM in CFD are restricted to flows of highly viscoelastic polymer melts. The main issue of FEM is that it does not ensure mass conservation, which may be a major problem for flow simulation.

- The finite volume method (FVM) assumes that all unknown parameters are constant in a cell (except velocity components) as in the small, perfectly mixed element of a plug flow reactor. This circumvents the problem of mass, momentum, and energy conservation as these are exactly satisfied. This approach has gained a general acceptance and has been adopted by almost all commercial and free CFD codes. The main limitation of FVM is that it is strongly adapted but also limited to Navier–Stokes equations. Conversely, it is less effective than FEM to analyze fluid–structure interactions.

Only FVM will be considered in the following paragraphs. A very simple physical presentation of this methodology can be found in Patankar (1980).

6.5.2 Numerical Strategies

While discretization methodology is imposed by the CFD software package, the choices offered to CFD users can be summarized as follows:

1. The discretization scheme of the convective terms
2. The choice between transient and steady simulation
3. The discretization of the transient term
4. The solution algorithm
5. The linear equation solver
6. Convergence monitoring

These six points are discussed successively in the following paragraphs.

Convective terms: The discretization scheme of the convective terms corresponds to the approximation of flux $\rho v_i \phi$ at each surface of the cell in Equations 6.3 and 6.4. This can be chosen and modified during calculations. In practice, the well-known upwind scheme is useful to start simulations. It has low-order accuracy and leads to numerical diffusion (i.e., a special type of truncation error that corresponds to an additional diffusion term in Equation 6.4 but is more stable than higher-order schemes. Numerical diffusion is the main weakness of FVM, but it may be circumvented by higher-order schemes. The common strategy consists in using higher-order schemes, such as QUICK or MUSCL (Ranade 2002), once a coarse solution has been obtained (typically a few hundred iterations). These improve solution accuracy but at the expense of higher computational times.

Transient versus steady simulations: Contrary to expectations, the use of transient solvers has become the rule even for steady problems involving industrial continuous bioreactors, using a pseudo-transient approach. This circumvents convergence failure with steady solvers when initialization is too far from the actual flow field. Transient simulations can be used to obtain a more accurate guess for solution initialization. The key difference between true transient and pseudo-transient simulations involves the time step Δt defined by users as the same solver is used. While transient simulations require very small time steps (Δt about 10^{-4} s), pseudo-transient simulations do not need to capture a time-accurate flow field, but only to approach numerically the steady flow, which allows higher time steps. It must, however, be mentioned that multiphase flows are intrinsically unsteady flows in which oscillating behaviors are often observed. In this case, only transient solvers must be used. This is particularly true for bubble column and airlift bioreactors for which only transient simulations provide accurate solutions for the flow field (Soklochin and Eigenberger 1999; Krishna et al. 2000). Similarly, flow is intrinsically transient in the impeller region of a stirred-tank reactor, which means that all steady solutions constitute only an approximation of the actual transient flow field of stirred tank bioreactors.

Temporal discretization: For temporal discretization, a first-order scheme corresponds to

$$a_P \frac{\phi_P(t+\Delta t) - \phi_P(t)}{\Delta t} = \theta_1 \left[\sum_{i,j,k \neq P} a_{ijk}\phi_{ijk} \right]_{t+\Delta t} + \theta_2 \left[\sum_{i,j,k \neq P} a_{ijk}\phi_{ijk} \right]_t + b_P \quad (6.11)$$

which is mathematically identical to $a_P\phi_P = \sum_{i,j,k} a'_{ijk}\phi_{ijk} + b'_P$ in which modified coefficients account simultaneously for values known at time t. This equation corresponds to an explicit scheme when the approximation of time derivatives depends only on ϕ_{ijk} values at time $t(\theta_1 = 0)$; it is fully implicit when ϕ_{ijk} values are estimated only at time $t + \Delta t(\theta_2 = 0)$ and semiimplicit otherwise. The famous semiimplicit Crank–Nicholson scheme corresponds to $\theta_1 = \theta_2 = 1/2$. Second-order temporal discretization schemes in which ϕ_P value at time $t + \Delta t$ depends on ϕ_{ijk} and ϕ_P values at times t and $t - \Delta t$ are now available but useful only when true transient simulations are required. Similarly, recent commercial solvers often propose to introduce auto-adaptive time step values, but these are not robust yet when multiphase flows have to be treated.

Solution algorithms: For CFD, all of them mainly derive from the SIMPLE algorithm (Patankar 1980), that is, the semiimplicit pressure-linked equations algorithm that is specific to Navier–Stokes equations and transforms the continuity equation (Equation 6.1) into a pressure-correction equation. For the algorithm, choices are, therefore, limited. Common improvements are the SIMPLER (revised SIMPLE) and the PISO (pressure implicit with splitting of operators) algorithms that may (sometimes) improve convergence speed. In Equation 6.3, a wrong estimation of pressure in the momentum equations obviously leads to mass imbalance in the continuity equation. For a guessed pressure field, velocity may be estimated, and Equation 6.1 may be used to estimate mass imbalance from the estimated velocity values. A continuity equation cannot directly predict pressure but only estimate a pressure correction to apply on the guessed pressure so as to reduce imbalance. This leads to an iterative procedure in which one alternatively solves the momentum and pressure correction equations (Figure 6.8). Coupled-phase versions of the SIMPLE algorithm are used for multiphase flows. Historically, the SIMPLE was only a segregated algorithm. This means that the three momentum equations on the one hand and the continuity equations on the other hand were solved sequentially before solving all other generic scalar transport equations, including energy, turbulence, concentrations (Figure 6.8), etc. Even though coupled SIMPLE algorithms that simultaneously solve velocity, continuity, energy, and concentration equations are readily available, their interest remains limited to aerodynamic simulations.

Linear equation solvers: In each step of the segregated SIMPLE algorithm, from one (general) to three (velocity components) unknown parameters must be solved on the total number of cells of the domain. For the general case, this constitutes a set of linear equations $a_P \phi_P = \sum_{i,j,k} a'_{ijk} \phi_{ijk} + b'_P$ involving a sparse squared matrix (i.e., populated primarily with zeros) for which special storage and inversion procedures can be applied. Matrix inversion is based on iterative methods. In transient simulations in which flow predictions are estimated as a function of time with a time step Δt, this means that three levels of iterating procedures are embedded (Figure 6.8):

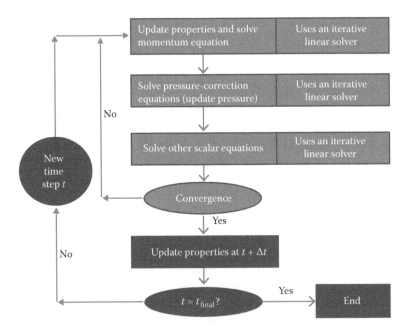

FIGURE 6.8 Description of SIMPLE algorithm for transient flow simulations with three levels of iterations.

namely, iterations in the linear solver for matrix inversion on each unknown parameter, iterations between pressure correction and velocity in the SIMPLE algorithm, and iterations on flow time from t to $t + \Delta t$ when convergence has been achieved at time t. Now, iterative multigrid methods have replaced Gauss–Seidel and Krylov subspace methods (such as conjugate gradients) in most CFD codes. They accelerate convergence but are rather complex to parameterize. Using default parameters is the rule except for under-relaxation factors that must be small enough to avoid divergence but not too low because they could slow down convergence rates. Inadequate choices for linear solver parameterization or for time step selection can, indeed, lead to divergence or increase computational time by more than a factor of 10. This is usually the more critical numerical issue for bioreactor modeling when multiphase systems are involved. CPU time may, therefore, increase from days to weeks or from weeks to months.

Convergence: A key point among numerical issues is to be able to define whether convergence is achieved when a steady state solution is desired or at each time step for transient simulations. Convergence criteria are commonly based on scaled residuals R_ϕ. These quantify the imbalance in the equations. Scaled residuals can be defined locally or, for the whole computational domain, as follows:

$$R_\phi = \frac{\left| a_P \phi_P - \sum_{i,j,k} a_{ijk} \phi_{ijk} + b_P \right|}{\left| a_P \phi_P \right|} \ \text{(local)} \qquad R_\phi = \frac{\sum_{\text{all cells}} \left| a_P \phi_P - \sum_{i,j,k} a_{ijk} \phi_{ijk} + b_P \right|}{\sum_{\text{all cells}} \left| a_P \phi_P \right|} \ \text{(global)}$$

$$(6.12)$$

Local residuals can be used to determine the zones in which convergence is difficult, but in practice, convergence criteria are only based on global residuals. As a rule of thumb, convergence criteria are R_ϕ lower than 10^{-3}, 10^{-4}, and sometimes less. For example, 10^{-5} was used by Mousavi et al. (2008) for bubble columns, but these guidelines may be adapted as a function of the problem. Rather than the value, the evolution of residuals is generally more meaningful. Convergence should, indeed, be characterized by constant residuals or slow variations for transient simulations (Figure 6.9a). This means residuals that still decrease when convergence criteria have been reached usually denote nonconverged steady solutions. Conversely, residuals may stabilize above the specified criteria for a converged steady solution if those have been underestimated. Knowing when a solution is converged is not always an easy task and sometimes requires monitoring other quantities than the global residuals, for example, surface-average velocities, pressures, temperatures, or concentrations. For mechanically stirred tanks, torque estimated on the shaft of the impeller is a key parameter to be monitored. Figure 6.9b illustrates the case of a food mixer for rheological complex fluids under laminar flow conditions in which torque evolves very slowly and stabilizes when global residuals are about 10^{-7} or smaller.

As a conclusion, CFD is not a monolithic tool in which a simple comparison of physical models can explain differences between simulations. The quality of the simulations depends mainly on the quality of the preliminary analysis of the problem,

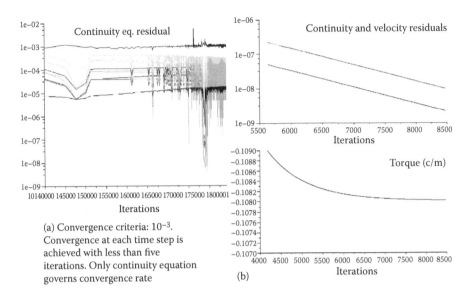

FIGURE 6.9 Example of convergence monitoring; (a) global residuals on pseudo-transient simulations of a multiphase airlift reactor; (b) torque monitoring to assess convergence in a food mixer used for highly viscous dough.

of the mesh generated based on this analysis, on physical models retained among those available, and also on the way simulations have been conducted, that is, on the numerical strategy and the experience of CFD users. In comparison to physical modeling, this last information often lacks in the literature because it constitutes the art of CFD expertise. Even for commercial CFD software packages, it must also be admitted that the quality of the simulations may also depend on the CFD code: Benchmarking between CFD codes has also exhibited differences in performance. In the 1990s, CFX was reported to better predict multiphase flows than Fluent. The situation is continuously evolving as new models and algorithms are implemented in each new version of CFD software packages, but each commercial code includes specific models that cannot be found in other ones. This may explain why CFX and Fluent codes still exist, although they both belong to the same company and are embedded in the same workbench now (Table 6.2).

6.6 APPLICATION TO BIOREACTORS THROUGH THE EXAMPLE OF BUBBLE COLUMNS

6.6.1 GENERALITIES

In Section 6.3, the main steps of CFD methodology have been established. The main physical issues have been detailed in Section 6.4, focusing on multiphase flow modeling that is still nonstandard, even today. The numerical issues have been discussed in Section 6.5. The objective of this section is to define how CFD methodology has been applied and how the main issues are treated on bioreactors. As mentioned in the Introduction, bioreactors entangle a wide variety of size, geometry, and operating conditions. These include conventional aerobic and anaerobic stirred fermenters and aerobic pneumatically agitated aerobic bioreactors, but also more recent photobioreactors (PBR) or membrane bioreactors (MBR). CFD analysis of all these kinds of reactors can be found in the literature now. A summary of various examples is summarized in Sharma et al. (2011). An additional interesting example is the CFD analysis of mixing and oxygen mass transfer in shake flasks for the culture of fungal and bacterial microorganisms and mammalian cells, as these are used for routine screening tests (Zhang et al. 2005). However, the aim of this section is not to cover all the possible bioreactor types and geometry. It is also not to establish a complete review of published CFD applications on any type of bioreactor as their number has too rapidly increased in the last decade. The objective is only to pinpoint, through a particular case that covers the most common situations, how the recent advances in CFD have been successfully applied to improve the design of bioreactors and where the remaining issues are. Bubble column reactors have been selected because they combine nearly all the possible issues of bioreactor modeling.

6.6.2 Preliminary Analysis of Bubble Column Bioreactors

Bubble columns are gas–liquid and gas–liquid–solid reactors often used in industry to develop and produce chemicals and fuels for use in chemical, biotechnological, and pharmaceutical processes. They are commonly utilized to carry slow gas–liquid reactions with a poorly soluble gas phase, such as chemical and biochemical oxidations, including aerobic fermentations. Industrial applications of bubble column bioreactors also include wastewater treatments (Joshi 2001), but their applications as PBRs for algal culture (Bitog et al. 2011) and bioleaching of mineral ores (García-Ochoa et al. 1999) have been known for more than 20 years. As bioreactors, they are used with microorganisms in order to produce industrially valuable products, such as enzymes, proteins, antibiotics, etc. (Kantarci et al. 2005). They offer many advantages over other kinds of multiphase bioreactors, especially stirred tanks: simple construction without internal or moving parts (i.e., no mechanical stirrer), good heat and mass transfer capabilities, high thermal stability, and effective mixing in both phases with low energy consumption. They also exhibit low construction and operation costs and low stress for stress-sensitive microorganisms. The first reason is that bubble columns often consist of a simple cylindrical tube with a larger section at the top to promote gas disengagement. The gas phase is introduced at the bottom using a gas distribution device that may consist of a porous plate, a perforated plate, or a gas sparger, such as a single-orifice or a multiple-orifice nozzle (Figure 6.10a). The second reason is that mixing is only induced by the dispersion of the gas phase in the form of tiny bubbles that simultaneously promote mass transfer for reaction purposes and heat transfer. Consequently, the flow pattern depends widely on the gas flow rate, which constitutes the key operating condition. Another major parameter is the geometry of the gas distribution device that controls how the gas is spatially distributed and determines the primary bubble size distribution. For continuous steady reactors, liquid flow rate is usually low and only slightly modifies reactor behavior. Conversely, key liquid properties are liquid viscosity and the coalescence behavior

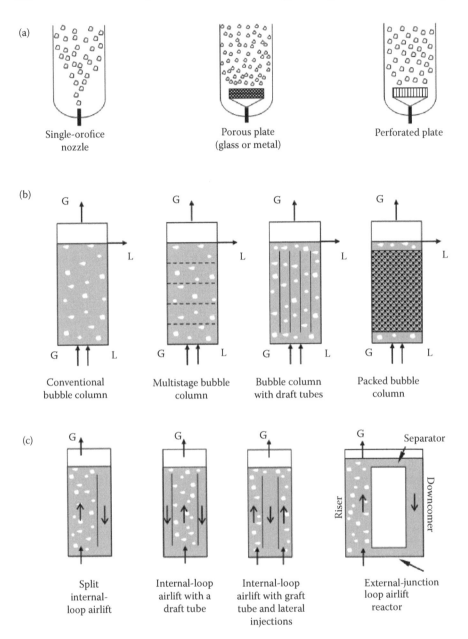

(a)

Single-orofice nozzle

Porous plate (glass or metal)

Perforated plate

(b)

Conventional bubble column

Multistage bubble column

Bubble column with draft tubes

Packed bubble column

(c)

Split internal-loop airlift

Internal-loop airlift with a draft tube

Internal-loop airlift with graft tube and lateral injections

External-junction loop airlift reactor

FIGURE 6.10 Typical bubble column bioreactors; (a) various gas distributor designs, (b) different classes of bubble columns; (c) different classes of airlift bioreactors.

of the liquid phase that governs the evolution of the bubble size distribution along the column.

As the geometry of bubble columns is simple, the only other geometry parameters are height and diameter. Although more complex classes of bubbles column can be found (Figure 6.10b), internals are usually to be avoided in bioreactors, and the conventional design prevails. The only class of nonconventional bubble columns that has found a wide range of applications is airlift reactors. These include an internal or external loop in which a global liquid recirculation is induced by a density difference between two regions: namely, the riser, in which the gas phase is sparged and the liquid moves upward, and the downcomer, in which the liquid flows downward, and the gas volume fraction is lower than in the riser (Merchuk and Gluz 2003). Several classes of internal-loop airlift reactors can be distinguished; some of them are presented in Figure 6.10c together with an external-loop device that is characterized by an external downcomer section. This figure is not exhaustive as activated sludge channel reactors are also considered to be horizontal airlift reactors (Le Moullec et al. 2011). Airlift reactors have the advantage of being more flexible than bubble columns because of the global liquid recirculation to minimize mechanical stress applied to microorganisms. For gas–liquid bubble columns and airlift reactors, the key design parameters, including liquid and geometry properties, and operating and nonoperating parameters, are summarized in Table 6.6. As a rule of thumb, CFD should be able to predict the nonoperating design parameters as a function of the operating parameters and system properties. Gas–solid–solid cases only introduce an increased level of complexity.

Even though bubble columns are ranging among the most popular gas–liquid contacting devices in the process industries, the present design practice is much closer to an art than a science (Joshi 2001). Contrary to stirred tanks in which fluid flow is clearly driven by impellers, hydrodynamics is only controlled by buoyancy forces in bubble columns, which leads to complex flow regimes and flow structures (Wild et al. 2003). Consequently, the performance of bubble column reactors depends largely on the interplay between reactor geometry (mainly height and diameter), gas volume fractions, the gas–liquid interfacial area, interfacial mass transfer, bubble size, bubble rise velocity, and bubble–bubble interactions (Joshi 2001). Bubble columns have been recognized, therefore, as good candidates for testing the ability of CFD on gas–liquid bioreactors as the simplicity of their design is counterbalanced by the complexity of their local hydrodynamic, mixing, and mass-transfer properties.

6.6.3 Physical Modeling

Hundreds of papers have now, been published on CFD applications involving bubble columns and airlift reactors. Most of them, except for the most recent, are limited to the analysis of hydrodynamics, and it is therefore difficult to distinguish whether they are dedicated to chemical or biochemical bubble column reactors as only cold flow without reaction is analyzed. In commercial CFD codes, multiphase flow models including species concentrations and interfacial mass transfer started in the middle of the 2000s. Now, they can also include heat transfer although this is rarely accounted for in bubble column simulations. In this section, the analysis of physical

TABLE 6.6

Key Parameters in the Design of Gas–Liquid Bubble Column and Airlift Reactors

Operating Parameters	
Gas flow rate	Key importance
Liquid flow rate	Low importance
Species concentrations, pH	Key importance on reaction kinetics
Temperature	Major importance on reaction kinetics
Pressure	Affects interfacial mass transfer
Mechanical power input	Irrelevant but key in stirred tanks
Geometry Parameters	
Height	Key importance
Diameter	Key importance
Distributor geometry	Key importance for primary bubble size distribution
Airlift: Geometry of the downcomer	Affects friction and global liquid recirculation
Airlift: Geometry of the junction	Affects friction and global liquid recirculation
Airlift: Geometry of the separator	Determines the efficiency of gas disengagement
Physicochemical Properties	
Reaction kinetics	Key parameter
Thermodynamic data	Key parameters of thermal effects
Liquid viscosity	Key parameter
Foamability	Difficult to quantify (surface tension is usually used but does not accurately predict foamability)
Species diffusion coefficients	May be important
Nonoperating Parameters	
Flow regime	Key parameters
Gas volume fraction (gas hold-up)	Key parameter
Local distribution of gas fraction	May be important to avoid anoxic regions
Local liquid velocity profiles	Influences stress on microorganisms
Airlift: Global recirculation velocity	Key parameter
Dispersion and mixing	Key parameters for reaction
$k_L a$	Mass transfer between the gas and the liquid phase
Interfacial area a	Mass transfer between the gas and the liquid phase
Local bubble size distribution	
Reaction yield and selectivity	Key parameter
Heat to evacuate	Key parameter of thermal effects

models will be limited to hydrodynamics as reaction kinetics is too specific to be compared, and the literature usually agrees on mass-transfer modeling. These points will be addressed in Sections 6.6.4 and 6.6.5.

A rapid analysis of the recent literature shows that the Euler–Euler frame of reference remains the rule, as gas volume fraction is usually high. The same stands for stirred-tank bioreactors. Exceptions involve the modeling of bubble formation or the coalescence of a pair of bubbles in which VOF can be used (Figure 6.7). A recent paper on this topic that reviews literature data is reported by Ma et al. (2011). Contrary to expectations, the Euler–Lagrange approach has received only little attention since the publication by Lapin and Lübbert (1994) even when gas-flow rates are low. Recent contributions include Hu and Celik (2008) that applied LES for turbulence modeling in the continuous phase but also Vikas et al. (2011) who used a modified Euler–Lagrange approach in which statistical groups of particles are monitored using PBE methodology. Only the Euler–Euler approach will be discussed in further developments. Another major problem was the modeling of the free surface at the top of bubble columns for which the Euler–Euler approach failed until the middle of the 2000s: Only a static liquid–gas–free surface could be modeled. This limitation has been circumvented, and now a continuous liquid phase with a continuous gas phase above can be simulated at the expense of slightly higher computational times. A key paper on this topic is reported by Talvy et al. (2007a), which describes the numerical strategy to handle this issue and highlights differences in flow predictions with the assumption of fixed and free surface, respectively.

For turbulence modeling, the standard k–ε model remains the rule although there have been recent attempts to introduce RSM and LES (Joshi 2001; Rafique et al. 2004; Tabib et al. 2008). These models not only imply an additional computational effort, but conclusions remain unclear and the subject of controversy. RSM and LES seem able to improve simulation accuracy as they account for turbulence anisotropy in flows in which turbulence anisotropy has been experimentally observed for more than 20 years by all the available measuring techniques (hot film anemometry, laser Doppler velocimetry, particle image velocimetry, radioactive particle tracking, etc.). However, the gain mainly concerns turbulent kinetic energy and transient phenomena while the average flow fields do not differ significantly (Tabib et al. 2008). According to Laborde-Boutet et al. (2009), the RNG k–ε model outperforms the standard and realizable versions, but their conclusions still need further validation on other experimental data sets and reactor geometries. A simple reason for which the standard k–ε still prevails is that alternative multiphase turbulence models have been implemented only recently in the multiphase solvers of some CFD commercial codes. Now, both mixture and dispersed versions of this standard model are used (Table 6.5), usually coupled with the turbulent dispersion force and bubble-induced turbulence in k and ε scalar equations. This applies both to bubble columns and airlift reactors despite the presence of a global liquid circulation (Talvy et al. 2007a). Conversely, the per phase approach has found little use because it is too costly. Only Laborde-Boutet et al. (2009) preferred the per phase framework, but data to confirm this opinion is lacking. The key role of turbulence modeling is illustrated by Figure 6.11. This shows that the radial dispersion of the gas phase strongly depends on the turbulent dispersion force, and the dispersed turbulence model predicts sharper radial gas volume fraction gradients. This could be expected as the bubbles move toward regions where turbulence is higher and pressure lower, which means that bubble-induced turbulence opposes radial gas dispersion.

Even for cylindrical or large aspect-ratio rectangular bubble columns, it has been known since the 1990s that the assumptions leading to 2D Euler–Euler simulations

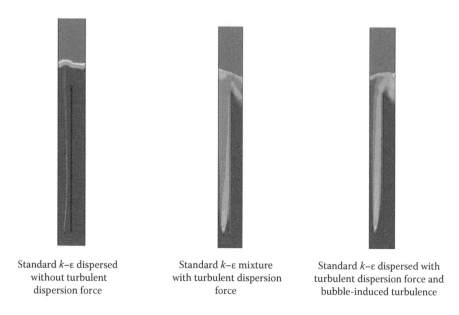

Standard k–ε dispersed without turbulent dispersion force	Standard k–ε mixture with turbulent dispersion force	Standard k–ε dispersed with turbulent dispersion force and bubble-induced turbulence

FIGURE 6.11 Influence of turbulence modeling on local gas holdup in 2D transient simulations of bubble column.

induces erroneous estimations of turbulence and mixing properties (Sokolichin and Eigenberger 1999; Pfleger and Becker 2001). The bubble plume is known to oscillate pseudo-periodically, which means that flow is 3D and transient. This behavior has been partially attributed to the presence of a bubble size distribution and could be captured by CFD simulations by Sokolichin and Eigenberger (1999) when fine 3D grids were used and also by Krishna et al. (2000). A more recent example of comparison between 2D and 3D simulations is reported by Ekambara et al. (2005) as 3D CFD modeling remained too costly for practical applications until the middle of the 2000s. This is the reason why the historical Zehner's cross flow model (Zehner and Schuh 1985) that empirically describes 2D mixing in bubble columns was validated only recently by Staykov et al. (2009). However, even today, 2D simulations can still be found in the literature (Ali et al. 2011).

A major physical issue in bubble column reactors remains the description of the interfacial forces between bubbles and the liquid phase as the liquid phase flow is mainly driven by the rate of momentum transfer from the bubbles that move upward because of buoyancy forces. Even for drag force, many models have been proposed to estimate the drag law coefficient. Comparisons between these models are common in the literature (see, e.g., Talvy et al. (2007a) for airlift reactors and Tabib et al. (2008) for bubble columns). The discrepancies between them are mainly a result of the fact that bubbles present a large variety of topologies (among which are cylindrical, ellipsoidal, spherical cap shapes, etc.) as a function of the gas distributor, the primary bubble size, and the coalescence behavior of the liquid phase (Clift et al. 1978). The selection, therefore, should be preferably based on experimental data involving the actual gas–liquid system. For other forces, no general rule can be established as opposite conclusions are drawn in the literature. For lift force, it emerges clearly that a negative lift force coefficient must be used (Tomiyama et al. 2002). The added-mass force is usually reported to be negligible in the literature. A key point is that these three forces have a qualitatively similar effect although they derive from different physical phenomena: They enhance the radial dispersion of the gas phase (Figure 6.12). In practice, two of them are frequently neglected when one is accounted for. In the steady simulations of Figure 6.12, one can notice that lift force overestimates the radial dispersion of the bubbles as it does not predict the experimental maximum of gas holdup at the column center.

A major issue of the Euler–Euler frame of reference is that it assumes only one unique bubble size. Many attempts have been developed to account for the bubble size distribution. Krishna and Van Baten (2001) proposed a two-bubble-class model that treated small and large bubbles as two dispersed phases but without mass transfer between them. In parallel, CFD solvers able to couple multiphase RANS equations to PBE emerged at the same time, especially with the MUSIG model used by Olmos et al. (2001) and more recently by Dìaz et al. (2008). Today, some commercial codes include MOC and MOM approaches. MOM is relatively attractive because it is less computationally intensive than MOC and because it reduces the number of additional scalars to a few moments of the distribution (see, e.g., Selma et al. 2010b; Vikas et al. 2011). MOM still requires, however, high computational power, and its validity is not assessed yet. The difficulty lies in the selection of the proper models for describing bubble coalescence and breakup. Many models have been proposed but are not mature enough. Consequently, CFD-MOM coupled solvers constitute a promising solution that will probably gain maturity in the next decade. At the

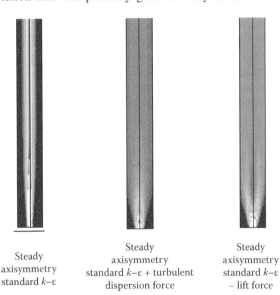

Steady
axisymmetry
standard k–ε

Steady
axisymmetry
standard k–ε + turbulent
dispersion force

Steady
axisymmetry
standard k–ε
– lift force

FIGURE 6.12 Comparison between local gas volume fraction predictions in 2D axisymmetric steady simulations and 3D transient simulations for 10 cm diameter cylindrical bubble column.

moment, it should be limited to fundamental research on hydrodynamics as a key issue remains the analysis of turbulence-induced breakup. In fact, the standard k–ε model is able to predict accurately the mean flow field but not the actual level of turbulent kinetic energy. This impairs the prediction of bubble breakup induced by turbulence in the MOM or MOC methodology.

As can be seen from this rapid picture, no general rule can be found for physical modeling. In the author's opinion, a robust choice that maintains reasonable computational times is the dispersed standard k–ε model in which only drag force and turbulent dispersion force have to be accounted for with bubble-induced turbulence. On the contrary, PBE is promising but remains in the future. This methodology should be able to capture the mean flow field, which should be sufficient to predict the local hydrodynamics and mixing properties and to analyze the influence of the operating parameters of Table 6.6. A major problem that remains is the validation of local CFD data. While global hydrodynamic parameters have been extensively studied, local data still remains scarce in the literature.

6.6.4 Numerical Strategy

For numerical aspects, commercial CFD codes now prevail, and among them is Fluent from Ansys, Inc., but simulations involving CFX can also be found when the previously mentioned MUSIG model is used. Recently, interesting developments with the free package OpenFoam have also been reported (Selma et al. 2010a, 2010b). Actually, little information is reported on numerical strategies and numerical issues in the literature. Under-relaxation factors are usually not mentioned except in a few papers (see, e.g., Mousavi et al. 2008), it is only a bit more frequent for convergence criteria (see, e.g., Mousavi et al. 2008; Laborde-Boutet et al. 2009). Also, little information is available on time-step size, but this information is of lesser importance as it may widely vary with reactor size, mesh size, and gas-flow rates. On the contrary, information on mesh size is always available, and the same stands for discretization schemes. For example, Tabib et al. (2008) summarized the physical models, reactor size, and grid size of the main 3D simulations on bubble columns published since the middle of the 1990s. The same comparison lacks, however, for airlift reactors that have been far less studied and present more various geometries. For time-step size, the common strategy is to start with a very low Δt, such as 10^{-4} s to avoid divergence in the first time steps and then to increase it as soon as possible.

The numerical strategy must be adapted to objectives. For example, three main methods have been applied for the scale-up of bioreactors (Marques et al. 2010). They are, respectively, based on the following:

- The volumetric mass transfer coefficient $k_L a$
- Power consumption
- Mixing time

All these parameters are defined at the reactor scale but can be deduced from local CFD data. In stirred tanks, power consumption may be inferred from surface-average torque estimations on the shaft of the impeller and also from local turbulent dissipation rate ε.

Similarly, local $k_L a$ values can be estimated from conventional hydrodynamic models for k_L and from local bubble size and gas volume fraction data for interfacial area a (see, e.g., Talvy et al. 2007b). Local ε and k data are also self-sufficient when enzyme deactivation or microorganism lethality resulting from hydrodynamic stress have to be investigated. Ghadge et al. (2005) illustrated this kind of analysis on enzyme deactivation in a bubble column although the correlation between experimental data and flow field is not easy to establish. A detailed analysis of mass transfer requires, in addition to hydrodynamic simulations, the introduction of species transport equations. A typical example is dissolved oxygen in the liquid phase or oxygen concentrations in both phases to account for oxygen depletion in the gas phase. This may be carried out on a frozen steady velocity field or using transient and far more computationally-intensive simulations that account for the instantaneous flow field. In bubble columns, the method for scale-up is a bit simpler than in stirred tanks as $k_L a$, power consumption, and mixing time only depend on gas-flow rate. This implies that the three previously mentioned scale-up methods are closer than in stirred tanks in which there is a complex interplay between gas-flow rate and impeller speed. On the contrary, transient simulations are the rule for the flow field and for mass transfer in bubble columns. Typical examples of mass transfer analysis can be found in Talvy et al. (2007b) in an airlift reactor and in Kerdouss et al. (2008) in stirred tanks.

An example of numerical strategy can be detailed as follows. The authors investigated the 63 L split-rectangular internal-loop airlift reactor from Gourich et al. (2006). It consists of a square column (0.20 m × 0.20 m) of 2 m height, divided into a riser and a downcomer section by a 1.3 m height Plexiglas baffle. The riser-to-downcomer cross-sectional area ratio is equal to one. The baffle is located 0.115 m from the bottom of the reactor. The gas distributor at the bottom of the riser is a single-orifice nozzle of 3.5 mm diameter. Air was used as the gas phase and tap water as the liquid phase. The superficial gas velocity was varied from 0.01 up to 0.08 m s^{-1}, and a batch liquid phase was used. The clear liquid height was always 1.59 m; the dispersion height above the baffle was, therefore, always higher than 0.175 m. Gourich et al. (2006) reported only global hydrodynamic parameters. Simulations were conducted with the Ansys-Fluent software package using the Euler–Euler frame of reference with a unique bubble size (5 mm). A dispersed standard k-ε turbulence model coupled with bubble-induced turbulence was retained. Only buoyancy, drag, and turbulent dispersion forces were accounted for. Three-dimensional transient simulations were carried out for 80 s flow time on the hybrid mesh shown in Figure 6.3 (60,398 cells, similar to Talvy et al. 2007a). A time step of 10^{-4} s was used in the first 10 s of flow time and then increased progressively to 10^{-3} s. The upwind scheme was used to start the simulation but was replaced by the second-order QUICK scheme after 100 time steps. Defaults were maintained for under-relaxation factors. Convergence was assumed when scaled residuals were lower than 10^{-4} at each time step for all variables. A maximum of 200 iterations per time step were allowed. Height variations of the free gas–liquid surface at the top of the reactor were taken into account as in Talvy et al. (2007a). Preliminary simulations showed that the major numerical problem was the high-velocity values near the single-orifice nozzle. As a result, CPU time requirements increased drastically with superficial gas velocity: They ranged from 10 days on one processor for 1 cm/s to three weeks on four processors in parallel for 8 cm/s. The simulations have been compared to the global hydrodynamic data available in Gourich et al. (2006). A good agreement has been obtained for both the overall liquid recirculation velocity and the volume–average gas fractions in the riser and in the downcomer (Figure 6.13). Then, ε and $k_L a$ values can be estimated and analyzed as explained previously. This illustrates the robustness of the basic strategy suggested in Section 6.6.3.

In practice, industrial strategies may be more complex than has been found in the scientific literature. A very interesting example is developed by Ranade (2002) on an industrial loop reactor. Because of the large size of this reactor, only rough simulations of the whole reactor were carried out. On the contrary, four separate detailed flow models were developed to analyze successively the following:

1. The bottom of the reactor with a Euler–Euler approach, including a dispersed gas phase
2. The free surface in the separation section using a VOF approach
3. The vapor space of the separator using a Euler–Lagrange frame of reference with dispersed droplets in the continuous vapor phase
4. The region where the liquid phase is fed into the reactor

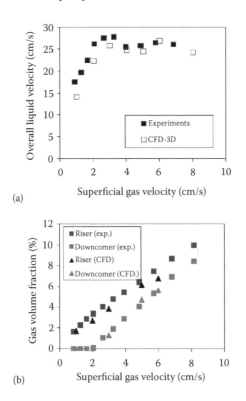

(a)

(b)

FIGURE 6.13 Comparison between experimental and CFD data; (a) for surface-average overall liquid circulation velocity; (b) for volume-average gas volume fraction in riser and in downcomer.

This constitutes an interesting use of CFD that accounts for the limits in available CPU and engineering times and attempts to capture flow details only where it is necessary. It highlights that commercial software packages are only tools that cannot be used systematically and require smart strategies that combine knowledge of the physics and the chemistry of bioreactors and know-how on the numerical issues of CFD models and on the practical limitations of CFD codes.

6.6.5 BEYOND HYDRODYNAMICS

As already mentioned in Sections 6.6.3 and 6.6.4, most CFD studies in the literature are limited to hydrodynamic data, which already gives access to mixing, power consumption, and k_La properties. Investigations including mass transfer or RTD data are few, and the same stands for CFD studies that directly include biochemical reactions in the local equations coupled to local mass transfer rates. This constitutes the most promising application of CFD to bioreactors. Table 6.7 summarizes the main contributions that have already coped with this topic to illustrate the potentiality of CFD to account for simple or structured bioprocess modeling in bubble column reactors. A review of CFD applications to PBRs is also reported by Bitog et al. (2011); bubble columns and airlift reactors indeed play a key role in this evolving field of bioenergy. While only a weak coupling was assumed when only local k_La, shear stress, or oxygen concentration values were monitored, CFD has the potential to strongly couple microorganism growth and product formation in multiphase gas–liquid and gas–liquid–solid biochemical systems with the abiotic environment. This should take into account flow and concentration heterogeneity in the two or three phases, local gas–liquid mass transfer and also mass transfer to microorganisms.

Table 6.7 highlights that two-phase and three-phase reactors with immobilized microorganisms can already be studied using CFD, of course, at the expense of highly time-consuming simulations (Jia et al. 2010; Wang et al. 2011). In these papers, bubble-size distribution was modeled using the MUSIG approach. To limit computational requirements, only oxygen, pollutant, and biomass concentrations were monitored. The two- and three-phase Euler–Euler approach was always retained, but the same also stands for the contributions of Table 6.7 even for the weakly aerated regions of the activated sludge channel reactor of Le Moullec et al. (2008). Similarly, the dispersed version of the standard k–ε model was always used for turbulence modeling except by Mousavi et al. (2008), but bubble-induced turbulence modeling was code-dependent. If the description of the gas phase was simpler in Le Moullec et al. (2010a,b, 2011) as only a unique bubble size was assumed, these authors introduced the ASM1 model for activated sludge that includes 12 different components and eight kinetic processes in their physical modeling. Their simulations accurately predicted the presence of anoxic and aerobic regions. Of course, their comparison with simpler models showed that CFD is still too time-consuming and cannot reasonably calculate transient phenomena over long periods. However, CFD is still an interesting approach to validate on a physical basis their simpler networks-of-zones model. For PBRs, the challenge over the next few years is to be coupled with flow equations and biochemical reactions with a radiation model able to predict how light transfer and radiation distribution affects the photosynthetic reaction (Bitog et al. 2011).

TABLE 6.7
Typical Examples in which CFD is Coupled with a Biochemical Reaction in Bubble Column and Airlift Reactors

Reference	Code	Objective	Reactor	Model
Wang et al. (2009)	CFX	Biodegradation of toluene on free microorganisms	Bubble column	Two-phase Euler–Euler with MUSIG (dispersed turbulence, 3D)
Wang et al. (2010)	CFX	Biodegradation of toluene on free microorganisms	Airlift reactor	Two-phase Euler–Euler with MUSIG (dispersed turbulence, 3D)
Jia et al. (2010)	CFX	Biodegradation of phenol on supported microorganisms	Bubble column	Three-phase Euler–Euler with MUSIG (dispersed turbulence, 3D)
Wang et al. (2011)	CFX	Biodegradation of toluene on supported microorganisms	Airlift reactor	Three-phase Euler–Euler with MUSIG (dispersed turbulence, 3D)
Mousavi et al. (2008)	Fluent	Ferrous bio-oxidation	Bubble column	Two-phase Euler–Euler (mixture turbulence, 3D)
Le Moullec et al. (2010a)	Fluent	Comparison CFD vs. experimental data	Activated sludge channel	Two-phase Euler–Euler (dispersed turbulence, 3D)
Le Moullec et al. (2010b)	Fluent	Comparison of CFD with simpler models	Activated sludge channel	Two-phase Euler–Euler (dispersed turbulence, 3D)
Le Moullec et al. (2011)	Fluent	Comparison of CFD with simpler models	Activated sludge channel	Two-phase Euler–Euler (dispersed turbulence, 3D)

6.7 CONCLUSIONS

Since the early 1990s, computational fluid dynamics (CFD) has been successfully applied to improve the understanding and the design methodology of single-phase reactors of the chemical and biochemical industries. Thus, it has become a key analysis tool for designing and optimizing the physical and chemical processes that may be limited by hydrodynamics and mixing or mass transfer, using commercial codes that avoid complex programming. In the last decade, the ability of CFD to predict quantitatively multiphase flows with multiple species in each phase, heat and mass transfer, chemical reactions, etc. has rapidly improved. Although CFD cannot provide exact solutions, robust simulations can be obtained now and assist engineers so as to maximize conversion yield and modify process equipment at lower cost and risk. In less than 10 years, the accuracy of the simulations in the multiphase reacting flows of bioreactors has so strongly improved that CFD is clearly the most promising tool for the intensification of bioreactors. In the near future, CFD will be able to capture accurately bubble-size distribution in aerated bioreactors and the heterogeneity of microorganism population. However, the use of CFD, even with commercial software packages, requires expertise that combines a solid understanding of physical phenomena and numerical issues. This explains why CFD is not yet fully integrated in the design methodology of biochemical reactors: There are still only a few CFD experts in multiphase reacting flows. This chapter will help understand how CFD works, what it can do and also its limitations and their multiple origins. It is expected also that it will develop an interest in this rapidly evolving field of chemical and biochemical engineering. Readers interested in the application of CFD in other fields of biotechnology than bioreactors can find additional information in Sharma et al. (2011).

REFERENCES

Ahmed, S.U., P. Ranganathan, A. Pandey and S. Sivaraman. 2010. Computational fluid dynamics modeling of gas dispersion in multi impeller bioreactor. *Journal of Bioscience and Bioengineering* 109:588–597.

Ali R.M., C. Jamel, B. Ghazi and A. Liné. 2011. Gas dispersion in air-lift reactors: Contribution of the turbulent part of the added mass force. *AIChE Journal* 57:3315–3330.

Anderson J.D. 1995. *Computational Fluid Dynamics—The Basics with Applications*. New York. McGraw-Hill.

Ansys, Inc. 2009. *Ansys Fluent 12.0/12.1 Documentation. User's Guide*. Canonsburg (PA). Ansys, Inc.

Azzopardi B., D. Zhao, Y. Yan, H. Morvan, R.F. Mudde and S.Lo. 2011. *Hydrodynamics of Gas–Liquid Reactors: Normal Operation and Upset Conditions*. Chichester (UK). John Wiley & Sons.

Bannari R., F. Kerdouss, B. Selma, A. Bannari and P. Proulx. 2008. Three-dimensional mathematical modeling of dispersed two-phase flow using class method of population balance in bubble column reactors. *Computers and Chemical Engineering* 32:3224–3237.

Bitog J.P., I.-B. Lee, C.-G. Lee, K.-S. Kim, H.-S. Hwang, S.-W. Hong, I.-H. Seo, K.-S. Kwon and E. Mostafa. 2011. Application of computational fluid dynamics for modeling and designing photobioreactors for microalgae production: A review. *Computers and Electronics in Agriculture* 76:131–147.

Cabral J.M.S., M. Mota and J. Tramper. 2001. *Multiphase Bioreactor Design*. London (UK). Taylor & Francis.

Clift R., J.R. Grace and M.E. Weber. 1978. *Bubbles, Drops and Particles*. London (UK). Academic Press.

Cockx A., Z. Do-Quang, J.M. Audic, A. Liné and M. Roustan. 2001. Global and local mass transfer coefficients in waste water treatment process by computational fluid dynamics. *Chemical Engineering and Processing* 40:187–194.

Dìaz M.E., A. Iranzo, D. Cuadra, R. Barbero, F.J. Montes and M.A. Galán. 2008. Numerical simulation of the gas–liquid flow in a laboratory scale bubble column: Influence of bubble size distribution and non-drag forces. *Chemical Engineering Journal* 139:363–379.

Ekambara K., M.T. Dhotre and J.B. Joshi. 2005. CFD simulations of bubble column reactors: 1D, 2D and 3D approach. *Chemical Engineering Science* 60:6733–6746.

Ferziger, J.H. and M. Perić. 2001. *Computational Methods for Fluid Dynamics* (3rd ed.) Berlin. Springer.

García-Ochoa J., S. Foucher, S. Poncin, D. Morin and G. Wild. 1999. Bioleaching of mineral ores in a suspending solid bubble column: Hydrodynamic, mass transfer and reaction aspects. *Chemical Engineering Science* 54:3197–3205.

Ghadge R.S., K. Ekambara and J.B. Joshi. 2005. Role of hydrodynamic flow parameters in lipase deactivation in bubble column reactor. *Chemical Engineering Science* 60:6320–6335.

Gourich B., N. El Azher, C. Vial, M. Belhaj Soulami and M. Ziyad. 2006. Study of hydrodynamics, mixing and gas–liquid mass transfer in a split-rectangular airlift reactor, *Canadian Journal of Chemical Engineering* 84:539–546.

Hinze J.O. 1975. *Turbulence* (2nd ed.). London (UK). McGraw-Hill.

Hu G.S. and I. Celik. 2008. Eulerian–Lagrangian based large-eddy simulation of a partially aerated flat bubble column. *Chemical Engineering Science* 56:253–271.

Huang W., C. Wu and W. Xia. 2009. Oxygen transfer in high-speed surface aeration tank for wastewater treatment: Full-scale test and numerical modeling. *Journal of Environmental Engineering* 135:684–691.

Jia X., X. Wang, J. Wen, J. Feng and Y. Jiang. 2010. CFD modelling of phenol biodegradation by immobilized *Candida tropicalis* in a gas–liquid–solid three-phase bubble column. *Chemical Engineering Journal* 157:451–465.

Joshi J.B. 2001. Computational flow modeling and design of bubble column reactors. *Chemical Engineering Science* 56:5893–5933.

Kantarci N., F. Borak and K.O. Ulgen. 2005. Bubble column reactors (review). *Process Biochemistry* 40:2263–2283.

Kerdouss F., A. Bannari, P. Proulx, R. Bannari, M. Skrga and Y. Labrecque. 2008. Two-phase mass transfer coefficient prediction in stirred vessel with a CFD model. *Computers and Chemical Engineering* 32:1943–1955.

Krishna R., J.M. Van Baten, J.M. and M.I. Urseanu. 2000. Three-phase Eulerian simulation of bubble column reactors operating in the churn-turbulent regime: A scale up strategy. *Chemical Engineering Science* 55:3275–3286.

Krishna R. and J.M. Van Baten. 2001. Eulerian simulations of bubble columns operating at elevated pressures in the churn turbulent flow regime. *Chemical Engineering Science* 56:6249–6258.

Laborde-Boutet C., F. Larachi, N. Dromard, O. Delsart and D. Schweich. 2009. CFD simulation of bubble column flows: Investigations on turbulence models in RANS approach. *Chemical Engineering Science* 64:4399–4413.

Lapin, A. and A. Lubbert. 1994. Numerical simulation of the dynamics of two-phase gas–liquid flows in bubble columns. *Chemical Engineering Science* 49:3661–3674.

Launder B.E. and D.B. Spalding. 1974. The numerical computation of turbulent flows. *Computer Methods in Applied Mechanics and Engineering* 3:69–289.

Laín S., D. Bröder, M. Sommerfeld and M.F. Göz. 2002. Modelling hydrodynamics and turbulence in a bubble column using the Euler-Lagrange procedure. *International Journal of Multiphase Flow* 28:1381–1407.

Le Moullec Y., O. Potier, C. Gentric and J.-P. Leclerc. 2008. Flowfield and residence time distribution simulation of a cross-flow gas–liquid wastewater treatment reactor using CFD. *Chemical Engineering Science* 63:2436–2449.

Le Moullec Y., C. Gentric, O. Potier and J.P. Leclerc. 2010a. CFD simulation of the hydrodynamics and reactions in an activated sludge channel reactor of wastewater treatment. *Chemical Engineering Science* 65:492–498.

Le Moullec Y., C. Gentric, O. Potier and J.P. Leclerc. 2010b. Comparison of systemic, compartmental and CFD modelling approaches: Application to the simulation of a biological reactor of wastewater treatment. *Chemical Engineering Science* 65:343–350.

Le Moullec Y., C. Gentric, O. Potier and J.P. Leclerc. 2011. Activated sludge pilot plant: Comparison between experimental and predicted concentration profiles using three different modelling approaches. *Water Research* 45:3085–3097.

Leon-Becerril E., A. Cockx and A. Liné. 2002. Effect of bubble deformation on stability and mixing in bubble columns. *Chemical Engineering Science* 57:3283–3297.

Ma D., M. Liu, Y. Zu and C. Tang. 2011. Two-dimensional volume of fluid simulation studies on single bubble formation and dynamics in bubble columns. *Chemical Engineering Science* 72:61–77.

Marques M.P.C., J.M.S. Cabral and P. Fernandes. 2010. Bioprocess scale-up: Quest for the parameters to be used as criterion to move from microreactors to lab-scale. *Journal of Chemical Technological Biotechnology* 85:1184–1198.

Marshall E.M. and A. Bakker. 1995. Computational Fluid Mixing. Lebanon (NH). Fluent Inc.

Merchuk J.C. and M. Gluz. 2003. Fermentation, biocatalysis, bioseparation, In: M.C. Ficklinger and W.D. Stephen, Eds., *Encyclopedia of Bioproccess Technology*. New York. J. Wiley & Sons.

Mousavi S.M., A. Jafari, S. Yaghmaei, M. Vossoughi and I. Turunen. 2008. Experiments and CFD simulation of ferrous biooxidation in a bubble column bioreactor. *Computers and Chemical Engineering* 32:1681–1688.

Mousavi S.M., S.A. Shojaosadati, J. Golestani and F. Yazdian. 2010. CFD simulation and optimization of effective parameters for biomass production in a horizontal tubular loop bioreactor. *Chemical Engineering and Processing: Process Intensification* 49:1249–1258.

Olmos E., C. Gentric, C. Vial, G. Wild and N. Midoux. 2001. Numerical simulation of multiphase flow in bubble column reactors: Influence of bubble coalescence and break-up. *Chemical Engineering Science* 56:6359–6365.

Patankar S.V. 1980. *Numerical Heat Transfer and Fluid Flow*. New York. McGraw-Hill.

Pfleger D. and S. Becker. 2001. Modelling and simulation of the dynamic flow behaviour in a bubble column. *Chemical Engineering Science* 56:1737–1745.

Rafique M., P. Chen and M.P. Dudukovíc. 2004. Computational modeling of gas–liquid flow in bubble columns. *Reviews in Chemical Engineering* 20:225–375.

Ranade V.V. 2002. *Computational Flow Modeling for Chemical Reactor Engineering*. San Diego (CA). Academic Press.

Sanyal J., D.L. Marchisio, R.O. Fox and K. Dhanasekharan. 2005. On the comparison between population balance models for CFD simulation of bubble columns. *Chemical Engineering Science* 44:5063–5072.

Scragg A.H. 1991. *Bioreactors in Biotechnology: A Practical Approach*. Hemel Hempstead (UK). Ellis Horwood, Ltd.

Selma B., R. Bannari and P. Proulx. 2010a. A full integration of a dispersion and interface closures in the standard *k–ε* model of turbulence. *Chemical Engineering Science* 65:5417–5428.

Selma B., R. Bannari and P. Proulx. 2010b. Simulation of bubbly flows: Comparison between direct quadrature method of moments (DQMOM) and method of classes (CM). *Chemical Engineering Science* 65:1925–1941.

Sharma C., D. Malhotra and A.S. Rathore. 2011. Review of computational fluid dynamics applications in biotechnology processes. *Biotechnology Progress* 27:1497–1510.

Simonnet M., C. Gentric, E. Olmos and N. Midoux. 2008. CFD simulation of the flow field in a bubble column reactor: Importance of the drag force formulation to describe regime transitions. *Chemical Engineering and Processing* 47:1726–1737.

Sokolichin A. and G. Eigenberger. 1999. Applicability of the standard turbulence model to the dynamic simulation of bubble columns. Part I. Detailed numerical simulations. *Chemical Engineering Science* 54: 2273–2284.

Staykov P., M. Fialova and S.D. Vlaev. 2009. Bubble-bed structural models for hybrid flow simulation: an outlook based on a CFD generated flow image. *Chemical Engineering Research and Design* 87:669–676.

Tabib M.V., S.A. Roy and J.B. Joshi. 2008. CFD simulation of bubble column: An analysis of interphase forces and turbulence models. *Chemical Engineering Journal* 139:589–614.

Talvy S., A. Cockx and A. Liné. 2007a. Modeling hydrodynamics of gas–liquid airlift reactor. *AIChE Journal* 53:335–353.

Talvy S., A. Cockx and A. Liné. 2007b. Modeling of oxygen mass transfer in a gas–liquid airlift reactor. *AIChE Journal* 53:316–326.

Tomiyama A., H. Tarnai, I. Zun and S. Hosokama. 2002. Transverse migration of single bubbles in simple shear flow. *Chemical Engineering Science* 57:1849–1858.

Van't Riet K. and J. Tramper. 1991. *Basic Bioreactor Design*. New York (NY). Marcel Dekker.

Vial C. (2000). Application of fluid mechanics tools to the analysis of gas–liquid bubbly reactors: Experiments and numerical simulation. PhD dissertation thesis (in French). Institut National Polytechnique de Lorraine. Nancy (France).

Vikas V., C. Yuan, Z.J. Wang and R.O. Fox. 2011. Modeling of bubble column flows with quadrature-based moment methods. *Chemical Engineering Science* 66:3058–3070.

Wang X., J. Wen and X. Jia. 2009. CFD modelling of transient performance of toluene emissions biodegradation in bubble column. *Biochemical Engineering Journal* 48:42–50.

Wang X., X. Jia and J. Wen. 2010. Transient modeling of toluene waste gas biotreatment in a gas–liquid airlift loop reactor. *Chemical Engineering Journal* 159:1–10.

Wang X., X. Jia and J. Wen. 2011. Transient CFD modeling of toluene waste gas biodegradation in a gas–liquid–solid three-phase airlift loop reactor by immobilized *Pseudomonas putida*. *Chemical Engineering Journal* 172:735–745.

Wild G., S. Poncin, H-Z. Li and E. Olmos. 2003. Some aspects of the hydrodynamics of bubble columns. *International Journal of Chemical Reactor Engineering* 1–36.

Wu B. 2010. CFD simulation of gas and non-Newtonian fluid two-phase flow in anaerobic digesters. *Water Research* 44:3861–3874.

Zadghaffari R., J.S. Moghaddas and J. Revstedt. 2009. A mixing study in a double-Rushton stirred tank. *Computers and Chemical Engineering* 33:1240–1246.

Zahradník J., R. Mann, M. Fialová, D. Vlaev, S.D. Vlaev, V. Lossev and P. Seichter. 2001. A networks-of-zones analysis of mixing and mass transfer in three industrial bioreactors. *Chemical Engineering Science* 56:485–492.

Zehner P. and G. Schuh. 1985. A concept for the description of gas phase mixing in bubble columns. *German Chemical Engineering* 5:282–289.

Zhang H., W. Williams-Dalson, E. Keshavarz-Moore and P.A. Shamlou. 2005. Computational-fluid-dynamics (CFD) analysis of mixing and gas–liquid mass transfer in shake flasks. *Biotechnology and Applied Biochemistry* 41:1–8.

7 Laboratory and Industrial Bioreactors for Submerged Fermentations

Parvinder Kaur, Ashima Vohra, and Tulasi Satyanarayana

CONTENTS

7.1 INTRODUCTION

Fermentation technology is the oldest of all biotechnological processes, dating back thousands of years. It was the means by which bread, wine, beer, and cheese were made. The term "fermentation" is derived from the Latin verb *fervere*, which means "to boil." This is because of the frothy appearance of fruit extracts or malted grain acted upon by yeast during the production of alcohol as gas bubbles of CO_2 are continuously released. It was only in the 1850s, when Louis Pasteur concluded that yeasts were responsible for fermenting sugar into ethanol and carbon dioxide in anaerobic conditions, that research interest in employing microbes in fermentations in the food and beverage industries and therapeutics was initiated. Biochemists consider fermentation to be an energy-generating process in which organic compounds act both as electron donors and acceptors and an anaerobic process where energy is produced without the participation of oxygen or other inorganic electron acceptors. Industrial microbiologists consider fermentation to be any process (aerobic or anaerobic) for the production of products by means of mass culture of microbes even if the final electron acceptor is not an organic compound.

Three modes of operation of fermenters are very common: batch, continuous, and fed-batch (Rani and Rao 1999). All industrial processes are described as one of these, depending upon the number of times the entire medium or a single component of the fermentation medium is added to the fermenter during the fermentation process as well as the change in net volume of the fermentation media in the fermenter. The choice of the fermentation mode is dependent upon the relationship of consumption of substrate to the biomass and products (Raj and Karanth 2006).

Batch fermentation is a closed or discontinuous system, which implies that all the nutrient components are added at the beginning of the fermentation process. The sterile medium is inoculated with the appropriate microorganisms, and the fermentation proceeds without the addition of fresh media or the removal of spent media. Thus, no net change in the volume of fermentation media occurs during the fermentation. The growth rate of the organisms will eventually proceed to zero resulting from either diminishing nutrients or accumulation of toxic waste products. The cells growing in a finite volume of liquid nutrient medium follow a sigmoid pattern of growth with lag, log, stationary, and death phases, and all cells are harvested at the same time. Most industrial fermentations are of this type for the production of organic acids, enzymes, and several others. The basic advantage is a lesser chance of contamination although it is difficult to maintain uniformity in the culture variables.

Fed-batch fermentation is a modification of the batch process; whereby nutrients are added periodically to the bioreactor to augment depletion of nutrients, but no medium is removed until the end of the process. This is called controlled feeding and results in a net increase in volume of the fermentation media. Overall, the system, however, remains closed, and there is no continuous flow. A fed-batch culture has the advantage of avoiding substrate overfeeding, which can inhibit the growth of microorganisms. Major disadvantages include difficulty in controlling cultural parameters, a greater chance of contamination, and higher cost. Commercially, baker's yeast is produced by this method.

In contrast, continuous fermentation is an open system where fresh medium is added continuously during fermentation, but there is also concomitant removal of an equal volume of spent medium containing suspended microorganisms. Thus, organisms and nutrients can continuously enter and leave the fermenter, and the number of cells removed with the outflow is exactly balanced by the number of newly synthesized cells, resulting in no net change in the volume of fermentation media. This method prolongs the exponential growth phase of microbes as nutrients are continuously supplied and metabolites and other wastes are continually removed, thus promoting continual growth of the microorganisms. Two control methods are used in continuous fermentations: chemostat and turbidostat. In the chemostat method, the medium contains an excess of all but one of the nutrients that determines the rate of growth of the microorganism. At steady state, the amount of biomass remains constant although medium is added to the fermenter and removed from it at the same rate while maintaining the same level of the medium. On the other hand, in the turbidostat method, the medium contains an excess of all nutrients so that the microbial growth is at its maximum specific growth rate. The system consists of a photoelectric cell, which is a turbidity sensor that detects changes in turbidity of the contents in the fermenter and then controls the amount of medium fed to the fermenter.

Continuous fermentation is advantageous because of its higher productivity, higher yield, lower cost, and better maintenance of cultural parameters. Major disadvantages include an increased chance of contamination and loss of stability resulting from strain degeneration during long-term cultivation. The continuous fermentation processes are not commonly used in industrial processes because they require sophisticated and automated instrumentation setup, and any snag in the system disturbs the process. They have, however, been used for the production of organic solvents, such as acetone butanol fermentation, and in understanding the effects of various variables on growth and product formation by microbes.

Fermenters of various types, shapes, and sizes made of glass, wood, concrete, steel, or stainless steel have been used for conducting microbial fermentations, each having its own advantages and disadvantages. This chapter deals with various types of fermenters or bioreactors employed in submerged fermentations with special reference to their utility in the food industry.

7.2 SUBMERGED FERMENTATION

Submerged fermentation (SMF) involves submersion of the microorganisms in an aqueous solution containing all the nutrients needed for growth. It is the most popularly used technique for the production of a large number of products using a wide range of microorganisms. Usually, highly processed ingredients are used in the medium for submerged fermentation. Asepsis needs to be maintained as the high water activity of the medium makes it prone to contamination. A limiting factor is mass transfer from the gas to the liquid phase, but better mixing can prevent diffusional limitation of nutrients in submerged fermentation. Online sensors help in better bioprocess control of the fermentation process. The main advantages are better media utilization resulting from uniform mixing and the lower cost of downstream processing, that is, extraction of product. Submerged fermentation has been employed extensively in industries for large-scale production of alcohol, organic acids, enzymes, vitamins, and amino acids.

Solid-state fermentation (SSF), on the other hand, involves cultivation of microorganisms on solid, moist substrates in the absence of a free aqueous phase (Pandey 2003). Hence, it is a microbial process in which a solid or semisolid material is used as the substrate or the inert support on which microorganisms grow. The medium used in SSF is usually a solid substrate (e.g., rice bran, wheat bran, or grain), which requires no processing. The low water content results in fewer problems resulting from contamination. The power/energy requirements are lower than those of submerged fermentation. SSF is generally applicable for low-value products requiring less monitoring and control because of problems of inadequate mixing, limitations of nutrient diffusion, metabolic heat accumulation, and ineffective process control.

7.3 FERMENTERS

Fermentation is carried out in vessels known as fermenters or bioreactors. A fermenter is defined as a vessel for the growth of microorganisms, which, while not permitting contamination, enables the provision of conditions necessary for the maximal production of the desired products (Okafor 2007). The fermentation unit in industrial microbiology is analogous to a chemical plant in the chemical industry with the exception that a fermentation process is a biological process. Hence, fermenters are also known as bioreactors as they are reactors used for biological reactions. A bioreactor is a reactor system used for the cultivation of microorganisms. They vary in size and complexity from a 10 mL volume in a test tube to computer-controlled fermenters with liquid volumes greater than 100 m³. The capacity of various fermenters varies as follows: laboratory-scale fermenter: 1–25 L; pilot-scale fermenter: up to 2000 gal.; production-scale fermenter: 5000–100,000 gal. They may vary in cost from a few cents to a few million dollars.

In the fermenters, the industrial microorganisms are grown under controlled conditions with the aim of optimizing the growth of the organism or production of a target microbial product. The basic function of a fermenter is to provide a suitable environment in which an organism can efficiently produce a target product, which may be cell biomass, a metabolite, or a bioconversion product. The type of fermenter ranges from a simple tank to a complex, integrated system of automated control. Some well-known manufacturers of fermenters are B. Braun Biotech-Sartorius (Germany), New Brunswick Scientific (USA), Applikon (Netherlands), Bioengineering (Switzerland), and INFORS (Switzerland).

The important components in most of the fermenters are the following:

- Body of construction: Lab fermenters are usually made up of either glass or stainless steel, and fermenters with higher capacities are generally made up of only stainless steel. In aseptic fermentation processes, fermenters constructed of wood or concrete may be used; for example, some alcohol manufacturers are using fermenters made of wood. The most common height-to-diameter ratio of the fermenters is between 1 and 2.
- Stirrers: The main function of the stirrers is to mix the gases and liquid media. Stirrers consist of a hollow or solid shaft and impellers. There are two types of stirrer shaft seals that are widely in use to arrest the leakage from the entry of the stirrer shaft into the vessel. They are gland packing (stuffing the box with PTFE threads) and mechanical seals.
- Impellers: The presence of impellers on the agitator shaft brings about uniform mixing of microorganisms and nutrients and dispersion of air in the nutrient solution, resulting in efficient mass and heat transfer. There are several types of impellers used in fermentation, such as the disc turbine, vaned disc, open turbine, marine propeller, Rushton turbine, and others. Recent developments in impeller design have led to the emergence of high-efficiency impellers, such as the Mig (counterflow impellers) and Intermig (a modified version of Mig), which require 25% and 40% less power input to get the same degree of mixing as a turbine impeller (Raj and Karanth 2006).

- Baffles: Baffles are mainly used to prevent a vortex and to improve the aeration efficiency of the stirrers. These are the blades with approximately 1/10 the diameter of the fermenter and mounted inside the fermenter vessel along the walls. Baffles should be installed so that there is a gap between the baffle and the vessel wall to minimize the microbial growth on the baffles and walls.
- Sparger: A sparger is a device for introducing air below the liquid level in the fermenter vessel. There are four types of spargers in general use: porous sparger, orifice sparger, nozzle sparger, and combined sparger. The use of spargers with very small orifices is more efficient than a single orifice delivering the same volume of air.
- Jacket or coils: There is a jacket on the outer side of the fermenter to facilitate the cooling or heating of the media inside the fermenter. In large-scale fermenters, internal and external cooling coils are used for this purpose.
- Valves and steam traps: Several types of valves are available, each playing different roles in different areas, for example, globe valve, gate valve, needle valve, ball valve, piston valve, diaphragm valve, pinch valve, butterfly valve, nonreturn valve, pressure-reducing valve, safety valve, etc. For the sterile operations in fermentation, diaphragm valves and pinch valves are recommended.
- Probes: Various kinds of probes are connected to measure and control various process parameters. These include probes for monitoring pH, foaming, temperature, and dissolved oxygen.

7.4 TYPES OF FERMENTERS

Laboratory-scale submerged fermentations are carried out in shaker flasks, and large-scale fermentations are carried out in glass or stainless-steel tank fermenters. The prerequisites for a good fermentation vessel are that it must be inexpensive, not allow contamination of the contents, be nontoxic to the microorganism used for the process, be easy to sterilize, be easy to operate, be robust and reliable, be leak proof, allow visual monitoring of the fermentation process, and allow sampling. The scale-up of the fermentation process is usually the final step in any research-and-development program, leading to the large-scale industrial manufacture of products by fermentation (Einsele 1978). It needs to be understood that the process of scaling up a fermentation system is not simply a matter of increasing culture and vessel volume, but it is frequently governed by a number of important engineering considerations. Therefore, usually a large-scale process does not perform as well as a small-scale laboratory process. It is often observed that the biomass yield and any growth-associated products are often decreased with the scale-up of an aerobic process (Enfors et al. 2001).

7.4.1 SHAKE FLASKS

These are conical vessels made of glass and are available in different sizes. The typical volume of these flasks is 250 mL. Erlenmeyer flasks with a cotton plug are

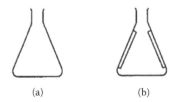

(a)   (b)

FIGURE 7.1   Shake flasks. (a) Standard (Erlenmeyer flask) and (b) with baffles.

the classic small-scale liquid fermentation system (Figure 7.1a). When the vessel is round-based and the motion is orbital, mixing is poor. Some modifications in the basic shake-flask system have been made to increase mixing and mass transfer with indentations (baffles) and corners (Figure 7.1b). Baffles have been used mainly to increase the turbulence of mixing to increase the liquid surface area and, therefore, the gas transfer. The upper limit on the volume of small-scale batch fermentations is typically set by the oxygen-uptake requirements of the culture to be fermented. Generally, the gas transfer rate at the gas–liquid interface is lesser as the volume gets larger. Shaker beds or shaker tables are used to allow oxygen transfer by their continuous rotary motion. Although higher oxygen transfer rates can be achieved with shake flasks than with standing cultures, oxygen transfer limitations will still be unavoidable particularly when trying to achieve high cell densities. The rate of oxygen transfer in shake flasks is dependent on the shaking speed, the liquid volume, and the shake-flask design.

7.4.2   Stirred Tank Fermenter

Stirred tank fermenters (STF) are some of the most commonly used fermenter types because of their flexibility (Chisti 2010). They are cylindrical vessels with a motor-driven central shaft with impellers or an agitator to stir the contents in the tank (Figure 7.2). The shaft supports three to four impellers placed approximately one impeller-diameter apart (Chisti 2010). The agitator may be top driven or bottom driven, depending on the scale of operation and other operational aspects. The top-entry–stirrer (agitator) model is most commonly used because it has many advantages, such as ease of operation, reliability, and robustness, and the bottom-entry–stirrer (agitator) model is rarely used. The fermenter has an aspect (working height to weight) ratio of 2:1 and 6:1. The choice of impeller depends on the physical and biological characteristics of the fermentation broth. Usually, a ring-type sparger with perforations is used to supply air to the fermenter. There are four equally spaced vertical baffles that extend from near the walls into the vessel to avoid vortex formation and improve mixing. The STR offers the advantages of high oxygen transfer rates required for high biomass productivity coupled with low investment and operating costs, which form the basis for any successful aerobic fermentation process.

Laboratory-scale STRs are made of borosilicate glass with a stainless-steel lid and top-entry stirrer. The typical volume of these fermenters is 1 to 100 L. Stainless-steel fermenters are also used in laboratories and have special requirements. They should be made of high-grade stainless steel, have an internal surface that is polished

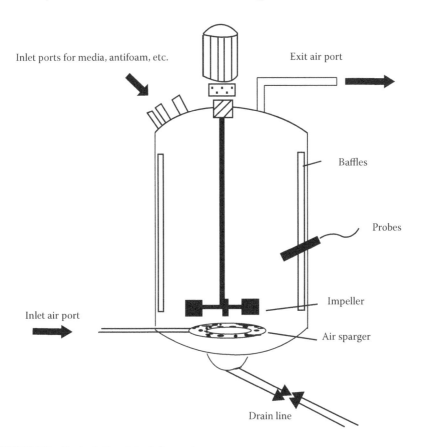

Inlet ports for media, antifoam, etc.

Exit air port

Baffles

Probes

Impeller

Inlet air port

Air sparger

Drain line

FIGURE 7.2   Typical stirred-tank fermenter.

to reduce adhesion of the contents to the walls of the fermenter, and have joints that are smooth and free from pinholes.

Traditionally, fermentation of grape juice was conducted in large wooden barrels or concrete tanks, but most modern wineries now use sophisticated stainless-steel tanks with temperature control and various other features for process management (Divies 1993).

In a study by Ali et al. (2002), a laboratory-scale stirred fermenter of 15-L capacity with a working volume of 9 L was used for citric acid production using *Aspergillus niger* GCBT7. Batch fermentation was employed in a New Brunswick bioreactor for obtaining probiotic biomass from *Lactobacillus plantarum* BS1 and BS3 strains. The conditions of fermentation were 37°C, pH 5.5, stir 100 rpm once at every 12 h of fermentation in order to homogenize the medium for 15 min (Vamanu 2009). The production of α-amylase by *Bacillus amyloliquefaciens* was performed in 5-L stirred-tank bioreactor (Biostat B-5, B. Braun Biotech-Sartorius, Melsungen, Germany). It is a baffled cylindrical acrylic vessel with a working volume of 3 L and a working volume to space ratio of 1:1.66, having an internal diameter of 160 mm and a height of 250 mm with dual impellers mounted on the shaft. The baffles with a width of 12 mm were placed perpendicular to the vessel. The system was equipped with a six-bladed Rushton turbine impeller for agitation with a diameter of 64 mm, a blade height of 13 mm, and a width of 19 mm. The spacing between the impellers was maintained at 110 mm, and the lower impeller was located at a distance of 80 mm from the bottom of the vessel. The sparger was located at a distance of 5 mm from the bottom of the vessel through which air was sparged to the tank. The ring sparger was 52 mm in diameter and had 16 symmetrically drilled holes 1 mm in diameter. The flow rate of sparged air was fixed at 1.5 vvm. The fermentation was carried out at 37°C and monitored by a temperature probe that was controlled by circulating the chilled water. Foaming in the fermentation broth was monitored by a ceramic-coated antifoam probe, and coconut oil was used as an antifoaming agent. The dissolved oxygen (DO) was maintained at a 100% saturation level, which was continuously monitored by a sterilized polarographic electrode (Mettler-Toledo InPro 6000 Series, Greifensee, Switzerland). The experiments indicated a requirement of high rates of aeration to enhance the enzyme yield (Gangadharan et al. 2011). The production of rennet was carried out in submerged fermentation by *Rhizomucor miehei* NRRL-3420 using two types of media for 40 h at 380 rpm, 1 vvm aeration, and 30 ± 1°C (De Lima et al. 2008).

7.4.3   TOWER FERMENTER

Tower fermenters are modified stirred-tank reactors, which are simple in design and easy to construct. They consist of a long, cylindrical vessel with an inlet at the bottom, an exhaust at the top, and a jacket to control the temperature. They do not require agitation; hence, there are no shafts, impellers, or blades. Tower fermenters are used for continuous fermentation of beer, yeast, and Single Cell Protein (SCP). In brewing, cylindroconical fermenters have become the vessels of choice since the 1970s (Maule 1986). Traditionally, open rectangular tanks of 2–3 m depth were used (Hough et al. 1982). An enclosed vessel reduces the risk of contamination over open vessels and also helps to capture the carbon dioxide produced. The cylindroconical fermenter has a conical lower section that facilitates the sedimentation and recovery of bottom yeast, which settles out late in the fermentation process. The shape also encourages a vigorous mixing of yeast cells in the fermentation wort, resulting in a faster fermentation. No aeration is required as the gas produced by the yeast cells contributes to mixing.

7.4.4   AIRLIFT FERMENTER

An airlift fermenter is a cylindrical fermentation vessel in which the cells are mixed by air introduced at the base of the vessel and that rises through the column of culture medium. The working aspect ratio of these fermenters is six (height/weight of 6:1) or more. These fermenters do not have mechanical agitation systems (motor, shaft, impeller blades), but the contents are agitated by injecting air from the bottom. The cell suspension circulates around the column as a consequence of the gradient of air bubbles in different parts of the reactor. Thus, the fluid of volume is divided into two interconnected zones by means of a baffle or draft tube. Only one of the two zones is sparged with air or gas. The sparged zone is known as the riser, and the zone that receives no gas is the downcomer (Chisti and Moo-Young 2001). These come in two models: the internal-loop and external-loop designs. In the internal-loop configuration (Figure 7.3a), the aerated riser and the unaerated downcomer are contained in the same shell, and in the external-loop design (Figure 7.3b), the riser and the downcomer are separate tubes linked near the top and the bottom (Chisti 1999). The external loop design has not been used frequently in industry.

Sterile atmospheric air is used if the microorganisms are aerobic and inert gas is used if the microorganisms are anaerobic. This is a gentle method of mixing the contents and is most suitable for fermentation of molds because the mechanical agitation produces high shearing stress that may damage the cells. An airlift fermenter differs from bubble column bioreactors by the presence of a draft tube, which provides better mass and heat transfer efficiencies and low shear conditions. Other advantages include increased oxygen solubility and easier maintenance of sterility. The major disadvantages of airlift fermenters are high capital cost, high energy requirements, excessive foaming, and cell damage resulting from bubbles bursting.

Airlift fermenters have been used for citric acid production with 900 m³ volume using *A. niger*. The fermenters must be resistant to acidity and are made of stainless steel because ordinary steel can be dissolved at pH 1–2, inhibiting the fermentation. In fermenters with a capacity of less than 1 m³, because of increased surface-to-volume ratio, corrosion becomes significant, and even steel chambers must be protected with a layer of plastic material. An aeration rate of 0.2–1 vvm is often used during the production phase to avoid dissolved oxygen levels lower than 20%–25% of the saturation (Honecker et al. 1989). The production of phytase and glucoamylase by *Sporotrichum thermophile* (Singh and Satyanarayana 2008) and *Thermomucor*

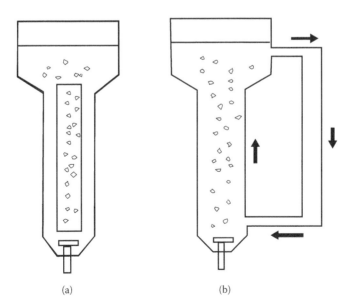

(a)   (b)

FIGURE 7.3   Airlift fermenter (a) with internal draft tube and (b) with external draft tube.

indicae-seudaticae (Kumar and Satyanarayana 2007) have been recently reported in the airlift fermenter, respectively.

Quorn is the leading brand of mock meat mycoprotein food product in the United Kingdom and Ireland. The mycoprotein used to produce Quorn is extracted from a fungus, *Fusarium graminearum*, which is grown in a very large airlift fermenter in a continuous-culture mode. *F. graminearum* A315 biomass is produced in an airlift or pressure-cycle fermenter at Billingham. A continuous-flow culture system is chosen for the process because the growth conditions in such cultures, unlike those in batch cultures, can be maintained as constant throughout the production phase and because much higher productivities can be achieved in continuous culture than in batch culture (Trinci 1994).

The technology of creating SCP from methanol has been well studied, and the most advanced process belongs to Imperial Chemical Industries (ICI, UK). The fermentation was carried out in a big airlift fermenter with the bacterium *Methylophilus methylotrophus*. This organism was selected from among other methanol utilized after screening tests for pathogenicity and toxicity. As a nitrogen source, ammonia was used and the product was named Pruteen. The pruteen contained 72% crude protein and was marketed as feed that contains a source of energy, vitamins, and minerals as well as a highly balanced protein source. The methionine and lysine content of Pruteen compared very favorably with whitefish meal. The ICI commissioned a 60,000 ton/year plant utilizing the single largest fermenter in the world (2 × 10,000,000 L) (Nasseri et al. 2011).

7.4.5 Bubble Column Reactor

A bubble column reactor is basically a cylindrical vessel with a gas distributor at the bottom. It was first applied by Helmut Gerstenberg. The gas is sparged in the form of bubbles into either a liquid phase or a liquid–solid suspension (Figure 7.4). The

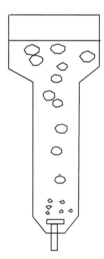

FIGURE 7.4   Bubble column reactor.

introduction of gas takes place at the bottom of the column and causes a turbulent stream to enable an optimum gas exchange. It is built in numerous forms of construction. The mixing is done by the gas sparging, and it requires less energy than mechanical stirring (Nigar et al. 2005). These reactors are generally referred to as slurry bubble column reactors when a solid phase exists. Bubble columns are intensively utilized as multiphase contactors and reactors in chemical, petrochemical, biochemical, and metallurgical industries (Degaleesan et al. 2001). They are used especially in chemical processes involving reactions, such as oxidation, chlorination, alkylation, polymerization, and hydrogenation, in the manufacture of synthetic fuels by gas conversion processes, and in biochemical processes, such as fermentation and biological wastewater treatment (Shah et al. 1982; Prakash et al. 2001).

Bubble column reactors have a number of advantages as compared to other reactors. They have excellent heat and mass transfer characteristics, require little maintenance, have low operating costs because of their lack of moving parts and compactness, and the durability of the catalyst or other packing material is high (Degaleesan et al. 2001). Moreover, online catalyst addition and withdrawal ability and plug-free operation are other advantages that render bubble columns an attractive reactor choice (Prakash et al. 2001).

Bubble column bioreactors have been used extensively to produce industrially valuable products, such as enzymes, proteins, antibiotics, and others. The production of thienamycin was carried out using *Streptomyces cattleya* in a continuously operated bubble column bioreactor (Arcuri et al. 1986). Federici et al. (1990) reported production of glucoamylase by *Aureobasidium pullulans*. Also, the production of acetic acid in a bubble column by *Acetobacter aceti* was investigated by Sun and Furusaki (1990). Rodrigues et al. (1992) reported cultivation of hybridoma cells in a bubble column reactor for attaining a high monoclonal antibody productivity of 503 mg/L/day. Chang et al. (2001) cultivated *Eubacterium limosum* on carbon monoxide to produce organic acids in a bubble column reactor. The study by Ogbonna et al. (2001) was based on the potential of producing fuel ethanol from sugar-beet juice in a bubble column. In this study, yeast cells (*Saccharomyces cerevisiae*) were used in order to investigate the feasibility of scaling up the process.

7.4.6 FLUIDIZED BED REACTOR

These are similar to bubble columns with an expanded cross section near the top so that fresh or recirculated liquid is continuously pumped into the bottom of the vessel at a velocity that is sufficient to fluidize the solids or maintain them in suspension (Chisti 2010). In these bioreactors, mixing is assisted by the action of a pump, and the cells or enzymes are immobilized in and/or on the surface of light particles (Figure 7.5). A pump located at the base of the tank causes the immobilized catalysts to move with the fluid. The pump pushes the fluid and the particles in a vertical direction. The upward force of the pump is balanced by the downward movement of the particles because of gravity to prevent washout of solids from the bioreactor, thus resulting in good circulation. Sparging is used to improve oxygen transfer rates for aerobic microbial systems. Fluidized bed bioreactors are one method of maintaining high biomass concentrations and, at the same time, good mass transfer rates in

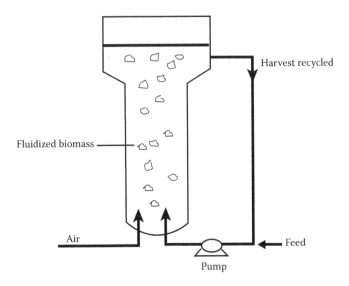

FIGURE 7.5   Fluidized bed reactor.

continuous cultures. This advantage has been observed in ethanol production from *S. cerevisiae* (Margaritis et al. 1983) with superior mass and heat transfer characteristics, very good mixing between the three phases, relatively low energy requirements, and low shear rates.

7.4.7   MEMBRANE BIOREACTOR

Membrane bioreactors comprising hollow fiber systems have been developed and tested for the immobilization of bacteria, yeast, and enzymes. The hollow fibers can be made from cellulose acetate with a uniform wall matrix or from acrylic copolymers or polysulfone fibers with asymmetric wall configurations (Raj and Karanth 2006). Among the several advantages of using a hollow-fiber reactor for microbial systems include high density of cell growth, using a perfusion system for simultaneous separation of product and biomass, and biocatalyst regeneration. The difficulty in monitoring and controlling the growth and metabolism of the culture is a major disadvantage along with low oxygen transfer rates at high cell density and blockage and rupture of the membranes resulting from excessive growth of the microorganisms. The metabolic activity of the cell system in the hollow fiber might also be inhibited by accumulation of toxic products. Membrane bioreactors have been used extensively for several microbial cell cultivations (Chang and Furusaki 1991). A few investigations have been reported on the probiotic cell production in membrane reactors; these systems supported high cell yields and volumetric productivities. Taniguchi et al. (1987) reported sevenfold higher *Bifidobacterium longum* biomass levels in a membrane bioreactor than that attained in free-cell batch fermentations. Similarly, Corre et al. (1992) attained high cell yields and a 15-fold enhancement in volumetric productivity as compared to free-cell batch cultures of *B. bifidum*.

TABLE 7.1

Comparison of Bioreactors with Mechanical and Pneumatic Agitation

	Mechanically Agitated Fermenters	Pneumatically Agitated Fermenters
Design	Complex (shaft, impeller, bearings, etc.)	Simple
Shear	High	Low
Cleaning	Difficult	Easy
Mixing	Nonuniform as controlled by impellers	More uniform
Operational flexibility	Better	Limited

The technique has also been used for the production of lactic acid (Moueddeb et al. 1996). A high-performance membrane bioreactor had been employed in ethanol (Cheryan and Mehaia 1983) and organic acid (Endo 1996) fermentations. The productivity was enhanced by continuous fermentation in a membrane bioreactor performed at a very high dilution rate (Boyaval and Corre 1987). A double vessel–membrane bioreactor used for the production of wine from grape juice supported a higher wine production than a single vessel in continuous fermentation with a low residual sugar level (Takaya et al. 2002).

7.5 COMPARISON OF MECHANICALLY AGITATED AND PNEUMATIC FERMENTERS

The mechanically stirred fermenters have traditionally been used extensively. They, however, suffer from several limitations when compared with fermenters agitated by gas or air injection, that is, pneumatic bioreactors (e.g., airlift and bubble columns) (Table 7.1). The main disadvantage of mechanically agitated fermenters is their structural and mechanical complexity, leading to difficulties in cleaning and greater chances of contamination. Moreover, the high shearing forces lead to problems in the culturing of molds; immobilized cells; and fragile, genetically engineered microorganisms. Pneumatic agitation promotes uniform mixing as compared to the turbulence provided mainly in the zone of impellers in mechanically stirred reactors, especially in viscous media. However, one major advantage of mechanical agitation is the operational flexibility of these bioreactors, which can be controlled by regulating the impeller speed and gas-flow rate (Chisti and Moo Young 1991).

7.6 CONCLUSIONS

The primary function of a fermenter or bioreactor is to provide a suitable environment in which an organism can grow and efficiently produce the desired product. Submerged fermentation is a method of choice for most microbial fermentations. The fermenters have been classified into various types, depending mainly on the method of agitation and aeration. The stirred-tank bioreactor is commonly used for the production of various industrially important products, such as organic acids, enzymes, and others. The pneumatically agitated fermenters are better for cultivating molds and immobilized cells because of reduced shear, and are fast becoming the bioreactors of choice, despite their high cost of operation. Novel types of fermenters, such as membrane bioreactors, have been designed with improved features to maximize their productivity while lowering production costs.

REFERENCES

Ali, S., Haq, I., Qadeer, M.A. and Iqbal, J. (2002) 'Production of citric acid by *Aspergillus niger* using cane molasses in a stirred fermentor,' *Electronic J. Biotechnol.*, 5(3):258–271.

Arcuri, E.J., Slaff, G. and Greasham, R. (1986) 'Continuous production of thienamycin in immobilized cell systems,' *Biotechnol. Bioeng.*, 28:842–849.

Boyaval, P. and Corre, C. (1987) 'Continuous fermentation of sweet whey permeate for propionic acid production in a CSTR with UF recycle,' *Biotechnol. Lett.*, 11:801–806.

Chang, H.N. and Furusaki, S. (1991) 'Membrane bioreactors: Present and prospects,' in A. Fiechter (ed.) *Advances in Biochemical Engineering Biotechnology*, Vol. 44, Berlin: Springer-Verlag, pp. 27–64.

Chang, I.S., Kim, B.H., Lovitt, R.W. and Bang, J.S. (2001) 'Effect of partial pressure on cell-recycled continuous cofermentation by *Eubacterium limosium* kist612,' *Process Biochem.*, 37:411–421.

Cheryan, M. and Mehaia M.A. (1983) 'A high performance membrane bioreactor for continuous fermentation of lactose to ethanol,' *Biotechnol. Lett.*, 5:519–524.

Chisti, Y. (1999) 'Fermentation (industrial): Basic considerations,' in R. Robinson, C. Batt and P. Patel (eds.) *Encyclopedia of Food Microbiology*, London: Academic Press, pp. 663–674.

Chisti, Y. (2010) 'Fermentation technology,' in W. Soetaert and E.J. Vandamme (eds.) *Industrial Biotechnology: Sustainable Growth and Economic Success*, Germany: Wiley VCH, pp. 149–171.

Chisti, Y. and Moo-Young, M. (1991) 'Fermentation technology, bioprocessing, scale-up and manufacture,' in V. Moses and R.E. Cape (eds.) *Biotechnology: The Science and the Business*, New York: Harwood Academic Publishers, pp. 167–209.

Chisti, Y. and Moo-Young, M. (2001) 'Bioreactor design,' in C. Ratledge and B. Kristiansen (eds.) *Basic Biotechnology*, 2nd edition, UK: Cambridge University Press, pp. 151–171.

Corre, C., Madec, M.N. and Boyaval, P. (1992) 'Production of concentrated *Bifidobacterium bifidum*,' *J. Chem. Technol. Biotechnol.*, 53:189–194.

De Lima, C.J.B., Cortezi, M., Lovaglio, R.B., Ribeiro, E.J., Contiero, J. and De Araújo, E.H. (2008) 'Production of rennet in submerged fermentation with the filamentous fungus *Mucor miehei* NRRL 3420,' *World Appl. Sciences J.*, 4(4):578–585.

Degaleesan, S., Dudukovic, M. and Pan, Y. (2001) 'Experimental study of gas induced liquidflow structures in bubble columns,' *AIChE J.*, 47:1913–1931.

Divies, C. (1993) 'Bioreactor technology and wine fermentation,' in G.H. Fleet (ed.) *Wine Microbiology and Biotechnology*, Harwood Academic, pp. 119–175.

Einsele, A. (1978) 'Scaling-up of bioreactors,' *Proc. Biochem.*, 13:13–14.

Endo, I. (1996) 'A membrane bioreactor,' *Membrane*, 21:18–22.

Enfors, S.O., Jahic, M., Rozkov, A., Xu, B., Hecker, M., Jurgen, B., Kruger, et al. (2001) 'Physiological responses to mixing in large bioreactors,' *J. Biotechnol.*, 85:175–185.

Federici, F., Petruccioli, M. and Miller, M.W. (1990) 'Enhancement and stabilization of the production of glucoamylase by immobilized cells of *Aureobasidium pullulans* in a fluidized bed reactor,' *Appl. Microbiol. Biotechnol.*, 33:407–409.

Gangadharan, D., Madhavan Nampoothiri, K. and Pandey A. (2011) 'α-Amylase produced by *B. amyloliquefaciens*,' *Food Technol. Biotechnol.*, 49(3):336–340.

Honecker, S., Bisping, B., Yang, Z. and Rehm, H.J. (1989) 'Influence of sucrose concentration and phosphate limitation on citric acid production by immobilized cells of *Aspergillus niger*,' *Appl. Microbiol. Biotechnol.*, 31:17–24.

Hough, J.S., Briggs, D.E., Stevens, R., and Young, T.W. (1982) *Malting and Brewing Science*, Vol. 2, London: Chapman and Hall.

Kumar, P. and Satyanarayana, T. (2007) 'Optimization of culture variables for improving glucoamylase production by alginate-entrapped *Thermomucor indicae-seudaticae* using statistical methods,' *Bioresour. Technol.*, 98:1252–1259.

Margaritis, A., te Bokkel, D. and Kashab, M.E. (1983) 'Pilot plant production of ethanol using immobilized yeast cells in a novel fluidized bioreactor system,' 18th ACS Meeting, Washington, DC.

Maule, D.R. (1986) 'A century of fermenter design,' *J. Inst. Brew.*, 92:137–145.

Moueddeb, H., Sanchez, J., Bardot, C. and Fick, M. (1996) 'Membrane bioreactor for lactic acid production,' *J. Membr. Sci.*, 114:59–71.

Nasseri, A.T., Rasoul-Amini, S., Morowvat, M.H. and Ghasemi, Y. (2011) 'Single cell protein: Production and process,' *American J. Food Technol.*, 6(2):103–116.

Nigar, K., Fahir, B. and Kutlu, O.U. (2005) 'Bubble column reactors,' *Process Biochem.*, 40:2263–2283.

Ogbonna, J.C., Mashima, H. and Tanaka, H. (2001) 'Scale up of fuel production from sugar beet juice using loofa sponge immobilized bioreactor,' *Bioresour. Technol.*, 76:1–8.

Okafor, N. (2007) *Modern Industrial Microbiology and Biotechnology*, USA: Science Publishers.

Pandey, A. (2003) 'Solid-state fermentation,' *Biochem. Engineering*, 13:81–84.

Prakash, A., Margaritis, A. and Li, H. (2001) 'Hydrodynamics and local heat transfer measurements in a bubble column with suspension of yeast,' *Biochem. Eng. J.*, 9:155–163.

Raj, E.A. and Karanth, G.N. (2006) 'Fermentation technology and bioreactor design,' in K. Shetty, G. Paliyath, A. Pometto and R.E. Levin (eds.) *Food Biotechnology*, 2nd edition, USA: CRC Press.

Rani, K.Y. and Rao, V.S.R. (1999) 'Control of fermenters: A review,' *Bioprocess Engineering*, 21:77–88.

Rodrigues, M.T.A., Vilaca, P.R., Garbuio, A. and Takagai, M. (1992) 'Glucose uptake rate as a tool to estimate hybridoma growth in a packed bed bioreactor,' *Bioprocess Eng.*, 21:543–556.

Shah, Y.T., Godbole, S.P. and Deckwer, W.D. (1982) 'Design parameters estimations for bubble column reactors,' *AIChE J.*, 28:353–379.

Singh, B. and Satyanarayana, T. (2008) 'Phytase production by *Sporotrichum thermophile* in a cost-effective cane molasses medium in submerged fermentation and its application in bread,' *J. Appl. Microbiol.*, 105:1858–1865.

Sun, Y. and Furusaki, S. (1990) 'Effects of product inhibition on continuous acetic acid production by immobilized *Acetobacter aceti*: Theoretical calculations,' *J. Ferment. Bioeng.*, 70(1):196–198.

Takaya, M., Matsumoto, N. and Yanase, H. (2002) 'Characterization of membrane bioreactor for dry wine production,' *J. Biosci. Bioeng.*, 93:240–244.

Taniguchi, M., Kotani, N. and Kobayashi, T. (1987) 'High-concentration cultivation of lactic acid bacteria in fermenter with cross-flow filtration,' *J. Ferment. Technol.*, 65:179–184.

Trinci, A.P.J. (1994) 'Evolution of the QuornB mycoprotein fungus, *Fusarium graminearum* A315,' *Microbiology*, 140:2181–2188.

Vamanu, E. (2009) 'Studies regarding the production of probiotic biomass from *Lactobacillus plantarum* strains,' *Archiva Zootechnica*, 12(4):92–101.

8 Laboratory and Industrial Bioreactors for Solid-State Fermentation

*José Angel Rodríguez-León,
Daniel Ernesto Rodríguez-Fernández,
and Carlos Ricardo Soccol*

CONTENTS

8.1 INTRODUCTION

Solid-state fermentation (SSF) is an old fermentation technique that has been employed by mankind since food was first preserved or transformed into new food products.

Submerged fermentation (SMF) processes have commonly been employed since ancient times in the manufacturing of wine, beers, fermented milk, and other food products. SSF has been employed in the production of bread and cheese, in the processing of fish, and in other processes. Indeed, there is no evidence to suggest SMF or liquid fermentation was a normal practice employed before SSF to process food.

SSF is already present in nature as a mechanism by which microorganisms survive and develop. The process occurred naturally, during fruit deterioration, wood attack by microorganism, and soil pollution, even before humans created permanent procedures or techniques to exploit microbial metabolism.

Both SMF and SSF processes have undoubtedly been developed through observations and trial and error. Initially, techniques for food preservation or transformation were developed through artisanal procedures until satisfactory results were achieved. These techniques—often called "the art of…"—were transmitted from generation to generation through descriptions rather than modern scientific explanations. Science, in contrast, is a discipline in which "the art of…" is transformed into "the knowledge of…." Currently, there are several different scientific approaches to biotechnological procedures, including, in simple terms, microbiology, biology, biochemistry, genetics, and biochemical engineering.

8.2 FERMENTATION CONCEPT

Fermentation is the materialization of the reproductive capacity (growth) that a microorganism has in a favorable medium or substrate. This capacity is determined by the metabolic patterns of the microorganism.

In the nineteenth century, Louis Pasteur demonstrated and described the production and acidification of wine resulting from the activities of yeast or detrimental bacteria. The concept of fermentation was applied to the anaerobic (absence of air) process, which is referred to as anaerobic fermentation today. Currently, the term "fermentation" has been broadened to include the process in which air is present, that is, the aerobic (presence of air) process or aerobic fermentation.

The metabolic pattern of a microorganism is represented by the multiple reactions that occur during the fermentation process, which are primarily based on the accessibility of the organism to an available energy supply.

The manner in which microorganisms obtain the energy required for reproduction leads to a general classification as follows:

- Autotrophic microorganisms: those microorganisms that employ an energy source from an external source, generally photoelectric.
- Heterotrophic microorganisms: those microorganisms that obtain the energy required for growth by degrading reduced organic substances by oxidation–reduction reactions.

The microorganisms employed in the food industry or in biotechnological processes in general are largely heterotrophic. Therefore, some generalization can be made when dealing with the aerobic fermentation process:

- At the beginning of the process, organic substances are present that provide the energy required for growth via oxidation depending, or not, on the availability of oxygen.
- The metabolic pathway represents the process that produces substances that are more oxidized than the substances that were initially present.
- The oxidative process developed in the presence of air is more energy efficient (ATP synthesis) than the anaerobic process.
- In the case of aerobic fermentation, the final products are CO_2, H_2O, biomass, and extracellular metabolites, which cannot be further oxidized by this pattern.
- In the case of anaerobic fermentation, the final products are normally CO_2, H_2O, biomass, and extracellular metabolites in fairly high concentrations.

These metabolites have greater reduction grades than the metabolites obtained by an aerobic fermentation, v. gr., ethanol production.

An overall description of an aerobic pattern (not considering a possible secondary metabolic pathway) shows the following:

$$\text{Organic material} + O_2 \rightarrow \text{Biomass} + CO_2 + H_2O + \text{heat} \tag{8.1}$$

Simple sugars, such as glucose or sucrose, are commonly employed in fermentation processes as the organic materials that supply energy. It can therefore be postulated that

$$\text{Glucose} + O_2 \rightarrow \text{Biomass} + CO_2 + H_2O + \text{heat} \tag{8.2}$$

Equations 8.1 and 8.2 can be considered to represent the following:

- A redox reaction.
- A combustion reaction.
- Any aerobic fermentation that takes place, regardless of whether the fermentation is submerged or solid. This means that the metabolic pathway is not determined by the fermentation conditions, whether submerged or solid.
- The stoichiometric relationship between the rates at which glucose or any other organic compound is consumed is related to biomass synthesis, O_2 consumption, or CO_2 production.

8.3   FERMENTATION KINETICS

The term "kinetic" indicates that there is a change or variation in some variable over time. In this sense, any available (measured) variable can be chosen to determine kinetic characteristics. The identification of variables and their relationships with other factors eventually leads to the establishment of appropriate models describing processes or parts of processes. This identification also allows one to propose studies for process optimization and to determine control criteria. In the case of fermentation, the most commonly employed variable is biomass synthesis.

Considering biomass variation during process running time as a measure of its kinetics, it can be defined that this variation, in general, depends on several factors; thus,

$$\frac{dx}{dt} = f(x,s,T,\text{etc.}) \tag{8.3}$$

where x is the biomass concentration (g/L), t is the time (h), s is the substrate concentration (g/L), and T is the temperature (°C).

Monod (1949) was one of the first scientists to mathematically describe a fermentation process. He proposed a model that related biomass synthesis with substrate. The model is as follows:

$$\mu = \mu_{max} \frac{S}{K_s + S} \tag{8.4}$$

where μ is the specific growth rate (h^{-1}), μ_{max} is the maximum specific growth rate (h^{-1}), S is the substrate concentration (g/L), and K_s is the affinity constant (g/L).

In the model proposed by Monod (Equation 8.4), time is not explicitly represented but rather is included in a term called "specific growth rate" (μ). The specific growth rate in Equation 8.4 is expressed as follows:

$$\mu = \frac{1}{X} \frac{dX}{dt} \tag{8.5}$$

where X is the biomass concentration at a particular time t (g/L), t is the time (h), and μ is the specific growth rate (h^{-1}).

The specific growth rate, as its name indicates, is related to a velocity term or rate (dx/dt) with the quantity of biomass present at a particular time in the fermenter being an intensity term. The parameters μ_{max} and K_s determine the values for a particular process.

The definition of the specific growth rate (μ) has permitted the representation of the kinetic pattern of microbial growth through different phases known as growth phases, which are expressed as follows:

Lag phase: In this phase, the specific growth rate is practically null ($\mu \approx 0$). It is a stage in which microbial biomass has not yet started to multiply. The lasting time of this phase depends on such factors as the physiology of the microorganism, the environment, and the type of inoculum, among other factors.

Accelerated growth phase: This phase is characterized by the initiation of growth (as a concept of cell multiplication) by the synthesis of new cells. In this phase, $\mu > 0$. The lasting time of this phase is relatively short compared to the entire process.

Logarithmic or exponential growth phase: This phase is characterized by a constant growth rate during the entire phase. The logarithmic-specific growth rate is, at the same time, the highest specific growth rate that can be attained in a fermentation process. This is the parameter that first characterizes fermentation from a kinetic point of view.

Decelerated growth phase: In this phase, the biomass synthesis rate starts to diminish in the process, and it can be postulated that $\mu_{max} > \mu > 0$.

Stationary growth phase: In this phase, an equilibrium is established between the quantity of cells that are synthesized and the quantity of cells that are no longer viable or have lost their capacity to reproduce. In this phase, it can be postulated that $\mu = 0$.

Negative growth rate phase: In this phase, there are more cells that have lost
their capacity for reproduction or viability than the cells that can reproduce.
In this phase, growth can be represented by $\mu < 0$.

A study of the microbial growth phases can be performed through the application
of Equation 8.5, using different available mathematical tools. Readers should keep in
mind that the definitions for the different phases might differ depending on the author,
but the significance of the terms and the proper mathematical definitions do not differ.

Although biomass synthesis was the first variable used to determine the kinetics
of fermentation processes, other variables are used to complete kinetic analysis or
even substitute for biomass, including the following:

- Substrate consumption
- Synthesis of a particular metabolite (product)
- O_2 consumption or CO_2 production
- Temperature or pH variation

Obviously, any variable or variables chosen to determine the kinetic parameters
of a fermentation process require the use of a device to measure the variable or variables in an acceptable way. This device is a reactor.

A reactor, in the broad sense, could be an Erlenmeyer flask or even a Petri dish,
both of which serve as reactors (fermenters) because they provide environments for
the development of microorganisms, in particular, isolated conditions. A fermenter
is considered to be a more advanced, developed device because it is a reactor with
more possibilities for taking measurements and controlling process variables. By far,
the most commonly employed fermenter at the laboratory scale for the study of SMF
is the stirred-tank reactor (STR).

The major advancements of this reactor include the following features:

- The system is homogenous.
- There is an improvement in the heat and mass transfer processes when rotation
velocity (rpm) of a proper axis is employed because of the breaking of gradients
that exist in static reactors, v. gr., gradients of dissolved oxygen supplied to the
microbial cells or temperature gradients in different locations in the reactor.
- The system is relatively simple to handle and clean.
- It is easy to operate the reactor in different operational patterns such as
batch, fed-batch, continuous, and cell-recycle.
- Important factors, such as temperature, pH, air supply, rpm, and foam, can
be controlled.

Kinetic representation is important in the fermentative process because these
studies allow us to determine important parameters, such as the following:

- The specific growth rate
- The process yield
- The process productivity

- The heat evolved in the process
- Oxygen demand in aerobic fermentation
- CO_2 production
- Process control criteria
- Strategies for the production of particular metabolites
- Scale-up considerations

It is worth emphasizing that the concept of fermentation kinetics holds true for any fermentation process, and the proper representation of such kinetics depends only on improvements in the measurement of the variables.

8.4   SSF CONCEPT

SSF, according to the accepted definition, is a process by which a microorganism grows in or on a solid substrate (matrix) at a determined humidity (moisture or water activity) level in the absence of free water. The presence of a solid matrix is the characteristic that distinguishes SSF from SMF.

SSF is characterized mainly by the following:

- The existence of solid particles, which may or may not be porous.
- Solid particles with dissimilar shapes, varying from perfect solid spheres, in the case of fermentation media supported in an inert solid matrix, to particles with different shapes and forms, in the case of agricultural residue employed as direct sources for microorganism growth.
- A heterogeneous system.
- A heterogeneous characteristic that may produce gradients (temperature, O_2, CO_2, biomass, etc.). Therefore, the mechanisms of heat and mass transfer phenomena are quite different from those in SMF.
- Biomass produced by microorganism growth that can barely be isolated from the remaining solid matrix.
- Biomass growth that generates constant changes in system properties, such as the porosity of the bed, modification of the particle shape, and the production of air channeling in the bed.
- Kinetics of the system that cannot be determined by direct measurement of biomass but must be determined by indirect methods related to biomass synthesis, such as those based on microbial cell composition, by considering substances that are not present in the initial fermentation media, temperature and heat evolution, O_2 consumption, and other proper variables.
- Difficulties encountered when determining the kinetics of SSF that lead to very poor representations of the processes, which are limited, in many cases, to a more or less qualitative explanation of what is occurring.
- Moisture levels in the solid matrix (water activity) that are relevant to the development of the process.

All of these considerations need to be taken into account when attempting to design a proper bioreactor in which a SSF could properly be described.

8.5 LABORATORY SSF BIOREACTOR

When considering an effective laboratory SSF bioreactor, a bioreactor that in some way could be equivalent to the STR of the SMF would be beneficial. The principal features that this reactor should have are as follows:

- A proper system with surrounding isolation that guarantees an uncontaminated process.
- A system for temperature control.
- A system for air supply, with air quality (% saturation) regulation to allow the air to serve as a possible heat transfer medium.
- An agitation system to be employed continuously or intermittently to disrupt the formation of gradients and to enhance the air supply if the microorganism that is cultured is not sensitive to the shear forces generated by mechanical agitation.

One of the first pieces of SSF equipment that better represented the process through corresponding kinetic parameters was the column reactor called the Raimbault (ORSTOM) reactor, which is depicted in Figure 8.1. The principal characteristics of this system are as follows:

- The system is composed of several parts.
- The column can be placed in a water bath at the selected temperature to allow a certain amount of temperature control.
- The air supply to each column may or may not be saturated, depending on whether the air has been passed through a humidifier (water reservoir) at the bottom of each column.
- The O_2 consumption and CO_2 production can be established by connecting each column to a gas chromatograph with an automated sampler that routinely samples the exhausted air that exits in each column.

It is, therefore, possible to establish the kinetic characteristics of a particular process, particularly biomass synthesis, by direct or indirect methods.

When using direct methods, each column is assumed to represent a particular point in the process at a determined time and is processed as a whole. It must be assured that, when protein synthesis is used to estimate biomass, the initial solid matrix contains an insignificant amount of protein. The same considerations need to be taken into account when biomass is determined using any component of the microorganism, not only protein.

Following this method avoids the problem of poor sample representation resulting from the heterogeneous nature of the system, assuming that each column was filled in the same way from an initial solid substrate already prepared and inoculated. All columns are therefore equal.

The former methods can be considered indirect methods in the sense that the biomass is not directly estimated by counting cells, dry weight, or other factors as in SMF because of the impossibility, in most cases, of achieving the separation of the biomass from the solid substrate.

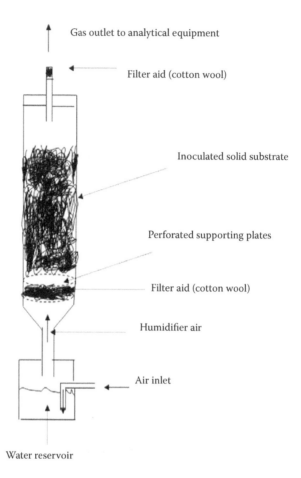

Gas outlet to analytical equipment

Filter aid (cotton wool)

Inoculated solid substrate

Perforated supporting plates

Filter aid (cotton wool)

Humidifier air

Air inlet

Water reservoir

FIGURE 8.1   Raimbault (ORSTOM) system for SSF studies at lab scale.

There is an indirect method that provides a satisfactory representation of the kinetics of the process, the gas balance method, which allows the determination of O_2 consumption and CO_2 production, which are directly related to biomass synthesis according Equation 8.1.

The advantage of this method is that the samples are not handled, thus avoiding sampling errors that can statistically alter the results resulting from the heterogeneous characteristics of the system, so the results obtained represent the entire process.

Using the gas balance method, Equations 8.6 and 8.7 were obtained. These equations allow us to determine the volumes of O_2 consumed and CO_2 produced during fermentation:

$$V_{O_2 \, cons} = \left(0.209 - \frac{0.791\%O_2}{(100 - \%O_2 - \%CO_2)} \right) F_e \qquad (8.6)$$

$$V_{CO_2\,prod} = \frac{0.791\%CO_2}{(100 - \%O_2 - \%CO_2)}F_e \tag{8.7}$$

where $V_{O_2\,cons}$ is the volumetric oxygen consumed (L/h), $V_{CO_2\,prod}$ is the volumetric carbon dioxide produced (L/h), $\%O_2$ is the O_2 concentration of the exit gas (%), $\%CO_2$ is the CO_2 concentration of the exit gas (%), and F_e is the volumetric air flow at the fermenter entrance (L/h).

In this equation, it is assumed that the CO_2 concentration in the air supply to the system is worthless and that the air is composed of only 20.9% oxygen and 79.1% nitrogen.

Once the volumetric flows for O_2 consumption and CO_2 production are determined, the corresponding molar concentrations or gravimetric flows for respective gases in moles/L or in g/L are readily obtained by applying the ideal gas law.

The mass balance for oxygen consumption states that

$$O_2 \text{ consumed} = O_2 \text{ employed in biomass synthesis}$$

$$+ O_2 \text{ employed in endogenous process} \tag{8.8}$$

Equation 8.8 can be expressed as follows:

$$\frac{\Delta_{O_2}}{\Delta t} = \frac{1}{Y_{x/o}}\frac{\Delta X}{\Delta t} + mX \tag{8.9}$$

where $\dfrac{\Delta_{O_2}}{\Delta t}$ is the oxygen uptake rate (OUR) in the time period considered (g O_2/h), $Y_{x/o}$ is the yield based on consumed O_2 for biomass synthesis at the time interval considered (g biomass/g O_2), ΔX is the biomass synthesized during the time interval considered (g), m is the maintenance coefficient (g O_2/g biomass h), and Δt is the time interval considered (h).

In Equation 8.9, the OUR or $\dfrac{\Delta_{O_2}}{\Delta t}$ is related to the yield coefficient that is based on oxygen consumption ($Y_{x/o}$) and the maintenance coefficient (m). This yield can be defined as $\Delta x/\Delta_{O_2}$, considering ΔO_2 as a substrate that is consumed for the synthesis of a particular biomass (Δx). The maintenance coefficient (m) represents the amount of energy that is employed by an individual cell to maintain its activity.

The differential equation 8.9 does not have an analytical solution, so it was solved by applying a numerical method, the trapeze rule, and the result obtained is as follows:

$$X_n = \left[Y_{xo}\Delta t \left(\frac{1}{2}\left(\left(\frac{dO_2}{dt}\right)t=0 + \left(\frac{dO_2}{dt}\right)t=n \right) + \sum_{i=1}^{i=n-1}\left(\frac{dO_2}{dt}\right)t=i \right) \right.$$

$$\left. + \left(1 - \frac{a}{2}\right)X_o - a\sum_{i=1}^{i=n-1}X_i \right) \bigg/ \left(1 + \frac{a}{2}\right) \tag{8.10}$$

where $\left(\dfrac{dO_2}{dt}\right) t = i$ is the OUR or O_2 consumed at the time interval $t = i$ (g O_2/h), n is the last time interval, Δt is the time interval considered for the estimation (h), X_0 is the biomass at the beginning of the time interval Δt (g), and X_n is the biomass at the end of the time interval Δt (g).

Equation 8.10 was obtained by considering the yield based on the amount of O_2 that is consumed and a maintenance coefficient (m) that was considered to be constant during the time interval, that is, $Y_{x/o} = k_1$ and $m = k_2$. Additionally, the term a, which emerges in Equation 8.10, is a result of a simple grouping of terms that corresponds to the relationship $a = Y_{x/o}\, m\Delta t$, which appears when Equation 8.10 is solved. The term a is also a constant if the yield, based on oxygen consumption and the maintenance coefficient, are constant, too. Equation 8.10 allows us to estimate the biomass content from the data obtained from oxygen consumption. It must be taken into account that the calculations are performed step by step; that is, first x_1 is determined, then x_2 and x_3 are determined, and so on.

Another extremely important parameter that can be determined by the gas balance method is the respiration quotient (Q):

$$Q = \frac{CO_2 \text{ produced}}{O_2 \text{ consumed}} \tag{8.11}$$

Obviously, the value of Q is dimensionless and equal to 1 according to Equation 8.1 if O_2 and CO_2 are measured in moles/L, and Q has another constant value if measured in another unit.

Based on this reasoning, it is valid to postulate the following strategy when dealing with SSF when the O_2 and CO_2 patterns from these fermentations are available:

- If it is observed that, in a specific time interval, the respiration quotient is approximately 1 when the substrates are simple sugars or related substances, it is reasonable to assume that the yield based on oxygen consumption ($Y_{x/o}$) and the maintenance coefficient (m) can be presumed to be constant during the entire time interval when the process is proceeding via aerobic fermentation, according to Equation 8.2.
- If the data shows a definite tendency to produce a respiration quotient of $Q < 1$, it is valid to assume that one or more nonconstitutive or newly induced metabolites have possibly been produced. It would then be necessary to calculate new values for $Y_{x/o}$ and m because previous values would no longer be valid.
- When the values for the respiration quotient are greater than 1 in aerated fermentation systems, it must be considered that the higher production of CO_2 relative to O_2 corresponds to the presence of anaerobic zones present in the fermenter. These anaerobic zones might be caused by a malfunction in the aeration system and the possible channeling of the air supply, system compaction resulting from microbial growth, system shrinkage resulting

from excess water evaporation, or other problems. In these cases, the system cannot be considered aerobic and must be corrected.

It must be kept in mind that all of these considerations are based on the employment of substrates comprising simple sugar or related carbohydrates or polysaccharides. When dealing with other substrates, the substrate properties and the overall metabolic pathway must be taken into account to obtain a proper value for the respiration quotient.

Undoubtedly, measuring the amount of O_2 consumed or the amount of CO_2 evolved during SSF serves to represent the kinetic pattern of SSF in a relatively noninvasive way, thereby not allowing the introduction of changes into the system. It would, therefore, be convenient for a basic laboratory SSF bioreactor to have a system for gas analysis at the exit of the reactor.

Another major problem faced in SSF is describing a way in which the energy evolved in the process could be evacuated. Undoubtedly, addressing this problem could not only determine the course of development of the process, but it could also have economic implications. As expected, the energy produced during fermentation could affect such factors as the rise in temperature in the system, the air demand, the specific growth rate of the microorganism employed, and even the type of reactor to be used. In this sense, it is extremely important to perform an overall energy balance that may allow us to establish the relationships among the factors already discussed.

The temperature control of the system is complicated by two factors:

1. The heterogeneity of the system
2. The quality of air supplied

The heterogeneity of the system could cause the existence of gradients that affect energy and mass transfer in the reactor. The existence of gradients could be avoided by sporadic or continuous agitation that, in some way, may also provide a certain homogeneity. This feature does not appear in the system depicted in Figure 8.1.

The air supply in aerobic SSF plays three important roles:

1. It provides the required oxygen.
2. It transfers part of the heat that is produced during fermentation to the gas phase.
3. It removes CO_2 and other volatiles that are produced from the solid matrix (substrate).

The oxygen supply to the cells is a vital factor in aerobic fermentation. It is so important that submerged fermenter design and utility are based on the efficient mass transfer of oxygen from the air to the liquid medium. This transference is represented by the overall gas–liquid mass transfer coefficient ($K_L a$), which was established as one of the principal criteria for process scale-up for SMF. In this sense, and assuming some sort of equivalence in SSF, several attempts were made to determine $K_L a$ in SSF. For SSF, however, the air supply not only brings the oxygen necessary for microbial growth, but could also serve as a medium to evacuate the heat produced during the process. Until now, it has not been demonstrated that K_La has a greater influence on the development of an aerobic SSF than the rise in temperature produced by the heat that is evolved during the process. As a matter of fact, the amount of air required during SSF to cool the process is approximately three times the amount needed to supply oxygen to the cells.

From the macro balance of energy in the process, we can determine the relationship between the amount of heat evolved and the amount of air that is required. Equation 8.12 describes this relationship:

$$W_g = \frac{Q_m - hA(T_{out} - T_{in})}{0.24(T_{out} - T_{in}) + \lambda(H_{out} - H_{in})} \tag{8.12}$$

where W_g is the mass velocity of dry air (kg dry air/h), Q_m is the metabolic heat evolved during the process (kcal/h), h is the heat transfer coefficient of the fermenter wall (kcal/m²/h/°C), A is the wall surface area of fermenter (m²), T_{out} is the temperature of the outlet air (°C), T_{in} is the temperature of the entering air (°C), λ is the latent heat of vaporization of water (kcal/kg water), H_{out} is the absolute humidity of air at the exit of the fermenter (kg water/kg dry air), and H_{in} is the absolute humidity of air at the entrance of the fermenter (kg water/kg dry air).

Equation 8.12 represents two mechanisms of heat transfer:

1. The transfer of metabolic heat as heat transfer from the solid substrate to the air that flows through the bed
2. The heat transfer from the wall of the reactor to the surroundings

For the first mechanism, it is very important to consider the conditions in the air, including humidity, temperature, and airflow. The second mechanism is directed by the heat transfer coefficient (h) through the fermenter wall and the temperature gradient between the system and the surroundings. Evaporative cooling enhances the removal of heat, but at the same time, it causes drying of the solid matrix, leading to a reduction in bioreactor performance. The heat transfer from the wall of the reactor to the surroundings could be improved by designing a reactor with a more effective surface for heat interchange, using agitation during the process and even controlling the surrounding conditions (temperature) by running cooling water through the system.

Figure 8.1 shows that the manipulation of air quality, air humidity, and the surrounding environment can all be performed for SSF. Heat transfer from the wall of the reactor to its surroundings, however, cannot be enhanced. Agitation could help enhance this transfer mechanism.

Equation 8.12 illustrates that the demand for air depends not only on the microorganism's growth requirements but also on the cooling of the reactor. If the value of the heat transfer coefficient through the fermenter wall, that is, heat transfer across the reactor wall, is not enough to evacuate the metabolic heat and keep the temperature in the medium constant, water evaporation becomes a controlling mechanism.

Equation 8.12 also demonstrates that if temperature control is attempted, the amount of air employed by the process will be higher than the value that corresponds only to metabolic air demand.

The quality of air employed in the system can range from saturated air to air with lower humidity levels by employing proper air humidifying or air-drying devices. The use of nonsaturated air affects the system because of a decrease in water activity or humidity, so saturated air is employed to bypass this effect. The use of saturated air, however, leads to a rise in temperature that cannot be avoided. Therefore, a contradiction exists concerning the optimal quality of air to employ. The humidity drop in the system could be prevented by spraying water onto the solid substrate coupled with appropriate control mechanisms. Obviously, particular procedures correspond to each SSF process.

Paying attention to the types of standard equipment is a first step in studying an SSF process that corresponds somewhat to a STR in a SMF. The column reactor may constitute such equipment when it is complemented with a gas analyzer and, eventually, an agitator. However, at the bench scale, other reactors besides the column packed bed reactor illustrated in Figure 8.1 are employed, primarily including the following:

• Erlenmeyer flasks
• Trays
• Drum reactors with or without agitation or rotation

The use of Erlenmeyer flasks in SSF is commonly reported in the scientific literature, but unfortunately, these reports present simple initial descriptions of the development of particular products. Tray and drum reactors have been employed in larger-scale fermentations, and in some cases, these reactors employ proper controls or registrations for temperature or airflow. Additionally, systems for large-scale SSF include silage-processing equipment, bags with holes that permit some oxygen supply and carbon dioxide to exit, and even bags with forced airflow. However, in these systems, there is almost no way to control or guide the fermentation process.

The handling of solids in SSF is particularly important. Unfortunately, solids cannot be manipulated by pumping like liquids can, and therefore, solid manipulation can lead to contamination. This factor can be lessened when:

• The solid matrix can be sterilized in situ.
• The inoculums can be introduced into a sterilized solid substrate.

The sterilization of the solid substrate in situ is easier to perform when it is possible to agitate the system during sterilization to produce efficient homogenization. Sterilization could be obtained by employing saturated hot air during a proper time interval until all of the contaminating microorganisms present in the initially nonsterilized solid matrix are destroyed. Dry hot air could then be employed until the desired humidity level in the solid matrix is achieved. The inoculation is uncomplicated if the inocula are in liquid form and the reactor can be agitated as previously stated.

Figure 8.2 illustrates a system that uses a drum reactor and employs the features that were previously discussed.

Additionally, several reactors were designed based on column or tray reactors for pilot or large-scale levels of SSF, and the authors claimed that good results were obtained.

A different reactor employed in fungal spore production is the Zymotis. This reactor contains heat exchanger sheets (with inner cooling water flow). The reactor is composed of vertical plaques. The substrate is introduced between the plaques, and each plaque has an inner surface through which cooling water flows, allowing better temperature control to occur in the reactor. Figure 8.3 shows a schematic diagram of the Zymotis SSF reactor.

Additionally, several reactors were designed based on column or tray reactors.

The major problem associated with this reactor is the difficulty in handling solids on a large scale.

Another new reactor or system, the air pressure pulsation solid-state bioreactor or APP-SSF, has recently been proposed. The aim of this reactor is to enhance mass and heat transfer in the SSF process in such a way that mycelia from fungi are not altered by agitation, and the production of metabolites, such as enzymes, is enhanced.

The APP-SSF reactor is based on a reactor that contains several trays with a control system that includes a pressurized air inlet, a regulated compartment to hold air in the reactor, and an air outlet. Using desired air conditions and the pulsation of pressurized entrance air, the desired level of water evaporation is obtained without affecting the process. However, the humidity levels or water activity of the solid matrix must be taken into account. It is worth noting that the proposed system may be effective using other reactor shapes in addition to tray reactors when suitable control devices, such as those used in the APP-SSF, are employed.

More consideration must be taken into account when considering or designing a proper SSF reactor. For example:

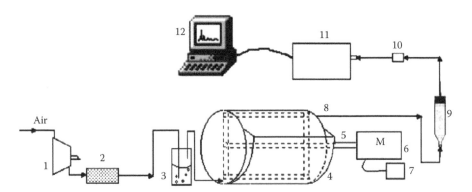

FIGURE 8.2 Schematic representation of drum SSF reactor with data collection system. (1) Compressor, (2) filter, (3) humidification, (4) bioreactor, (5) agitator, (6 and 7) motor, (8) gas outlet from the bioreactor, (9) filter, (10) CO_2 and O_2 sensors, (11) control system, and (12) computer.

- The reactor must be as simple as possible. Adding inner parts for air distribution and agitation may produce complications when, for example, sterilization is required.
- A scale-up criterion needs to be considered. In fact, the development of reactors for SSF, regardless of the scale in which they will be employed, is based on intuition or trial and error.

8.6 SCALE-UP CRITERIA

When a particular process developed at the bench is scaled up, it is crucial to know how to proceed to obtain the best results at the larger scale. Several factors need to be considered to guarantee that the fermentation process proceeds in a manner that is as close to the bench-scale process as possible. These factors are relevant to both the design and the employment of the reactor.

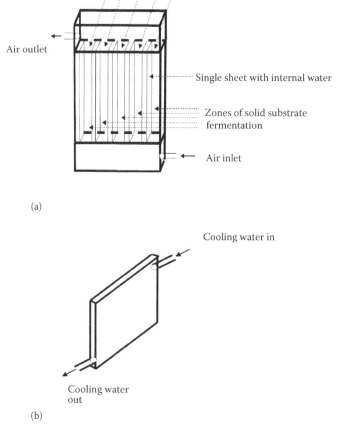

FIGURE 8.3 Scheme of SSF Zymotis reactor. (a) Ensemble zymotis SSF reactor. (b) Characteristics of one sheet in Zymotis SSF reactor.

For many SSF processes, the kinetic models reported in the scientific literature are limited to describing product synthesis in a semiqualitative manner (e.g., statistical empirical models or graphic representation). Therefore, well-instrumented reactors on the bench scale are very important. The paucity of information concerning mathematical models that describe SSF processes correctly might imply that the models are inadequate when applied to semiindustrial or industrial applications. The construction and employment of large-scale equipment similar or equivalent to their laboratory counterparts creates several problems, ranging from heat transfer and adequate aeration to shear forces produced by eventual agitation.

Attempts to scale up the SSF process are based on what is called the art of..., which implies a certain amount of subjectivity and the lack of a more accurate foundation.

Several methods have been reported for the SMF process to establish scale-up strategies that attempt to keep subjectivity to a minimum. These strategies highlight three different, greatly important phenomena: thermodynamics, transport phenomena, and kinetics.

Scale-up criteria in submerged fermentations are primarily based on parameters involved in mixing; aeration; oxygen and heat transfer, including specific power inputs; P/V; the volumetric mass transfer coefficient, $K_L a$; impeller tip speed of the agitator or shear; and dissolved oxygen concentration. These factors are considered alone or in combination with each other, and they are not necessarily conveyed as dimensionless numbers but are preferably portrayed as numbers, such as the N_p (power number) or N_{Re} (Reynolds number). A dimensionless number is defined as the relationship between several factors or quantities that have dimensions, but through this relationship, the final result is a dimensionless expression. The implication of such dimensionless associations is readily obvious; as a dimensionless liaison, the expression is independent of scale.

For scale-up, it is also worthwhile to consider important parameters that are intensive, that is, independent of the scale that is considered.

In the case of SSF, air supply (L/h) has a significant value as was previously demonstrated, not only for the delivery of oxygen to the microorganism cells, but also for the evacuation of the heat evolved during the process. However, air supply is an extensive factor, so this factor depends on the scale that is considered. Nevertheless, if air supply is related to the initial solid matrix or substrate (wet or dry), then an intensive factor is obtained called VKgM or airflow intensity (L air supply/mass of substrate/min.), so a factor that does not depend on the scale employed is obtained. Unfortunately, only a few scientific publications report this factor. However, the factor, which is very simple to calculate, takes the air supply and the initial solid substrate into consideration. Because VKgM is an intensive factor, it could initially be considered to be a scale-up criterion.

Furthermore, a dimensionless number could be derived from the airflow intensity (VKgM) multiplied by the bed density (ρ), taking into account that bed density (kg/L) is the ratio between the solid substrate and the real fermentation volume. After this multiplication, the corresponding dimension is min^{-1}. If the inverse of the specific growth rate (μ) is also considered, then the following relationship of these three factors is obtained $\left(\dfrac{\rho VKgM}{\mu} \right)$ and gives a dimensionless number.

This proposed dimensionless number implies that not only air intensity, VKgM, but also bed density (and therefore the porosity of the bed and particle distribution) must be considered in the scale-up of SSF. It should be noted, however, that agitation and mixing with minimal shear damage was not initially considered, but it may eventually be included to complement these scale-up criteria. Additionally, it may be important that, for the first time, a dimensionless number, including a biological parameter (μ) that represents microbial activity, has been presented.

Consequently, the airflow intensity appears to represent a good reference as criterion to scale up SSF processes when one also considers the density of the bed and the specific growth rate (μ) related in a dimensionless number that may be important in future studies of scale-up or the manipulation of SSF fermenters of any size.

In summary, there are some characteristics that a laboratory fermenter, a pilot plant, or an industrial fermenter should have. Those characteristics are resumed below

- Must be constructed from suitable materials, such as glass or polished steel (as for a submerged fermenter).
- Could have a jacket for circulated water cooling.
- Could possess the possibility for movement and agitation, which would be appreciated.
- Should properly contain a suitable seal to isolate the reactor.
- Should possess a particular improvement for sterilization in situ. This improvement would be required for pilot or industrial fermenters because of the requirement for the complex manipulation of solids.
- Should possess an air supply designed to take into account not only air sterilization, but also air quality.
- Should contain a system for determining the concentrations of O_2 and CO_2, which are needed to determine kinetic parameters.
- Should ideally be coupled to a computer containing suitable programs for data acquisition and processing.

It is noteworthy that, a reactor that fits this description, independent of form (column, drum-like, or tray), allows the determination of parameters that permit the use of concepts described in this report, including the work with dimensionless numbers proposed for scale up and handling of larger-scale fermenters.

8.7   FUTURE DIRECTIONS

Future works related with SSF reactors should be considered the following aspects.

- For bench scale reactors, it would be very desirable to produce standard reactors for SSF with proper controls that allow the study and characterization of SSF as is already the case for submerged fermentation.
- The design and development of future SSF reactors should focus on controls at a suitable scale to enable performance similar to that observed for bench-scale reactors.

- More appropriate studies need to be developed that consider transport phenomena, primarily including heat and mass transfer in a system characterized by a gas–solid interface.
- The need to study and practice handling solids, regardless of whether they are humid, is essential when considering large-scale reactors. This factor might determine whether a larger scale can be employed in SSF.
- Other considerations for the scale up of SSF would enhance the criteria described in this report, particularly if parameters related to transport phenomena that occur in SSF could be addressed.

SUGGESTED READING

ARTICLES

Ali, H., Kh, Q., and Zulkali, M.M. (2011) Design aspects of bioreactors for solid-state fermentation: A review, *Chemical and Biochemical Engineering Quarterly* 25(2):255–266.

Andre, G., Moo-young, M., and Robinson, C.W. (1981) Improved method for the dynamic measurement of mass transfer coefficient for application to solid substrate fermentation, *Biotechnology and Bioengineering* 23:1611–1622.

Durand, A. (2003) Bioreactor designs for solid state fermentation, *Biochemical Engineering Journal* 13:113–125.

Fujian, X., Hongzhang, C., and Zuohu, L. (2002) Effect of periodically dynamic changes of air on cellulase production in solid-state fermentation, *Enzyme and Microbial Technology* 30:45–48.

Garcia-Ochoa, F. and Gomez, E. (2009) Bioreactor scale-up and oxygen transfer rate in microbial processes: An overview, *Biotechnology Advances* 27:153–176.

Gowthaman, M.K., Raghava Rao, K.S.M.S., Ghildyal, N.P., and Karanth, N.G. (1995) Estimation of $K_L a$ in solid-state fermentation using a packed-bed bioreactor, *Process Biochemistry* 30(1):9–15.

Liu, J., Li, D.B., and Yang, J.C. (2007) Operating characteristics of solid-state fermentation bioreactor with air pressure pulsation, *Applied Biochemistry and Microbiology* 43(2):211–216.

Monod, J. (1949) The growth of bacterial cultures, *Annual Reviews in Microbiology* 3:371–394.

Narahara, H., Koyama, Y., Yoshida, T., Attahasampunna, P., and Tagushi, H. (1984) Control of water content in a solid-state culture of *Aspergillus oryzae*, *Journal of Fermentation Technology* 62(5):453–459.

Raghavarao, K.S.M.S., Ranganathan, T.V., and Karanth, N.G. (2003) Some engineering aspects of solid-state fermentation, *Biochemical Engineering Journal* 13:127–135.

Rodríguez-Fernández, D.E., Rodríguez-León, J.A., de Carvalho, J.C., Sturm, W., and Soccol, C. R. (2011) The behavior of kinetic parameters in production of pectinase and xylanase by solid-state fermentation, *Bioresource Technology* 102:10657–10662.

Rodríguez-León, J.A., Sastre, L., Echevarría, J., Delgado, G., and Bechstedt, W. (1988) A mathematical approach for the estimation of biomass production rate in solid state fermentation, *Acta Biotechnologica* 8(4):299–302.

Rodriguez-Leon, J.A., Torres, A., Echevarria, J., and Saura, G. (1991) Energy balance in solid state fermentation processes, *Acta Biotechnologica* 11(1):9–14.

Roussos, S., Raimbault, M., Prebois, J.-P., and Lonsane, B.K. (1993) Zymotis, a large scale solid state fermenter: Design and evaluation, *Applied Biochemistry and Biotechnology* 42:37–52.

Sato, K., Nagatani, M., Nakamura, K., and Sato, S. (1983) Growth estimation of *Candida lipolítica* from oxygen uptake in solid state culture with forced aeration, *Journal of Fermentation Technology* 61(6):623–662.

Schmidt, F.R. (2005) Optimization and scale up of industrial fermentation processes, *Applied Microbiology and Biotechnology* 68:818–820.

Sebastián Lekanda, J. and Ricardo Pérez-Correa, J. (2004) Energy and water balances using kinetic modeling in a pilot-scale SSF bioreactor, *Process Biochemistry* 39:1793–1802.

Suryanarayan, S. (2003) Current industrial practice in solid state fermentations for secondary metabolite production: The Biocon India experience, *Biochemical Engineering Journal* 13:189–195.

Zeng, W. and Chen, H.Z. (2009) Air pressure pulsation solid state fermentation of feruloyl esterase by *Aspergillus niger*, *Bioresource Technology* 100:1371–1375.

BOOKS

Mitchell, D.A., Krieger, N., and Berovic, M. Eds. (2010) *Solid-state Fermentation Bioreactors: Fundamentals of Design and Operation*, Heidelberg: Springer.

Pandey, A., Soccol, C.R., and Larroche, C. Eds. (2007) *Current Developments in Solid-State Fermentation*, New Delhi: Asiatech Publishers, Inc.

Pandey, A., Soccol, C.R., Rodríguez-León, J.A., and Nigam, P. (2001) *Solid-State Fermentation in Biotechnology: Fundamentals and Applications*, New Delhi: Asiatech Publishers, Inc.

9 Downstream Operations of Fermented Products

Júlio César de Carvalho,
Adriane Bianchi Pedroni Medeiros,
Daniel Ernesto Rodríguez-Fernández,
Luiz Alberto Júnior Letti,
Luciana Porto de Souza Vandenberghe,
Adenise Lorenci Woiciechowski,
and Carlos Ricardo Soccol

CONTENTS

9.1 INTRODUCTION

Most fermented products must be further processed after biotransformation and before packaging; these are called downstream operations. These may range from simple sedimentation to sophisticated drying of frozen products at low pressures. One useful scheme for looking at downstream operations of fermented products is the RIPP scheme (removal of insolubles, isolation of the product, purification, and polishing; Table 9.1) (Belter 1988). Removal of insolubles is usually necessary for fermented beverages, such as wine and beer, where biomass and protein precipitates may add cloudiness; also, if the product is the biomass itself, as in starter cultures, the solid-removal step is essential. Isolation of the product stands for the concentration, which may be done by distillation in fermented beverages or membrane separations in some dairy products.

Purification and polishing (which is essentially a high-resolution purification step) are uncommon in fermented foods but are very important for the production of fermented additives. Finally, a finishing step for many fermented foods is stabilization by drying, sometimes after adequate formulation.

This chapter describes the most important operations for solid–liquid separation, concentration, and drying of fermented foods.

TABLE 9.1

RIPP Scheme and Most Common Unit Operations for Fermented Foods Downstream Processing

Removal of Insolubles (Sometimes after Flocculation)	Isolation of Product (Concentration)	Purification and Polishing	Finishing (Sometimes after Formulation)
Filtration	Evaporation	Uncommon for	Drying solids
Sedimentation	Distillation	fermented foods;	Spray drying of liquids
Centrifugation	Ultrafiltration	important for food	Thermal processing
		additives	(sterilization)

9.1.1 FORMULATION AND DRIVING FORCES IN SEPARATION

Although the development of a separation step may be done based on experience, and will certainly depend on experiments, it is useful to take a look at the physical principles behind these separations. Most separation processes require three kinds of equations to be solved: a material balance, a transport equation, and an equilibrium condition. The material balance derives from the amount of materials to be processed, for example, the amount of liquid that must be evaporated from a wet material in order to give a desired product. The transport equation describes how the flow of the component of interest in the mixture is affected by a driving force and a resistance; for example, the flow of liquid through a porous bed depends on the pressure applied and the resistance of the medium. The equilibrium condition concerns the distribution of a component of interest between phases; for example, the chemical potential of components in liquid and gaseous phases must be equal in equilibrium. Actually, most separations do not reach equilibrium, and complex formulations must account for the kinetics of transport. Furthermore, fermentation processes produce complex mixtures that defy rigorous modeling of the operations; the equations presented in each section of this chapter should be regarded as guides that must be complemented by experimental data.

9.2 SEDIMENTATION, COAGULATION, FLOCCULATION, AND CENTRIFUGING

9.2.1 SEDIMENTATION

Sedimentation is a process where a solid suspended in a fluid or two immiscible fluid phases separate on the basis of a difference in density. The mixture must be maintained at rest or at low velocities, and the denser phase (usually a solid) will descend and accumulate at the bottom. The time needed for the phases to separate depends on several parameters, and while batch sedimentation of small particles will require longer times than centrifugation, in several cases, a large gravity settler may operate continuously; this is the case for some liquid–liquid separations with immiscible fluids with a relatively large density difference as in water and oil mixtures.

Sedimentation relies on gravity for the separation and will result in two product phases: a top phase (supernatant) with a lower density and a bottom phase (precipitate) with a higher density.

The phases frequently need further processing, but the reduction of volumes may compensate for the low resolution of this first separation step. For example, suppose that a biomass suspension has 0.5% of cells and gives, after decanting, a bottom phase with 5% solids with a separation efficiency of 95% regarding solids. In such a process, the volume would have been reduced to roughly 1/10 of the initial volume; this is advantageous both if biomass is the product (in which case it is concentrated in a smaller volume for the next processing step) or if the supernatant is the product (in which case it has a lower suspended solids load and may be centrifuged of filtered).

Sedimentation is a useful operation for recovery of cells, proteins, antibiotics, and a variety of molecules and aggregates produced in fermentative processes.

9.2.2 SEDIMENTATION MECHANISM

Sedimentation occurs as a result of the effect of gravity on two different phases. For example, imagine a solid particle suspended in a liquid with a lower density. On a static particle, the weight and buoyant forces will result in a combined force, which will bring the particle down. Because it depends only on gravity and the properties of the mixture, this force is constant and will accelerate the particle until it reaches the bottom:

$$F_R = F_g - F_b = \text{weight of the particle} - \text{weight of the liquid dislocated}$$

$$F_R = (1/6)\pi d^3 \rho_s g - (1/6)\pi d^3 \rho_l g \tag{9.1}$$

where ρ_s is the density of the solid, ρ_l is the density of the liquid, g is the acceleration of gravity (9.81 m s^{-2}), and $(1/6)\pi d^3$ is the volume of a sphere of diameter d.

However, as soon as the particle starts moving, a reactive force comes into play: the viscous drag, which is caused by the interaction between the particle and the liquid. This drag depends on the particle size and shape, but for small spherical particles, it may be expressed using Stoke's law (Belter 1988):

$$F_D = 3\pi d\mu v \tag{9.2}$$

where F_D is the viscous drag force, d is the particle diameter, μ is the viscosity of the liquid, and v is the velocity of the particle. Because of this viscous resistance, the particle that accelerates by action of gravity also suffers a growing viscous resistance; F_D will eventually equipoise F_R, and from this moment on, the particle velocity will be constant, that is, the terminal velocity of the particle under action of gravity, often denominated v_g:

$$F_R = F_D = (1/6)\pi d^3(\rho_s - \rho_l)g = 3\pi d\mu v_g$$

Isolating v_g,

$$v_g = (1/18\mu)d^2(\rho_s - \rho_l)g \tag{9.3}$$

Although it is developed for spheres, Equation 9.3 should also work for other particles with a form-factor correction. For particles that are less dense than the liquid, v_g will be negative. This approach also works for mixtures of immiscible liquids.

Equation 9.3 holds only for laminar flow with a Re (Reynolds number) smaller than 0.1. That is the case for microorganisms and particles below 250 μm even if their density is as high as 2 g/cm^3. For larger particles, flocs, or in centrifugation at high speeds, the higher velocity causes turbulence, and the drag is partly caused by eddies. Finally, the movement of bubbles or droplets of a fluid suspended in another fluid is more complex because of the movement inside the droplets. Modeling all these parameters and determining the settling time or particle velocity in order to design a decanter is difficult, and experimental data are always necessary; however, the formulation shows which variables involved in settling velocity could be tweaked in order to improve sedimentation.

Sedimentation is favored when the viscosity is reduced, the particle size is increased, or the density difference increases. While changing the particle composition or liquid density may be difficult, decreasing the medium viscosity or enlarging the particles via aggregation is possible and may significantly improve the process (Leme 1979; Belter 1988; Atkinson and Mavituna 1991). These two strategies are discussed in the following sections.

9.2.3 CHANGING THE VISCOSITY

The first constraint that can be changed in order to enhance the sedimentation is the fluid viscosity, which is inversely proportional to the settling velocity. A reduction of viscosity of 35% for a fluid such as water from 1 cP at 20°C to 0.65 cP at 40°C causes a 54% increase in the settling velocity. This may seem to be a small increment at the expense of energy and product stability at a higher temperature; however, for viscous solutions, this may be a real advantage: Honey would reduce its viscosity from an average 200 cP at 20°C to approximately 25 cP at 40°C, causing an eightfold increase in velocity. Another way to reduce the viscosity of highly viscous suspensions is to dilute the fluid phase; normally dilutions of 1:2, 1:3, or 1:4 are enough to allow an efficient sedimentation or centrifugation. Higher dilutions are not recommended because they will greatly increase the final volume of the liquid medium, increasing the costs of pumping, transfer, storage, and further processing. Dilution may be useful to separate cells from a viscous fermented broth where the viscosity of the fermented medium was produced by a soluble product of the fermentation. Finally, an important way to reduce the viscosity caused by polymers, such as pectin or starch, is their cleavage using enzymes as is the case in starch liquefaction or in juice processing. Because the objective here is to make random cuts on the macromolecule in order to reduce the average molecular size, small loads of enzyme have a dramatic effect on the viscosity. For example, a 2% solution of methyl cellulose has a viscosity at 20°C given by the equation μ (cP) = $10^{-5}DP^{3.2}$, where DP is the degree of polymerization; the viscosity may be reduced ninefold by simply cutting every polymer molecule in half—of course, a better estimate must use a DP distribution accounting for random cuts in the molecules.

9.2.4 CHANGING THE SIZE: PRECIPITATION AND FLOCCULATION

Small particles suspended in a fluid may sediment very slowly or even not sediment at all—being kept in suspension because of Brownian motion. An alternative to enhance the sedimentation rate is to aggregate the particles through flocculation (in which solid particles form aggregates) or precipitation (in which colloidal particles, such as proteins, form aggregates). In order to form aggregates, particles must come close enough for van der Waals and dipole–dipole forces to keep them together; however, that is difficult because these particles have a net electrical charge, usually negative, that is strong enough to repel them. This is actually the principle of some stabilizers, for example, phospholipids, which confer a charge to oil droplets in water. The electrical charge is a result of the chemical structure of its superficial molecules or specific sites in particles and proteins, such as carboxylic acids that will be ionized at a pH near to or above their pKs.

While total neutralization of surface charges works very well (e.g., using polymeric flocculants), just using salts to increase the ionic force of the solution may be enough to reduce the electric field, causing a salting-out effect where solvation layers around charged regions of the suspended particles will be reduced, and these particles may come close enough for interaction to occur. Likewise, protein aggregates of approximately 1 μm may form from proteins smaller than 0.01 μm if an adequate pH or salt concentration is used. There are several equations that successfully explain flocculation and precipitation phenomena, but because of the complexity and specificity of the particles, laboratory experiments are essential in order to determine adequate conditions for flocculation; as a starting point, lower temperatures, higher ionic forces, lower solvent dielectric constants, and flocculants with polyvalent ions (especially regarding cations, such as Fe^{+3} or Al^{+3} salts; Harrison 2003) should be tested.

After the addition of a flocculating agent, the mixture must be gently agitated in order to enhance the aggregates' structure; flocs tend to be less dense than the original suspended material (because of the amorphous structure and solvent entrainment into the floc), but that is compensated by a larger diameter: as sedimentation velocity is proportional to the square of particle diameter, the formation of a 50 μm yeast cell aggregate from 5 μm cells represents a hundredfold increase, turning an impractical sedimentation at $v_g = 0.5$ cm/h into a feasible process that takes 2 h to clarify a column of 1 m of liquid.

9.2.5 Applying Force: Centrifugation

Centrifugation is an operation used when the settling velocity of solids in a suspension is so low that sedimentation is impractical. Centrifuges use rotors to create a centrifugal acceleration that is usually thousands of times greater than the acceleration of gravity. The acceleration in a centrifuge is sometimes divided by gravity acceleration, $g = 9.81$ m s^{-2}, and referred to as relative centrifugal force, (RCF). That force depends on the rotation speed and the distance of the particle to the center of rotation (radius):

$$\text{RCF, (×}g\text{)} = 11.18 \times 10^{-6} \times \text{rpm}^2 \times R \tag{9.4}$$

where rpm is the rotation speed of the centrifuge, in revolutions per minute, and R is the rotor radius in cm. For an estimate of RCF in common centrifuges, the average radius of the bowl may be used.

Centrifuges push the denser particles or liquids to the periphery of the rotor and may operate in batches (as is typical in laboratory centrifuges and in some types of process centrifuges) or continuously (in which case a solid removal system must be devised). Centrifugal separators are also very useful for separation of immiscible liquid phases after liquid–liquid extraction or oil (including essential oils) and water separation. Some centrifuges may even separate three phases: decanted solid, liquid, and lighter liquid phases. Centrifugation is particularly useful for removing cells from a fermented broth prior to the product recovery. Centrifugation produces a solid phase that is more concentrated and with a lower water content when compared with gravity settling, which may be a great advantage.

While the rotation and radius of a centrifuge define the RCF that it may achieve, the throughput of a centrifuge also depends on other dimensions, such as the inner radius of the rotor and the radius of the liquid layer. One useful approach for centrifuge comparison is to derive a sigma factor (Σ), which is the area of the gravity settler that could achieve the same separation that the centrifuge does. A general formula for the sigma factor is given below (Lander et al. 2005, p. 34):

$$\Sigma = (LR)\frac{(\omega^2 R)}{g} \text{ Geometric factors}$$

$$\Sigma = (\text{area}) (\text{settling relative to gravity})$$

(9.5)

where L is the length of the bowl, R is the radius of the bowl, ω is the angular velocity, and g is the acceleration of gravity. For a tubular centrifuge with a thin liquid layer, an approximation for R is the radius of the liquid layer, but when the layer is relatively large, an average radius must be used; more complex designs need further adaptations.

Centrifuges with a similar sigma value should be able to achieve the same separations with the same flow rate. The centrifuge throughput may be calculated by the following equation:

$$Q = v_g \Sigma$$

(9.6)

where Q is the flow rate in m³/h, and v_g is in m/h. For example, a tubular centrifuge with $\Sigma = 20$ m² should be able to process 1480 L/h of a yeast suspension with 5 μm cells with $v_g = 0.074$ m/h, not considering the solid load. This type of centrifuge operates in batches whose duration will depend on the volume of solids collected.

9.2.6 CENTRIFUGE TYPES

Centrifuges have several geometries, but there are four main types of process equipment: tubular or bowl, filtering or basket, disk-stack, and decanter or scroll centrifuges. Their shapes and dimensions will determine Σ values; for example, in tubular centrifuges, the radius and length of the bowl will determine the maximum flow rate that will permit enough time for particle settling. In disk centrifuges, the angle and number of disks is also important, and decanter centrifuges may be regarded as tubular centrifuges with an extra solid dewatering section, which greatly enhances the centrifuge throughput. Specific descriptions of centrifuges are listed here.

Tubular centrifuges: The simplest type of centrifuge is basically composed of a long, hollow cylinder that spins around its axis, usually in a vertical position. The liquid is fed into the bottom of the cylinder, and the centrifuge forces the denser phase to the internal wall while the liquid is removed at the top of the equipment. The level of the liquid is determined by the distance of the discharge orifice from the axis. The operation goes on until unclarified liquid is discharged at the top of the equipment, which means that the solid-holding capacity of the bowl has been reached. At this point, the system must be cleaned and the solid fraction removed mechanically. Speeds up to 20,000 g may be achieved in process centrifuges of this kind with excellent broth clarification and solid dewatering. The expression for the sigma factor in a tubular centrifuge is given by Lander et al. (2005, p. 34) as

$$\Sigma = \frac{\pi\omega^2 L}{g} \frac{\left(R_2^2 - R_1^2\right)}{\ln\left(\dfrac{2R_2^2}{\left(R_2^2 + R_1^2\right)}\right)} \tag{9.7}$$

Basket-based centrifuges: This equipment consists of a cylindrical bowl that spins either horizontally or vertically around its axis. The mixture is fed near the bottom of the bowl, and the separation may go on pretty much as in tubular centrifuges; however, it is usual to have perforated baskets or screened baskets, which turn this equipment into a filter driven by centrifugal force. The filter medium may be a metal grid or a cloth material that retains the solid particles and allows the liquid to pass. The solids can be removed manually; however, using moderate velocities (approximately $1000 \times g$) and scrapers or slightly conical baskets, the operation may be semicontinuous. Pusher centrifuges use a bottom dumper or membrane to push the solids toward the open end of the bowl, and peeler centrifuges use scrapers to remove solids periodically. Although these filtering centrifuges need relatively large particles (usually above 50 μm) to operate properly, they are built in diameters up to 2 m, giving relatively high throughputs. In a filtering centrifuge, one expression for throughput evaluation is given by Chen (1997):

$$Q = \frac{\pi\rho L\left(R_0^2 - R_L^2\right)\omega^2}{\mu\left(\left(\dfrac{1}{k_{av}}\right)\ln\left(\dfrac{R_0}{R_c}\right) + \dfrac{\Omega_m}{R_0}\right)} \tag{9.8}$$

where ρ_L is the liquid density, R_0 and R_L are the outer and liquid radius, k_{av} is the average cake permeability, R_c is the radius up to the cake surface, and Ω_m is the filtering medium resistance.

Disk-type centrifuges: These are one of most common designs and use a stack of cones inside the inundated part of the rotor to enhance separation. These cones are usually approximately 1 mm apart, forming thin channels. The heavier phase must travel a small distance in order to reach the inner surface of a disk, and at this surface, the force exerted by the counterflow of the light phase is smaller, enhancing the efficiency of the equipment. The solids may be discharged manually, but it is much more common to have nozzle dischargers (with a lower solid content) or ejector-type dischargers (with a periodic, drier solid discharge). The sigma factor of this kind of centrifuge is given by (Najafpour 2007, p. 178)

$$\Sigma = \frac{2\pi\omega^2(N-1)\left(r_2^3 - r_1^3\right)}{3g\tan\theta} \qquad (9.9)$$

Decanter or scroll centrifuges: This equipment is similar to a tubular centrifuge but uses an internal scroll to push the solids toward a conical section of the bowl. This has the effect of enhancing the drying capability of the centrifuge while turning it into continuous equipment: The mixture is fed at some point near the center of the centrifuge with the solids being carried to one direction while the liquid (whose level is regulated by the height of the output) leaves at the other side. The scroll rotates more slowly than the bowl. The sigma factor of this kind of centrifuge is given by (De Loggio and Letki 1994, p. 73)

$$\Sigma = 2\rho\omega^2 L_e R^2 g \qquad (9.10)$$

where R is the maximum radius, and L_e is the length at $0.75R$.

Because of the small size of most cells, biomass separation with centrifuges is best done using tubular or disk centrifuges; the other types are more useful for larger solids, such as modified starches and crystals.

9.3 FILTRATION

This is a technique where solid particles or very large molecules are separated from a fluid (a liquid or a gas) by a physical barrier (the filter medium), which retains relatively large particles and allows smaller particles and the fluid to pass through it. This is one of the most common steps in downstream processes, both in the beginning of the processing sequence (e.g., in order to separate cells and cell debris from the fermentation broth) and in later steps (e.g., in order to separate crystals from a mother liquor, for clarification, or sterilization).

Filtration may be divided into conventional or dead-end filtration, which will be discussed in this section, and tangential filtration (or membrane filtration), which has unique characteristics and shall be discussed in the next section. In conventional filtration, the liquid flow is perpendicular to the filter medium, and the driving force of this process is a differential pressure between the two regions of space separated by the medium. In gravity filtration, the pressure is the hydrostatic pressure resulting from the weight of the liquid layer above the medium and the cake; in vacuum filtration (common in laboratories), the pressure is the difference between the atmospheric pressure and that below the filter medium, usually indicated directly by a manometer; several filters use pumps to apply pressures beyond 1 atm.

9.3.1 PorOUS Beds

Porous beds are solid layers that let fluids pass through convection. They may have multiple chambers or channels interconnected (pores), as in some solid foams, or consist of a mass of compacted particles, as in sand beds. They occur in several processes, such as column adsorption and chromatography, heterogeneous catalysis, packed distillation and absorption columns, and drying, and appear in biological systems as biophysical components (such as alveoli and kidney tissues). Perhaps the most important occurrence of porous beds is in filtration processes: The physical barrier used in conventional filtration systems usually is a porous material, and the solids retained on the filter—usually referred to as a solids "cake"—is a porous bed. Several equations with variable complexity apply to liquid flow in porous beds, and these are useful both for filtration and other operations; the basic theory behind filtration is discussed in Section 9.3.2.

9.3.2 Filtration Theory

Filter systems are adapted or commissioned according to the demands of the process from experimental pilot scale or laboratory data. The most important things to know are the amount of solids that will be collected, the efficiency of the filtration, the time it will take, and the size of the filter that will do it. Of course, the process throughput depends on the filtering area but also on pressure, viscosity, and particle size. A useful approach to assessing filtration velocity is Darcy's law (Equation 9.11), which states that the velocity of the fluid through the bed (v) is directly proportional to the pressure drop (ΔP), inversely proportional to the thickness of the solids layer (l), and the viscosity of the fluid (μ). The proportionality constant (K) is called the permeability of the bed:

$$v = \frac{1}{A}\frac{dV}{dt} = \frac{K\Delta P}{\mu l} \tag{9.11}$$

This equation may be derived from Poiseuille equations considering laminar flow (which is the case in most filtration systems). From these equations, Kozenzy and later Carman (Badger and Banchero 1955) developed an expression for estimating K, the permeability of the bed, which, with some rearrangement, gives the following:

$$K = \frac{\varepsilon^3 d_p^2}{180(1-\varepsilon)^2} \tag{9.12}$$

where ε is the bed porosity, that is, the ratio between void volumes and total volume, and d_p is the particle diameter. Although the equation was developed for spherical particles, it works well for incompressible beds, such as packed columns.

Because ΔP is a driving force, $1/K$ may be seen as a resistance term. In filtration, two solid layers exist—the filtering medium and the cake—so that this resistance may be divided into a constant medium resistance, R_m, and a variable cake resistance,

R_c (variable because of the buildup of retained solids). Rewriting Equation 9.11 with resistances, we have the following:

$$v = \frac{1}{A}\frac{dV}{dt} = \frac{\Delta P}{\mu(R_m + R_c)}$$

(9.13)

The resistance offered by the cake depends on the nature of the solids and the thickness of the cake layer, which happens to vary throughout the filtration as the filter collects solids. One way to express cake resistance is to account for the buildup of solids and the nature of the cake separately:

$$R_c = \alpha V C_{SS}/A$$

(9.14)

where α is a specific cake resistance constant for incompressible cakes, V is the volume of suspension filtered, C_{SS} is the suspended solids concentration, and A is the cross-section area of the filter. As several biomass cakes are compressible, the factor α may be substituted by an expression accounting for the pressure effect on the cake (Choudhury and Dahlstrom 1957):

$$\alpha = \alpha'(\Delta P)^n$$

(9.15)

The higher the compressibility of the cake, the greater the exponent n will be, which, in extreme cases, can reach the value of 1.0.

Substituting the expression for R_c (Equation 9.14) into Equation 9.13, we obtain a differential equation that may be used to derive expressions for theoretical filtrations:

$$v = \frac{1}{A}\frac{dv}{dt} = \frac{\Delta P}{\mu\left(R_m + \dfrac{\alpha V C_{SS}}{A}\right)}$$

(9.16)

Equation 9.16 may be solved for constant-rate or constant-pressure filtrations. Constant-rate processes rely on pumps to raise the pressure as the resistance increases, thus maintaining v. This kind of process is less common, and Equation 9.16 will have all values constant except for V and ΔP. If the process is run at constant pressure—which is more common—the equation may be integrated from an initial time, when $V = 0$, to a final time t and volume V, giving the usual equation for batch filtration:

$$t = \frac{\mu\alpha C_{SS}}{2\Delta P}\left(\frac{V}{A}\right)^2 + \frac{\mu R_m}{\Delta P}\left(\frac{V}{A}\right)$$

(9.17)

The parabolic form of the equation is consistent with the reduction of velocity resulting from cake buildup during the filtration.

These equations should be used with care because the process conditions will affect the constants. Not only may α vary with pressure, but in some cases, R_m may by deformation with pressure or because of the entrainment of solids. That being said, it is easy to gather laboratory filtration data, which may be used for estimation of filtration time on a large scale. For that, a batch filtration apparatus (e.g., a Buchner funnel with a filter, a Kitasato, and a vacuum pump) may be used for recording the time and volume filtered, and these may be plotted as $t/(V/A)$ versus V/A, a suitable linearization of Equation 9.17:

$$\frac{t}{\left(\dfrac{V}{A}\right)} = \frac{\mu\alpha C_{ss}}{2\Delta P}\left(\frac{V}{A}\right) + \frac{\mu R_m}{\Delta P} \tag{9.18}$$

For example, say we want to filter yeast with an average size of 5 µm from a suspension with 20 g/L of solids. The desired product is dissolved in the clarified liquid, and it is being considered that a pressure filter with a perlite precoat over sintered metal could be used. It is thought that the resistance of the sintered basis will be low in comparison with the cake resistance, and in order to simulate the process, the following data were gathered using a 15 cm Buchner funnel coated with a 2 mm layer of perlite:

Time (min)	2	4	6	8	10
Filtrate volume (mL)	73	118	150	184	210

The liquid has a viscosity of 3 cP at 20°C, and the pressure differential used was 400 mmHg. We want to determine the filtration constants in order to estimate the time for filtering 2 m³ of fermented broth with a 0.78 m² pressure filter with a ΔP of 1100 mmHg.

From the experimental values, $t/(V/A)$ may be plotted against V/A, giving a straight line ($R^2 = 0.994$) represented by the following equation:

$$t/(V/A) = 4.6003\ (V/A) + 2.973 \tag{9.19}$$

with t in seconds and V/A in centimeters. Comparing Equations 9.17 and 9.18, with the given viscosity, solids concentration, and pressure, we determine the following:

$$\alpha = 1.6 \times 10^9\ \text{cm/g}$$

and

$$R_m = 2.07 \times 10^8\ \text{cm}^{-1}$$

Applying these values in Equation 9.16 or 9.17 but with the new area, volume, and pressure to be used in the process, we find that the time for filtration would be approximately 6 h. This filtration would form a cake layer of 10 cm if the porosity is approximately 0.5. The increasing resistance of the cake is reflected by the fact that in order to filtrate half the amount of suspension (1 m^3) the time would be only 1.7 h; one way to maintain the filtration efficiency in such a filter would be to use short cycles with periodic backwashing, therefore preventing thick cake buildup.

Porous cakes may easily hold up to 50% of liquid trapped among the solids; therefore, a washing step may be necessary either to recover dissolved substances of interest or to clean the solid collected in the filter. In theory, pumping a fluid through the cake would cause a displacement of the liquid, and one volume of liquid would be enough; in practice, there are diffusion, adsorption, and channeling effects in play, and the efficiency is smaller. According to Choudhury and Dahlstrom (1957), washing efficiency is usually between 35% and 86% with the following equation relating washing efficiency (E) and the residual fraction of solute within the cake (R):

$$R = (1 - E/100)^n \qquad (9.20)$$

where n is the ratio between the volume of washing liquid and the liquid trapped in the cake (or the void space). Washing also takes considerable time: If the resistances are considered constant at the end of filtration, and R_m is negligible in comparison with R_c, then the time for washing may be related to the overall filtration time by the following expression:

$$t_w = 2t_f(V_w/V_f) \qquad (9.21)$$

where t_w is the washing time, t_f is the time of filtration, V_w is the volume of water to be used in washing, and V_f is the total volume of filtrate prior to washing.

9.3.3 FILTER AIDS

Filter aids are solids with low compressibility and, usually, a high porosity, which facilitate the filtration by forming porous cakes and preventing clogging of the pores of the filter membrane. Although several particulate inert solids may be used as filter aids, including microcrystalline cellulose, calcium carbonate, or gypsum, they still have low porosities compared with the most common filter aids: diatomaceous earth (composed of microalgae exoskeletons) and perlite (a volcanic material composed mainly of aluminum silicate). Filter aids may be previously applied over the filter medium, forming a precoat, which will prevent the solids in the suspension under filtration to be deposited directly on the filter medium. Precoats are commonly used with rotary drum filters, facilitating cake separation. Filter aids may also be added directly to the suspension to be filtered, and then the mixture is forced through the filter media; this strategy enhances the cake quality of compressible materials. However, in both cases, the filter aid may adsorb the target compound or cause an excessive increase in the viscosity of the broth. Also, the use of filter aids makes the process more expensive.

9.3.4 FILTER TYPES

There are a variety of filter types, grouped according to the driving force (gravity, pressure or vacuum, mechanical pressure, and centrifugal force) (Chen 1997), to the type of process employed (batch, fed-batch, or continuous process), or even to the geometry. The simplest types operate in batch mode and can be mounted vertically or horizontally: candle, leaf, multiple tray, and Nutsche filters. This equipment operates under pressure or in a vacuum and can be easily contained, backwashed, and automated. Another common filter design is the plate-and-frame filter, which is compact and very efficient for the dewatering of cakes. Bag filters are useful for low solid concentrations, and the equipment most commonly used in large-scale bioprocesses are the rotary vacuum filter and belt filters.

Rotary vacuum filters consist of a rotating perforated drum whose round surface lies partially submerged in the suspension to be filtered. A vacuum is applied inside the drum while its surface may be covered with a filter medium or a precoat; the submerged section of the drum filters the material and slowly rotates while the cake builds up and is washed and removed with a "knife" that is slowly advanced. With thick filter aid layers, small knife advancements, and a rotation of 20–100 rpm, these large filters operate several hours between batches.

Belt filters have resistant filter media (such as polyester tissues), and are useful for separating solids that form porous cakes, such as crystals from suspensions: The material is applied to a moving filter medium, dewatered, and eventually washed and pressed with a membrane and cylinders with good dewatering.

9.4 MEMBRANE SEPARATIONS

Membrane separations usually consist of cross-flow filtration, that is, in feeding a fluid tangentially to a selective membrane; only part of the fluid passes through the membrane (the permeate flow), and the rest is pumped away from the membrane with a higher concentration of some solutes (the retentate flow). The membrane is selected regarding the size of the particles or molecules to be concentrated (or segregated in case of impurities): Smaller species pass through the membrane while larger ones are retained. This selectivity of the membrane is the main advantage of membrane separations. In addition, the processes with membranes do not require additives and can operate isothermally at low temperatures and with relatively low energy consumption. Also, upscaling and downscaling of membrane processes, as well as their integration into other separation or reaction processes, is easy (Ulbricht 2006).

Most applications of membrane technology refer to concentration, purification, and fractionation. These have also been applied to membrane reactors and bioreactors in order to obtain higher productivity. The utilization of the membrane separation process in the food and beverage industry is widespread. Many applications, nowadays, substitute traditional unit operations in such plants. Various applications of membrane technology in the industry are given in Table 9.2.

9.4.1 Transport in Membranes

In a cross-flow filtration, the feed stream moves parallel to the membrane surface, and purified liquid (or solvent) passes through the membrane (permeate). Most of the particulates and aggregates are carried away by the cross flow. Transport across the membrane can occur by a variety of driving forces, such as gradients in pressure,

TABLE 9.2
Selected Industrial Applications of Membrane Separation Processes

Industry	Membrane Technology	Application
Dairy	Reverse osmosis	Milk and whey concentration
	Ultrafiltration	Milk and whey fractionation
	Microfiltration	Clarification of cheese whey
Beverages	Microfiltration and ultrafiltration	Clarification of beer, wine, and vinegar
	Reverse osmosis	Alcohol content reduction in beers
Juices	Ultrafiltration, reverse osmosis, and microfiltration	Clarification and concentration of fruit and vegetable juices; sugarcane processing
Gelatin	Ultrafiltration	Concentration of soluble collagen before drying
Biotechnology	Microfiltration, ultrafiltration	Sterilizing process, concentration of enzymes, cell separation

concentration, electrical potential, or temperature as in the case of thermo-osmosis and membrane distillation. The separation of chemically different species results from the differences in transport rates through the membrane.

For porous membranes, the transport of solvent is directly proportional to the applied driving force, and models can be used to describe the convective flow. The existence of different pore geometries also describes the transport in different models. Presented here is the film model. The porous membrane has relatively low resistance, the fluxes of solvent are high, greater than 500 L m^{-2} h^{-1} atm^{-1} for microfiltration (MF) and about 100–500 L m^{-2} h^{-1} atm^{-1} for ultrafiltration (UF) (Mulder 1996, p. 46). The flux (J) through the membrane can be written as follows:

$$J \propto \frac{\text{driving force}}{\mu \times R} \tag{9.22}$$

where μ is the viscosity and R is the resistance.

The driving force for microfiltration and ultrafiltration processes is pressure. The total resistance includes all resistance to the permeate flow. In addition to the resistance of the membrane itself, another resistance arises as a result of the transport of components retained by the membrane, which are adsorbed on the surface of the membrane, block the pores, or form a gel layer—in all cases exerting a barrier to mass transfer through the membrane (which adds to the concentration polarization resistance). In gas separation and pervaporation, the resistance of gel layer, adsorption, and pore-blocking do not occur. For these processes with nonporous membranes, only membrane resistance and concentration polarization have to be considered.

9.4.2 Concentration Polarization

The concentration polarization is a reversible phenomenon different from fouling, which is often an irreversible process that can result from polarization phenomena.

When solutes retained by the membrane accumulate on the surface of the membrane, its concentration increases gradually over time. This concentration gives rise to a reverse convective flow to the bulk of the solution fed. When the steady state is established, the convective flow of solute ($J_c = JC$) toward the membrane surface is balanced by the diffusive flux ($J_d = -DdC/dx$), plus the flux of permeate through the membrane (JC_p):

$$JC + D\frac{dC}{dx} = JC_p \qquad (9.23)$$

A concentration profile is established in a boundary layer, where

$$x = 0, \ C = C_m$$

$$x = \Delta, \ C = C_b$$

where C_m is the concentration of solute in the membrane surface; C_b is its concentration in the bulk, and C_p is the solute concentration in the permeate.

The integration of Equation 9.23 results in the film model:

$$\frac{C_m - C_p}{C_b - C_p} = \exp J \frac{\delta}{D} \qquad (9.24)$$

where D is the diffusion coefficient, and δ is the boundary layer thickness. D/δ is the mass transfer coefficient k. C_m/C_b is called the concentration polarization modulus. This ratio increases with increasing J and with decreasing k. The mass transfer coefficient depends strongly on the hydrodynamics of the system and can be optimized. The concentration polarization can be reduced by manipulating the flux J and the mass transfer coefficient k. The cross-flow velocity is a very important variable to improve mass transfer.

The effect of concentration polarization is very severe in MF and UF compared to reverse osmosis because of its lower flux and higher mass transfer coefficients. The concentration polarization in gas separation and pervaporation can be neglected, or it is low. Gas molecules through a porous membrane (asymmetric or composite) tend to diffuse from the high-pressure side to the low-pressure side. The transport mechanism depends on the membrane structure and includes transport through a dense (nonporous) layer, Knudsen flow in narrow pores, viscous flow in wide pores, and surface diffusion along the pore wall (Mulder 1996, p. 226).

Separation with nonporous membranes is used when molecules (solute and solvent) have the same order of magnitude; the "molecular pores" in this case are described in terms of free volume. The transport of a gas or liquid through a nonporous membrane can be described in terms of a solution-diffusion mechanism. Permeability through the membrane is a function of solubility and diffusivity.

9.4.3 MEMBRANE TYPES

Mulder (1996) gives a definition of a membrane as a selective barrier between two phases; the term "selective" is inherent to the membrane process. Membranes can be classified according to their nature, morphology, or structure. Considering nature, two groups of membranes exist: biological and synthetic membranes. Biological membranes constitute every living cell, all of which are surrounded by a membrane. Biological membranes or cell membranes have very complex structures as they are responsible for specific functions. Current (first-generation) membrane polymers are biopolymers (mainly cellulose derivatives) or synthetic engineering polymers, which had originally been developed for other purposes. Synthetic membranes can be divided into organic (polymeric) and inorganic membranes.

9.4.3.1 Polymeric Membranes

All kinds of film-forming polymers can be used as membrane material; however, their physical and chemical characteristics differ substantially, and this makes only a few polymers useful in practice. Polymeric membranes can be classified as porous and nonporous. According to the definitions adopted by the IUPAC (Mulder 1996), pore size is classified into macropores (>50 nm), mesopores (2 nm < pore size < 50 nm), and micropores (< 2 nm).

Porous membranes have a well-defined pore structure, which, depending on the formation process, can be highly connected and tortuous or nonconnected and straight. This type of membrane is mainly applied in microfiltration and ultrafiltration. The flux and selectivity of the separation process depend essentially on the pore size and also the material of the membrane because the chemical and thermal stability and fouling tendency will affect the progress of the operation. For microfiltration membranes, pore sizes are in the range of 0.1–10 μm. Polycarbonate, poly(tetrafluorethylene) (PTFE), poly(vinylidene fluoride) (PVDF), and polypropylene (PP) are often used for microfiltration membranes. These four kinds of polymers are hydrophobic and have excellent chemical and thermal stability. On the other hand, hydrophilic polymers have fewer tendencies toward absorption of solutes, decreasing the resistance to mass transfer and facilitating the flow through the membrane. Cellulose and cellulose esters are the most common polymers of this type. The drawback of cellulose ester membranes is their susceptibility to degradation by chemical and biological agents. They constitute an important class of materials for membranes used in microfiltration, ultrafiltration, reverse osmosis, gas separation, and dialysis. Aliphatic and aromatic polyamides also find application as membrane materials. They have chemical stability and are used in microfiltration and ultrafiltration processes.

In addition to these materials, some less common polymers are used specifically for ultrafiltration membranes, where special techniques are used to obtain nanosized pores (2–100 nm). Polysulfone and polyether polymers are widely used as the basis for ultrafiltration membranes. Polyetherketones (PEEK) and polyacrilonitrile (PAN) are also examples of materials for ultrafiltration membranes.

Dense, nonporous membranes are applied in gas separation and pervaporation. Composite and asymmetric membranes are also used for these processes. The chemical nature and morphology of the polymeric membrane are the main factors to consider in transport across the nonporous membranes. The separation occurs because of differences in the solubility and/or diffusivity of the permeate compounds in the polymer. This is an important property; permeates of similar sizes may be separated if their solubility in the membrane differs significantly. Although highly selective, dense membranes have an inherently low rate of permeation.

Asymmetric membranes consist of two structurally distinct layers; one of which is a thin, dense selective skin or barrier layer, and the other is a thick, porous matrix whose main function is to provide physical support for the thin skin (Pandey and Chauhan 2001).

9.4.3.2 Inorganic Membranes

Inorganic materials have higher thermal and chemical stability comparing to polymeric membranes. Four different types frequently find applications in the field of MF and UF: ceramic, glass, metallic, and zeolitic. Ceramic membranes are the main class of material for inorganic membranes, which are prepared combining a metal (aluminum, titanium, or zirconium) and a nonmetal in the form of an oxide, nitride, or carbide. Zirconia and glass (silicon oxide or silica) are typical ceramic materials.

9.4.4 Pore Size versus Type of Operation: Membrane Process

Based on membrane material, flux density, or selectivity, the optimal separation performance can only be achieved by process conditions adapted to the separation problem and the membrane material.

Membrane separation technologies commercially established in industrial scale are

- Dialysis (D): for blood detoxification and plasma separation (medical devices)
- Reverse osmosis (RO): for the production of ultrapure water, including potable water
- Microfiltration (MF): for particle removal, including sterile filtration (various industries)
- Ultrafiltration (UF): for many concentration, fractionation, or purification processes (various industries)
- Gas separation (GS): for air separation or natural gas purification

Table 9.3 presents membrane processes according to their membrane type, pore size, and driving force.

9.4.5 Membrane Modules and Peripherals

Membranes for cross-flow filtration are available in a number of module designs. Membrane configurations are basically of two types: flat and tubular. Flat membranes are arranged in plate-and-frame or spiral-wound modules. Tubular, capillary,

TABLE 9.3
Membrane Processes and Their Characteristics

Membrane Process	Driving Force	Membrane Type/Pore Size
Microfiltration (MF)	Pressure	Symmetric porous/0.05–10 μm
Ultrafiltration (UF)	Pressure	Asymmetric porous/1–100 nm
Reverse osmosis (RO)	Pressure	Asymmetric or composite, microporous/<2 nm
Electrodialysis (ED)	Electrical field	Cation and anion exchange, nonporous
Dialysis (D)	Concentration	Homogeneous, microporous/ 2 nm or Mesoporous/2–50 nm
Nanofiltration (NF)	Pressure	Composite, microporous/ <2 nm
Pervaporation (PV)	Concentration	Composite, nonporous
Gas separation (GS)	Pressure	Asymmetric or composite nonporous or microporous/<1 μm

Source: Adapted from Mulder, M., *Basic Principles of Membrane Technology.* Dordrecht: Kluwer Academic Publishers, 1996.

and hollow-fiber modules are arranged in tubular modules. In general, a system does not consist of a single module but a number of modules arranged together as a system. The choice of module configuration is based on economical, mechanical, and hydrodynamic considerations (Harrison et al. 2003).

Plate-and-frame module: This design is composed of arrangements of two or more flat membranes, usually rectangular, arranged in parallel and separated by a spacer. Feed is held at one side of the membrane, and the separation occurs along the surface of the membranes. The permeate is collected in the opposite face of the membrane and the retentate on the other end of the module.

Spiral-wound module: This is a wrapped plate-and-frame system with a central collection pipe. The feed solution is fed into one end of the module and flows through the separator screens along the surface of the membranes. The retentate is collected at the other side, and the permeate spirals toward a central tube and exits at one end of the module.

Tubular module: Membrane tubes are placed inside porous supports with a diameter of more than 10 mm. These supports are made of porous stainless-steel, ceramic, or plastic material. The number of membrane tubes inside the module may vary. The feed solution flows inside the membrane tubes, and the permeate flows through the porous supports into the module shell and exits the module.

Capillary module: In a capillary module, a large number of capillary membranes are assembled together. The free ends of the fibers are potted with epoxy resins, polyurethanes, or silicone rubber (Mulder 1996, p. 470). Feed solution flows inside the capillaries (through the bore) and the permeate is collected on the outside. In a less-common arrangement, the feed solution enters the module on the shell side of the capillaries, and the permeate passes into the fiber bore.

Hollow-fiber module: The hollow-fiber modules consist of an array of narrow-bore, self-supporting fibers. As in capillary modules, the feed can flow inside or outside of the fiber. The difference between capillary and hollow-fiber modules is simply a matter of dimensions.

Peripherals: Membranes need several peripherals to be used: Beside valves and tubing, pumps are necessary for generating the flow and maintaining an adequate pressure drop across membranes; control valves and manometers are also necessary. For the regeneration of membranes at the end of an operational cycle, solutions of acids or bases, as well as enzymes, might be necessary, and because these processes are used for concentration and dialysis, holding tanks are essential. Modern systems are largely furnished as skid systems, where there is a set of pumps, manometers, control valves, and holding tanks connected to the filters. The systems are easily up-scaled, and membrane cartridges may be changed according to operational needs.

9.5   EVAPORATION AND DISTILLATION

The concentration of liquids during processing or as a final production step represent a significant reduction in transport, packing, and storage costs and increase the stability of foods (Bui et al. 2004). Distillation and evaporation are used to separate volatile fractions of liquid mixtures in order to increase their concentration—giving a concentrated juice or stream suitable for a subsequent spray drying step, for example—or to obtain the volatile fraction itself as it is done in the spirits industry. In both cases, the selective property is the difference in vapor pressures (volatilities) between substances, and heat is applied to cause the phase change for removal of the more volatile fraction.

9.5.1 DISTILLATION

Distillation is a unit operation that is very common in the food industry for separation of compounds present in a solution in two phases (liquid and vapor), which are at the same pressure and temperature but with different composition (Perry et al. 2008). The separation occurs because of the difference in volatility of the substances. Although rigorous modeling of the operation is complex, an approximation of the system may be done using vapor pressure expressions for the components of the mixture. Figure 9.1 illustrates the vapor pressure for six liquids.

It is clear that vapor pressures are quite different for different solvents. When a mixture of two or more liquids is heated, the vapor phase is enriched in the more volatile component (that with the higher vapor pressure); conversely, condensation of a mixture of vapors will result in a liquid enriched in the less-volatile component. These are the principles underlying distillation: Vapor generated by boiling liquid in one stage (or tray) goes up to cooler sections, losing the heavier components, which condense, while liquid from any stage goes down to warmer sections, losing the lighter components, which vaporize. In the column where distillation is carried out, any section is cooler than that below it; energy is usually supplied at the last (bottom) section, and a condenser is used for the vapor collected in the top of the column. Several details, such as the temperature and phase of the feed stream, the recycle ratio, etc., are carefully regulated in order to enhance the separation.

It is impossible to recover totally pure components by distillation, but the operation is routinely used to concentrate ethanol streams from 6%–15% to 95% ethanol

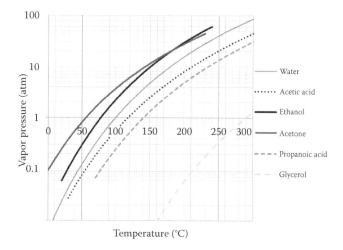

FIGURE 9.1 Vapor pressure of six selected solvents.

in the production of several distilled beverages, such as whiskey, rum, and vodka. Although the process may be carried out in stills, for large production systems, the process is carried out in columns divided into sections with plates or trays or packing materials, which guarantee an intimate contact between both phases (Treybal 1981). Figure 9.2 shows the equilibrium curve for an ethanol–water system and the

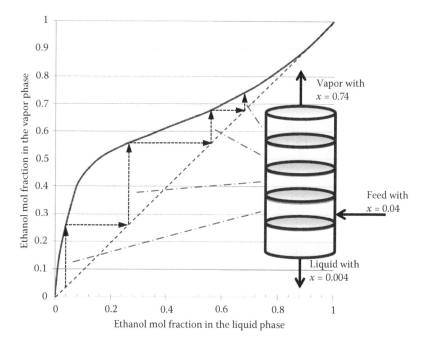

FIGURE 9.2 Liquid–vapor equilibrium for ethanol and water with a scheme of five ideal plates of distillation column.

corresponding plates in a distillation column without reflux. The arrows (start at $x = 0.04$, the ethanol molar fraction in the feed) show the liquid and vapor concentration in each plate; the vapor that leaves one plate is considered to have the same composition of the liquid in the next (upper) plate.

With just a few plates, the feed may be concentrated from an ethanol molar fraction of 0.04 to 0.74. There are important details that may be discussed regarding the operation: First, the liquid that is left at the bottom in Figure 9.2 is diluted but still has 10% of the ethanol from the feed; some more plates below the feeding stage could be used to enhance ethanol recovery in what is called the stripping section of the column. The section above the feeding plate is the rectifying section. Second, it seems that the top plates of the column concentrate the stream less and less. Adding more plates to the top of the column will enhance the concentration efficiency but just up to $x = 0.894$. That happens because ethanol forms an azeotrope with water, and further concentration requires a modification of the process. Third, real distillation columns return part of the vapor from the top, after condensation, to enhance the rectifying section efficiency, and that alters the equilibrium conditions in the top plates. Fourth, the volume of liquid in the stripping section is much smaller than that of the rectifying section (and vice versa). That will affect the column design. Tray efficiency, vapor velocities, and other factors will also affect distillation.

The design of distillation systems requires the following steps: a material balance based on the desired concentrations, an initial column design based on vapor–liquid equilibria, an energy balance accounting for vaporization enthalpies, and a final design considering heat transfer between the liquid and vapor phases. In spite of the complexity of real distillation systems, there is a great deal of knowledge about the operation, and several methods may be used from the classic McCabe-Thiele graphic design to modern process simulators. However, presenting such developments would demand a dedicated chapter and is beyond the scope of this text.

9.5.2   DISTILLATION FOR FERMENTED LIQUIDS

The composition of the feed stream determines the selection of the operating conditions, column, and packing materials. For example, terpene-based aromas may polymerize in high temperatures; lactones and esters may hydrolyze; and acids, such as lactic or propionic, are corrosive. One solution for thermal degradation is the use of a vacuum, which lowers the boiling point of the mixtures. An industrial vacuum may be economically obtained using compressors and steam ejectors. Vacuuming is mandatory for the separation of high boiling-temperature metabolites, such as free fatty acids in microbial oils. A disadvantage is the higher energy consumption and the high investment cost (Chanachai et al. 2010).

Although traditional distillation systems use perforated trays, structured packings are becoming popular. These have two major features: a low pressure drop (which is good for relatively viscous mixtures) and low height equivalent to theoretical plate (HETP) compared to multiple tray columns. Those features improve the effectiveness of the distillation process with structural packing for heat-sensitive materials (Zivdar et al. 2006a; Spiegel and Meier 2003), and it is preferred for high separation performance at low pressure (Zivdar et al. 2006b, Fischer et al. 2003; Bender and Moll 2003).

Membrane distillation is a relatively new technique: In this process, hydrophobic pores separate two aqueous streams (the feed and stripping streams) with the vapor pressure difference being the driving force. The difference in vapor pressure between both sides of the membrane may be a result of a temperature gradient or, in osmotic distillation, because of the higher concentration in the feed. Distillation across membrane systems consists of three main steps (Mansouri and Fane 1999):

- Evaporation at the feed-side membrane
- Transport of the vapor through the pores of the hydrophobic membrane
- Condensation of the vapor at the permeate side of the membrane

These techniques are promising because the process takes place at ambient pressure and temperature; however, the process still has relatively low fluxes.

9.5.3 Evaporation

Evaporation is the partial removal of water (or another solvent) from liquid feeds. It differs from distillation in that the solute of interest is nonvolatile and remains in the concentrate; it differs from drying in that the feed and concentrate are liquid streams. Evaporation considerably reduces the volume and weight for storage and transportation (Prodanić et al. 2008). It is common in the food industry—used mainly to concentrate fruit juices, milk, and coffee—frequently before a drying or freezing operation. Evaporation is also applied for production of concentrated soups, jellies, and ice creams. Maximal concentrations achieved by evaporation are approximately 80%, higher than other concentration techniques, such as ultrafiltration and reverse osmosis (30%) and freeze concentration (40%), but at higher energy consumption by kilogram of water removed.

Evaporators are essentially heat exchangers where the energy supplied to the feed stream is enough to convert part of the liquid to vapor; the phase change may occur inside the evaporator, which will consist of a phase separation chamber and a boiler section, or outside the chamber with forced circulation through the heat exchanger. Several designs of these evaporators exist, including equipment adapted for crystallization. In order to economize energy, the vapor produced in an evaporator may be used to heat another evaporation stage, provided that the boiling temperature is lower (using a vacuum) so as to maintain a suitable temperature difference for the heat exchange.

The sugar industry has been the most important industry related to the development of evaporators; multistage evaporation (where the vapor generated in one stage is used in the next) has been applied since 1844 to produce sugar from sugar cane and was later adopted in Europe in the beet-sugar industry. These multistage systems can occupy large areas, and vapor is supplied at 3.5 atm (Urbaniec 2004) in three main designs: falling films in tubes, falling films in plates, and climbing films in plates, operating at 125°C–140°C. The concentrated molasses that are separated after crystallization may have thermal degradation products from this step.

9.5.4 Evaporator Types

There are several evaporator types; all of them have a feed tank (eventually with a preheater), a heat exchanger, a large separation chamber, and a condenser for the vapor (Lavis 1994). For the design of the operation, it is necessary to perform a material balance and energy balances for the heater and condenser and to define separation parameters, such as the use of mist separators. Most evaporators fall into three classes: film evaporators, forced-circulation evaporators, and wiped-film evaporators. The first class is by far the most common with rising or falling films forming over the heat-exchanging surface while the liquid vaporizes. These evaporators have high heat transfer capacities with low residence times but are suitable only for low-viscosity fluids (up to 200 cP). Viscous liquids may be processed in forced-circulation evaporators when positive displacement pumps guarantee the flow of fluid. Because a pump will give enough backpressure to suppress boiling in the heater, good heat-exchange coefficients may be obtained in this process. Wiped-film evaporators use blades for the removal of liquid from the surface of tubes, and although the equipment design is complicated by moving parts inside the evaporator, the design is adequate for very viscous liquids.

9.5.5 Vacuum-Evaporative Cooling

Several fermented products are heat sensitive and cannot be concentrated at high temperatures, losing part of their nutritional and flavor properties during processing. Again, the use of a vacuum enhances the process quality, allowing the evaporation to take place at low temperatures.

Vacuums may also be used to cool materials (evaporative cooling), a process which traditionally has been applied to remove field heat after the harvest of vegetables, increasing their shelf life (Sun and Zheng 2006). Vacuum cooling evaporation has several advantages, such as short-time evaporation, reduction of energy consumption, and minimization of microbial growth, especially in cooked meats.

TABLE 9.4
Examples of Vacuum Evaporative Cooling in Food Processing

Food Process	References
Tortillas	Taylor et al. (1998)
Precooked chicken breast	Huber and Laurindo (2005)
White mushrooms	Tao et al. (2006)
Cooked beef	Jackman et al. (2007)
Bread	Primo-Martín et al. (2008)
Cabbage	Han et al. (2009)
Lettuce	Ozturk and Ozturk (2009)
Stuffed bun	Deng et al. (2011)
Vegetables	Yu et al. (2011)

In recent decades, the evaporation under vacuum method has been extended to other products, such as beef, dairy, and baked products. Table 9.4 shows some examples of vacuum cooling (Taylor et al. 1998).

9.6 DRYING

Drying is the removal of solvent from moist solids or from a liquid, giving a product a solid, dry material. Although the term applies to the removal of any solvent, water is by far the most common solvent in drying. The removal of water from gases is also called drying although this operation usually relies on adsorption into solids.

Drying is indispensable in most processes in order to reduce weight, increase concentration, increase stability, recover solvents, and control the solubilization characteristics of a product. It is also an essential step for converting liquids or moist solids into free-flowing powders, which is the desired product form for use in several food formulations.

There are many types of driers, but in all of them, thermal energy is transferred to the material in order to vaporize the liquid; therefore, heat transfer and the properties of the solid matrix and of the solvent are paramount. We will focus on water as a solvent and describe some properties that affect drying and, subsequently, describe principles of drying theory and practice.

9.6.1 PSYCHROMETRICS

Psychrometrics is the study of the thermodynamic properties of solvents in gaseous systems. In drying, the solvent of interest is usually water, and the gas system is air. The properties that are more important in drying are the vapor pressure of the solvent, which indicates the tendency of the liquid phase to change into vapor, and the vaporization enthalpy, which tells us how much energy would be required for that change.

For water in food systems, a good approximation (1.5% average deviation from experimental data) for vapor pressure is given by the equation of Antoine:

$$\text{Log } P_v = 8.0600 - 1725.3280 \, (- 40.0199 + T)^{-1} \tag{9.25}$$

Equation 9.25 gives the vapor pressure (P_v) of pure water, in mmHg, at a temperature T, in Kelvin. The vaporization enthalpy may be approximated using the Watson correlation, based on vaporization enthalpy (ΔH_{vap}) at the normal boiling point (Rao 1997, p. 261):

$$\Delta H_{vap} = 3128.35(1 - T/Tc)^{0.38} \tag{9.26}$$

where Tc is the critical temperature of water, 647.3 K; T is the temperature in Kelvin; and ΔH_{vap} is in kJ/kg.

Because drying may also require heating air and water vapor, it is also necessary to use an expression for air enthalpy in order to solve energy balances. Air and water heat capacities ($Cp_{Air} = 1$ kJ kg^{-1} K^{-1}, $Cp_{Vapor} = 1.8541$ kJ kg^{-1} K^{-1}) are essentially constant at the range of temperatures of driers; an adequate expression for the gas enthalpy would be concerned with the mixture composition. With reference enthalpies H_{air} and H_{water} equal to zero at 0°C, an adequate expression would be the following:

$$H_{Mixture} = T + w(\Delta H_{vap} + 1.854T) \qquad (9.27)$$

where $H_{Mixture}$ is in kJ/kg dry air, w is in kg_{Water}/kg_{DryAir}, and T is in degrees Celsius.

These values are frequently aggregated in a psychrometric chart, a graphic that relates temperature and absolute humidity to enthalpy, relative humidity, and wet-bulb temperature as shown in Figure 9.3. The wet-bulb temperature is a convenient way to assess the relative humidity and, consequently, the drying capacity of air.

Finally, the properties of the drying solid are important, but widely variable; an approximate value is given by the following equation (Sokhansanj and Jayas 2006):

$$Cp = 0.34Xc + 0.37Xp + 0.4Xf + 0.2Xa + 1Xm \qquad (9.28)$$

where Xc, Xp, Xf, Xa, and Xm are the mass fractions of carbohydrates, proteins, lipids, ashes, and moisture in the material, respectively.

In drying, it is not necessary to remove all the water from a material but just enough water to stabilize it. This stabilization should be looked at regarding the water activity (a_w), rather than the absolute moisture content. Water activity is the ratio of the vapor pressure of the water in the material vis à vis pure water and

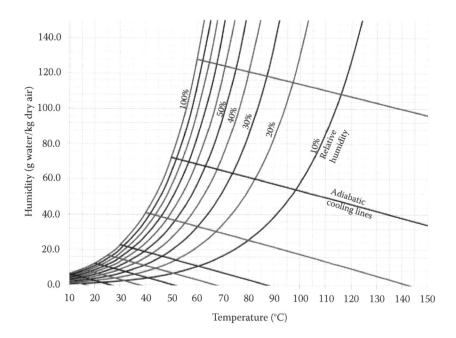

FIGURE 9.3   Absolute and relative humidity × temperature for system at 1 atm.

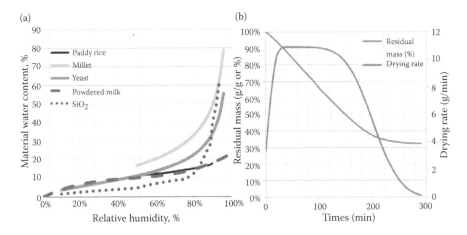

FIGURE 9.4 (a) Water activity as function of absolute humidity for five materials at 22°C; (b) drying curve for yeast cake from filter press at 60°C with the drying rate.

may be easily determined using equilibrium moisture or dew point measurements. Growth of bacteria usually requires an a_w higher than 0.9, and fungi usually require an a_w higher than 0.8. The relationship between water activity and absolute water content in a solid may be represented by curves, such as those shown in Figure 9.4. The shape may be explained by the affinity of the material for water: At low absolute humidities, the water is tightly bound to proteins and other molecules, and at higher humidities, there is water adsorbed in pores and on the surface—enough absolute humidity gives rise to unbound or free water, rising the material's a_w to 1. At the other side, every material eventually reaches equilibrium with the surrounding air with hygroscopic materials having high equilibrium humidities. In order to protect food from spoilage, the low a_w values obtained by drying should be maintained by adequate packaging.

9.6.2 MATERIAL AND ENERGY BALANCES, DRYING RATES

Using the equations and material properties and defining the final humidity of the material, it is possible to determine the amount of water to be removed and the minimum amount of energy required for that. However, nothing can be said at this point about the drying velocity, which depends on heat and mass transfer and, thus, on the material and the type of drier itself. A typical drying curve is shown in Figure 9.4 (b). After an initial heating phase, the material loses water at a constant rate; in this phase, drying is controlled by heat transfer. At a later stage, the water that is more tightly bound is lost but at a rate that diminishes steadily; it is the falling-rate period. This phase follows a first-order kinetic, controlled by mass transfer or equilibrium between the humidity in the solid and gas phases and may be represented by the following:

$$DR = k(C_{eq} - C_{bulk}) \tag{9.29}$$

where DR = drying rate, for example, in g/min; k is an overall mass transfer coefficient; C_{eq} is the equilibrium concentration in the gas at the material surface; and C_{bulk} is the concentration in the drying air.

Most driers may be classified, by way of the energy-transfer mode, into contact driers and indirect driers. In contact driers, the heat is transferred to the material directly by the heating fluid (air) and includes tray, tunnel, and spray driers. In indirect driers, heat is transferred to the material through a contacting interface, such as a tray heated by vapor; some examples are spray, tray, and pan driers and most designs that treat viscous pastes. In direct driers, the material temperature is maintained near the wet-bulb temperature of the gas during the constant-rate phase of drying, and in indirect driers, the material is kept near the boiling temperature of the liquid (Kimball 2001). Therefore, direct driers are more adequate for heat-sensitive materials.

Considering that drying rate will depend on the type of drier and the material to be processed, data should be gathered through experiments and the optimization through the modeling of the operation. An easy way to obtain essential data for drying is to dry small amounts of the material of interest at different temperatures and different layer thicknesses, computing the water loss by periodic weighing. The data obtained may be used to build a set of drying curves and determine the effect of temperature on the drying rate and product quality.

Example

We wish to dry 500 kg of a pressed yeast cake with 70% humidity, initially at 25°C, down to 10% humidity. The maximum allowed temperature for this material is 60°C, and we intend to use a cabinet drier heated indirectly with hot water; the process will be done at atmospheric pressure. We wish to estimate the drying time and the amount of energy and air required; the cabinet has 10 shelves of 2 × 1 m, and the yeast cake's apparent density is 0.8 g/cm³. Previous drying data in a small drier at 60°C with a layer of 4 cm of yeast cake were used to plot the drying curve in Figure 9.4.

MATERIAL AND ENERGY BALANCES

The water to be removed may be calculated considering that the solids are the same in both wet and dry material:

1000 kg – water to be removed = 1000 – [1000*(1 – 0.7)/(1 – 0.1)] = 666.7 kg

We could just calculate the amount of energy necessary to evaporate the water (666.7 kg of water times 2376.9 kJ/kg, the enthalpy of vaporization of water at 60°C from Equation 9.26), which is 1584.6 MJ. However, the atmospheric drying process depends on air carrying the moisture out of the drier; if air is not blown into the drier, the internal atmosphere will soon be saturated, and the process will be inefficient.

The minimum amount of air necessary depends on the moisture content of saturated air at 60°C, which can be read in the psychrometric chart to be approximately 124 g water/kg dry air. This air does not enter the drying cabinet free of moisture and may carry up to 16 g of water/kg dry air if the air outside the cabinet is saturated. Therefore, the minimum amount of moisture that 1 kg of air carries entering at 25°C and leaving at 60°C is 124 – 16 = 108 g or 0.108 kg.

Therefore, the amount of air necessary for drying is 666.7 kg water/0.108 kg water/kg dry air = 6172.9 kg dry air.

Now, we may use Equation 9.27 to estimate the enthalpy of the air entering at 25°C and leaving the drier at 60°C:

$$H_{25} = 25 + 0.016 \times (2474.2 + 1.854 \times 60) = 65.33 \text{ kJ/kg dry air}$$

$$H_{60} = 60 + 0.124 \times (2376.9 + 1.854 \times 60) = 368.52 \text{ kJ/kg dry air}$$

The difference, 303.2 kJ/kg dry air, multiplied by the amount of air necessary gives 1872 MJ or 18% extra energy (compared to that required for evaporation) just for air heating.

Drying time: This depends on the conduction of heat through the solids, the mass transfer rate of water from a drying front to air, and the diffusion of water through the dry material into the surrounding air. Both may limit the overall drying rate; however, that depends on the layer of solids deposited on the trays, and therefore, experimental data are always important. The drying curve in Figure 9.4 was plotted from drying with a cake layer of 4 cm; the amount of cake that will be dried may be calculated using the mass, density, and drier dimensions:

volume of solids = number of trays × tray area × solid layer thickness

(500 kg × 1000 g/kg)/0.8 g cm³ = 10 × 100 cm × 200 cm × layer thickness

625,000 cm³ = 200,000 cm² × layer thickness; the layer has 3.125 cm.

Considering that the residual mass will be 33% of the initial mass, the drying time would be 250 min with a layer of 4 cm; with a smaller layer, the process will be more efficient. Considering that the duration of the heating and constant-drying rate phase depends linearly on the amount of material fed to the drier, the reduction of the layer might proportionally reduce these phases from 120 to (120 × 3.125 cm/4 cm) = 93.8 min. In the extra 130 min of the falling-rate phase, drying is controlled by diffusion and should change less. This phase could be modeled using Fick's second law (Harrison et al. 2003, p. 300); however, because the material drying is a porous solid, the thickness used would be that of the pores—which did not change—and not that of the macroscopic layer. Therefore, a safer initial consideration for drying time would be 130 + 94 = 224 min.

9.6.3 Types of Driers

There are dozens of types of driers with the most common designs being belt, flash, fluid bed, rotary, spray, and tray driers, which are direct, convection, or adiabatic driers, and thin film, contact, drum, tray, bed, and steam-jacketed rotary driers, which are indirect or nonadiabatic driers. Some driers, such as the tray drier, may be operated both as adiabatic or convection driers. The selection of the drier depends highly on the process volume, the temperature to be used, and the need for containment.

The solid exposure to heat depends of the type of drier: Most driers need 10 to 60 min to dry solids, but some (drum, flash, and spray) need less than 30 s (Kimball 2001) while tray driers and lyophilizers may need several hours.

Considering that biological solids are heat sensitive, drying at low temperatures is desirable. This may be achieved using a vacuum, which may be easily obtained using steam ejectors or vacuum pumps. One recurrent problem is the formation of dry films, which shrink and reduce the drying rate. This problem may be avoided using carriers, selecting adequate temperatures (i.e., starting with a lower temperature), and changing the type of drier (e.g., using a dispersing drier, such as a flash drier).

Indirect driers: One of the most common types of drier is the tray drier. In this equipment, the material to be dried is heated by the tray (which is itself heated using vapor or electricity) or by the drying air. In both cases, there are two concurrent transfer phenomena: the conduction of heat from the tray (or air) to the material and the diffusion of water vapor through the material to be dried. This kind of drier is easy to maintain and operate and is flexible and low cost. Several trays are used in a single cabinet, which gives this equipment a high area-to-volume ratio. However, this equipment is not adequate for pastes and may require grinding of the product.

For viscous solids, such as polysaccharide pastes, an agitated, indirect contact drier may be effective. In this type of drier, although the heat-exchange area is comparatively smaller than in tray driers, the scraping of the surface with helicoidal agitators enhances good homogeneity and the overall drying rate.

Another popular drier design is the rotary drum drier where a film of the material to be dried is formed on the surface of a rotating cylinder that is heated internally. The material dries quickly because of the small thickness of the solids layer and is scraped away from the surface. However, the final temperature reached by the solids in this type of drier is high.

Adiabatic driers: For loose solids and liquid suspensions or solutions, adiabatic driers offer short contact times, which is frequently an advantage for biological solids.

A popular adiabatic drier is the spray drier: In this equipment, a liquid solution of the material is heated and sprayed into a large chamber with hot air. The drops from the spray fall and lose water and are collected at the bottom of the drier.

Some key parameters that may be controlled are the liquid temperature at the inlet, the drop size, and the air temperature and humidity. These parameters will control both the drying and falling time. Drying time must be greater than falling time; otherwise wet solids could cling to the walls of the drying chamber. Most of the energy necessary for the water to vaporize is provided by the hot air; if all the water in the liquid (usually with 50% solids) were to be evaporated solely based on the inlet enthalpy, the inlet temperature would have to be too high.

There are several geometries and operating modes for spray driers; the most common are cocurrent (when the dried solids leave in equilibrium with the air at the wet-bulb temperature) and countercurrent (when the solids leave in contact with the hottest air, which ensures a higher drying rate, but it is not suitable for heat-sensitive solids). The resulting solids are a powder with low bulk density and even with a hollow core (when pellicles form during drying). The size distribution depends on the dispersion system but tends to be narrow. Considering that there are spray designs adequate for several liquids (including viscous materials), this kind of drying is a good option whenever solutions of fine suspensions are to be dried. Because the product size distribution and internal structure are affected by the process, it is possible to fine-tune the product properties in spray driers.

Another type of adiabatic drier is the flash drier, which is able to handle solids, such as starches or amino acid crystals, which are transported into the equipment (using a screw transporter, for example) and spread on hot air at high speed. The suspension is conducted to a cyclone separator, and the dry solids are collected at the bottom.

If higher contact times are necessary, one effective design is a fluidized bed drier: In this type of drier, the material to be dried is spread in a chamber with a high amount of solids in suspension in the air; the solids dry and eventually leave the drier to a cyclone separator while heavy (wet) solids are maintained in the drier because of its buoyancy. The gas generated for adiabatic drying may be air or inert gas heated using vapor coils or air heated directly in a combustion chamber. Both flash and fluidized bed driers are fed with solids and do not change the powder characteristics (Shaw 1994, p. 78).

Freeze dryers: The process of drying frozen solids is called freeze-drying or lyophilization, a useful method for preserving cell and tissue structures. These are contact driers, which work at very low pressures. The triple point of water (the condition where solid, liquid, and vapor phases coexist at equilibrium) is 273.16 K and 611.73 Pa (or 0.01°C and 0.0060373 atm). Therefore, giving heat to solid water below 611 Pa will not melt it as usual, but rather it will sublimate it. The same will happen with solutions and moist frozen solids.

In freeze-drying, there are usually three steps: freezing, primary drying, and secondary drying. A solution or wet solid is frozen usually below –20°C, and then it is subjected to a pressure below 0.006 atm. From this point on, the material must be warmed so that the energy provided is used for water sublimation. However, the heating must be very gentle, matching the slow diffusion of vapor through the solid matrix. If the material is heated too quickly, the local water pressure could surpass the triple-point pressure, thereby creating conditions for fusion rather than sublimation. It is usual to maintain temperatures below –20°C throughout the primary drying; in the secondary drying, the matrix is usually rigid enough to support higher temperatures for evaporation of bound water.

Because the materials are solidified before freezing, this type of drying is very gentle, generating porous, easy-to-rehydrate products. Although the process is slow, relying on diffusion, the rehydrated solids have flavor and texture closest to the original material and maintain biological activity, which makes lyophilization the best method for long-term preservation of microorganisms.

9.6.4 PREPROCESSING FOR DRYING

Drying foods or biological materials frequently involves some sort of preprocessing in order to stabilize or concentrate the material.

Concentration for spray drying: Spray driers work best with streams containing a high solid concentration (20%–50%). In order to achieve this concentration, a previous evaporation or membrane concentration may be used.

Osmotic dehydration: This process removes water by contacting the wet solid with a hypertonic solution, usually of NaCl or a simple sugar. The process removes part of the water and is useful for the preconcentration of fruits, vegetables, and meats, lowering the water content. It is very gentle and reduces the time during which the material will be heated. A reduction of water content from 92% to 80% in mushrooms, for example, represents the removal of 60% of the water before the drying process without heat. However, this method is useful only for materials with high moisture contents and which have a barrier for the loss of water without important loss of other solutes; also, the inclusion of salt or sugars in the material is not always desirable.

Enzyme activity: During the drying of fermented foods, the activity of enzymes may cause important modifications before the inactivation or the removal of water. This is especially important in sliced or grated foods that will be dried, such as mushrooms or cheese. These reactions may be prevented by drying in an inert atmosphere, by using an additive, or by blanching the material prior to drying. Additives may be antioxidants, such as ascorbic acid, or sequestrants, such as EDTA, and blanching is the quick immersion of the food in hot water (the idea is to heat just the surface), which both washes away substrates and inactivates enzymes that catalyze reactions, such as enzymatic browning, a classic cause of mushroom darkening during drying.

Shrinking: If the control of drying rates through temperature is not enough to prevent the shrinkage of liquids that tend to form gels, an alternative is the use of an inert bulking agent, such as simple sugars, starches, cellulose, talcum, gypsum, calcium carbonate, or silica. The liquid can be sprayed over the solid, which will then be mixed. Another possibility is the use of mixed-mode spray driers (countercurrent with a fluidized bed section below the nozzle).

Cryoprotective agents: In lyophilized solids, it is often the freezing step that causes the most damage to cells and tissues. This damage reduces the viability of microorganisms, which is undesirable for starter cultures, for example. There are several substances capable of enhancing the cell viability after lyophilization, usually sugars, such as trehalose or lactose. The action of these agents is not completely understood, but one of the enhancing effects of these solutes is the formation of glasses, avoiding the formation of large crystals, which could cause mechanical damage to cells. Another effect is the substitution of water interacting with phospholipids in membranes, thus avoiding phase transitions in the membranes (Nounou et al. 2005). Finally, the extra solids may avoid collapse and ensure that a highly hydrophilic porous solid will be created, facilitating rehydration.

Wetting agents: The drying rate depends on the capillary permeation of liquid in porous solids. The effect of tensoactive molecules, such as sorbitan monooleate, in drying processes is to reduce the surface tension, enhancing wettability and vaporization.

9.6.5 OTHER DRYING METHODS

Considering that drying involves heat transfer to the wet material, there are drying methods that rely on radiative energy transfer; also, drying operations may be allied to other processes in order to give the desired final product.

Electromagnetic radiation: Driers may be equipped with microwave magnetrons or infrared sources as heating elements. Microwave ovens operate at 2.45 GHz (or 5.8 GHz in some new equipment), using electromagnetic radiation to heat foods. Microwave driers use the same principle to heat the material to be dried; the microwaves partially penetrate the material, heating it efficiently. In low-fat materials, the absorption of energy is proportional to the moisture level; however, temperature control on the solid may be difficult, and the equipment should be routinely checked against microwave leakage. Because of the penetrating effect, microwave burns are insidious.

Infrared radiation has a very low penetration depth but is easily generated using resistors, lamps, or even flames; beside that, infrared is intrinsically safer than microwaves but less effective in the heating effect. Also, because the exposed surface will dry first and will be kept heated, care must be taken with the drying of sensitive materials. Both infrared and microwave sources are useful for tunnel driers or processes where the heat-exchange surface is small or the solid has a poor conductivity.

Foam drying: In this process, a liquid is dried in foam form; this may be done by forming a foam with air or other gas and then drying the foam in an indirect contact drier or using low pressure evaporation inside a lyophilizer to create foam with continuous collapsing until the dispersion thickens and solidifies. Controlling the temperature and drying rate, it is possible to obtain materials with variable surface area, lower than that obtained in spray drying or lyophilization with enhanced stability. The drying rates are superior to lyophilization.

Explosion puffing: In this process, the material to be dried is heated directly with high-pressure saturated steam and then released into a chamber with atmospheric pressure. The water inside the material instantly expands, exploding the material and creating a drier, porous mass. This process is useful for structured materials that would form a paste when milled or when expansion of the material is important as in puffed cereals. This process may be applied to mushrooms, fruits, and vegetables; other gases or very rapid heating may be used to generate the pressures that expand the material.

Solar drying: In this process, radiant energy from the sun is used as source of heat. Although the process is subject to weather changes and may be difficult to control, it is very cheap and used for several agricultural products, which may be dried directly under sunlight (as is done for fermented cacao and some fermented cassava products). Other products, such as rice, may be dried under a roof, which protects the produce from rain but still radiates energy. The process uses relatively low temperatures (usually 40°C–50°C), relying on natural aeration. It is probably the most ancient drying process, and yet, with solar panels and proper control, it has a promising future in agro-industrial processing, especially in small and medium scales.

9.7 FUTURE DEVELOPMENTS

Although the operations described in this chapter are mature, constant improvement follows the understanding of the processes behind the separations and the availability of new technologies. That is the case in membrane development, where new polymeric materials and construction techniques increase the efficiency and throughput of the operations; the availability of more efficient vacuum systems, which enhance lyophilization and evaporation; or the use of new heat sources, such as microwave radiation, just described in the drying section.

One may guess the goal of these developments; beside the cost reduction desired in any process, we expect that separation processes will focus on the following:

- Greener processes; reducing energy intensity by pretreatment steps, such as osmotic dehydration prior to drying; or membrane distillation prior to traditional distillation or evaporation
- Reduction of emissions through recycling of streams
- Processing with one eye on formulation, such as in foam and spray drying
- Better integration of operations aimed at the reduction of separation steps, especially for higher value-added products, such as natural food additives
- The use of smaller, highly controlled operations, especially for relatively small batches of niche products, such as probiotic and prebiotic foods

REFERENCES

Atkinson, B. and Mavituna, F., *Biochemical Engineering and Biotechnology Handbook*, 2nd Edn. London: Macmillan Publishers, Ltd., 1991.

Badger, W. L. and Banchero, J. T., *Introduction to Chemical Engineering*, 1st Edn. Tokyo: McGraw-Hill, Inc., 1955.

Belter, P. A., *Bioseparations: Downstream Processing for Biotechnology*. John Wiley & Sons, Inc., 1988.

Bender, F. and Moll, A., Modifications of structured packings to increase their capacity, *Trans IChemE, Part A, Chem. Eng. Res. Des.*, (81A) 2003: 58–67.

Bui, V. A., Nguyen, M. H. and Muller, J., A laboratory study on glucose concentration by osmotic distillation in hollow fibre module, *J. Food Eng.*, (63) 2004: 237–245.

Chanachai, A., Meksup, K. and Jiraratananon, R., Coating of hydrophobic hollow fiber PVDF membrane with chitosan for protection against wetting and flavor loss in osmotic distillation process, *Sep. Pur. Technol.*, (72) 2010: 217–224.

Chen, W., Solid liquid separation via filtration, *Chemical Engineering*, 104 (2) 1997: 66–72.

Choudhury, A. P. R. and Dahlstrom, D. A., Prediction of cake washing results with continuous filtration equipment, *AIChE J.*, (3) 1957: 433.

De Loggio, T. J. and Letki, A. G., New directions in centrifuging, *Chemical Engineering*, 101 (1) 1994: 70–76.

Deng, Y., Song, X. and Li, Y., Impact of pressure reduction rate on the quality of steamed stuffed bun, *J. Agr. Sci. Tech.*, (13) 2011: 377–386.

Fischer, L., Buhlman, U. and Melcher, R., Characterization of high performance structured packing, *Trans IChemE, Part A, Chem. Eng. Res. Des.*, (81A) 2003: 79–84.

Han, Z., Xie, J., Pan, Y., Effect of pressure fluctuation on vacuum cooling performance of cabbage. Nongye Gongcheng Xuebao/Transactions of the Chinese Society of Agricultural Engineering (25) 2009: 313–317.

Harrison, R. G., Todd, P., Rudge, S. R. and Petrides, D., *Bioseparations Science and Engineering*, 1st Edn. New York: Oxford University Press, 2003.

Huber, E. and Laurindo, J.B., Weight loss of precooked chicken breast cooled by vacuum application, *J. Food Processing Eng.*, (28) 2005: 299–312.

Jackman, P., Sun, D. W. and Zheng, L., Effect of combined vacuum cooling and air blast cooling on processing time and cooling loss of large cooked beef joints, *J. Food Eng.*, (81) 2007: 266–271.

Kimball, G., Direct vs. indirect drying: Optimizing the process, *Chemical Engineering*, (108) 2001: 5.

Lander, R., Daniels, C. and Meacle, F., Efficient, scalable clarification of diverse bioprocess streams, *Bioprocess International*, (11) 2005: 32–40.

Lavis, G., Evaporators: How to make the right choice. *Chemical Engineering*, April 1994: 93–102.

Leme, F. P., *Teoria e Técnicas de Tratamento de água*. São Paulo: CETESB, 1979.

Mansouri, J. and Fane, A. G., Osmotic distillation of oily feeds, *Journal of Membrane Science*, (153) 1999: 103–120.

Mulder, M., *Basic Principles of Membrane Technology*. Dordrecht: Kluwer Academic Publishers, 1996.

Najafpour, G., *Biochemical Engineering and Biotechnology*, 1st Edn. Oxford: Elsevier Kindle Edition, 2007.

Nounou, M. M., El-Khordagui, L., Khallafallah, N. and Khalil, S., Influence of different sugar cryoprotectants on the stability and physico-chemical characteristics of freeze-dried 5-fluorouracil plurilamellar vesicles, *DARU Journal of Pharmaceutical Sciences*, (4) 2005: 13.

Ozturk, H. M. and Ozturk, H. K., Effect of pressure on the vacuum cooling of iceberg lettuce, *Int. J. Refrig.*, (32) 2009: 402–410.

Pandey, P. and Chauhan, R. S., Membranes for gas separation, *Prog. Polym. Sci.*, (26) 2001: 853–893.

Perry, R. H. and Green, D. W., *Perry's Chemical Engineers' Handbook*, Eighth Edn. McGraw-Hill, 2008.

Primo-Martin, C., de Beukelaer, H., Hamer, R.J., van Vliet, T., Fracture behaviour of bread crust: Effect of bread cooling conditions *J. Food Eng.* (89) 2008: 285–290.

Prodanić, B. B., Jokić, A. I., Marković, J. D. and Zavargo, Z. Z., Improving the economic performances of the beet-sugar industry, *Acta Periodica Tecnologica*, (39) 2008: 55–61.

Rao, Y. V. C., *Chemical Engineering Thermodynamics*, 1st Edn. Mumbai: Universities Press, 1997.

Shaw, F. V., Fresh options in drying, *Chemical Engineering*, (101) 1994: 7.

Sokhansanj, S. and Jayas, D. S., Drying of foodstuffs, in *Handbook of Industrial Drying*. CRC, 2006.

Spiegel, L. E. and Meier, W., Distillation columns with structured packings in the next decade, *Chem. Eng. Res. Des.*, (81) 2003: 39–47.

Sun, D. W. and Zheng, L., Vacuum cooling technology for the agri-food industry: Past, present and future, *J. Food Eng.*, (77) 2006: 203–214.

Tao, F., Zhang, M., Hangqing, Y., and Jincai, S. Effects of different storage conditions on chemical and physical properties of white mushrooms after vacuum cooling. *Journal of Food Engineering* (77) 2006: 545–549.

Taylor, T. A., Heldman, D. R., Chao, R. R. and Kramer H. L., Simulation of the evaporative cooling process for tortillas, *J. Food Process Eng.*, (21) 1998: 407–425.

Treybal, R. E., *Mass Transfer Operations*, Third Edn. McGraw-Hill, 1981.

Ulbricht, M., Advanced functional polymer membranes, *Polymer*, (47) 2006: 2217–2262.

Urbaniec, K., The evolution of evaporator stations in the beet-sugar industry, *J. Food Eng.*, (61) 2004: 505–508.

Yu, K.C., Chang, H.S., and Lai, S.M., Discussion on the vacuum pre-cooling rate of vegetables. *Appl. Mechan. Mat.* (44) 2011: 1633–1640.

Zivdar, M., Fard, M. H. and Prince, R. G. H., Evaluation of pressure drop and mass-transfer characteristics of a structured packing for production and separation of food flavours, Part I: Pressure drop characteristics, *Food and Bioproducts Processing*, 84 (C3) 2006a: 200–205.

Zivdar, M., Fard, M.H. and Prince, R.G.H., Evaluation of pressure drop and mass transfer characteristics of structured packing for production and separation of food flavors, Part II: Mass transfer characteristics, *Food and Bioproducts Processing*, 84(C3) 2006b: 206–212.

10 Instrumentation and Control of Industrial Fermentative Processes

Alexandre Francisco de Moraes Filho,
Alysson Hikaru Shirai, Wilerson Sturm,
and Dario Eduardo Amaral Dergint

CONTENTS

10.1   INTRODUCTION

The use of automated systems, such as sensors, actuators, and controls, specifically in the food industry involves some peculiarities. In most other industrial processes, chemicals are inorganic. But the food industry will generally be processing organic materials. In some cases, the control system must maintain the conditions necessary for the life of microorganisms; in others, the reverse process, for example, the deactivation of microorganisms or enzymes, is true. In this context, all the elements of the system should be capable of supporting, for instance, the temperatures of sterilization and the presence of liquid for washing and cleaning, which may be at high pressure. Thus, the materials used in casings must be in accordance with the standards and also achieve mechanical and chemical resistance when faced with aggressive substances. These materials cannot provide conditions for the proliferation of bacteria and other pathogenic organisms; generally used materials are stainless steel or glass. These characteristics affect the quality and price of system elements.

An automated system is basically formed by sensors, control systems, actuators, and an industrial network that interconnects all these elements. This chapter presents some of the most common devices in the industry.

10.2 SENSORS

These are the sensorial components of the system, feeding back to it, so that an automatic or human intervention will keep it under control. A sensor is composed of one or more elements in order to convert mechanical, thermal, magnetic, optical, chemical, or even electrical energy into electrical signals. The device that directly converts a form of energy into something else that can be presented by a sensor is called a transducer. Primary sensing elements can be constructed so that one of their electrical characteristics, such as resistance, capacitance, or inductance, changes under the action of a physical quantity of specific interest. This section will present some physical quantities that are very important for food industries and also for several others and their respective sensors.

10.2.1 Temperature

The need to measure temperature is present in virtually any industrial process whether for security, for the equipment's protection, or by the process conditions themselves. In bioprocesses, this need is reflected in the condition of the microorganisms' survival. In fermentation processes, the heat generation or the maintenance of a certain temperature value is essential for determining their behavior, for instance. Following are some solutions for the measurement of temperature.

10.2.1.1 Thermocouples

One way of converting thermal energy into electricity was discovered in 1821 by Thomas J. Seebeck. He realized that with a couple of different metal wires attached at two points (or junctions) forming a loop there was an electrical current if such ends were subjected to different temperatures. If this circuit is opened at one of the joints, it generates a voltage proportional to the temperature gradient and dependent on the metals used (the Seebeck effect) (Webster 2009; Sturm 2004). A thermocouple is a sensor based on the Seebeck effect where metals or alloys are chosen to meet certain characteristics, such as high sensitivity, linearity, and resistance to oxidation (Pallas-Areny and Webster 2001).

The response of the thermocouple voltage can be approximated by the equation $V = kT$ where k is the temperature coefficient or Seebeck and T is the difference in temperature between junctions. It happens that such response is not linear; that is, the coefficient k is not constant over the entire range of operation. Thus, such sensors have a relative linearity, which is subject to certain temperature ranges. To facilitate its use, there are standard tables according to the type of thermocouple, indicating the output voltages for each hot-junction temperature relative to $0°C$. The value of k in a given range can be considered constant and defines the sensitivity of the sensor. What is desired is that for small variations of temperature great variations of electric tension occur. The higher the coefficient, the greater the sensitivity and, in turn, the higher the resolution of the device. The thermocouples have coefficients on the order of tens of microvolts per degree Celsius, and Seebeck voltage on the order of tens of millivolts are common (Sturm 2004; Pallas-Areny and Webster 2001). Following are the compositions, ranges of use, and maximum output voltages for some of the most used types (Pallas-Areny and Webster 2001):

Type S: Platinum–rhodium–platinum, 0°C to 1538°C, 16 mV
Type K: Chromel–alumel, –184°C to 1260°C, 56 mV
Type J: Iron–constantan, 0°C to 760°C, 43 mV
Type T: Copper–constantan, –184°C to 400°C, 26 mV

Although the cited sensors are the most common, specifically in the area of food production, the types T and J are the most used because their temperature range is the most suitable for these applications.

A thermocouple generates a voltage corresponding to the difference in temperature between the hot junction (point of measurement) and the reference junction (reading point). Thus, it is necessary to measure the reference junction temperature to obtain the absolute temperature of the hot junction. Nowadays, an electronic circuit with a thermistor, for example, compensates for this reference temperature, remembering that this device indicates the difference in temperature between junctions originally (Norton 1998).

Regarding the installation of the thermocouple, some care must be taken. For instance, the thermocouple has polarity. In addition, if it is necessary to connect cables that are too long (which would not be cost-effective), extension cables (same material) or compensating cables (cheaper, but with the next thermal response of the original materials) can be adapted; other cables would cause errors in the Seebeck voltage because of the formation of new joints. Physically, the hot junction is usually protected by a rod (stainless steel). This junction may or may not be in electrical contact with such a rod, but it will always be thermally coupled to it (the rod will be filled with a compacted ceramic, such as magnesium oxide) (Sturm 2004; Pallas-Areny and Webster 2001). Figure 10.1 shows one example of external protection. Despite the small output voltages and low sensitivity afforded by the thermocouple, advantages, such as ease of fabrication, fast response, robustness, a wide range of use, and low cost, justify the fact that they are the most common temperature sensors currently used in industry (Sturm 2004; Norton 1998).

10.2.1.2 Resistive Temperature Detectors

The conductors, when heated, present an increase in resistivity because of the higher vibration of the atoms because the free electrons inside collide more if an electric current is established. Thus, its electrical resistance is directly proportional

FIGURE 10.1 External temperature sensor protection.

to temperature. Resistive temperature detector (RTD) sensors are based on this physical principle, and they are produced from a metal wire wound on a ceramic or glass base or from a thin metal layer (film) deposited by so-called "sputtering" on a ceramic substrate (because of their constructive aspect, they are sensitive to mechanical vibrations). The resistance of this sensor is relatively low and, according to the distance of installation, may require physical links with three or four wires to reduce the influence of the resistance of its terminals when compared to that affected by temperature (Sturm 2004). The metals most commonly used to obtain RTDs are platinum, nickel, and copper. Their work ranges are −200°C to 850°C for platinum, −200°C to 260°C for copper, and −80°C to 320°C for nickel. Platinum is a metal that is especially suitable for this application because of its high linearity between resistivity and temperature, high repeatability, and wide range of use. The platinum sensor, also known as platinum resistance thermometer (PRT), has the best accuracy and does not require periodic calibration. The industrial model Pt100 (a platinum sensor whose resistance is 100 Ω at 0°C with a sensitivity of 0.4 Ω/°C), after the thermocouple, is the temperature sensor most commonly used today. This type of sensor is one of the most often used in the food industry also because of its temperature range, easy installation, and price. Versions in nickel and copper do not exhibit the same performance as the platinum, but they are often employed because of their low cost (Pallas-Areny and Webster 2001; Norton 1998).

10.2.1.3 Thermistors

The electrical resistance of a semiconductor is more dependent on the amount of electric charge it carries than the vibration of the atoms that constitute it. Thus, the natural behavior of a semiconductor is to reduce its electrical resistance with increasing temperature because it will have a greater number of free electrons or holes resulting from this increase in thermal energy. Thermistors are sensors that explore this feature of the semiconductor and are divided into two categories: negative temperature coefficient (NTC) and positive temperature coefficient (PTC) (Pallas-Areny and Webster 2001). For NTC thermistors, the electrical resistance decreases with increasing temperature. They are made from semiconductor materials, such as nickel oxide, cobalt, manganese, and copper, whose more usual application is for temperature measurement in the range of −55°C to 150°C. They have a high sensitivity, greater than that of thermocouples and RTDs, and, therefore, a high resolution for temperature measurement. However, their nonlinear response (exponential) limits their range of use, and precise values are always close to a certain temperature (such a disadvantage can be minimized if linearization circuits are used) (Pallas-Areny and Webster 2001; Sturm 2004).

Thermistors can be manufactured with a very small physical size and a relatively high electrical resistance. Thus, they have some advantages, such as long-distance, two-wire connections, fast response, and easy incorporation with other elements or equipment. These sensors are, by far, the cheapest for temperature measurement, but they also have low linearity, are less stable than other (self-heating) sensors, are more fragile, and require greater care in relation to the calibration (Norton 1998; Sturm et al. 2008). Because of these characteristics, they are more commonly used for indication only. In other words, they are often used in refrigeration chambers for monitoring, for example, in the dairy processing industry, where temperature monitoring is crucial and its value does not exceed the limit of the sensor, or also in the meat-processing industry, both in storage and in the process itself.

10.2.1.4 Integrated-Circuit Temperature Sensors

Some electronic component manufacturers provide temperature sensors based on semiconductors. The advantage of such devices is to integrate the sensor and its signal conditioning circuits (such as amplifiers, linearization circuits, and other adjustments) into one housing. Such sensors have high linearity and accuracy and usually do not require calibration. They are inexpensive compared to other types previously seen but have relatively limited ranges of operation. Their use is practically the same as for the thermistors, considering their temperature range. The following are three components that already exist (National Semiconductor Corporation 2006; Sturm 2004):

LM35: Manufactured by National Semiconductor, has a range of −55°C to 150°C, an accuracy of 1°C, and a sensitivity of 10 mV/°C

LM235A: Manufactured by National Semiconductor, has a range of −40°C to 125°C, an accuracy of 1°C, and a sensitivity of 10 mV/K

ADQ590: Manufactured by Analog Devices, has a range of −55°C to 150°C, an accuracy of 1°C, and a sensitivity of 1 μA/K

10.2.2 Pressure

Pressure sensors are required in the food industry for several processes. They appear directly in sterilization applications but also in machines that transform the products.

10.2.2.1 Piezoelectric Pressure Transducers

Some natural crystals, such as quartz and tourmaline, when subjected to tensile or compressive forces (mechanical elastic deformation), generate a voltage proportional in amplitude and polarity. The reverse effect also occurs, that is, materials that stretch or compress depending on the external electric fields applied to them. This phenomenon is known as the piezoelectric effect and was first observed by Pierre and Jacques Curie in 1880–1881 (Pallas-Areny and Webster 2001). The pressure sensors based on this effect transform the amplitude of the force of deformation into an electric voltage, as the area can be previously determined; the result is the measure of an electrical signal proportional to the pressure applied (Sturm 2004). The conversion of mechanical energy into electricity by a piezoelectric transducer is limited. If a constant force is applied to the strain element, the internal electric particles responsible for the generation of electrical voltage are neutralized. Thus, the transducer is limited to measuring only variations in pressure (dynamic pressure); it is not possible to measure static pressures (Webster 2009). This way, their use is limited, for example, to presses, which are present in certain segments of the food industry.

10.2.2.2 Strain-Gauge Pressure Transducers

Conductors and semiconductors, when subjected to mechanical stresses, have their electrical resistance changed. This is justified not only by the dimensional variations suffered by the element when compressed or stretched but also by the change in resistivity because of mechanical stress (a piezoresistive effect) (Pallas-Areny and Webster 2001). The sensors based on this effect consist of a piezoresistive wire, which is firmly fixed on a base element (aluminum or other metal) with a grid-shaped layout that is as compact as possible. Such a device is known as a strain gauge or extensometer and is able to detect extremely small deformations (caused by mechanical forces) because of proportional variations in electrical resistance. Some pressure sensors use strain gauges so that these forces are converted into pressure. These elements allow the measurement of static and dynamic pressures (Sturm 2004).

10.2.2.3 Capacitive Pressure Transducers

The capacitance of a parallel plate capacitor depends on the area of overlapping plates, the distance between them, and the dielectric constant (the insulating material between the plates). Capacitive pressure sensors are based on the variation of the distance between the capacitor plates, depending on the pressure to which they are subjected. Normally, such a capacitive element determines the oscillation frequency of an electronic circuit, so frequency variations correspond to variations of pressure. These sensors have high sensitivity and resolution, having some advantage over the strain-gauge sensors because they are much less dependent on temperature (this dependence is higher or lower depending on the dielectric material) (Pallas-Areny and Webster 2001). This type of sensor is one of the most common solutions to measure liquid or gaseous material pressure in the food and beverage industry, for instance, as carbon dioxide addition in soft drinks or pasteurization by steam injection.

10.2.3 Level

The level measurement is the food industry's main function to indicate the quantity, both in stock and in the dosage and preparation processes.

10.2.4 Discrete Level

The following sensors include some principles for the detection of discrete level; that is, they are able to detect the presence or absence of a liquid or solid material in a given height of a tank, also known as point-level sensors.

10.2.4.1 Photoelectric Level Sensing

This type of sensing is based on the emission and subsequent detection of a beam of light. The light source is usually pulsating and obtained with the use of light emitting diodes (LEDs), which are more commonly used in the infrared range. Once emitted, such light sometimes goes through a converging lens to increase its reach, directly or by reflection, to get to the receiver circuit. This circuit consists of photodiodes or phototransistors, which are sensitive to incident light and tuned according to the emitted light pulses (frequency of the transmitter). The photoelectric sensor detects the presence of a liquid or solid material when the return of light is insufficient. A model widely used to detect the sensor level is a retro-reflected beam. It has a transmitter and a receiver in one housing assembly and a tip made of glass or polymer, which serves as a probe. Figure 10.2 shows the sensor elements. The light reflects

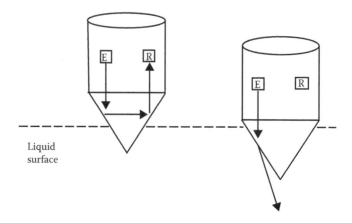

FIGURE 10.2 Photoelectric level sensor elements.

inside the probe so that when it is immersed in a liquid, for example, the internal beam will suffer refraction, and the sensor will be triggered. The electrical insulation between the sensor and the environment allows the application of this sensor, including the detection of flammable liquids. They also have a fast response, have high noise immunity, and can be installed horizontally or vertically, permitting the detection of several different levels (Sturm 2004). It can be used, for example, to detect the level of beer because it does not detect the foam, only the liquid.

10.2.4.2 Conductivity Level Sensing

In applications involving the detection of conductive liquids, sensors based on vertical measuring rods can be a very practical solution, commonly for the detection of water because of its extensive use in the food and beverage industries and because of their cost-effective relationship. Basically, they are conductive rods of different lengths isolated from each other, and the electrical contact between them is given by the liquid when it reaches a certain level. Most are made of stainless steel and polarized with alternating current to avoid electrolysis. Figure 10.3 shows the sensor basics. These sensors are easy to install, have no moving parts, and usually require frequent cleaning. Sensors are provided with movable rods, fixed with multiple (six) rods, adjustable stems in situ, and versions coated against corrosion, among other possibilities (Krishnaswamy and Vijayachitra 2005; Sturm 2004).

10.2.4.3 Vibrating Fork Level Sensing

These are sensors with a rod shaped like a tuning fork. These devices vibrate at a natural frequency when in contact with air from an electronic system used to generate the oscillations. When the rod comes into contact with a solid or liquid, there is a change of frequency and detection takes place. They do have restrictions, such as materials that can become embedded in the rod, indicating continuous contact. On the other hand, they have advantages, such as immunity from product power,

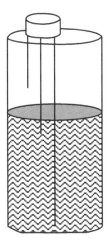

FIGURE 10.3   Conductivity level sensor basics.

robustness, and ease of installation (Krishnaswamy and Vijayachitra 2005; Sturm 2004). They also have interesting characteristics when there is need for detection of powder materials, such as color additives and other ingredients.

10.2.5   CONTINUOUS LEVEL

These are able to measure the level continuously, providing a metric measure related to a given volume of material contained in a tank.

10.2.5.1   Capacitive Level Sensing

These are based on a capacitive element whose dielectricity depends on the level of a liquid or solid. Thus, a rod is introduced into a tank to form the plate of a capacitor (when the tank is conductive, it can be used as one of the plates). The material fills the space between the plates (the dielectric medium) according to its level, thus varying the capacitance of the sensor. This capacitance determines the oscillation frequency of an electronic circuit, so variations in frequency are the corresponding level. One advantage of such sensors is that they can be used in tanks with high pressure and temperature as well as flammable materials (Krishnaswamy and Vijayachitra 2005; Sturm 2004). They have features that allow their use in showing the quantity of a very large group of materials, such as milk, sugar, wheat flour, color additives, sweeteners, acidulants, thickeners, and others.

10.2.5.2   Ultrasonic Level Sensing

The operating principle of the ultrasonic sensor is the emission of an ultrasonic wave (sound above the audible range) and the measure of its flight time, that is, the time between wave emission and reception of its echo from the shock with some material.

FIGURE 10.4 External ultrasonic sensor housing.

Because the speed of wave propagation in the medium is known (usually air) with the measured time, we get the distance, and then the level can be determined. Most sensors use a piezoelectric crystal as a sender and receiver, which receives a pulse train for a short period of time to send the wave and then is kept in standby mode to capture the echo (Norton 1998). Figure 10.4 shows one example of an external housing. These sensors have the great advantage of not coming into contact with the material and may be used for liquids or solids. However, they are sensitive to air currents and movements and susceptible to temperature changes (Krishnaswamy and Vijayachitra 2005; Sturm 2004).

10.2.5.3 Level Measurement by Weighing

Pressure sensors can be used for continuous level measurement. One method employed is the positioning of a sensor in the bottom of a container in order to capture the column pressure of a given liquid, converting it to level indication (Norton 1998). It is practically independent of the material to be measured.

10.2.6 Flow

The flow measurement is associated with determination of the quantities being used in a process. This amount is related to the elapsed time. In fermentation processes, these quantities may determine various parameters associated with their actual operation, as respirometric data, for example.

10.2.6.1 Thermal Dispersion

An RTD can be used indirectly to measure the flow of a gas or liquid. The element is heated and controlled by an electronic circuit and the fluid flowing through it; there is a drop in temperature in proportion to the speed of the fluid (thermal convection). Usually, such sensors use a temperature compensation circuit (Wheatstone bridge) with a referenced element at room temperature. They are compact, have no moving parts, and are resistant to vibration with the one inconvenience of the mechanical resistance to fluid flow, which can cause loss of pressure or turbulence (Norton 1998; Sturm 2004).

10.2.6.2 Electromagnetic Flowmeters

According to Faraday's law of induction, when a conductor of length L is going through an orthogonal magnetic field with density B (N and S as shown in Figure 10.5), there is an induced voltage proportional to the displacement velocity

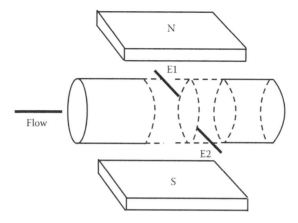

FIGURE 10.5   Electromagnetic flowmeter elements.

u ($e = BLu$). Electromagnetic flow sensors are constructed so that the moving fluid acts as a conductor. A nonconductive section of the tube is subjected to a magnetic field. A voltage is produced proportional to the velocity of the fluid and detected by electrodes placed on the sides of the tube, E1 and E2, as shown in Figure 10.5. The fluid needs to be a conductive liquid, and the magnetic field's generating coils are usually excited in AC to avoid electrolysis. This sensor has the advantage of being minimally invasive by electrodes and has small pressure loss in addition to not having moving parts (Webster 2009; Sturm 2004). They are widely used for the measurement of water as well as beverages and other liquid components, although they are relatively expensive.

Following are covered sensors that measure the flow from the generation of a pressure difference between two points (dynamic pressure).

10.2.6.3   Pitot Tubes

A Pitot tube is a device inserted into the duct of a fluid; the format, for a better understanding, can be considered U-shaped. One of its sides is positioned facing the flow and measures the so-called impact pressure. The other end opens perpendicular to the flow by measuring the static pressure. The difference between these pressures is proportional to fluid velocity and therefore flow. Pressure sensors previously mentioned can be installed in these tubes, and from an appropriate electronic circuit, an electrical signal proportional to the flow can be obtained directly (Krishnaswamy and Vijayachitra 2005; Sturm 2004).

10.2.6.4   Venturi Tube

A Venturi tube is a gradual restriction in a pipe installed in series with a duct. The fluid flow by constriction leads to an increase in its speed and the consequent reduction of its static pressure. The pressure difference in this region of bottlenecks and beyond is proportional to fluid velocity and, hence, its flow. These elements have the advantage of a reduced pressure loss and do not impose barriers to the passage of fluid (not retaining particles in suspension) (Norton 1998; Sturm 2004).

10.2.6.5 Orifice Plate

An orifice plate consists of a plate with a hole, which is inserted into the duct and, because of narrowing, causes a differential pressure. The shape and size of the hole depend on the application, characteristics of fluid, and flow value. The plate is mounted between two rings containing holes for taking the pressure. As in two previous cases, the pressure difference measurement can be made by something as simple as a manometer, or it may be something more sophisticated, such as electronic sensors. Such a device has advantages, such as simplicity and relatively low cost. Its main limitation is to generate considerable pressure loss in addition to mechanical wearing on the board caused by the passage of fluid (Norton 1998; Sturm 2004).

10.2.7 HUMIDITY

Concerning to fermentation processes, humidity control can be as critical as temperature control, depending on the bioreactor type, to assure adequate performance because of its influence on oxygen transfer, which can cause a reduction in the growth rate (Pandey 2004). Two types of humidity sensors are more common for this subject; their operation is based on a sensitive polymer coating over an electrode baseplate (Soloman 1998). One type is the capacitive sensor; depending on the humidity, the capacitance value varies, and thus the output signal changes. The other type is the resistive output sensor, which changes electrical resistance according to the humidity variation. In comparison with the capacitive sensor, this sensor must have a slower output response. The capacitive humidity sensor is faster than the resistive; however, it is normally more expensive (Sturm et al. 2008).

10.2.8 WATER ACTIVITY

In most processes, it is important to know that water activity is closely related but not equal to the water content exactly. Water content can be normally determined off-line by dry weight measurement, but this amount is not equal to water available for microorganism activity. In order to measure water activity, one of the usual methods is the capacitive sensor, designed to regulate the circulating air humidity (Christen et al. 2003).

10.2.9 pH

The pH variations indicate changes in metabolic activity in the process (Christen et al. 2003). Normally, it is measured off-line; flat-ended electrodes are used to determine pH directly on the surfaces of solid substrates. It can be determined by measuring the pH of aqueous suspensions or extracts of the solid sample (Scheper and Berovic 2000). New pH sensors have been developed, which have no necessity of temperature compensation and filling of liquid solution (Myrvoll 2003). But pH value is still difficult to measure in some kinds of processes because of the lack of free water.

10.2.10 O_2 AND CO_2

In fermentation processes, the O_2 consumption and CO_2 production allow the estimation of the kinetics of microorganism growth (Pandey et al. 2001). Instead of many other gases, oxygen has a strong affinity for magnetic fields (Khandpur 2003). In the fermentation process, it is the only gas that is paramagnetic; hence, this is the characteristic used to measure it. In a deflection sensor, the paramagnetic behavior is compared to other gasses, both involved in a magnetic field. The more concentrated the oxygen sample, the more unbalanced the system, which is measured by mechanical device deflection. This kind of sensor is often quite expensive but very accurate. Another kind of oxygen sensor works like a metal/air battery (Jardine and McCallum 1994). When in contact with the cathode, oxygen is reduced into hydroxyl ions with a balancing reaction of lead oxidation at the anode. Through this chemical reaction, an electrical current is generated proportionally to the rate of oxygen consumption; this property determines the sensor's lifetime because of the availability of lead, and it will stop working when all the lead is oxidized. The thermal-type sensor is also based on the oxygen paramagnetic property; its flow changes the thermal balance of a temperature sensor, proportionally to gas concentration, which is turned into an electrical output signal (Christen et al. 2003). When facing the necessity of continuous measuring of CO_2, a correct choice could be a nondispersive infrared (NDIR) gas sensor. It is based on a single-path, dual-wavelength infrared detection system. Compared to other devices, it could be a low-cost and low-maintenance analyzer (Singh 2010).

10.3   CONTROL

A controlled variable must maintain its value, that is, stay at a stable value or as close as possible considering the tolerances, and when facing a disturbance, it must return to the previous state as soon as possible. In the study of control strategies, some terms will appear very often, such as set point (SP), which is the desired value of some variable under control to maintain; manipulated variable (MV), which is the variable that is changed by the control strategy to act most directly in the process; and present value (PV), which is the variable value at some particular time. To achieve these objectives, some control strategies are very common: on–off; proportional; proportional and derivative; and proportional, integral, and derivative. But there are some unusual problems, which demand updated solutions, such as neural network, genetic algorithm, and fuzzy logic.

10.3.1   ON–OFF CONTROLLER

One of the most rudimentary control methods is the so-called on–off control (Montague 1997). One example of this strategy could be the temperature control used in home electric ovens (Svrcek et al. 2006). The heating power is either fully switched on when the temperature is below the SP or fully turned off whenever it is above. Resulting from this behavior, the temperature oscillates around the SP; the amplitude and the time period of the oscillation depend on the specific process, or the thermal lag, that could exist between the heating energy source and the temperature sensor. To minimize the output switching frequency, as the measured temperature crosses the SP, the controller does not turn on and off at precisely the same point; it introduces a small gap known as hysteresis, the value of which depends on the process behavior to maintain the SP at the average value of the total oscillating curve. Into the hysteresis area, or dead band, the controller does not switch the output. On–off control is satisfactory only for the noncritical process applications where some oscillation in the present value is permissible (Sturm et al. 2008).

10.3.2 PROPORTIONAL, INTEGRAL, AND DERIVATIVE CONTROLLER

When facing a critical process application, on–off control would not be the correct choice because it could not achieve all the control objectives, such as stability, for instance. In this case, the proportional, integral, and derivative (PID) method should be the best option; it is widely used in industrial plants. To assure the global comprehension, this method will be analyzed in parts as follows.

10.3.2.1 Proportional Controller

To illustrate proportional controller behavior, a temperature process will be used as an example. In this case, the controller applies power to the heater, proportional to the difference in temperature variable between the process instantaneous value and the SP. This difference is usually called error, and the parameter P is known as the controller proportional gain (Svrcek et al. 2006). When the proportional gain increases, the system tends to react faster with new values in the SP or when dealing with a disturbance, but it becomes progressively faded and eventually unstable. The final process temperature value normally stabilizes below the SP because of the equilibrium between the energy supplied to the system and the resistances from it; the proportional amount of energy is too weak to reach the SP because the error value close to the SP used to be very small. This lasting difference is normally called steady-state error or offset.

10.3.2.2 Proportional and Derivative Controller

When using only a proportional strategy with a high gain adjust, the PV may stabilize above the SP, and some problems could appear, depending on the process. In order to minimize these problems, some terms should be added to the output signal or control signal (Omstead 2000). For example, a term proportional to the time derivative of the error signal can be added. It is the rate of error variation in relation to the time, and it is usually called derivative time, expressed in seconds (Sturm et al. 2008). Its response is faster than the proportional control alone because the derivative term's purpose is to anticipate the control action when facing a large error change in a narrow time gap. The amount of this anticipation can be adjusted through the derivative time parameter (Smith 2002). This technique is known as PD control, but it is not used very often by itself. It normally appears in association with an integral term. Nevertheless, it was applied in batch pH control processes (Svrcek et al. 2006). It must be kept in mind that the derivative term, used alone, is not a control strategy because its action is about the error-changing rate and not the error value itself; thus, one process could be stabilized at another point even if it is too far from the SP as long as the difference between SP and PV was not changing in time reference, so the control objective would not be reached.

10.3.2.3   Proportional and Integral Controller

Practically speaking, most processes cannot have an offset or steady-state error (Smith 2002). In order to eliminate this error in the proportional controller, an integral term can be used, which is sometimes called reset action. This term will increase the output signal, or control signal, as long as an error exists (King 2011). Depending on the manufacturer, the integral time can be expressed in seconds or minutes and indicates the amount of this term's influence over the control output signal. The integral action can cause a kind of delay compared with the proportional action strategy used alone.

10.3.2.4   Proportional, Integral, and Derivative Controller

In order to achieve the objectives of control strategy with some velocity, error antici-pation, and no offset, it is possible to combine the three terms in one output function. Figure 10.6 shows the probable behavior of this kind of control. The PID control can deal properly with the overshoot and stability problems in comparison with the pro-portional control working alone. It is also able to solve the problem with the steady-state error. The output control function becomes the following:

$$Sc = Pe + \frac{G}{Ti} \int_0^T e \, dt + Td \frac{de}{dt} + B$$

where Sc is the control output signal, P is the proportional band, G is a constant, Ti is the integral time, e is the error (SP-PV), Td is the derivative time, and B is a kind of bias (Bryce and Mansi 1999).

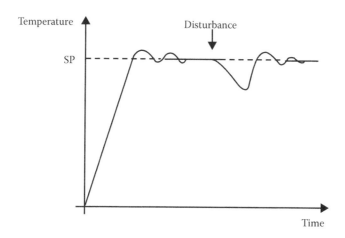

FIGURE 10.6   PID control behavior example.

10.3.2.5   Advanced PID Control

Over the years that it has been used, the PID technique has had some upgrades because of continuous development of new control strategies and facing new application challenges. Thus, some manufacturers offer more-specialized control algorithms. One of these novelties is the fuzzy logic algorithm working together with the usual PID (Sturm et al. 2008). This controller is able to solve some PID limitations, such as the problem of controlling highly nonlinear and time-variant plants, which is one example where fuzzy logic would offer certain advantages, as well as during the auto-tuning function that adjusts the PID parameters.

10.3.3   OTHER CONTROL TECHNIQUES

10.3.3.1   Fuzzy Logic

The concept of fuzzy logic was not first presented as a control technique but as a new way to process data (Zadeh 1965). Previously, classical logic was able to consider only sets where only membership or nonmembership options were possible. But Professor Lotfi A. Zadeh presented the possibility of partial set membership, explaining that people do not require precise numerical information input and, even when this information is dispersed, are capable of highly adaptive control (Zadeh 1990). Human beings are capable of understanding imprecise expressions, such as "tall," "expensive," or "not so far" (Sturm 2005), but a computer-processing mode is reduced to a one-or-zero, everything-or-nothing, or true-or-false mode of thinking. Professor Zadeh proposed the method of modeling linguistic variables, which suggests thinking as linguistic objects or words instead of numbers. This approach allows designing a whole process control strategy instead of just helping a PID system as exposed in the advanced PID example. It is often able to handle problems with poorly defined variables or when they are not exactly known, considering the process behavior as well as the nonlinearities that normally appear in bioprocesses.

10.3.3.2   Neural Networks

Most human abilities were learned by example (Norgaard et al. 2000). This is the principle of neural network control design. It is supposed to emulate the real human brain neural connections, so it needs a training process (Anthony and Bartlett 2009) where examples are inserted and supposed to be learned by the control system. This approach consists of a large number of neurons, simple program parts, richly interconnected with each other. Each neuron often has a very simple function to execute, but the number of connections will define its complexity. The synaptic weight, the parameter that links the neurons, together with the connections between them, will determine the network behavior—actually the pattern of synaptic weight values in association with all the connections in the whole network that carries real information. During the training process, the synaptic weight values are automatically adjusted (Priddy and Keller 2005). According to Norgaard et al. (2000), this approach is able to work with nonlinear systems and has the advantage of a learning capacity. Sometimes designers know the behavior of a specific bioprocess, but not each variable influences itself. Neural networks can be used through learning practical examples. The disadvantage is that the mathematical model will not be clearly understandable as an equation, for instance (Pandey et al. 2008).

10.4 ACTUATORS

An actuator is one of the control system devices that is closely in contact with the process itself and, thus, needs to be designed to facilitate sterilization, asepsis, insertion, and removal (Omstead 2000). In the specific case of food industry fermentation or similar processes, certain types of actuators can be installed outside the process. This section will expose the more complex and common actuators and will not discuss elements that are electrically and logically simpler, such as microswitches, limit switches, and contactors.

10.4.1 Motors

The field of industrial application of electric motors is wide and is normally used to allow any kind of movement. For instance, ventilation is indirectly or directly performed by a motor attached to an air pump or compressor as well as the rotating mechanical parts to achieve the proper mixing of the fermenting substrate (Pandey et al. 2008). Motors can be supplied by alternating current (AC) or direct current (DC), depending on the application, power, and availability, so in the industrial application, the AC three-phase supplied motors are more often used (Mittle and Mittal 2009). It is possible to change three-phase motor speed using an electronic device called a frequency inverter, which is nowadays more cost-effective; in other words, it allows the control of some process variables by changing the motor velocity. The use of a servomotor is generally prohibitive because of its costs. However, it has advantages, such as controlling speed and position with high accuracy.

10.4.2 Valves

A valve is an electromechanical device that is able to control a large scope of fluid components that are normally present in industrial processes, such as acid and base for pH control, for instance. Understanding this device's function is indispensable for instrumentation design, and there are many types of valves (Sturm et al. 2008). In order to automate the control loop, motorized and proportional valves could be the correct choice (Janocha 2004). These valves are able to change their status from completely closed to wide open, depending on the input signal, which is normally 4 to 20 mA as a control signal for proportional models. Proportional valves can be used, for example, to control the air inlet flow, component pressure, and other substance volume to be added to the process, achieving automatic control requirements, such as accuracy and resolution.

10.4.3 Heater Power Control

Instead of just turning the heater source on and off, there are two common ways to control the heater power in industrial applications: by phase angle and pulse-width modulation (PWM) (Ibrahim 2002). Phase angle is performed when the start of each main power supply half-cycle, applied to a resistance heater, is delayed by an angle. Concerning the electricity case, this angle means some amount of time, and it is important to keep in mind that each full cycle lasts approximately 16.6 ms, considering a frequency of 60 Hz. Electronic components are responsible for turning the heater on and off but do so very quickly, in other words, each main cycle; thus, the power that is delivered can be fully controlled. The PWM method, also known as the duty cycle, is able to turn on and off very quickly but, in this case, cutting a certain number of full cycles and leaving the other cycles directly applied to the heater element. These methods are often used in temperature control with electrical resistance heaters because they are able to apply between zero and the full available power to the process, depending on the controller signal. While allowing a more accurate and reliable control, these methods generate electrical noise, which can cause malfunction in some equipment (Rexford and Giuliani 2004).

10.4.4 PERISTALTIC PUMPS

The peristaltic motion of a pump chamber allows fluid to flow in the correct direction by squeezing it. Normally, peristaltic pumps need three or more pump chambers with actuating membranes (Nguyen and Wereley 2002). Because of their operation, peristaltic pumps can be used for all liquid materials and, in specific cases, even gases. Depending on the construction, the flow can vary greatly. This kind of pump is usually able to maintain the flow, so it is the correct choice for dosing components, such as acids and bases for pH control, sweeteners, color additives, or moisture control, for instance. These pump models are leak-free, depending on the flexible elastomeric element strength, facing mechanical stresses. This element should be checked and replaced regularly to avoid failure. To prevent mistakes in selecting the correct elastomer and casing material, it is indispensible to consider the chemical properties and temperature of the liquid (Nesbitt 2006).

10.5 SENSING NETWORKS

Nowadays, the selection of the communication protocol and the interconnection technology is as important as the selection of a sensor. Thus, the selection of the process control is more dependent on the communication systems and the control protocols than it used to be in the past, when the process control was basically defined on the sensor and its control algorithm. The varieties of network and distributed control systems that exist now are the result of two main factors. The first one is a result of the natural evolution of electronic devices, which made possible the construction of devices with higher processing capabilities and lower costs and power consumption to operate. The other reason is the increasing complexity of the control system that is required to manage the processes. There is interdependence between these two factors, but this discussion is not in the scope of this chapter. With the evolution of electronics and their dissemination, there was a popularization of sensors, but the same has not happened with communication protocols. Even in recent years, there has not been a consensus in terms of fieldbus-based networks. In this context, this section aims to present the alternatives of modern, industrial fieldbus networks, but it is worth noting that there is a market competition between different types of industrial networks. But even in simple projects, it is interesting to choose fieldbus-based communication given its inherent advantages, which allow the systems to be more flexible, scalable, and evolutionary. This chapter also illustrates the different approaches to sensing networks in industrial scope, where some well-known standards developed for use in industrial networks are discussed. The ZigBee protocol is then presented, which is a modern sensing methodology based on the formation of wireless sensor networks; it is a pattern that has shown great potential in this application area.

10.5.1 INDUSTRIAL COMMUNICATION NETWORKS

The selection of the communication system depends on many factors. Often the technical constraints are such that the system is already defined. Generally, the network definition depends on the composition of technical and economic aspects and also on political and cultural aspects of the company. The goal here is to present an overview of the main networks and their technical characteristics. Another problem that the designer can face is the diversity of control systems and industrial networks available on the market. There are systems that perform adaptation through a hybridization process (HART, for example, where 4–20 mA systems are endowed with digital communication) or an evolution process from simple buses (CAN, ASI, etc.), and there are native fieldbus systems that were born with the goal of providing complete network solutions for complex process control (PROFIBUS, WordFIP, etc.), even though many protocols work with the same physical level, that is, the same electrical configuration, and show some standardization at the superior layer, also known as the user layer. This superior layer permits interoperability between different manufacturers and vendors as will be discussed here. Thus, the systems differ primarily in the intermediate layers of communication, which generally comprise a data-link layer (medium access control) and an application layer (high-level methods to control the communication). Also, two other situations should be considered when designing a solution. The first is when there is no working control system implemented, so the project can be freely analyzed from technical, economic, cultural, and political aspects. The second situation is the case where there is already a working control system. In this situation, it is usually not feasible to replace the existing system with a new one because of cost and infrastructure. Normally in these cases, there is the possibility of using hybrid technology, such as HART protocol, for example, that allows an immediate replacement of 4- to 20-mA–based sensors and also offers the advantages of digital communication. Finally, a wireless alternative should also be considered. Although receiving critics regarding its reliability and the security of data transmission, these protocols are increasingly being used in industrial applications, especially in systems that have some kind of mobility.

10.5.1.1 Standardization and Trends

The vast majority of fieldbus standards are based on the International Organization for Standardization–Open Systems Interconnection (ISO–OSI) layers communication model. However, not all layers specified in this model are present in the fieldbus structure. In general, only layers 1, 2, and 7 are adopted. In addition to them, the user layer (layer 8) is also presented in many protocols, and it is the layer that is aimed at the interoperability of equipment from different manufacturers (Mackay et al. 2004). This last layer is not presented as an OSI model because it describes mechanisms of control and is not related to communication aspects. The other intermediate layers of the OSI model are not present in the fieldbus protocols because of the critical characteristics of fieldbuses: cost and response time. Adopting the OSI model does not mean that the systems are interchangeable. It simply means that it is possible to use devices as gateways, linking the communication protocols of two different networks, and is able to perform the connection of these networks. In this situation, the interconnectivity is guaranteed. Regarding the fieldbus, because it has only three layers, these devices would not be visible at different levels of the network except through artifices that complement the missing layers. In other words, this architecture needs intermediate systems that have all OSI layers implemented and can communicate with the rest of the system.

There are some standardization proposals for networks that have emerged with efforts conducted by international organizations, such as the International Society of Automation (ISA) and the International Electrotechnical Commission (IEC), to accelerate the convergence of adopting a chosen interconnection technology. In terms of standardization, the main interest is a system that meets the needs and, at the same time, does not become dependent on only one vendor. In this context, with this evolutionary process, some convergences have arisen. In terms of physical layers, the standards RS-485 and IEC 61158-2 show a strong presence. In the data-link layer, there is not a great convergence, although the solutions derived from FIP and PROFIBUS protocols are proclaimed as trends because of their utilization.

Other discussions related to the protocols includes the context of their application, which can be basically divided into manufacturing automation and process control. Manufacturing automation generally employs several subsystems that are responsible for a part of the overall process. In this sense, many materials are moved from one machine to another, showing some discontinuity in the activity. It is also logic control–intensive and often requires high-speed features. An example would be the automobile industry. Regarding process control, it can be understood as activities that show continuity, where the response time of the systems is not so fast. Typically, the transformation of raw materials through chemical reaction or physical changes is involved to make the final product. An example of this group is the chemical industry in general (Siemens Energy and Automation 2007).

Contrasting with industrial process control, the characteristics of the projects in the manufacturing environment are as follows: a fast response time, availability of a local power supply, usually not potentially explosive environments, and small distances between sensors or actuators and the control center. In this context, the PROFIBUS (DP version) has been largely employed as have simpler networks, such as CAN and ASI, which are interesting alternatives. Compared with the manufacturing environment, industrial process control has the following characteristics: large distances between sensors or actuators and the control center, a lower data sampling rate, and environments with difficulty having the power supply near the sensor elements. In this context, the WordFIP network is claimed to be more appropriate, mainly because of its medium access mechanism. However, PROFIBUS (PA version) is also specific to this context. Thus, it is up to the designer to evaluate the political and economic context and choose the network that best meets their needs. The time required to perform the exchange of the messages between sensors and actuators, the system's flexibility, and its scalability are some other criteria that can guide the designer when choosing the network for the system.

10.5.1.2   Interchangeability and Protocol Standards

The fieldbus systems normalize the transmission of data and the interconnection of equipment from several manufacturers in a given standard protocol. The user has the possibility of choosing among the manufacturers and the devices that best fit in their application. However, the user will remain linked to a specific protocol.

The interchangeability is usually achieved using the same protocol. Connections to communicate between different networks are not recommended, although there may be bridges or gateways (systems for interconnecting heterogeneous networks) that carry out this interconnection. Generally, communication is done horizontally on a network where there should be an intensive exchange of data. But on different networks, in theory, it is better that the central control makes the distribution of tasks and data, being transparent as the network and subnet devices are, thanks to devices that make the link to the networks with different protocols. According to the OSI model, an open system is modeled containing seven abstraction layers to perform the communication with other systems. Logically, if the protocols are different, there also have to be bridges. Each layer has to have its match, allowing a transparent communication. The point is that industrial networks do not support all the seven layers because of the processing required. Thus, the industrial fieldbus network has only the strictly necessary layers to perform its function (layers 1, 2, and 7). Even with a reduced number of layers, communication is possible, but not with the same flexibility as the seven-layer systems. Thus, in industrial networks, the communication is done with intermediate equipment, and this equipment communicates with the reduced stack devices (sensors and actuators).

Analyzing the current automation's needs, normally, the homogeneity of the standards is preferred. Regarding the protocols, there are protocols from particular vendors, others that are open standards, and there are also standards elaborated from normalizing associations. The following sections will point out the main standards presented at industrial contexts. It should be clear that there are dozens of others.

10.5.1.2.1   The 4–20 mA Standard and HART Protocol

The 4–20 mA standard was a consolidated protocol in industries, and it is still widely used. This analog standard allows long-distance connections and offers high immunity to electromagnetic interference. It makes use of current levels in the 4- to 20-mA range to transmit the information. The initial range of the current level is used to supply the circuit, which can also be used for fault indication, such as power fault or line interruption (Pereira 2004). In order to create a digital communication protocol compatible with the 4–20 mA infrastructure for transmitting information to and from sensor and actuator components, in 1986, Rosenmount created the so-called HART protocol (Highway Addressable Remote Transducer).

The HART became an open standard in 1990, and later, the protocol was organized by the HART Communication Foundation. Equipped with digital communication features, it was possible to interface HART devices with microprocessed systems. An important characteristic is that the HART signal does not cause interference on a 4–20 mA signal, making it possible to send both signals simultaneously. Thus, HART is considered a hybrid technology, that is, a transition between analog point-to-point systems and digital fieldbus systems. Given the wide variety of 4–20 mA devices that are still on the market, the HART protocol is still largely used (Pereira 2004; Helson 2004).

10.5.1.2.2 WorldFIP

The World Factory Instrumentation Protocol (WorldFIP) is a standard developed from the French standard NFC 46-600, usually known as FIP. The WorldFIP is completely defined as a fieldbus protocol and is part of the European fieldbus standard EN 50170. This protocol takes into account the real-time constraints imposed by a large number of applications at the factory-floor level. Regarding the OSI model, this protocol implements the following layers: physical, data link, and application. The physical layer follows the IEC 61158-2 standard. The data-link layer performs medium access control through the producer–consumer method where the data sent by the transmitter can be used by one or more receivers. The application layer is divided into three: ABAS (bus arbitrator application services), MPS (manufacturing periodical/aperiodical services), and subMMS (subset of messaging services). The WorldFIP can be structured as a centralized, decentralized, or master–slave system. Distributed applications can be synchronous or asynchronous. It allows distribution of control and data where, for example, an algorithm can be located in a single processing unit or can be distributed across multiple devices. As WorldFIP is a widespread pattern, it is possible to choose between different suppliers (de Azevedo and Cravoisy 1998).

10.5.1.2.3 PROFIBUS

Process fieldbus (PROFIBUS) is a very popular fieldbus standard. It is used in process applications as well as in manufacturing and automation applications in general. This pattern follows the EN 50170 and EN 50254 standards. Since January 2000, the PROFIBUS standard is regulated by IEC 61158. There are basically three PROFIBUS variants: PROFIBUS-DP, PROFIBUS-FMS, and PROFIBUS-PA. PROFIBUS-DP is the high-speed solution of PROFIBUS applications focused on the manufacturing sector. This variant uses the RS-485 or fiber-optic transmission at the physical layer, being capable of transmitting 1 kb of data in less than 2 ms. The PROFIBUS-FMS offers a wide set of control and supervision functions. This architecture is based on PROFIBUS-DP and is capable of managing complex controls at the production-unit level as the full implementation of programmable logic controllers (PLCs) and distributed control systems (DCSs). The PROFIBUS-PA was developed in cooperation with the NAMUR (control and process industry). It was designed for processes automation, showing as strong points its support for use at potentially hazardous areas, its intrinsically safe feature, and its powering and transmission of data based on IEC 61158-2 technology, using the same bus, in similar way to 4–20 mA technology (PROFIBUS 2010; Smar 2011).

10.5.1.2.4   ASI

In 1990, through a group of companies in Germany, a protocol standard for distributed sensors and actuators named actuator sensor interface (ASI) was developed. ASI is an open standard normalized by the EN 50295/IEC 62026-2. Its main feature is its simplicity, which allows easy and fast interfacing between sensors and actuators with their respective controllers. The bus is simple and inexpensive and allows powering and communicating at the same time. As in other fieldbus standards, the data are transmitted cyclically. The ASI system establishes a master–slave relationship in a tree-topology bus. The bus supports up to 31 slaves, and the master can interrogate and update all the inputs and outputs in less than 5 ms (AS-Interface 2011; Schneider Electric 2000).

10.5.1.2.5   CAN

The controller area network (CAN) was developed in the 1980s by the Robert Bosch Company, initially aimed at the automotive industry. The CAN offers high reliability and a high data transfer rate and, nowadays, is also used in other areas (CIA 2011). This protocol is standardized by ISO 11898, which openly describes the physical and data-link layers. For example, this standard does not specify aspects relating to cables and connectors or the meaning of the data sent or requested (Bosch 2011). The CAN is based on multimaster architecture, where all controllers are masters and have autonomy to access the bus. This architecture is, therefore, fault tolerant because there is not only one central node. As with WorldFIP, CAN also adopts the use of the producer–consumer model to transmit the information where multiple nodes receive the data sent by the station, and each node decides whether to use the received data or not (IXXAT Automation 2011).

10.5.1.3   Wireless Networks and ZigBee

Solutions based on wireless communication are interesting because they do not require any data cabling. It is becoming increasingly feasible, thanks to advances in wireless technology and also in the power consumption of these devices (Iyer et al. 2008; Schneider Electric 2007). In some cases, a simple battery can be used as a power supply as well as a combination of other sources, such as solar cells, for example. The ZigBee wireless communication protocol is a compromise between low power operating and cost of deployment with the rate of data transmission around 250 kbps. Its physical and data-link layers are normalized by the IEEE 802.15.4 standard. Its network and application layers are normalized by the ZigBee Alliance. In addition to the star and cluster tree topologies, mesh networks can also be built using ZigBee, where node devices relay data from other nodes so that multiple paths can connect each node (Ergen 2004).

REFERENCES

Anthony, M., Bartlett, P. L. (2009) *Neural Network Learning: Theoretical Foundations.* 403 pp. pp. 1, 5. Cambridge: Cambridge University Press.

AS-Interface (2011) *Welcome to the Virtual Academy of AS-International Association.* Online. Available at http://as-interface.net/academy/content/sys/start/start.en.html (accessed October 15, 2011).

BOSCH. (2011) *What is CAN?* Online. Available at http://www.semiconductors.bosch.de/en/ipmodules/can/whatiscan/whatiscan.asp (accessed October 16, 2011).

Bryce, C. F. A., Mansi, E. M. (1999) *Fermentation Microbiology and Biotechnology.* pp. 203–236. London: Taylor & Francis.

Christen, P., Veronique, B. M., Orliac, O. (2003) "Sensors and measurements in solid-state fermentation: A review," *Process Biochemistry,* 38, 881–896.

CiA. (2011) *CAN history.* Online. Available at http://www.can-cia.de/index.php?id=522 (accessed October 16, 2011).

de Azevedo, J., Cravoisy, N. (1998) *WorldFIP Technology.* Online. Available at http://www.dte.us.es/docencia/etsii/itis/emc/docs/bc/wfip/at_download/file (accessed October 15, 2011).

Ergen, S. C. (2004) "IEEE 802.15.4 Summary," Technical report, Advanced Technology Lab of National Semiconductor, August.

Helson, R. (2004) *The Benefits of HART Protocol Communication in Smart Instrumentation Systems.* Online. Available at http://www.smar.com/hart.asp (accessed September 20, 2011).

Ibrahim, D. (2002) *Microcontroller-Based Temperature Monitoring and Control.* pp. 203–204. Oxford: Newnes.

IXXAT Automation (2011) *Controller Area Network (CAN)—Introduction.* Online. Available at http://www.ixxat.com/can-controller-area-network-introduction_en.html (accessed October 16, 2011).

Iyer, A., Kulkarni, S. S., Mhatre, V., Rosenberg, C. P. (2008) "A taxonomy-based approach to design of large-scale sensor networks." In: Li, Y., Thai, M. T., Wu, W., ed. *Wireless Sensor Networks and Applications.* pp. 3–6. New York: Springer.

Janocha, H. (2004) *Actuators: Basics and Applications.* pp. 167–170. New York: Springer.

Jardine, F. M., McCallum, R. I. (1994) *Engineering and Health in Compressed Air Work.* pp. 334–335. London: Spon Press.

Khandpur, R. S. (2003) *Handbook of Biomedical Instrumentation.* New Delhi: Tata McGraw-Hill.

King, M. (2011) *Process Control: A Practical Approach.* p. 33. West Sussex: John Wiley and Sons.

Krishnaswamy, K., Vijayachitra, S. (2005) *Industrial Instrumentation.* New Delhi: New Age International Publishers.

Mackay, S., Wright, E., Reynders, D., Park, J. (2004) *Practical Industrial Data Networks: Design, Installation and Troubleshooting.* pp. 200–201. Oxford: Newnes.

Mittle, V. N., Mittal, A. (2009) *Basic Electrical Engineering.* New Delhi: Tata McGraw-Hill.

Montague, G. ed. (1997) *Monitoring and Control of Fermenters.* pp. 89–92. London: Institution of Chemical Engineers (IChemE).

Myrvoll, F. ed. (2003) *Field Measurements in Geomechanics.* Lisse (UK): Taylor & Francis.

National Semiconductor Corporation. (1996) *The Practical Limits of RS-485.* Online. Available at http://www.national.com/an/AN/AN-979.pdf (accessed September 15, 2011).

National Semiconductors Corporation. (2006) *LM35 Data Sheet.* Online. Available at http://www.national.com/ds/LM/LM35.pdf (accessed September 12, 2011).

Nesbitt, B. (2006) *Handbook of Pumps and Pumping.* Burlington: Elsevier.

Nguyen, N. T., Wereley, S. T. (2002) *Fundamentals and Applications of Microfluidics.* Norwood: Artech House.

Norgaard, M., Ravn, O., Poulsen, N. K., Hansen, L. K. (2000) *Neural Networks for Modelling and Control of Dynamic Systems: A Practitioner's Handbook.* 246 p. Great Britain: Springer-Verlag London Limited.

Norton, H. N. (1998) *Handbook of Transducers.* 1st ed. New Jersey: Prentice Hall Professional Technical Reference.

Omstead, D. R. (2000) *Computer Control of Fermentation Processes*. Boca Raton, Florida: CRC Press.

Pallàs-Areny, R., Webster, J. G. (2001) *Sensors and Signal Conditioning*. 2nd ed. New York: John Wiley and Sons.

Pandey, A. ed. (2004) *Concise Encyclopedia of Bioresource Technology*. New York: Haworth Press.

Pandey, A., Soccol, C. R., Rodriguez-Leon, J. A., Nigam, P. (2001) *Solid-State Fermentation in Biotechnology: Fundamentals and Applications*. New Delhi, India: Asiatech Publishers, Inc.

Pandey, A., Soccol, C. R., Larroche, C. org. (2008) *Current Developments in Solid-State Fermentation*. v. 1. pp. 146–168. 1 ed. New York: Springer Science + Business Media, LLC.

Pereira, J. M. D. (2004) "A fieldbus prototype for educational purposes," *Instrumentation and Measurement Magazine, IEEE*, 7, 24–31.

Priddy, K. L., Keller, P. E. (2005) *Artificial Neural Networks: An Introduction*. Washington: SPIE Bellingham.

Rexford, K. B., Giuliani, P. R. (2004) *Electrical Control for Machines*. New York: Thomson Delmar Learning.

Scheper, T., Berovic, M. (2000) *New Products and New Areas of Bioprocess Engineering*. Berlin: Springer.

Schneider Electric. (2000) *Actuator-Sensors Interface AS-i-bus*. Online. Available at http://www.dee.hcmut.edu.vn/vn/ptn/sch/download/Network_Architecture/AS_i_bus.pdf (accessed September 18, 2011).

Siemens Energy and Automation. (2007) *DCS or PLC? Seven Questions to Help You Select the Best Solution*. Online. Available at http://leadwise.mediadroit.com/files/7405DCS_PLC_WP.pdf (accessed October 31, 2011).

Singh, M. (2010) *Introduction to Biomedical Instrumentation*. New Delhi: PHI Learning.

Smar. (2011) *Profibus Protocol*. Online. Available at http://www.smar.com/profibus.asp (accessed October 15, 2011).

Soloman, S. (1998) *Sensors Handbook*. New York: McGraw-Hill Professional.

Smith, C. A. (2002) *Automated Continuous Process Control*. Volume 1. p. 50. New York: John Wiley and Sons.

Sturm, W., Dergint, D. E. A., Soccol, C. R., Pandey, A. (2008) "Instrumentation and Control in SSF." In: A. Pandey, C. R. Soccol, C, Larroche, org. *Current Developments in Solid-State Fermentation*. v. 1. pp. 146–168. 1 ed. New York: Springer Science + Business Media, LLC.

Sturm, W. (2004) *Sensores Industriais: Conceitos Teóricos e Aplicações Práticas*. Rio de Janeiro: Papelvirtual Editora.

Sturm, W. (2005) "Avaliação do Potencial de Uso da Lógica Fuzzy para a Identificação de Indicadores de Competências no Currículo Lattes." Dissertation. p. 104. April 25, 2005. Centro Federal de Educação Tecnológica do Paraná, Curitiba.

Svrcek, W. Y., Mahoney, D. P., Young, B. R. (2006) *A Real-Time Approach to Process Control*. p. 93. West Sussex: John Wiley and Sons.

Webster, J. G. (2009) *Medical Instrumentation Application and Design*. 4th ed. New York: John Wiley and Sons.

Zadeh, L. A. (1965) "Fuzzy sets." In: *Information and Control* 8. pp. 338–353.

Zadeh, L. A. (1990) "The birth and evolution of fuzzy logic." In: Turksen, I. B. Proceedings of NAFIP '90 (June 6–8, 1990).

11 Fermented Foods and Human Health Benefits of Fermented Functional Foods

Juliano De Dea Lindner, Ana Lúcia Barretto Penna,
Ivo Mottin Demiate, Caroline Tiemi Yamaguishi,
Maria Rosa Machado Prado, and José Luis Parada

CONTENTS

11.1 INTRODUCTION

The majority of plant and animal material that is used as a primary food source is also a good substrate for microbial growth and may be subject to different fermentative processes.

Methods for the natural fermentation of milk and cereals have been used to transform fresh raw materials, and the earliest records of these methods date back to 6000 BC. Yeasts have been used in bread making and alcoholic beverage production since the development of ancient Egyptian civilization more than 5000 years ago. The fermentation process evolved with the Jewish culture and later in Greece and Rome. Through the Middle Ages to the present, the quality and development of starter productions have improved [1].

Many fermented foods are well known and form a part of universal gastronomy and regional diets. These foods are obtained by the fermentation of milk, meat, fish, soy, rice, corn, and tubercles. The aroma and flavor compounds produced include alcohols, acids, esters, aldehydes, and ketones. The flavor of cheese is derived from milk components, which are modified chemically by specific enzymes of bacteria and molds. Bread most frequently prepared from wheat is leavened by *Saccharomyces* metabolism on starch or by some heterofermentative lactic acid microorganisms that are able to produce CO_2. Beers, wines, and spirits are also produced by the alcoholic fermentation of yeasts with cereals and fruits and vinegar by acetic acid bacteria (e.g., *Acetobacter* and *Gluconobacter*).

Meat, fish, and vegetables are protected from spoilage microorganisms through the use of spices, salt, and microbial fermentations using lactic acid bacteria (LAB) that promote acidification to help in preservation. The combined action of salt and the acid produced by the LAB lowers the activities of the enzymes responsible for the deterioration of vegetable cells (sauerkraut, olives, and silage), reduces oxidative changes in tissues, and inhibits the growth of spoilage and pathogenic microorganisms [1]. Starter cultures may behave as transformers, as functional probiotics, and/or as biopreservatives (nisin) [2]. LAB in fermented milks may also produce biopeptides that have therapeutic activities.

Some microorganisms are able to grow in the cherries of coffee, chocolate, and tea, and they contribute to their conservation and typical flavor. *Theobroma cacao* beans are liberated from their pulp by a fermentative process. Orient tempeh is produced with soybeans through the action of molds (*Rhizopus oligosporus*); with peanuts fermented by yeasts (*Neurospora*); and angkak, the red yeast rice, fermented by

Monascus purpureus. Other fermented products include soybean foods produced by fermentation (soy sauce and miso) in Japan.

In addition to new flavor characteristics and textures, fermented foods may also have increased shelf lives, compared to their original raw materials, and be more nutritious. Traditional methods for fermented food production have been scientifically studied to improve their productivity, quality, safety, and nutritional value by introducing modern scientific techniques (molecular biology, proteomics, and genetic engineering) and novel technologies, such as automation and microencapsulation.

Artisanal methods devised over the centuries have been adapted to the production of fermented foods on an industrial scale with the systematic selection of strains and introducing controlled fermentation processes in large and modern bioreactors, which make it possible to meet the demands of the increasing population. In this chapter, the emphasis will be on meat and vegetable fermentation, the human health benefits of consuming fermented foods, functional fermented foods, and new technologies and trends for food fermentation.

11.2   FERMENTED MEAT AND VEGETABLE PRODUCTS

Additional topics covered in this book, including the sections on the production of nondairy probiotic beverages, the production of dairy products, and industrial fermentations for the production of alcoholic beverages, describe other fermented foods that are common in the diets of consumers around the world. Important fermented foods, such as meat and vegetables, will be covered in this section, which provides current scientific and technological information on fermentation.

11.2.1   FERMENTED MEAT

The consumption of animal products has been present throughout human evolution, and meat has served as a source of protein with a very high nutritional value for millions of years. The ingestion of meat and its derivatives is a very common part of alimentary diets in all world regions, which is mainly a result of its large sensorial appeal and the satiety it provides. Recently, an increase in meat consumption has been observed, mainly in European countries and China, which is the first place with 25% consumption of all animal-based foods consumed on the planet [3,4].

However, meat and its derivatives suffer from deterioration very quickly and require special care for storage. These deterioration processes are associated with intrinsic characteristics, such as a high water activity (A_w), a high pH, and the disposal of nutrients and minerals, which supply ideal conditions for microbial growth [5,6].

Meat industrialization, originating from its derivatives, has the objective of increasing shelf life, developing different flavors, using parts of the carcass that are difficult to commercialize to form some subproducts, as well as the addition of condiments, spices, and authorized additives and the use of starter cultures [6,7]. The use of these ingredients, in combination with physical and thermal treatments, promotes physicochemical modifications in meat to improve its sensory characteristics and increase the stability of the product [3,6,7].

Because meat and its derivatives display a high perishability and must be preserved using conservation techniques, fermentation appears to be an efficient conservation method. In general, meat preservation is obtained using a combination of fermentation, dehydration, and salting processes [3,8].

11.2.1.1 Embedded Fermentation

There have been reports of the use of fermentation, drying, and curing processes for meat conservation for centuries; the Mediterranean is the cradle of fermentation where the climate favors the fermentation process. These techniques have been improved, and, currently, there are different fermentation processes that employ diverse microorganisms, which are responsible for the characteristics of the meat products and produce several original types of commercialized, embedded fermentation products [9,10].

The embedded elaboration began as a simple process of salting and drying to conserve fresh meat that could not be immediately consumed and when there was no other way to conserve it. The production characteristics of embedded fermentation are related to geographical region, and they provide specific textures and flavors that appeal to the local palate; furthermore, many products are known by their region of provenance [6,9].

A significant factor in the development of different ways to acquire meat derivatives is related to climatic conditions. In northern Europe, cooked embedment appeared because the colder weather allowed for better conservation and storage. Dry embedment was established in southern Europe where the mild temperatures provided a higher stability to the product [9,11,12].

The countries that present the largest production and consumption per capita of embedded fermentation are Germany, Italy, Spain, and France. The European country that consumes the most fermented products, specifically sausages, is Germany, where 330 different types are manufactured. In Italy, sausage diversity is about a thousand, and, because of fermentation processes, these products are stable to environmental temperatures. Currently, an increase in the manufacture and consumption of embedded fermentation products can be observed in the United States, Australia, Great Britain, Brazil, and Japan [4,11,12].

11.2.1.2 Fermented Sausages

The main representative of an embedded fermentation product is sausages, which are internationally classified into two large groups based on the manufacturing technology used and the final pH of the product. Northern sausages use cow or steer meat and are more acidic with a pH of under 5, and sausages from the Mediterranean commonly use pork meat and have a pH of over 5 and a smoother taste [11,12].

There is a large variety of sausages that are manufactured industrially or handmade; however, they are defined as meat products produced by the mixing of minced meat, pieces of fat, salt, cure agents (nitrite, nitrate), sugar, and flavors embedded in a wrapper and submitted to the fermentation and posterior drying process. Sausages are stable at environmental temperatures and commonly consumed without heating [8,9].

The manufacturing process of fermented sausages is described in Figure 11.1, which, in general, shows the steps involved in sausage production. The differences

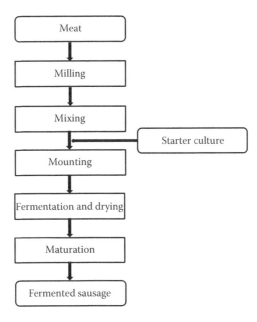

FIGURE 11.1   Manufacturing flowchart of fermented sausages.

in the processes consist of the raw meat material composition, the starter culture composition, curing, mold usage, and drying time, among other factors [11–13].

During sausage manufacturing, the fermentation and drying processes aid in the development of the texture and slice ability, the inhibition of pathogenic microorganisms, the formation of aromatic compounds and characteristic flavors, the formation of a typical red tinge, and the extension of shelf life. All of these are possibly due to physical, chemical, and microbiological reactions [14,15].

The addition of a starter culture composed of different microorganisms provides specific characteristics to the products, and LAB are the main factors responsible for the pH reduction to nearly 5, which prevents the development of undesirable bacteria and acts to improve food safety, product stability, and the drying process. However, catalase positive cocci, such as *Staphylococcus* and *Kocuria*, may contribute to aroma formation and the inhibition of rancidity [13,14,16]. Yeasts (*Debaryomyces*) synthesize proteolytic and lipolytic enzymes and metabolize lactic acid, which reduces product acidity and helps to improve taste. In this case, the taste tends to be smoother and the aroma more intense [13,15,17].

The application of mold on a product after the mounting step enzymatically alters the fermented product's flavor, regulates fast dehydration, and partially hampers oxygen penetration. These alterations consequently prevent the occurrence of undesirable oxidative processes.

The starter culture used in sausage manufacturing varies according to the cure agent applied. When nitrite is used as the unique curing agent, the usage of LAB and *Staphylococcus* is normally recommended. However, the addition of nitrate during the curing process requires the inclusion of nitrate-reducing bacteria from genus *Micrococcus* in the starter culture [8,16].

During sausage production, the reduction of nitrate to nitrite is a fundamental step in color development. This reaction is catalyzed by the enzyme nitrate reductase, which is produced by members of the family *Micrococcaceae* [8,16,18]. Later, the nitrite is transformed by a series of reactions to nitrous oxide, which reacts with myoglobin to form the complex nitrosomyoglobin. This complex is responsible for the typical coloration of the cured products and limits lipid oxidation [8,18,19].

The metabolism of *Micrococcaceae* liberates enzymes responsible for proteolysis and lipolysis, which promote protein and fat catabolism to improve texture and flavor because of the formation of aromatic substances and organic acids [8,19]. Among the *Micrococcaceae*, the predominance of *Staphylococcus* is observed, and the species *S. saprophyticus*, *S. xylosus*, and *S. carnosus* and the subspecies *carnosus* and *utilis* are commonly part of a starter culture [16,19]. The capacity to produce antimicrobial compounds increases interest in their usage in food because they can reduce deterioration and inhibit pathogenic microorganisms, such as *Listeria monocytogenes* [10,17].

The use of a starter culture in the manufacturing of meat derivatives assures product innocuousness; reduces fermentation time; and favors texture, flavor, aroma, and stability, which improve the final product's quality.

11.2.1.3   Use of Probiotics in Meat Products

There is an association between sausage consumption and the development of some diseases, such as cardiomyopathy. In an attempt to minimize this problem and increase the beneficial characteristics for consumer health, new technologies, such as the addition of probiotic microorganisms, have been used [15,17,20].

The probiotic bacteria used in embedded fermentations need to be viable under conditions that are, many times, inadequate because embedded fermentation products possess competitor microorganisms, the presence of curing salts, and a low pH, which provide a poor environment for probiotics [12,20,21]. The presence of nitrite, nitrate, and sodium chloride can inhibit probiotic cultures [15,18,19].

Potential probiotics, such as *Lactobacillus rhamnosus*, *Lactobacillus casei*, *Lactobacillus plantarum*, *Bifidobacterium lactis*, and *Lactobacillus reuteri*, can be mixed into starter cultures to ensure that the probiotics have the desired beneficial effects [15,21].

11.2.1.4   Other Applications of Microorganism Cultures

Different types of cultures are available for fermented and aged meat product manufacturing. The most common use is in the manufacturing of fermented sausages. However, there is another growing area for the application of catalase-positive cocci that are nonpathogenic, such as *Staphylococcus* and *Kocuria*, specifically in whole muscles, such as coppa and raw ham. These microorganisms are responsible for enhancing the development of aromas, colors, and stability [15,21].

One technology being used is the application of lactic cultures as a protecting culture to improve the safety and quality of fresh meat products (e.g., sausages). In this case, the lactic cultures will compete with other, undesirable, microorganisms, and this prevents the proliferation of undesirable organisms in the product [21,22].

Culture application to cooked meat product surfaces to avoid cross contamination during handling is also another new possibility. The advantage is that because of competitive exclusion, the culture abolishes the development of deteriorating bacteria, such as native lactic bacteria that produce acetic acid or gas. Furthermore, the growth of some pathogenic bacteria, such as *L. monocytogenes*, can also be reduced [17,21,22].

11.2.2 FERMENTED VEGETABLES

The principle methods of traditional vegetable processing are canning, drying, freezing, fermenting, and chemical preservation. Canned or frozen foods are considered expensive or not readily available for the majority of people living in developing countries where acid fermentation combined with salting remains one of the most practical methods of preservation and often enhances the organoleptic and nutritional quality of fresh vegetables [23,24]. Biopreservation refers to extending the shelf life and enhancing the safety of vegetables using their natural or a starter microflora and their antibacterial products [25,26]. Vegetables may also be processed and preserved by direct acidification or a combination of these techniques.

At present, cabbage (sauerkraut and kimchi), pickles (cucumber), olives, and peppers account for the majority of vegetables commercially fermented. In 2006, more than 800 million kg of fermented vegetables were produced annually in the United States. Germans consume approximately 1.8 kg of sauerkraut per capita per year, and Syrians eat nearly 6 kg of fermented olives. Notably, Koreans consume more than 43 kg of kimchi per year [27]. Cauliflower, carrots, artichokes, beans, and other products are also fermented but lack any real economic importance [28]. In this section, we will review the manufacturing procedures involved in the production of some popular fermented vegetables and illustrate their microbiological and economic aspects.

11.2.2.1 Production Techniques

The development of vegetable technology actually began more than 2000 years ago [27]. Vegetable fermentation procedures consist of varied technological steps directly dependent on the starter culture inoculum and the adjustment of operations based on the fermentation environment [29].

The manufacturing of fermented vegetables is very similar, and, in a general sense, the technology is based on the same principles as other lactic acid fermentations in that sugars are converted to acids and the finished product acquires new characteristics. Unlike the dairy industry, for example, the fermented vegetable industry still relies on nonstarter lactic microflora (autochthonous) to conduct the fermentation. Given the diversity of microflora present in raw vegetables and the logarithmic disparity between the LAB and others, measures must be adopted to establish the environment (salt and oxygen concentrations and temperature) necessary for lactic acid fermentation.

The pretreatment step is the stage that involves common operations for the selection and cleaning of the raw vegetables. Specific treatments, such as blanching, peeling, and/or cutting, are commonly used in industrial settings.

A suitable environment around the vegetable must be established to guarantee that the desirable microflora can proliferate and predominate. Salt addition is necessary in most types of vegetable fermentations to inhibit the growth of spoilage micro-flora, to enhance the purging of water from the vegetable, and to salt and flavor the fermented vegetables. Differences in the salt concentrations used during production accounts for the difference in the biotypes of LAB that grow in each fermentation environment [30,31]. The use of a starter culture is another method of facilitating the enrichment of desirable microflora in the fermenting vegetables. The LAB used for this purpose include *Lactobacillus* and *Lactococcus* species and *Leuconostoc mesenteroides*. The starter cultures are able to colonize the substrate rapidly, compete under environmental conditions, and ferment the vegetable.

The major factors that influence vegetable fermentation are the temperature, the pH, and low oxygen concentrations. Mesophilic temperatures ranging from 10°C to 38°C are generally used in the process. The optimal temperature for the process depends on the predominant organisms present during the fermentation. The buffer systems promote the utilization of sugars by the starter culture. Buffering the pH and acid neutralization during fermentation with a pH controller has also been used to guarantee complete sugar utilization [32]. To preserve the shelf life of fermented vegetables, the pH should be kept below 4. To maintain anaerobic conditions, the vegetable must be totally covered by the brine in the vessels. Industrial anaerobic tanks provide more suitable anaerobic conditions than vessels.

11.2.2.2 Cabbage Fermentation

Sauerkraut or kraut is a product resulting from the lactic acid fermentation of fresh shredded, salted, white cabbage (*Brassica oleracea* var. *Capitata* for *alba* L.). Historically, sauerkraut was an essential food for naval forces and seafarers, who had little access to fresh vegetables and fruits that normally would have served as a source of vitamin C to prevent scurvy [28].

The production procedures for sauerkraut are shown in the flowchart given in Figure 11.2. The cabbage delivered to the factory is transported to a coring machine, and the cored heads are conveyed to a trunning table where they are properly trimmed and the bad spots are removed. The cabbage is then shredded and transported to the fermentation vat. Two to three percent salt is added evenly as the shreds are distributed in the vat. Juice is released from the cabbage almost immediately after the addition of salt. After the vat is filled, it is covered with a heavy lid and left to ferment at below 15.5°C for at least a month. The fermentation is complete when the shreds are cured and the titratable acidity, expressed as lactic acid, has reached 1.5%. The final step is the transfer of the kraut and brine into retail packages. The product may be canned, thermally processed (at approximately 75°C) (pasteurized sauerkraut), or packaged for nonpasteurized, refrigerated (fresh sauerkraut) marketing in sealed plastic bags (polybags) with or without the addition of approved preservative ingredients [33].

Sauerkraut fermentation is a complex microbiological process that has been reported to occur in two stages: an initial heterofermentation followed by a homo-fermentation where the microorganisms originating from the raw cabbage are the main participants [34]. The fermentation is started by the conversion of sugar to lactic acid, acetic acid, alcohol, and CO_2 by *L. mesenteroides*. Carbon dioxide helps maintain the anaerobic conditions necessary for fermenting cabbage. Afterward, *L. mesenteroides* is inhibited by the CO_2 accumulation in the vat, and fermentation

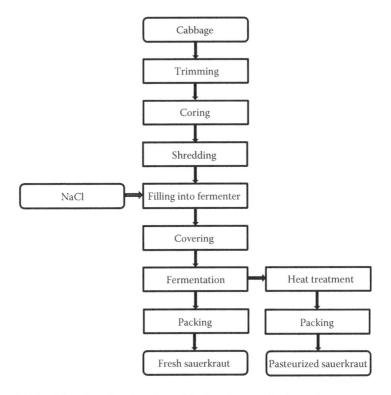

FIGURE 11.2   Manufacturing flowchart of fresh and pasteurized sauerkraut.

continues with *Lactobacillus brevis*, *Pediococcus cerevisiae*, and finally *Lb. plantarum* [34].

Kimchi is a traditional Korean fermented cabbage and vegetable product. In the Far East, kimchi has become the most popular of all fermented vegetables. Kimchi production uses fresh cabbage that is cut in half or shredded, soaked in brine of approximately 10% of salt overnight, then washed and drained. The kimchi is packed in a jar and pressed to keep the ingredients immersed in the juice. *L. mesenteroides* is the dominant microorganism before ripening. *Lactobacillus* species may be dominant in the later stages of kimchi fermentation depending on the temperature [36]. The diversity and richness of bacterial communities vary depending on the type of kimchi, and these differences could largely be explained by differences in the major ingredients and the manufacturing process of each type of kimchi [37].

The main difference between sauerkraut and kimchi is that the latter product contains ingredients other than cabbage and salt. Kimchi is usually made from cabbage, radishes, and cucumbers. Spices and flavoring agents are also commonly added to kimchi, depending on the particular type of kimchi being produced.

11.2.2.3   Cucumber Fermentation

The combination of acid, spices, and sugar with cucumbers (*Cucumis sativus* L.) creates the acidic food product known as pickles. Pickled cucumbers are made in

Africa, Asia, and Latin America. Pickled products are classified based on the ingredients used and the method of preparation: fermented pickles or brined pickles that undergo a natural curing process and fresh-pack or quick-process pickles (the direct addition of vinegar to an equilibrated pH of 4.6 or below).

In fermented pickles, selected cucumbers are brined with approximately 5% salt. The brine strength is gradually increased during the fermentation to 16%. Procedures have been developed for brining cucumbers in closed anaerobic tanks at substantially lower salt concentrations [38]. This technique may allow for cucumber fermentation and storage at sufficiently low salt concentrations that do not require desalting.

L. mesenteroides, P. cerevisiae, Lb. brevis, and *Lb. plantarum* are involved in the pickling process at approximately day 7 to day 14. When the pH is lowered to 3.2, the metabolism of *Lb. plantarum* is inhibited, and fermentation is complete [35].

The product may be preserved by fermentation, pasteurization, or refrigeration and may contain sweeteners, flavorings, spices, and other permissible ingredients.

11.2.2.4 Olive Fermentation

The practice of olive fermentation has become a large-scale production worldwide. The world production of table olives surpasses 2.2 million tons per year, and the Mediterranean countries are the main producers. The 27 members of the European Union, Egypt, Turkey, Syria, and Morocco are the major fermented olive-producing countries with approximately 83% of the worldwide production. The Spanish-style green olive is the most important industrial preparation with approximately 60% of the production [39]. There has been an increase in demand for fermented green and black table olives in recent years in all regions of the world because of their nutritional and functional food proprieties [40].

The olive is the fruit of the olive tree (*Olea europaea* L.) and is consumed only after fermentation because it makes the olive more digestible and reduces the bitterness and toxicity of its phenols. Table olives are prepared from specifically cultivated fruit varieties harvested at a predetermined stage of maturation [41].

There are three major table olive production methods. Spanish-style green olives are treated with sodium hydroxide (1.6% to 2.0%) and fermented. Greek-style black olives are just fermented. The fermentation for both types of olive is mediated by the indigenous olive microflora. The third type is the Californian-style black olive, which is lye-treated but not fermented. A low pH treatment is necessary to remove oleuropein, a bitter component in olives. Afterwards, the olives are rinsed in fresh water to completely remove the lye and brined in containers at a brine concentration ranging from 5% to 15%, which depends on the variety and size of the olives. They also undergo an aeration treatment that promotes the oxidation of pigments and conversion of a green color to black [40].

The entire fermentation process may take 2 weeks to 10 months, and the acid content of the final product varies from 0.18% to 1.27% [33]. *L. mesenteroides* and *P. cerevisiae* are the first LAB to become prominent during the olive fermentation, and they are followed by *Lb. plantarum* and *Lb. brevis*. The lye treatment may affect the microbial flora. Inoculation with *Lb. plantarum* as a starter promotes the production of bacteriocins that better control lactic acid fermentation [42].

11.3 HUMAN HEALTH BENEFITS OF FERMENTED FOODS

Many foods chosen by humans are rich in nutrients and very perishable in their natural form. They commonly have high moisture content as well as high water activity values, and their chemical compositions favor microbial development. Microbiological spoilage is a major concern as a result of these and other factors, and humans have learned how to halt food deterioration and take advantage of foods modified by fermentative processes, which result in nutritious and tasty products.

It has taken a very long time for us to understand the presence of microscopic organisms that are involved in the processes of both food and beverage spoilage or fermentation, even though these products were already present in the daily human diet.

The presence of microorganisms in food products is almost always associated with worry and distress as people think about hazardous bacteria, toxins, and other frightening subjects that are commonly reported in the press. There is a large portion of the population that has never before heard about the benefits of microorganisms in foods and that become surprised that yogurts, cheeses, salami, pickles, and many other foods contain live microorganisms that are ingested in their daily diets.

The presence and activity of microorganisms in foods suggest that their enzymes hydrolyze complex food compounds, including oligosaccharides and polysaccharides, proteins, and many other minor components that, in some cases, may be associated with toxicity or low digestibility if consumed by humans. Several types of seeds, such as pulses, may present this type of problem, and cassava roots are toxic if used without the proper processing, which includes the action of microorganisms.

Several types of microorganisms, including many bacteria and fungi, play a large role in fermented foods. They can be considered to be nutritious biomass when consumed together with foods, such as in moldy cheeses, salami, yogurts, pickles, fermented cassava flours, and many other products that are a part of diets worldwide. The enzymatic action of these microorganisms in vegetable- and animal-based foods also generally increases their digestibility. In vegetable-based foods, microorganisms also increase nutritional quality, as they synthesize vitamins and other compounds as well as high-quality proteins.

Fermented foods and beverages are part of ancient and current human diets. Traditional Oriental dishes are prepared with fermented foods and ingredients derived from soybeans, fishes, and many other raw materials in the same manner as in Western countries with pickled vegetables, cheeses, yogurts, and so on. Alcoholic fermented beverages and dairy fermented foods are by far the most important examples of fermented products that present good acceptability as well as health interest. Although the consumption of ethanol is considered hazardous, other components present in wine appear to be very important for a healthy lifestyle. In the case of artisanal beers, there are reports of their importance in several African countries because of the presence of B-complex vitamins. The fact that fermentation helps to increase the shelf life of very perishable foods should also be considered as this kind of quasi fresh food is rich in partially digested components that are of great interest in human diets.

Even though many advantages have been described in terms of the nutritional aspects of fermented foods, it is obvious that some difficulties also exist other than the presence of ethanol in fermented alcoholic beverages. The high level of sodium in some pickled vegetables and in fermented meat products is an example of these constraints that are related with this type of processed food.

In this section, some of the nutritional benefits of fermented foods are discussed, and consideration is given to selected types of fermentation that exemplify some well-known raw materials and foods.

11.3.1 INDIGENOUS FERMENTED FOODS

Fermentation is one of the oldest and most economical ways of producing and pre-serving foods worldwide [43]. Microbial and enzymatic transformations occur and change the texture, appearance, flavor, and nutritional value of the raw material. If the transformed product is repugnant, off-flavor, or toxic, the food is considered spoiled. However, if not, it is described as a fermented food [43].

Steinkraus [44] focused on indigenous fermented foods from different countries that have been classified based on fermentation and raw material types as follows: (a) protein-rich vegetarian meat substitutes, such as Indonesian tempeh; (b) fer-mented foods involving lactic acid fermentation, including acid-fermented vegeta-bles and acid-fermented milk; (c) alkaline-fermented foods, such as Japanese natto; (d) fermented foods in which ethanol is a major product, such as alcoholic beverages, vinegars, and acetic acid fermentations; (e) amino acid/peptide sauces and pastes with meat-like flavors, such as Chinese soy sauce and Japanese shoyu and miso; and (f) mushrooms (microbial protein) produced on agricultural lignocellulosic wastes. This well-known classification of fermented foods has been adopted worldwide.

From a food science and technological point of view, fermentation plays at least five roles [44]: (a) it develops flavors and aromas, changes food textures, and enriches diets; (b) it is responsible for food preservation through lactic acid, alcoholic, acetic acid, and alkaline fermentations; (c) it increases or enriches food substrates with protein, essential amino acids, essential fatty acids, and vitamins; (d) it detoxifies certain raw materials to make them edible; and (e) it decreases cooking times and fuel requirements.

Whereas in developed Western countries foods are enriched with synthetic vita-mins and other nutrients, in the developing world, the bio-enrichment promoted by fermentative processes plays an important role, mainly in the case of vitamins and essential amino acids [44].

Steinkraus [45] reviewed the worldwide industrialization of select indigenous fer-mented foods, which included, among others, Japanese shoyu and miso (soybean sauces and pastes), sake, Japanese natto, Malaysian tapai (made from glutinous rice or from cassava roots), Africa's indigenous brewed beers (from sorghum, millet, or corn), and gari (made from cassava roots). There are considerations given in terms of the success of some industrialized fermented foods, such as cheeses, yogurts, soy sauces and miso, brewing, winemaking, and vinegars, whereas others, such as tem-peh and indigenous fermented beverages, offer opportunities for extending village-level technology to large-scale industrialization.

As stated by Blandino et al. [46], fermented foods are produced worldwide using different microorganisms and raw materials, and fermentation processes can be grouped into four main types: alcoholic, lactic acid, acetic acid, and alkaline fermentation. The authors reviewed cereal-based fermented foods and beverages because of the importance of cereal grains as major sources of dietary nutrients worldwide. These primary products are deficient in some amino acids and vitamins, and fermentation may improve their nutritional value, sensory properties, and functional qualities at a low cost.

Legumes are important sources of protein in Egypt and many other developing countries because of the high prices of meat [47]. The fermentation processes associated with legume grains improve digestibility, nutritional value, and flavor and make the food safer. Considering that Indonesian tempeh is a very interesting fermented food that has a bland flavor, is protein rich, and has a high acceptability especially among vegetarians in Western countries, including the Netherlands and the United States. Nassar et al. [47] studied the small-scale production of tempeh in Egypt using mixtures of locally available legumes (green pea or *Pisum sativum*, broad bean or *Vicia faba*, chickpea or *Cicer arietinum*, and lupine or *Lupinus termis*). The authors succeeded in producing tempeh from alternative raw materials with an increased nutritional value and lower content of flatulence-producing oligosaccharides (stachyose and raffinose).

According to Nout and Kiers [48], fermented foods and beverages represent, on average, one-third of total food consumption. Tempeh is one of the best-known soybean foods with an attractive flavor, texture, and high digestibility; it is often employed as a meat substitute in burgers and salads and is valued for its potential health benefits. The most common raw material for tempeh production is yellow-seeded soybeans that are dehulled, soaked in water, and cooked and then inoculated with *Rhizopus* sp., which overgrows and holds the seeds together by the entanglement of their white hyphae. The nutritional advantages of tempeh over raw soybeans include a reduction of raffinose and stachyose, an increased digestibility, increased levels of B vitamins (mainly riboflavin, niacin, B6, and B12), and an increased sensory quality. As a result of all of these qualities, tempeh may also potentially protect against diarrhea. Therefore, fermented soybean foods should be considered in the treatment of chronic diseases [48].

Namgung et al. [49] described the quality of doenjang, a traditional fermented food (soybean paste) with a high consumption rate in Korea. Doenjang serves as an important source of protein and is rich in essential amino acids and vitamins. The authors detected chemical changes during the fermentation process of the product, which included increases in the levels of essential amino acids.

11.3.2　Lactic Acid-Fermented Foods

With very limited alternatives for preserving animal- and vegetable-based products, fermentation has played a major role in making good-quality and nutritious foods available for centuries, and traditional recipes were inherited from ancient generations in China [50]. As highlighted by the authors, fermentation increment the level of safety, nutritional value, and the flavors of foods. The importance of fermented products in developing countries with large populations must be recognized since it contributes to the supply of safe food and increases the food's sensory and nutritional qualities. LAB are involved in numerous food fermentations, are accepted as generally recognized as safe (GRAS), and are associated with the probiotic properties of some commercial products.

LAB have been reported as the fermentation agents in several food products, and four main types have been discussed by Liu et al. [50] based on the raw materials that are staples in Chinese diets: dairy-based fermentation, vegetable-based fermentation, soybean-based fermentation, and meat-based fermentation. The authors emphasize flavor enhancement as well as the release of active substances through the metabolic processes of the microorganisms that have additional benefits for human health.

Steinkraus [51] states that fermentations involving the production of lactic acid are generally safe and are based on the conversion of carbohydrates to lactic acid by several types of LAB. The increased shelf life of perishable raw materials provided by lactic acidification is reported as an important factor that provides nutritional benefits to consumers, mainly in developing countries.

van Hylckama Vlieg et al. [52] briefly described the importance of microorganisms in foods for human health by citing vitamin enrichment as well as the introduction of probiotics. Among the other benefits of fermentations of raw food materials, the authors cited the detoxification of cassava as it contains toxic cyanogens that are drastically decreased by the action of microbial enzymes. In their review, van Hylckama Vlieg et al. [52] concluded that there is a rapid increase in the understanding of the health benefits of microbial intake from food, and much of this evidence has come from emerging studies on the close relationship between host physiology and gastrointestinal microbiota metabolism. There is still a need to improve our knowledge of how food microorganisms interact with the gastrointestinal microbiota and host metabolism, and this will expand the possibilities for functional food innovation. Food fermentation can be regarded as an extension of digestion.

Fermented and sundried cassava starch or sour cassava starch presents a typical flavor and produces expanded salty biscuits that are very well appreciated in Brazil [53]. This indigenous fermented cassava starch is produced in Brazil and, on a smaller scale, in a few other South American countries and is known as *polvilho azedo* in Brazil or *almidón agrio* in Colombia [54]. Natural lactic acid fermentation occurs during a period of approximately 20 to 40 days, and organic acids are produced, which consequently decrease the pH. After this period, the sour starch is exposed to sunlight for drying for a period of 1 to 2 days [55]. This starch is of great technological interest because of its ability to generate tasteful expanded baked foods (salty biscuits and cheese breads, for example) without the need of extrusion. It can be produced at home, and the foods are completely gluten-free [56].

11.3.3 ALKALINE-FERMENTED FOODS

Alkaline-fermented foods constitute a group of less well-known food products but are very popular in Southeast Asia and African countries [43]. A relatively well-known product is Japanese natto, which is made from dehulled and partially cooked soybean seeds by fermentation with *Bacillus* spp. (mainly *Bacillus subtilis*), which causes the pH increase to values in the range of 8 to 9 resulting from the liberation of ammonia; however, the product presents a strong, pungent, ammonia-like smell.

Alkaline fermentation is also related with increased digestibility because of the extensive hydrolysis of proteins to amino acids and peptides, and there are reports of vitamins that are synthesized in the process, including riboflavin and vitamin B12, which are found in concentrations three to five times greater in natto compared to cooked soybeans [43]. Detoxification, or the decrease of toxic compounds, has also been described as an important contribution of the alkaline fermentation of African locust beans, which are poisonous to humans and become edible after this process, which results in a food known as dawa dawa.

11.3.4 NUTRITIONAL IMPROVEMENT IN FERMENTED FOODS

In several developing countries, there are still many starving people and nutritional deficiencies involving the lack of certain vitamins and amino acids as well as the low ingestion of protein. Fermented foods are able to minimize these problems by raising protein contents and improving the balance or availability of essential amino acids in diets. Especially for people living with diets based on cereals (corn or polished rice, for example), the biological enrichment of foods via fermentation is of great value because of the resulting increase in vitamins, such as niacin or nicotinic acid, riboflavin, and thiamine [51]. Another important contribution for human nutrition is the possibility of converting vegetable residues, such as straw, into edible mushrooms.

Steinkraus [51] published a review on fermented food classification and included a section about bio-enrichment by fermentation in terms of protein and essential amino acids in fermented cassava in Indonesia; vitamins (mainly niacin and B12) in tempeh, the Indonesian protein-rich meat substitute made by overgrowing soaked, dehulled, and partially cooked soybeans with *R. oligosporus*; and thiamine, riboflavin, niacin, and pantothenic acid in Mexican pulque, the oldest American alcoholic beverage produced with agave juice (a cactus plant). Another important subject presented by Steinkraus [51] is the reduction of cooking time (if necessary) for the culinary preparation of fermented foods, which, as a consequence, makes soybeans and other hard-to-cook foods easier to prepare with fuel.

Consumers are very interested in the health benefits related to their diets, and this fact is impacting food choices. Fermented foods containing elevated levels of B vitamins are a good example [57] and eliminate the need for fortification with these essential vitamins. Burgess et al. [57] reviewed the overproduction of B vitamins (riboflavin, folic acid, and cobalamin) by bacteria. Many of these bacteria are directly involved in food fermentation and have a GRAS status, and they include lactic acid- and propionic acid-producing bacteria. Fermentation with B vitamin-producing bacteria adds nutritional and commercial value to primary products, and research projects are in progress to further increase the synthesis of these essential nutrients by GRAS microorganisms. At the conclusion of their review,

Burgess et al. [57] claimed that the concept of in situ fortification by fermentation provides a method for the development of special foods targeted at specific groups in society, for instance, the elderly, adolescents, pregnant women, children, athletes, and vegetarians.

Bhatia and Khetarpaul [58] studied the reduction of phytic acid, an antinutritional factor frequently found in cereals and in indigenous fermented bread (doli ki roti). Although indigenous fermented foods from India seem to be very nutritious, scientific information on them is scarce, which justifies further research on this process. The authors concluded that indigenous natural fermentations improved the digestibility of starch and protein and reduced the content of phytic acid while producing a food that had an appealing and attractive color, flavor, and texture and an extended shelf life.

Arora et al. [59] studied the effect of germination and fermentation on the nutrient composition of pearl millet (*Pennisetum glaucum*)-based foods. The authors emphasize that this cereal is a staple food for a large part of the population of African and Asian countries and contributes a major portion of their dietary nutrients. On the other hand, millets have low digestibility and availability of minerals resulting from the presence of antinutritional factors. To overcome these problems, various processing methods, such as dehulling, cooking, germination, and natural fermentation, have been employed. The authors discuss how, in India, lactic acid-fermented foods have a special significance in the diet of vegetarians as the major source of vitamin B12, but they also emphasize that consumer interest has been renewed in health promotion and disease prevention by including probiotic bacteria in foods. Finally, Arora et al. [59] concluded that a combination of germination followed by fermentation with probiotic microorganisms of an indigenously developed food mixture is a potential method for the development of vegetable-based foods with improved nutritional quality that are also safe for human consumption.

Buckley et al. [60] considered the soybean as a functional food containing significant levels of biologically active compounds that are thought to reduce disease risk and, for that reason, would be of interest for the nutrition of astronauts. The authors studied the development and properties of fermented soymilk-containing probiotic bacteria in terms of the nutritional needs of astronauts carrying out long-term missions. Their interest was in the nutritional advantages of both soybeans and probiotic bacteria, and they found that the following health benefits could be provided for astronauts: (a) enhanced bone metabolism, (b) improved nutritional value, (c) immune system stimulation, (d) improved flavor and texture of soy foods, (e) reduced problems with flatulence, and (f) better preservation of foods. Soy fermentation is considered a way to increase the bioactive aglycone forms of isoflavones, which results in more rapid and efficient absorption (of genistein and daidzein, for example) and, consequently, makes the functionality of foods more evident. The increase in B-group vitamins from soy fermentation has been well documented, and specific levels of riboflavin, folate, niacin, B6, and B12 generated by soybean fermentation also have the potential to generate bioactive peptides.

In Table 11.1 some examples of food fermentation and their health or nutritional benefits are shown.

TABLE 11.1

Examples of Health and Nutritional Benefits from Selected Fermented Foods

Processing/ Fermentation Type	Raw Material and Microorganisms	Health/Nutritional Benefits	Regions and Countries	References
Germination followed by probiotic fermentation	Pearl millet (*Pennisetum glaucum*); *Lactobacillus acidophilus*	Increased *Lactobacilli* count; increased thiamine and niacin levels; increased lysine levels; increased in vitro availability of calcium, iron, and zinc	Africa and Asia	[59]
Indigenous fermented bread	Whole wheat, flour and spices; natural fermentation	Reduction of phytic acid content; improved starch and protein digestibility	India and Pakistan	[58]
Lactic acid fermentation	Soybeans (*Glycine max* L.); soy milk; *Streptococcus thermophilus*, *Bifidobacterium longum*, and *Lactobacillus helveticus*	Deconjugation of isoflavones increasing their bioavailability, reduction of stachyose and raffinose levels	USA	[60]
Lactic acid fermentation	Dairy-, vegetable-, meat-, and soybean-based foods; LAB	Increased food safety; improved protein digestibility; enrichment with vitamins; increased utilization ratios of calcium, phosphorus, and iron; adsorption of iron and vitamin D; lactose hydrolysis; beneficial effects of LAB on the intestinal microbiota (enhanced immune function, controlled serum cholesterol levels, reduced intestinal infections, and elimination of deleterious substances)	China	[50]

(*continued*)

TABLE 11.1 (Continued)
Examples of Health and Nutritional Benefits from Selected Fermented Foods

Processing/ Fermentation Type	Raw Material and Microorganisms	Health/Nutritional Benefits	Regions and Countries	References
Tempeh fermentation	Soybeans, green pea (*Pisum sativum*), broad bean (*Vicia faba* L.), chickpea (*Cicer arietinum*), and termis (*Lupinus termis*); *Rhizopus oligosporus* and LAB	Increased protein and fiber contents and decreased flatulence sugars (stachyose and raffinose); reduction of trypsin inhibitor activity and of phytic acid level; increased B-vitamin levels; modification of soybean isoflavones, increasing bioavailability; increased digestibility and nutrient bioavailability	Indonesia, Japan, The Netherlands, the USA, and other countries	[47,48]
Lactic acid and indigenous fermentation	Cassava roots	Detoxification of cassava by eliminating highly toxic cyanogens	Africa	[52]

11.4   FERMENTED FUNCTIONAL FOODS

The production of fermented foods is one of the oldest food-processing technologies known to man. Fermented foods are considered functional because they may offer health benefits. In general, functional foods are recognized as benefiting the health of the consumer through the improvement of general health conditions and a reduction in the risk of some diseases. The term "functional food" was introduced in 1984 when the Japanese began to evaluate the relationships among nutrition, sensory satisfaction, fortification, and the modulation of physiological systems [61]. Functional foods were described by Roberfroid [62] as "food similar in appearance to conventional food that is intended to be consumed as part of a normal diet, but has been modified to sub serve physiological roles beyond the provision of simple nutrient requirements" or "foods that may provide health benefits beyond basic nutrition."

11.4.1   CLASSES OF FUNCTIONAL FOODS

There are basically four types of functional foods on the market according to Spence [63]: (a) fortified products; (b) enriched products, including the use of fermented probiotics and prebiotics; (c) altered products, when existing components are replaced with beneficial components; and (d) enhanced commodities, when raw commodities suffer changes in nutrient composition (e.g., high lysine corn and golden rice).

11.4.2 PROBIOTICS AS FERMENTED FUNCTIONAL FOODS

Probiotic fermented foods have been one of the fastest-growing industrial segments in recent years. This may have been supported by the fact that probiotic products can be consumed on a daily basis and used as preventive medicine against diseases affecting all age groups [64].

Probiotics are defined by the FAO as "live microorganisms which, when administered in adequate amounts, confer a therapeutic or health benefit on the host" [65]. The most commonly employed microorganisms in the probiotic field are LAB and bifidobacteria, and *Bacillus*, *Aspergillus*, and *Saccharomyces* are less common genera.

Lactobacillus and *Bifidobacterium* have been used in the manufacturing of different dairy products (e.g., fermented milk, yogurts, buttermilk, and cheese), fermented meats, soft drinks, vegetables, and cereals. *Enterococcus*, *Bacillus*, *Aspergillus*, and *Saccharomyces* have already been used in feed-animal preparations [66].

One of the main claims of probiotic products is the prevention of inflammatory diseases, allergies, increase in cholesterol level, gastrointestinal disorders, and some types of cancer. Host immune modulation is one of the suggested benefits of the consumption of LAB probiotic functional foods containing *Bifidobacterium longum*, *B. lactis*, *Lactobacillus acidophilus*, and *L. rhamnosus*, which have been widely studied. Probiotics may also be used in cases of lactose intolerance, food-borne allergies, or inflammatory processes, such as inflammatory bowel diseases (IBD), and in the prevention of infections caused by pathogens, such as *Escherichia coli*, *Salmonella*, and *Helicobacter pylori* through receptor-site competition and bacteriocin production. *Saccharomyces boulardii* confers protection against toxin A, which is produced by *Clostridium difficile* and may cause intestinal injury and inflammation [67].

Several studies have suggested an anticarcinogenic effect of LAB probiotics by the induction of proinflammatory, antiinflammatory, or secretory responses, and the reduction of procarcinogenic factors has been described [68]. Other properties attributed to probiotics are protection against allergies and mycotoxin effects.

In Table 11.2, the principal characteristics of probiotic microorganisms that have been successfully used are listed.

11.4.3 HEALTH BENEFITS AND MODE OF ACTION OF PROBIOTICS PRODUCED BY FERMENTED FOODS

Probiotics have been claimed to possess various health benefits, including protection against infection, antitumor properties, improved immune stimulation, better digestion, lactose tolerance, cholesterol reduction, alleviation of atopic dermatitis and allergy symptoms, and positive effects on some bowel diseases and constipation. Some of the postulated mechanisms of action include the following: (a) competition with other microorganisms and the modulation of intestinal flora; (b) binding at receptor sites of pathogenic bacteria and rotavirus in diarrhea treatment and

TABLE 11.2
Probiotic Strains and Some of Their Relevant Claims

Microorganisms	Target Applications
Lactobacillus casei Imunitass	Immune response
Lactobacillus casei Shirota	Gut health, antimutagenic
Lactobacillus helveticus	Antihypertensive
Lactobacillus plantarum 299v	Digestive system
Lactobacillus plantarum PH04	Cholesterol-lowering effect
Lactobacillus rhamnosus GG	IBD, immune response, allergies
Bifidobacterium animalis subsp. *lactis* Bb12	Gut microbiota, immune system
Bifidobacterium lactis HN019	Infection diseases
Bifidobacterium longum BB 536	Gut microbiota, immune system
Enterococcus faecium CRL183	Cholesterol-lowering effects
Escherichia coli Nissle 1917	IBD, antimutagenic
Saccharomyces boulardii	Against diarrhea, prevents or limits mycotoxigenic *Clostridium difficile* infection

Source: From Soccol, C.R. et al., *Food Technology and Biotechnology* 48, 413–34, 2010. With permission; De Vuyst, L., *Meat Science* 80, 75–8, 2008. With permission.

prevention [70,71]; (c) antagonizing pathogens by the production of antimicrobial compounds, acids, H_2O_2, and bacteriocins; (d) modifying intestinal pH by lactic acid and short-chain fatty-acid production; (e) competing for available nutrients and producing some growth factors in the gut; (f) inducing immunomodulation by the activation of lymphocytes, immunoglobulins IgA, IgM, and IgE, the antiinflammatory cytokine IL-10, and interferon IFN-γ [72–74]; (g) reducing mutagenic activity and procarcinogenic factors in the colon, lowering fecal enzyme activities (glucuronidase, azoreductase, and nitratoreductase) [75]; (h) inducing metabolic effects by producing lactase and other hydrolytic enzymes, such as proteases and lipases; (i) reducing *H. pylori* infection as well as ulcer-inducing factors and protecting against induced gastritis by specific LAB strains [76,77]; and (j) reducing cholesterol levels.

Not all of the above-mentioned beneficial effects can be attributed to each or all probiotic strains as several of the positive responses are strain specific. A significant amount of evidence has been reported, and most of the effects were demonstrated mainly in diseased human populations. However, a demonstration of health benefits in the average (the generally healthy) population must be demonstrated.

Recent studies have reported that probiotics can convert milk protein into bioactive peptides, which can have an antihypertensive effect. This effect is associated with inhibition of enzymes, such as dipeptidyl carboxypeptidase, termed angiotensin I-converting enzyme (ACE). These enzymes catalyze the conversion of angiotensin I to the potent vasoconstrictor angiotensin II, inducing a release of aldosterone and an increased sodium concentration, which leads to an increase in blood pressure. In vivo studies have demonstrated the effect of the ACE inhibitor peptides in reducing blood pressure when administered intravenously or orally [78].

11.4.4   Food Matrices as Vehicles for Probiotics

Cheese, butter, powdered milk for infants, mayonnaise, meat, cereal, and vegetables can serve as vehicles for delivering probiotics [79]. Dairy products are regarded as ideal vehicles for the delivery of probiotic bacteria to the human gastrointestinal tract [68] because the dairy components buffer the bacteria in the stomach and contain functional ingredients that interact with the probiotics (e.g., milk proteins) to provide protection. Additionally, ice cream and frozen dairy desserts can be vehicles for delivering probiotics because they have the advantage of being stored at low temperatures and present a good viability at the time of consumption, being easily consumed by people of all ages.

Examples of nondairy matrices are fruits, vegetables, legumes, and cereals, which are relatively low in fat and free of lactose. An important quota of these products is represented by nondairy fermented beverages based on fruits and vegetables [80]. However, fruit juices have been demonstrated to have some limitations because the probiotics need to be protected against the acidic conditions. This drawback can be overcome through microencapsulation or nanoencapsulation technologies, which prevent contact between probiotic bacteria and the external environment by a protective coating. The food-grade polymers usually employed in microencapsulation are alginate, chitosan, carboxymethyl cellulose (CMC), carrageenan gelatin, and pectin [81,82].

In the manufacturing of probiotic products, the microorganisms can be added at multiple stages: (a) together with the starter cultures (direct vat inoculation); (b) in two batches separately, one containing the probiotic microorganism in milk and another with the starter cultures; or (c) as a starter culture after the fermentation stages are completed. In the last case, the fermentation time is generally higher than traditional processes using nonprobiotic starter cultures [83]. The standards that regulate functional foods establish that probiotics should remain viable during large-scale production, should be stable and viable during storage and use, and still should be able to survive in the intestinal ecosystem, and the host must be able to obtain the benefits offered by the ingestion of probiotics [80].

11.4.5   Synergy of Prebiotics with Probiotic Fermented Foods

According to Gibson and Roberfroid [84], prebiotics are "nondigestible food ingredients that beneficially affect the host by selectively stimulating the growth and/or activity of one, or a limited number of bacteria in the colon." Oligosaccharides, such as lactulose, galactooligosaccharides, inulin, fructooligosaccharides, and other food carbohydrates, are examples of prebiotics and may be added to fermented probiotic foods. The synergistic effect of probiotics and prebiotics promotes the growth and improves the activity of the microorganisms, making the host's bacterial gut flora healthier [85].

11.4.6   Principal Suppliers of Fermented Functional Foods

The industry of functional fermented foods is usually fragmented into different types of producers, which can be divided into the following categories: (a) multinational food companies (e.g., Nestlè: LC1 yogurt; Yakult Honsha Co. Ltd.: Yakult fermented milk; Unilever: Becel proactive margarine; Dannon: Activia, Actimel, and Densia yogurt); (b) pharmaceutical and/or dietary product-producing companies (e.g., Novartis Consumer Health: biscuits, cereal, cereal bars, and beverages with the brand Aviva; Johnson & Johnson: Lactaid lactose-free milk and Benecol cholesterol lowering foods); (c) national market leaders (e.g., Eckes: drinks fortified with A, C, and E vitamins [or ACE drinks]; Becker Fruchsäfte: ACE fruit juice); (d) small- and medium-sized food companies (SMEs) (e.g., Charlie Luciano Ltd.: cereal and grains from Familia A.C.E. balance); (e) food retail companies (e.g., Aldi: probiotic yogurt "Bi'Ac"); and (f) suppliers of food ingredients (e.g., Roche: vitamins; DSM: vitamins and enzymes; Südzucker: Orafti prebiotics) [86].

Multinational food companies are leaders in the functional food market. They spend billions of dollars on the development and marketing of these products in addition to research costs to prove the efficacy of the bioactive compounds in clinical trials.

11.5   NEW TECHNOLOGY FOR FOOD FERMENTATION

Recently, significant improvements have been made in fermentation technology for the production of various foods and ingredients, chemicals, nutraceuticals, and industrial enzymes. A number of innovations have been employed in classical fermentation processes to improve biological conversion and optimize the process: the selection of strains, metabolic engineering approaches to improve the production of biomolecules, the elimination or reduction of byproduct formation, and the improvement of yield or productivity [87]. Bioprocesses tend to have higher reaction specificities and milder reaction conditions and produce fewer toxic products, which are considered environmentally friendly [88].

In seeking clean or novel technologies for food biotechnology to obtain ultrapure and high-quality products or ingredients, a significant amount of scientific interest has been aroused over the past few years. For example, the technology of the supercritical extraction of bioactive compounds for use in functional foods has been applied in the food and nutraceutical industries for only the last three decades [89]. In addition to the extraction and purification of bioactive compounds from plants and animals, there are vast sources of microorganisms capable of producing biomolecules, which are important nutrients for the diets of humans or animals. In this respect, worldwide trends in the food industry and the recent advances in biotechnology have significantly increased the demands for high-quality, nutritious, healthier, and tastier foods with less fat, sugar, and salt that remain fresh for a long time and that are completely safe and less reliant on artificial additives [90].

It is known that diet plays an important role in the modulation of important physiological functions in the body. The biotechnological revolution currently in place promotes the use of food as a means of promoting welfare and health while reducing the risk of disease. Food processors face a major challenge in meeting this need, and, as a consequence, there has been a rapid increase in functional foods and beverages, especially fermented foods, available to consumers in many markets of the world. Among a number of functional compounds recognized so far, bioactive compounds from fermented foods and probiotics certainly take center stage because of their long tradition of safe use and established and postulated beneficial effects [91]. In addition, it is necessary to eliminate or replace more traditional chemical processes with faster, cheaper, and better enzymatic or fermentation methods [87].

At the same time, the interest in food biopreservation has also been increasing during recent years. The demands for minimally processed and fresher foods that do not contain chemical preservatives, ready-to-eat foods, or functional foods and nutraceuticals could be satisfied, at least in part, by the application of biopreservatives [92,93].

Notable advances in bioprocessing to obtain healthier ingredients and foods are appearing. The next sections will describe the production of some types of biomolecules by LAB, mainly vitamins, bioactive peptides, and antimicrobial compounds.

11.5.1 Vitamin Production

Vitamins play an important role in human health; their molecular functions can vary from their involvement in genetic regulation to their antioxidant properties. Vitamins are essential organic compounds that need to be obtained from the diet because of the inability of humans to synthesize them. Vitamin production by LAB has recently gained the attention of the scientific community. Although most LAB are auxotrophic for several vitamins, it is known that certain strains have the capability to synthesize group B vitamins (thiamine, folate, and riboflavin) and vitamin K (Figure 11.3). The use of selected starter cultures for fermentation potentially provides routes not only to enhance the nutritional food profile but also to deliver microorganisms to the gut where they can synthesize such vitamins in vivo [94,95].

It has been reported that two bifidobacteria strains (*Bifidobacterium infantis* and *B. longum*) can increase the levels of riboflavin and thiamine during a 48-h fermentation in soymilk [96]. Jayashree et al. [97] used the riboflavin-producing strain

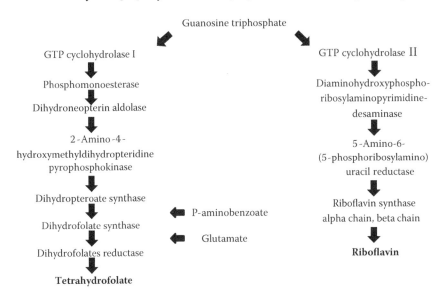

FIGURE 11.3 Schematic pathways involved in folate and riboflavin biosynthesis [94].

Lactobacillus fermentum MTCC8711 for the fermentation of different fermented milk products and obtained 2.29 mg/L of riboflavin after 24 h of growth in a chemically defined medium. The authors reported that this strain could be further exploited for the enhanced production of riboflavin and proposed that it could be used to replace the conventional strains that are being employed in LAB-based fermented products.

Recently, two natural riboflavin-overproducing strains were isolated from durum wheat samples. They are used for fermenting dough during the preparation of bread and pasta to enhance their vitamin B2 content [98].

Folate is an essential component of the human diet and is involved in nucleotide and cofactor biosynthesis in many metabolic reactions [99]. It has been reported that milk contains approximately 6 µg/100 g folate, and, in certain cheeses, relatively higher quantities are found [100]. However, the levels in fermented milk have been shown to be higher with folate concentrations in excess of 14 µg/100 g folate detected in yogurt fermented by *Streptococcus thermophilus* [101]. This increased level of folate is a result of the metabolic activity of LAB and *Bifidobacterium* spp. during the fermentation process. In contrast to bifidobacteria, it has been reported that strains of *Lactobacillus*, when used as both starters and probiotics, generally utilize more folate than they produce. However, there are exceptions, as a number of *Lactobacillus* strains, including *Lb. plantarum*, *Lb. acidophilus*, and *Lactobacillus delbrueckii* subsp. *bulgaricus*, are able to generate excess folate in the fermentation of dairy products. Propionibacteria have also been reported to produce bioactive fatty acid conjugated linoleic acid (CLA), vitamin B12, and folate [94].

The use of folate-producing probiotics has recently been proposed to efficiently confer protection against inflammation and cancer by both exerting the beneficial effects of probiotics and preventing the folate deficiency that is associated with premalignant changes in the colonic epithelia [102].

Vitamin K plays an important role in blood-clotting, tissue calcification, and atherosclerotic plaque and tissues, including the bones and kidneys. It is also essential for the synthesis of vitamins. This vitamin occurs in two forms: phylloquinone (vitamin K1) and menaquinone (MK, vitamin K2). Some genera of LAB, including *Lactococcus*, *Lactobacillus*, *Enterococcus*, *Bifidobacteria*, *Leuconostoc*, and *Streptococcus*, have been reported to produce MK [103]. Strains of *Lb. acidophilus*, *B. longum*, *Enterococcus faecium*, and *Lactococcus lactis* subsp. *lactis* can be used in multispecies probiotic fermented products [104].

Vitamin-producing LAB have led to the generation of novel fermented foods with increased and bioavailable vitamins. In addition, the use of genetic engineering strategies could be used to increase vitamin production or to create novel vitamin-producing strains, and the use of vitamin-producing LAB could be a cost-effective alternative to current vitamin fortification programs and be useful in the production of novel vitamin-enriched products [105].

11.5.2 PRODUCTION OF MILK BIOACTIVE PEPTIDES

Milk proteins (casein and whey proteins) and milk fat contain a wide range of biologically active compounds that have specific activities, and many milk protein-derived peptides have more than one functional role. For example, peptides from β-casein show immunostimulatory, opioid, and angiotensin-converting enzyme (ACE)-inhibitory activities; peptides from α_{s1}-casein show immunomodulatory and ACE-inhibitory activities; the opioid peptides α- and β-lactorphin also exhibit ACE-inhibitory activity; and the casein phosphopeptides possess immunomodulatory properties. Other examples of the physiological functions of food-derived bioactive peptides include antihypertensive, antithrombotic, antioxidative, antimicrobial, immunomodulatory, opioid, or mineral-binding activities [106–108], and these activities depend on their amino acid sequence [109]. These compounds present physicochemical and physiological properties that play an important role as health-promoting ingredients or nutraceuticals. Some of the biological activities of the milk protein component are latent and released only upon proteolysis. The activation of the latent biologically bioactive compounds usually occurs during the digestion of milk in the gut and during fermentation and processing [109]. Thus, milk bioactive peptides can be produced from milk proteins in different ways: enzymatic hydrolysis with digestive enzymes, the fermentation of milk with a proteolytic starter culture, and through the action of enzymes produced by proteolytic microorganisms. In comparison with digestive enzymes, microbial enzymes, either in the gut or in the food, use different cleavage sites. Thus, peptides released by exogenous enzymes may differ from those released by digestive enzymes [110]. The size of active sequences may vary from two to 20 amino acid residues, and many peptides are known to possess multifunctional properties [111]. Recently, Szwajkowska et al. [107] explained the influence of selected peptides released from bovine milk proteins on the immune system.

The enzymes produced by LAB are widely exploited by the dairy industry to hydrolyze milk proteins to produce the typical flavor, aroma, and texture of fermented products. *Lc. lactis*, *Lactobacillus helveticus*, and *Lb. delbrueckii* subsp. *bulgaricus* can release proteinase and peptidases that result in the production of bioactive peptides derived from milk proteins. In the gastrointestinal tract, digestive enzymes may hydrolyze milk proteins, which leads to a release of bioactive peptides. These compounds may provide protection against pathogenic microorganisms or pass through the intestinal wall into the blood circulation and reach a specific organ to modulate the neural, immune, vascular, or endocrine system [106].

Several studies have reported the production of ACE-inhibitory and immunomodulatory peptides using various proteolytic enzymes or proteolytic starter cultures. The *in vitro* incubation of milk proteins with gastrointestinal proteinase preparations (pepsin, trypsin, and chymotrypsin) results in the release of ACE-inhibitory peptides. Therefore, it is likely that ACE-inhibitory peptides can be generated during gastrointestinal transport. Commercial preparations of these proteinases are used in the large-scale manufacturing of bioactive peptides [112]. The antihypertensive and antimicrobial bioactive peptides derived from milk proteins have been reviewed by Haque and Chand [109].

Different combinations of proteinases (alcalase, chymotrypsin, pancreatin, pepsin, and bacterial or fungal enzymes) are used to produce bioactive peptides [106]. The proteolytic systems of lysed *Lc. lactis*, *Lb. helveticus*, and *Lb. delbrueckii* ssp. *bulgaricus* contain proteinase and intracellular peptidases, including endopeptidases, aminopeptidases, tripeptidases, and dipeptidases, which cause the degradation of casein into bioactive peptides in fermented dairy products [109].

ACE-inhibitory peptides can be released from α_{s1}- and β-casein and from α-Lactalbumin and β-lactoglobulin using different LAB. *Saccharomyces cerevisiae* and *Lb. helveticus* produce ACE-inhibitory peptides composed of Val-Pro-Pro and Ile-Pro-Pro, and *Kluyveromyces marxianus* var. *marxianus* releases a tetrapeptide composed of Tyr-Leu-Leu-Phe. Enzymes from *Lactobacillus* GG var. *casei* subsp. *rhamnonus* hydrolyzed α_{s1}-casein followed by treatment with pepsin–trypsin to simulate gastrointestinal conditions and release many bioactive peptides, such as immunosuppressive, ACE-inhibitory, immunomodulatory, and opioid peptides that present high activities.

A significant amount of research has shown the presence of bioactive peptides in fermented milk products. ACE-inhibitory peptides were found in ripened cheese (Manchego, Gouda, Emmental, Edam, Havarti, blue, and Camembert) and in Italian varieties. Peptides derived from α_{s1}- and β-casein presented strong antihypertensive effects [106,113,114].

Phosphopeptides have been identified in fermented milks and various types of cheese. These compounds are able to chelate various metals, such as Ca^{2+}, Zn^{2+}, Mn^{2+}, and Fe^{2+}, which enhance mineral solubility at an intestinal pH and the absorption of minerals [106].

11.5.3 PRODUCTION OF ANTIMICROBIAL COMPOUNDS

Antimicrobial compounds derived from natural sources have received increased attention over the last decade. Currently, more than 500 different antibacterial peptides have been isolated and characterized from a wide variety of organisms.

Several probiotic microorganisms, especially LAB with antimicrobial properties, have been commonly associated with foods. In addition to acid production, some probiotic LAB strains have the ability to produce a variety of other antimicrobial compounds, which function as a natural means of competition with other microorganisms that share the same niche. The main compound metabolized by those cultures is lactose to lactic acid, which lowers the pH and changes the environment into an unfavorable media for the development of pathogens and spoilage organisms [91,115,116].

LAB as competitive microbiota have a long history of application in fermented foods. Because of their metabolic properties, LAB are generally employed because of their positive contribution to the flavor, texture, and nutritional value in food products and their natural antimicrobial properties, which extend the product's shelf life.

Interest in food biopreservation—the extension of shelf life and enhanced safety of foods by the use of natural or controlled microbiota and/or antimicrobial compounds for food preservation [117]—has been increasing during recent years and has become increasingly popular for several reasons: natural food preservation methods that do not affect health are considered favorable for consumers and should have a smaller impact on food nutritional and sensory properties (as opposed to chemical or physicochemical treatments); these methods can reduce processing costs while simultaneously extending the shelf life of the product and do not require advanced technological equipment or skills and, therefore, can be exploited by smaller economies; these methods offer new possibilities for solving emerging issues, such as the increase of antibiotic resistance in the food chain, the need to improve animal productivity by natural means, and the control of emerging pathogens [118,119].

The preservative effect of LAB is a result of antimicrobial properties resulting from the production of one or more active metabolites, such as organic acids (lactic, acetic, formic, propionic, and butyric acids), which intensify their action by reducing the pH of the media, and other substances, such as ethanol, fatty acids, acetoin, hydrogen peroxide, diacetyl, antifungal compounds (propionate, phenyl-lactate, hydroxyphenyl-lactate, cyclic dipeptides, and 3-hydroxy fatty acids), bacteriocins (nisin, reuterin, reutericyclin, pediocin, lacticin, enterocin, and others), and bacteriocin-like inhibitory substances (BLIS). However, there are other mechanisms that have been suspected of being involved in the killing or inhibition of growth of other related species of bacteria and/or pathogens. Many bacteriocins are produced by LAB, and this offers the possibility of manipulating food microbial ecosystems in a deliberate fashion. As a result, although their potential application as biopreservatives has not been fully developed, a large number of bacteriocins produced by probiotic LAB have been identified [93,115,120].

Several strategies have been used to incorporate or use biopreservatives in food: the direct use of LAB strains with proven antimicrobial activity as starter cultures or starter adjuncts (probiotic concept); the use of biopreservative preparations in the form of a previously fermented product; or the use of semipurified, purified, or chemically synthesized bacteriocins [93,121]. Foods can be supplemented with ex situ produced bacteriocin preparations or inoculated with a bacteriocin-producer strain under conditions that favor the production of bacteriocins in situ [122,123].

A variety of studies of the ex situ production of bacteriocins in the form of immobilized preparations, in which the partially purified bacteriocin or the concentrated cultured broth is bound to a carrier, have been proposed [124]. A recent advance in this field is the use of immobilized bacteriocins in the development of antimicrobial packaging. A polyethylene film containing immobilized bacteriocin 32Y from *Lactobacillus curvatus* reduced viable counts of *L. monocytogenes* during storage in packaged pork, steak, and ground beef as well as frankfurters [125,126].

In situ bacteriocin production offers several advantages when compared with *ex situ* production in regards to both legal aspects and costs. The lower cost of biopreservation processes may be highly attractive, especially for smaller economies and developing countries, where food safety may be seriously compromised [127].

Several researchers have studied the application of bacteriocin and/or strains of bacteriocin producers in dairy products, and they have demonstrated effectiveness against pathogenic bacteria, such as *Staphylococcus aureus*, *Escherichia coli*, *Salmonella* spp., *L. monocytogenes* [122,128,129], *Clostridium perfringens*, and *E. faecalis* [130].

Bacteriocins can also be used to promote quality rather than simply to prevent spoilage or safety problems. For example, bacteriocins can be used to control adventitious nonstarter biota, such as nonstarter LAB (NSLAB) in cheese and wine. Bacteriocins can also be applied in other ways to enhance food fermentation. During semihard and hard cheese production, bacteriocins can control the lysis of starter LAB cultures, which results in the release of intracellular enzymes and ultimately accelerates ripening and even improves flavor [93].

11.6   FUTURE DIRECTIONS

There is an enormous commercial interest in this area for food innovation, and both industrial and research institutes are working hard to achieve advances in functional fermented foods. However, the development of functional foods is still a challenge because the claimed benefits need to be validated and confirmed by credible scientific researches.

A trend in biotechnology that has gained the attention of the scientific community is the use of selected starter cultures for food fermentation in order to produce biomolecules: vitamins, bioactive peptides, and antimicrobial compounds. These novel fermented foods present increased biologically active compounds that have specific activities and/or antimicrobial compounds for food preservation.

Therefore, it is desirable to continue to expand our understanding of the effectiveness of the use of microorganisms as a biotechnological tool for quality assurance of fermented food products while, at the same time, retaining the sensory qualities and nutritional value.

In order to provide consumers with a more appropriate diet for specific nutritional needs, a new trend called nutrigenomics or personalized nutrition is being evaluated. This science is based on the application of the human genome to nutrition and personal health, providing individual dietary recommendations. Functional foods in this context could be specifically designed for individuals who showed a genetic tendency to develop certain diseases, thus preventing their occurrence.

11.7   FINAL REMARKS

Fermented foods and beverages are consumed worldwide and have been for a long time. Although well documented, their benefits for health and welfare are still under investigation. Many scientific papers have demonstrated that fermentation increases digestibility and can lead to detoxification and nutritional enrichment, and have discussed other advantages, such as enhancing food shelf life and improving sensory quality by adding flavor and modifying textures.

The presence of functional foods containing probiotic microorganisms is quickly growing in the food market as consumers are looking for healthy foods that will extend life. The success of the marketing of fermented functional foods is related to information campaigns about the health benefits of a specific product, which must be communicated to the consumer in a simple and accessible mode. The price also influences the purchase of the product. Generally, functional food costs approximately 30% to 50% more than conventional food. Consumers have to be willing to pay for functional foods. Hence, it is important to include scientific evidence in information campaigns to justify the functionality of the products. Some other limitations that must be overcome are a loss of functionality during processing, storage, commercialization, and consumption; a lack of compatibility with the food matrix; the development of undesirable flavors; and control of texture.

Nowadays consumers demand safety and high quality with minimal processing of foods. The food industry has faced this challenge and has been applying existing emerging technologies and recent advances in food fermentation to the development of novel products, such as probiotic fermented dairy products, and the production of bioactive compounds and biopreservatives by LAB, which are desirable from a nutritional and/or health promotion perspective.

REFERENCES

1. F. Leroy and L. De Vuyst, Lactic acid bacteria as functional starter cultures for the food fermentation industry, *Trends in Food Science and Technology* 15, 2004, 67–78.
2. J.L. Parada, C.R. Caron, A.B.P. Medeiros and C.R. Soccol, Bacteriocins of lactic acid bacteria: purification, properties and use as biopreservatives, *Brazilian Archives of Biology and Technology* 50, 2007, 521–42.
3. G. Giraffa, Studying the dynamics of microbial population during food fermentation, *FEMS Microbiology Reviews* 28, 2004, 251–60.
4. FAO, *Meat consumption,* Online, available at http://www.fao.org/ag/againfo/themes/en/meat/background.html (accessed October 15, 2011).
5. S. Damodaran, K.L. Parkin and O.R. Fennema, *Química de Los Alimentos,* Zaragoza: Acribia, 2010, pp. 1116.
6. K. Arihara, Strategies for designing novel functional meat products, *Meat Science* 74, 2006, 219–29.
7. M.S. Ammor and B. Mayo, Selection criteria for lactic acid bacteria to be used as functional starter cultures in dry sausage production, *Meat Science* 76, 2007, 138–46.
8. W. P. Hammes, Metabolism of nitrate in fermented meats: the characteristic feature of a specific group of fermented foods, *Food Microbiology* 30, 2011, 1–6.
9. S. Marianski and A. Marianski, *The Art of Making Fermented Sausages*, New York: Bookmagic, 2009, pp. 274.
10. J.M. Fernandéz-Ginés, J. Fernández-López, E. Sayas-Barberá and J.A. Pérez-Alvarez, Meat products as functional foods: A review, *Journal of Food Science* 70, 2005, 37–43.
11. G. Comi, R. Urso, L. Iacumin, K. Rantsiou, P. Cattaneo, C. Cantoni and L. Cocolin, Characterization of naturally fermented sausages produced in the north east of Italy, *Meat Science* 69, 2005, 381–92.
12. R. Talon, I. Lebert, A. Lebert, S. Leroy, M. Garriga, T. Aymerich, E.H. Drosinos, et al. Traditional dry fermented sausages produced in small-scale processing units in Mediterranean countries and Slovakia, *Microbial Ecosystems of Processing Environments* 77, 2007, 570–79.
13. F. Leroy, J. Verluyten and L. Vuyst, Functional meat starter cultures for improved sausage fermentation, *International Journal of Food Microbiology* 106, 2006, 270–85.
14. T. Komprda, D. Smelá, P. Pechová, L. Kalhotka, J. Stencl and B. Lejdus, Effect of starter culture, spice mix and storage time and temperature on biogenic amine content of dry fermented sausages, *Meat Science* 67, 2004, 607–16.
15. L.D. Vuyst, G. Falony and F. Leroy, Probiotics in fermented sausages: A review, *Meat Science* 80, 2008, 75–8.
16. G. Mauriello, A. Casaburi and F. Villani, Isolation and technological properties of coagulase negative *Staphylococci* from fermented sausages of Southern Italy, *Meat Science* 67, 2004, 149–55.
17. C. Pennacchia, E.E. Vaughan and F. Villani, Potential probiotic *Lactobacillus* strains from fermented sausages: further investigations on their probiotic properties, *Meat Science* 73, 2006, 90–101.
18. M. Ruusunen and E. Puolanne, Reducing sodium intake from meat products, *Meat Science* 70, 2005, 531–41.

19. J. Gotterup, K. Olsen, S. Knochel, L.H. Stahnke and J.K.S. Moller, Colour formation in fermented sausages by meat-associated *Staphylococci* with different nitrite and nitrate-reductase activities, *Meat Science* 78, 2008, 492–501.

20. E. Papamanoli, N. Tzanetakis, E. Litopoulou-Tzanetaki and P. Kotzekidou, Characterization of lactic acid bacteria isolated from a Greek dry-fermented sausage in respect of their technological and probiotic properties, *Meat Science* 65, 2003, 859–67.

21. R. Rebucci, L. Sangalli, M. Fava, C. Bersani, C. Cantoni and A. Baldi, Evaluation of functional aspects in *Lactobacillus* strains isolated from dry fermented sausages, *Journal of Food Quality* 30, 2007, 187–201.

22. M. Gandhi and M.L. Chikindas, Listeria: a foodborne pathogen that knows how to survive, *International Journal of Food Microbiology* 113, 2007, 1–15.

23. K.H. Steinkraus, *Handbook of Indigenous Fermented Food*, New York: Marcel Dekker Inc., 1996, 2nd ed.

24. E. Caplice and G.F. Fitzgerald, Food fermentations: Role of microorganisms in food production and preservation, *International Journal of Food Microbiology* 50, 1999, 131–49.

25. L. Settanni and A. Corsetti, Application of bacteriocins in vegetable food biopreservation, *International Journal of Food Microbiology* 121, 2008, 123–38.

26. B. Tamang and J.P. Tamang, Traditional knowledge of biopreservation of perishable vegetable and bamboo shoots in Northeast India as food resources, *Indian Journal of Traditional Knowledge* 8, 2009, 89–95.

27. R.W. Hutkins, *Microbiology and Technology of Fermented Foods*, Ames: Blackwell Publishing, 2006, pp. 233–59.

28. J. Prakash Tamang and K. Kailasapathy, *Fermented Foods and Beverages of the World*, London: CRC Press, 2010, pp. 149–90.

29. Y.H. Hui, S. Ghazala, D.M. Graham, K.D. Murrell and W. Nip, *Handbook of Vegetable Preservation and Processing*, London: CRC Press, 2003, pp. 1–9.

30. M.A. Daeschel and H.P. Fleming, Selection of lactic acid bacteria for use in vegetable fermentations, *Food Microbiology* 1, 1984, 303–13.

31. R.K. Pundir and P. Jain, Change in microflora in sauerkraut during fermentation and storage, *World Journal of Dairy and Food Science* 5, 2010, 221–5.

32. H.P. Fleming, R.F. McFeeters, R.L. Thompson and D.C. Sanders, Storage stability of vegetables fermented with pH control, *Journal of Food Science* 48, 1983, 975.

33. J.M. Jay, *Modern Food Microbiology*, New York: Chapman and Hall, 1996, 5th ed., pp. 163–4.

34. V.F. Plengvidhya, F. Breidt and H.P. Fleming, Use of RAPD-PCR as a method to follow the progress of starter cultures in sauerkraut fermentation, *International Journal of Food Microbiology* 93, 2007, 287–96.

35. Y.H. Hui, L. Meunier-Goddik, J. Josephsen, W. Nip and P.S. Stanfield, *Handbook of Food and Beverage Fermentation Technology*, London: CRC Press, 2004, pp. 768–76.

36. C.H. Lee, Lactic acid fermented foods and their benefits in Asia, *Food Control* 8, 1997, 259–69.

37. E. Park, J. Chun, C. Cha, W. Park, C.O. Jeon and J.W. Bae, Bacterial community analysis during fermentation of ten representative kinds of kimchi with barcoded pyrosequencing, *Food Microbiology* 30, 2012, 197–204.

38. H.P. Fleming, R.F. McFeeters, M.A. Daeschel, E.G. Humphries and R.L. Thompson, Fermentation of cucumbers in anaerobic tanks, *Journal of Food Science* 53, 1988, 127–33.

39. International Olive Council, *Olivae,* Madrid: Advantia S.A., 2011, No. 115, pp. 29–31.

40. E.R. Farnaworth, *Handbook of Fermented Functional Foods*, Boca Raton: CRC Press, 2008, 413–27.

41. E.Z. Panagou, U. Schillinger, C. Franz and G.H. Nychas, Microbiological and biochemical profile of cv. Conservolea naturally black olives during controlled fermentation with selected strains of lactic acid bacteria, *Food Microbiology* 25, 2008, 348–58.

42. J.L. Ruiz-Barba, D.P. Cathcart, P.J. Warner and R. Jimenez-Diaz, Use of *Lactobacillus plantarum* LPCO10, a bacteriocin producer, as a starter culture in Spanish-style green olive fermentations, *Applied Environmental Microbiology* 60, 1994, 2059–64.

43. J. Wang and D.Y.C. Fung, Alkaline-fermented foods: A review with emphasis on Pidan fermentation, *Critical Reviews in Microbiology* 22, 1996, 101–38.

44. K.H. Steinkraus, *Handbook of Indigenous Fermented Foods*, New York: Marcel Dekker, 1996, pp. 776.

45. K.H. Steinkraus, *Industrialization of Indigenous Fermented Foods*, New York: Marcel Dekker, 2004, pp. 796.

46. A. Blandino, M.E. Al-Aseeri, S.S. Pandiella, D. Cantero and C. Webb, Cereal-based fermented foods and beverages, *Food Research International* 36, 2003, 527–43.

47. A.G. Nassar, A.E. Mubarak and A.E. El-Beltagy, Nutritional potential and functional properties of tempeh produced from mixtures of different legumes: Chemical composition and nitrogenous constituent, *International Journal of Food Science and Technology* 43, 2008, 1754–8.

48. M.J.R. Nout and J.L. Kiers, Tempeh fermentation, innovation and functionality: Update into the third millennium, *Journal of Applied Microbiology* 98, 2005, 789–805.

49. H. Namgung, H. Park, I.H. Cho, H. Choi, D. Kwon, S. Shima and Y. Kima, Metabolite profiling of doenjang, fermented soybean paste, during fermentation, *Journal of the Science of Food and Agriculture* 90, 2010, 1926–35.

50. S. Liu, Y. Han and Z. Zhou, Lactic acid bacteria in traditional fermented Chinese foods, *Food Research International* 44, 2011, 643–51.

51. K.H. Steinkraus, Classification of fermented foods: Worldwide review of household fermentation techniques, *Food Control* 8, 1997, 311–7.

52. J. van Hylckama Vlieg, P. Veiga, C. Zhang, M. Derrien and L. Zhao, Impact of microbial transformation of food on health from fermented foods to fermentation in the gastrointestinal tract, *Current Opinion in Biotechnology* 22, 2011, 211–9.

53. M.J.A. Marcon, G.C.N. Vieira, K.N. Simas, K. Santos, M.A. Vieira, R.D.M.C. Amboni and E.R. Amante, Effect of the improved fermentation on physicochemical properties and sensorial acceptability of sour cassava starch, *Brazilian Archives of Biology and Technology* 50, 2007, 1073–81.

54. I.M. Demiate and V. Kotovicz, Cassava starch in the Brazilian food industry, *Ciência e Tecnologia de Alimentos* 31, 2011, 388–97.

55. I.M. Demiate, A.C. Barana, M.P. Cereda and G. Wosiacki, Organic acid profile of commercial sour cassava starch, *Ciência e Tecnologia de Alimentos* 19, 1999, 131–5.

56. I.M. Demiate, N. Dupuy, J. Huvenne, M.P. Cereda and G. Wosiacki, Relationship between baking behavior of modified cassava starches and starch chemical structure determined by FTIR spectroscopy, *Carbohydrate Polymers* 42, 2000, 149–58.

57. C.M. Burgess, E.J. Smid and D. van Sinderen, Bacterial vitamin B2, B11 and B12 overproduction: An overview, *International Journal of Food Microbiology* 133, 2009, 1–7.

58. A. Bhatia and N. Khetarpaul, Development of an indigenously fermented Indian bread—doli ki roti—effect on phytic acid content and in vitro digestibility of starch and protein, *Nutrition and Food Science* 39, 2009, 330–6.

59. S. Arora, S. Jood and N. Khetarpaul, Effect of germination and probiotic fermentation on nutrient profile of pearl millet based food blends, *British Food Journal* 113, 2011, 470–81.

60. N.D. Buckley, C.P. Champagne, A.I. Masotti, L.E. Wagar, T.A. Tompkins and J.M. Green-Johnson, Harnessing functional food strategies for the health challenges of space travel: Fermented soy for astronaut nutrition, *Acta Astronautica* 68, 2011, 731–8.

61. I. Siró, E. Kápolna, B. Kápolna and A. Lugasi, Functional food, product development, marketing and consumer acceptance: A review, *Appetite* 51, 2008, 456–67.

62. M.B. Roberfroid, What is beneficial for health? The concept of functional food, *Food and Chemical Toxicology* 37, 1999, 1039–41.

63. J.T. Spence, Challenges related to the composition of functional foods, *Journal of Food Composition and Analysis* 19, 2006, 4–6.
64. C.T. Yamaguishi, M.R. Spier, J. De Dea Lindner, V.T. Soccol and C.R. Soccol, Current market trends and future directions, in M. Liong, *Probiotics: Biology, Genetics and Health Aspects*, London: Springer, 2011, pp. 299–319.
65. Food and Agriculture Organization/World Health Organization (FAO/WHO), Report of a joint FAO/WHO Expert Consultation on Evaluation of Health and Nutritional Properties of Probiotics in Food including Powder Milk with Live Lactic Acid Bacteria, Córdoba, Argentina. Online. 2001. Available at http://www.who.int/foodsafety/publications/fs_management/en/probiotics.pdf (accessed October 24, 2011).
66. S.M. Fox, Probiotics: Intestinal inoculants for production animals, *Veterinary Medicine* 83, 1988, 806–30.
67. X. Chen, E.G. Kokkotou, N. Mustafa, K. Ramakrishnan Bhaskar, S. Sougioultzis, M. O'Brien, C. Pothoulakis and C.P. Kelly, *Saccharomyces boulardii* inhibits ERK1/2 mitogen-activated protein kinase activation both in vitro and in vivo and protects against *Clostridium difficile* toxin A-induced enteritis, *Journal of Biological Chemistry* 281, 2006, 24,449–54.
68. C.R. Soccol, L.P.S. Vandenberghe, M.R. Spier, A.B.P. Medeiros, C.T. Yamaguishi, J. De Dea Lindner, A. Pandey and V.T. Soccol, The potential of probiotics: A review, *Food Technology and Biotechnology* 48, 2010, 413–34.
69. L. De Vuyst, G. Falony and F. Leroy, Probiotics in fermented sausages, *Meat Science* 80, 2008, 75–8.
70. E. Isolauri, M. Juntunen, T. Rautanen, P. Sillanaukee and T. Koivula, A human *Lactobacillus* strain (*Lactobacillus casei* sp. *strain* GG) promotes recovery from acute diarrhea in children, *Pediatrics* 88, 1991, 90–7.
71. H. Majamaa, E. Isolauri, M. Saxelin and T. Vesikari, Lactic acid bacteria in the treatment of acute rotavirus gastroenteritis, *Journal of Pediatric Gastroenterology and Nutrition* 20, 1995, 333–8.
72. A. Donnet-Hughes, F. Rochat, P. Serrant, J.M. Aeschlimann and E.J. Schiffrin, Modulation of nonspecific mechanisms of defense by lactic acid bacteria: Effective dose, *Journal of Dairy Science* 82, 1999, 863–9.
73. Y.H. Sheih, B.L. Chiang, L.H. Wang, L.K. Chuh and H.S. Gill, Systemic immunity enhancing effect in healthy subjects following dietary consumption of the lactic acid bacterium *Lactobacillus rhamnosus* HN001, *Journal of American College Nutrition* 20, 2001, 149–56.
74. E. Isolauri, T. Arvola, Y. Sutas, E. Moilanen and S. Salminen, Probiotics in the management of atopic eczema, *Clinical and Experimental Allergy* 30, 2000, 1604–10.
75. M. Hosada, H. Hashimoto, D. He, H. Morita and A. Hosono, Effect of administration of milk fermented with *Lactobacillus acidophilus* LA-2 on faecal mutagenicity and microflora in human intestine, *Journal of Dairy Science* 79, 1996, 745–9.
76. P.D. Midolo, J.R. Lambert, R. Hull, F. Luo and M.L. Grayson, In vitro inhibition of *Helicobacter pylori* NCTC 11637 by organic acids and lactic acid bacteria, *Journal of Applied Bacteriology* 79, 1995, 475–9.
77. C. Rodríguez, M. Medici, F. Mozzi and G.F. Valdez, Therapeutic effect of *Streptococcus thermophilus* CRL 1190-fermented milk on chronic gastritis, *World Journal of Gastroenterology* 16, 2010, 1622–30.
78. J.Y. Xu, L.Q. Qin, P.Y. Wang, W. Li and C. Chang, Effect of milk tripeptides on blood pressure: A meta-analysis of randomized controlled trials, *Nutrition* 24, 2008, 933–40.
79. J.L. Parada, C.R. Soccol, L.A. da Costa and A. Pandey, Potential of probiotics in the food industry, in C.R. Soccol, A. Pandey, V.T. Soccol and C. Larroche (eds.), *Advances in Bioprocesses in Food Industry*, New Delhi: Asiatech Publishers, 2011, pp. 94–120.

80. F.C. Prado, J.L. Parada, A. Pandey and C.R. Soccol, Trends in non-dairy probiotic beverages, *Food Research International* 41, 2008, 111–23.
81. W. Krasaekoopt, B. Bhandari and H. Deeth, Evaluation of encapsulation techniques of probiotics for yoghurt, *International Dairy Journal* 13, 2003, 3–13.
82. V. Chandramouli, K. Kailasapathy, P. Peiris and M. Jones, An improved method of microencapsulation and its evaluation to protect *Lactobacillus* spp. in simulated gastric conditions, *Journal Microbiological Methods* 56, 2004, 27–35.
83. A.Y. Tamime, M. Saarela, A. Korslund Søndergaard, V.V. Mistry and N.P. Shah, Production and maintenance of viability of probiotic micro-organisms in dairy products, in A.Y. Tamime (ed.), *Probiotic Dairy Products*, Oxford, UK: Blackwell Publishing, 2005, pp. 44–51.
84. G.R. Gibson and M.B. Roberfroid, Dietary modulation of the human colonic microbiota: Introducing the concept of prebiotics, *Journal of Nutrition* 125, 1995, 1401–12.
85. A. Zuleta, M.I. Sarchi, M.E. Rio, M.E. Sambucetti, M. Mora, S.V. de Fabrizio and J.L. Parada, Fermented milk-starch and milk-inulin products as vehicles for lactic acid bacteria, *Plant Foods for Human Nutrition* 59, 2004, 55–160.
86. K. Menrad, Market and marketing of functional food in Europe, *Journal of Food Engineering* 56, 2003, 181–8.
87. B.C. Saha, *Fermentation Biotechnology*, American Chemical Society Symposium Series, Boston: Oxford University Press, 2003, pp. 13–7.
88. T.H. Walker, C.M. Drapcho and F. Chen, Bioprocessing technology for production of nutraceutical compounds, in Shi, J., *Functional Food and Nutraceuticals: Processing Technologies*, Boca Raton: CRC Press, 2007, pp. 211–36.
89. J. Shi, L.S. Kassama and Y. Kakuda, Supercritical fluid technology for extraction of bioactive components, in Shi, J., *Functional Food and Nutraceuticals: Processing Technologies*, Boca Raton: CRC Press, 2007, pp. 3–44.
90. A.H. Soomro, T. Masud and K. Anwaar, Role of lactic acid bacteria (LAB) in food preservation and human health: A review, *Pakistan Journal of Nutrition* 1, 2002, 20–4.
91. T. Vasiljevic and N.P. Shah, Probiotics: From Metchnikoff to bioactives, *International Dairy Journal* 18, 2008, 714–28.
92. A. Robertson, C. Tirado, T. Lobstein, M. Jermini, C. Knai, J.H. Jensen, A. Ferro-Luzzi and W.P.T. James, Food and health in Europe: A new basis for action, *WHO Regional Publications, European Series* 96, 2004, 1–385.
93. P.D. Cotter, C. Hill and R.P. Ross, Bacteriocins: Developing innate immunity for food, *Nature Reviews Microbiology* 3, 2005, 777–88.
94. E.B. O'Connor, E. Barrett, G. Fitzgerald, C. Hill, C. Stanton and R.P. Ross, Production of vitamins, exopolysaccharides and bacteriocins by probiotic bacteria, in A.Y. Tamime, *Probiotic Dairy Products*, Oxford: Blackwell Publishing Ltd., 2005, pp. 167–94.
95. J.G. LeBlanc, M.P. Taranto, V. Molina and F. Sesma, B-group vitamins production by probiotic lactic acid bacteria, in F. Mozzi, R. Raya and G. Vignolo, *Biotechnology of Lactic Acid Bacteria: Novel Applications*, Ames: Wiley-Blackwell, 2010, pp. 211–32.
96. J.W. Hou, R.C. Yu and C.C. Chou, Changes in components in soymilk during fermentation with bifidobacteria, *Food Research International* 33, 2000, 393–7.
97. S. Jayashree, K. Jayaraman and G. Kalaichelvan, Isolation, screening and characterization of riboflavin producing lactic acid bacteria from Katpadi, Vellore district, *Recent Research Science and Technology* 2, 2010, 83–8.
98. V. Capozzi, V. Menga, A.M. Digesu, P. De Vita, D. van Sinderen, L. Cattivelli, C. Fares and G. Spano, Biotechnological production of vitamin B2-enriched bread and pasta, *Journal of Agriculture and Food Chemistry* 59, 2011, 8013–20.
99. E. Giovannucci, Epidemiologic studies of folate and colorectal neoplasia: A review, *Journal of Nutrition* 132, 2002, 2350–5.

100. P.F. Fox and P.L.H. McSweeney, *Dairy Chemistry and Biochemistry*, London: Blackie Academic & Professional, 1998, pp. 265–93.

101. E.J. Smid, M. Starrenburg, I. Mireau, W. Sybesma and J. Hugenholtz, Increase of folate level in fermented foods, *Innovations in Food Technology*, 2001, 13–5.

102. M. Rossi, A. Amaretti and S. Raimondi, Folate production by probiotic bacteria, *Nutrients* 3, 2011, 118–34.

103. T. Morishita, N. Tanura, T. Makino and S. Kudo, Production of manaquinones by lactic acid bacteria, *Journal of Dairy Science* 82, 1999, 1879–903.

104. H.M. Timmermann, C.J.M. Konning, L. Mulder, F.M. Rombouts and A.C. Beynen, Monostrain, multistrain and multispecies probiotics: A comparison of functionality and efficacy, *International Journal of Food Microbiology* 96, 2004, 219–33.

105. J.G. LeBlanc, J.E. Laiño, M. Juarez del Valle, V. Vannini, D. van Sinderen, M.P. Taranto, G. Font de Valdez, G. Savoy de Giori and F. Sesma, B-group vitamin production by lactic acid bacteria: Current knowledge and potential applications, *Journal of Applied Microbiology*, Online-version 10, 2011.

106. H. Korhonen and A. Pihlanto-Leppälä, Food-derived bioactive peptide: opportunities for designing future foods, *Current Pharmaceutical Design* 9, 2003, 1297–308.

107. M. Szwajkowska, A. Wolanciuk, J. Barlowska, J. Król and Z. Litwinczuk, Bovine milk protein as the source of bioactive peptides influencing the consumers' immune system: a review, *Animal Science Papers* 29, 2011, 269–80.

108. J.T. Ryan, R.P. Ross, D. Bolton, G.F. Fitzgerald and C. Stanton, Bioactive peptides from muscle sources: Meat and fish, *Nutrients* 3, 2011, 765–91.

109. E. Haque and R. Chand, Antihypertensive and antimicrobial bioactive peptides from milk proteins, *European Food Research Technology* 227, 2008, 7–15.

110. N.P. Möller, K.E. Scholz-Ahrens, N. Roos and J. Schrezenmeir, Bioactive peptides and proteins from foods: Indication for health effects, *European Journal of Nutrition* 47, 2008, 171–82.

111. H. Meisel and R.J. FitzGerald, Biofunctional peptides from milk proteins: Mineral binding and cytomodulatory effects, *Current Pharmaceutical Design* 9, 2003, 1289–95.

112. N. Yamamoto, M. Ejiri and S. Mizuno, Biogenic peptides and their potential use, *Current Pharmaceutical Design* 9, 2003, 1345–55.

113. T. Saito, T. Nakamura, H. Kitazawa, Y. Kawai and T. Itoh, Isolation and structural analysis of antihypertensive peptides that exist naturally in Gouda cheese, *Journal of Dairy Science* 83, 2000, 1434–40.

114. J.A. Gomes-Ruiz, M. Ramos and I. Recio, Angiotensin-converting enzyme-inhibitory peptides in Manchego cheeses manufactured with different starter cultures, *International Dairy Journal* 12, 2002, 697–706.

115. L. De Vuyst and F. Leroy, Bacteriocins from lactic acid bacteria: Production, purification, and food applications, *Journal of Molecular Microbiology and Biotechnology* 13, 2007, 194–9.

116. R.B.P. Oliveira, A.L. Oliveira and M.B.A. Glória, Screening of lactic acid bacteria from vacuum packaged beef for antimicrobial activity, *Brazilian Journal of Microbiology* 39, 2008, 368–74.

117. S. Annou, M. Maqueta, M. Martínez-Bueno and E. Valdivia, Biopreservation, an ecological approach to improve the safety and shelf-life of foods, *Applied Microbiology* 1, 2007, 475–86.

118. B. Kos, J. Suskovic, J. Beganovic, K. Gjuracic, J. Frece, C. Iannaccone and F. Canganella, Characterization of the three selected probiotic strains for the application in food industry, *World Journal of Microbiology and Biotechnology* 24, 2008, 699–707.

119. A. Gálvez, H. Abriouel, N. Benomar and R. Lucas, Microbial antagonists to food-borne pathogens and biocontrol, *Current Biotechnology* 21, 2010, 1–7.

120. A. Sobrino-Lopez and O. Martín-Belloso, Review: Use of nisin and other bacteriocins for preservation of dairy products, *International Dairy Journal* 18, 2008, 329–43.
121. S.F. Deraz, E.N. Karlsson, A.A. Khalil and B. Mattiasson, Mode of action of acidocin D20079, a bacteriocin produced by the potential probiotic strain, *Lactobacillus acidophilus* DSM 20079, *Journal of Industrial Microbiology and Biotechnology* 34, 2007, 373–9.
122. U. Schillinger, R. Geisen and W.H. Holzapfel, Potential of antagonistic microorganisms and bacteriocins for the biological preservation of foods, *Trends in Food Science and Technology* 7, 1996, 158–64.
123. M.E. Stiles, Biopreservation by lactic acid bacteria, *Antonie van Leeuwenhoek* 70, 1996, 331–45.
124. A. Gálvez, H. Abriouel, R.L. López and N.B. Omar, Bacteriocin-based strategies for food biopreservation, *International Journal of Food Microbiology* 120, 2007, 51–70.
125. G. Mauriello, D. Ercolini, A. La Storia, A. Casaburi and F. Villani, Development of polythene films for food packaging activated with an antilisterial bacteriocin from *Lactobacillus curvatus* 32Y, *Journal of Applied Microbiology* 97, 2004, 314–22.
126. D. Ercolini, A. La Storia, F. Villani, G. Mauriello, Effect of a bacteriocin-activated polythene film on *Listeria monocytogenes* as evaluated by viable staining and epifluorescence microscopy, *Journal of Applied Microbiology* 100, 2006, 765–72.
127. W.H. Holzapfel, Appropriate starter culture technologies for small-scale fermentation in developing countries, *International Journal of Food Microbiology* 75, 2002, 197–212.
128. M. Jamuna, S.T. Babusha and K. Jeevaratnam, Inhibitory efficacy of nisin and bacteriocins from *Lactobacillus* isolates against food spoilage and pathogenic organisms in model and food systems, *Food Microbiology* 22, 2005, 449–54.
129. J.G. Máqueza and C.E.G. Rojas, Efecto de la nisina sobre la microflora patógena del queso blanco artesanal tipo "telita" elaborado en una quesera de Upata, Estado Bolívar, Venezuela, *Revista de la Sociedad Venezolana de Microbiologia* 27, 2007, 108–11.
130. R. Bromberg, I. Moreno, R.R. Delboni and H.C. Cintra, Características da bacteriocina produzida por *Lactococcus lactis* ssp. *hordniae* CTC 484 e seu efeito sobre *Listeria monocytogenes* em carne bovina, *Ciência e Tecnologia de Alimentos* 26, 2006, 135–44.

12 Industrial Fermentation for Production of Alcoholic Beverages

Saurabh Jyoti Sarma, Mausam Verma, and Satinder Kaur Brar

CONTENTS

12.1   INTRODUCTION

The term "fermentation" is derived from the Latin word *fervere*, which means "to boil." Microbiologically, the term fermentation describes any process where microbial biomass is used for production of a product. However, from a biochemical point of view, it is an anaerobic energy-generating process where both the electron donor and acceptor are organic compounds, and neither oxygen nor any other inorganic compound acts as an electron acceptor (Stanbury 1988). Fermented foods and beverages have been produced since the Neolithic age without knowing the role of microorganisms in this process. In the mid-1600s, A.V. Leeuwenhoek (of The Netherlands) observed "animalcules" in a sample of fermented wine and beer (Fleet 1998).[1] Nearly 200 years after Leeuwenhoek's observation, in 1861, Louis Pasteur (of France) discovered that yeast is involved in fermentation, and it can convert sugar to ethanol and carbon dioxide. In 1883, Emil Christian Hansen from the Carlsberg brewery (in Denmark) first isolated yeast and used it for the fermentation of beer (Fleet 1998; Dequin 2001). A few years later, in 1890, a pure culture of wine yeast was isolated by Müller-Thurgau from Geisenheim (Germany) (Dequin 2001). In continuation with these achievements, application of pure yeast cultures for alcoholic beverage fermentation has been considered. For industrial production of alcoholic beverages, application of selected yeast strains as inoculum has certain advantages. First, it can dominate the indigenous yeast or other fungus present in raw materials, which may otherwise be involved in spoilage or production of undesirable compounds. Second, selected yeast with high production efficiency can be applied, and the final product quality can be predicted. Further, using recent tools for molecular microbiology and genetic engineering, novel and more-efficient yeast strains have been developed. Successful commercial exploitation of such novel microbial systems and other recent technological developments in this field have helped the global alcoholic beverage industry to grow rapidly. Presently, the global alcoholic beverage market is US$900 billion, and it is still growing with active participation of developing regions, such as China, India, and Russia.[2]

The word "alcohol" was derived from Arabic *al-kuhul*, and the type of alcohol found in alcoholic beverages is ethanol or ethyl alcohol, which is produced by the fermentation of fruits or grains.[3] For different alcoholic beverage types, different production processes are used, and hence, there is a variation in their final alcohol content. Besides alcohol content, flavor, color, CO_2 content, and mouthfeel may also vary among different beverage types. Even the overall quality of two different batches of the same beverage type or the same beverage from two different producers may vary significantly. There may be a number of reasons for such variations; first of all, the chemical makeup of the raw material influences the beverage quality. In turn, the quality of the raw material depends on its genetic variety, geographical origin and cultivation, and harvesting techniques followed. Further, maturation is an important part of the alcoholic beverage fermentation process, and it is mainly responsible for imparting typical color and flavor to the final product. Traditionally, maturation is achieved by using special wooden vessels, and the physical as well as the chemical properties of the wood surface play a vital role in determining the overall quality of the beverages. Thus, the purpose of this chapter is to provide a brief outline on the different varieties of alcoholic beverages, the technical aspects of their production processes, and recent advances in this field.

12.2 TYPES OF ALCOHOLIC BEVERAGES

In general, alcoholic beverages can be divided into two major groups: fermented beverages and distilled beverages. Each of these groups can be further divided into two subdivisions: wine and beer, and spirits and liqueurs.

A generalized classification of alcoholic beverages is presented in Figure 12.1. Wine is a naturally fermented fruit juice commonly produced from grapes. Grapes are first crushed and then yeasts are applied for natural fermentation without the external addition of sugar, acid, enzymes, or a nutrient solution. Depending upon the variety of grape or yeast strain applied, the quality and name of the wine may be different. Besides grapes, wine can also be produced from other fruit juices, such as apple or berry, and is then known as fruit wine or according to the name of the respective fruit, such as apple wine. Usually, the alcohol content of wines may vary from 10% to 14% (v/v), which may be 14% to 20% (v/v) in some special types, such as dessert wine.

Similarly, beer is another type of fermented beverage produced by the fermentation of sugar derived mainly from malted barley or malted wheat. Hops are used as a flavoring agent and a natural preservative, and they give beer its typical bitter taste. Sometimes, certain herbs and some fruits may also be used as flavoring additives. Usually, the alcohol content of beer is 4% to 6% (v/v); however, it may be as much as 20% (v/v) in some rare cases.

Spirits are distilled beverages where the alcohol concentration is further increased by distilling a fermented product of grain, fruits, or vegetables. The alcohol content of spirits is usually more than 20% (v/v), and whiskey, rum, gin, vodka, and brandy are some well-known spirit types. Liqueurs are also a kind of distilled beverage containing added sugar and flavor from fruits, flowers, nuts, herbs, and spices. They are

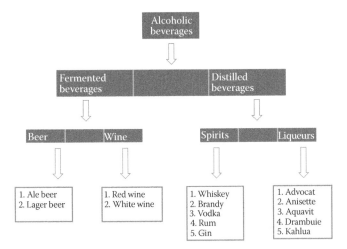

FIGURE 12.1 Different types of alcoholic beverages.

sweet in nature, and the alcohol content is usually 15% to 30% (v/v), which is lower than most spirits.

Similarly, each of these four subdivisions of alcoholic beverages has a number of varieties based on variations in the raw material (substrate) used for fermentation, alcohol content, flavor, geographic origin, or nature of postfermentation processing.

12.3 MICROBIOLOGY OF ALCOHOLIC BEVERAGE FERMENTATION

The production of ethanol by the fermentation of sugar is a relatively simple technique. However, for production of alcoholic beverages such as beer or wine, the main aim of the fermentation process is to obtain balanced flavor as well as a high ethanol yield (Linko et al. 1998). As mentioned earlier, yeast is the organism responsible for the fermentation process, and during alcoholic beverage fermentation, sugar present in grape juice or barley wort is converted to ethanol by the activity of yeast enzymes. Therefore, application of the proper yeast strain is the most important step in alcoholic beverage fermentation on which the success of the entire process depends. Yeasts are unicellular organisms, which are nonfilamentous fungi and reproduce either by budding or fission (Mohamudha and Begum 2005). Based on their behavior during the brewing process, yeasts are categorized as top-fermenting and bottom-fermenting yeasts (Cudlínová et al. 2011). Actually, the basis of this classification is the fact that top-fermenting yeasts are capable of forming a foam layer at the top of the liquid during fermentation; however, others do not. Top-fermenting yeasts are very active and grow rapidly at 20°C; however, bottom-fermenting yeasts grow slowly and are suitable for only low temperature (10°C–15°C) fermentation (Mohamudha and Begum 2005). *Saccharomyces cerevisiae* is an example of a top-fermenting yeast and is commonly used for alcoholic beverage production. It is also commonly known as wine yeast, brewer's yeast, or ale yeast (McKenzie et al. 1990; Jin and Alex 1998; Pretorius 2000). Bottom-fermenting yeasts are mainly used for production of lager-type beers, and *Saccharomyces pastorianus*, formerly known as *S. carlsbergensis*, is a good example of a bottom-fermenting yeast. Grape juice used for wine production contains wild yeast, which is a natural flora of grape skin. It is also capable of producing alcohol from grape juice; however, in such cases, the final product may also contain some unknown compound, and its quality will not be predictable. Further, growth of certain yeasts may lead to spoilage of beer or wine (Rodrigues et al. 2001). *Zygosaccharomyces bailii*, *Saccharomycodes ludwigii*, *Breflanomyces intermedius*, and some *Candida* species are such yeasts, and they can spoil alcoholic beverages. They may cause a secondary fermentation of wine and production of large amounts of CO_2, acetic acid, esters, and acetaldehyde. A film layer may develop as a result of flocculant masses, and volatile phenols may result in medicinal, phenolic, horsey, and barnyard taints (Toit and Pretorius 2000). Therefore, for reliable and predictable alcoholic beverage fermentation, a pure yeast culture is preferably used, one which can quickly dominate the fermentation so that wild yeasts are repressed. As mentioned above, the most commonly used yeast species is *S. cerevisiae*; however, all strains of the species are not suitable for fermentation. Different *S. cerevisiae* strains may be physiologically different from each other, and hence, their fermentative properties are also different (Dunn et al. 2005). Therefore, the quality of the final product is directly related to the strain of yeast selected for fermentation. *S. cerevisiae* (Lalvin QA23), *S. cerevisiae* (AWRI796), *S. cerevisiae* (Vin 13), and *S. cerevisiae* (VL 3) are some of the commercial strains used for wine production. Similarly, *S. cerevisiae* (Fosters O) and *S. cerevisiae* (Fosters B) are two commercial strains used for brewing (Borneman et al. 2011).

The presence of toxin-producing molds in cereals used for alcoholic beverage production is a serious problem. Naked cereal grains, such as maize, wheat, sorghum, and millet, which are not protected by husks, are more prone to such mycotoxin contamination (Mbugua and Gathumbi 2004). Molds, such as *Aspergillus flavus, Penicillium parasiticus, Fusarium graminearum, F. culmorum, F. roseum*, and *F. moniliforme*, are naturally present on grains; and during storage or malting, they can grow and produce different toxins, such as aflatoxins, trichothecenes, fumonisins, ochratoxins, and zearalenones (Mbugua and Gathumbi 2004; Schwarz et al. 1995; Usleber et al. 1994). Mycotoxins, such as DON (Deoxynivalenol), may be lost during the steeping of infected barley; however, the mold may still grow and produce mycotoxins during the rest of the malting process. Alternatively, different physical or chemical methods, such as irradiation and ozonation, are available for the treatment of barley during the malting process (Wolf-Hall and Schwarz 2002). Further, development of a genetically engineered yeast carrying a detoxifying gene may be a suitable option for the elimination of mycotoxin contamination during alcoholic beverage fermentation. Similarly, bacteria, such as *Acetobacter*, are capable of converting wine into vinegar within a night, and it can be a challenge for the alcoholic beverage manufacturer. However, these bacteria are sensitive to sulfur dioxide, and hence, their growth can be prevented before harm is caused.

12.4   INDUSTRIAL FERMENTATION OF BEER

12.4.1   Raw Material

Cereal grains, such as barley, wheat, corn, and rice, are raw materials that can be used for production of beer and distilled beverages. As mentioned earlier, barley (*Hordeum vulgare*) is the main raw material for beer production; however, all other aforementioned cereal grains can also be used for the purpose. Based on the number of spikelets, barley is broadly classified as either two-row barley or six-row barley. Two-row barley contains more sugar and, hence, is widely used for malting and brewing (Schwarz and Horsley 1996). High protein-containing six-row barley can also be used for beverage production; however, it is more suitable as feed. For production of alcoholic beverages, starch-containing raw materials are first processed to obtain fermentable sugars. Depending upon the nature of the raw material, prefermentation processing is different. Preprocessing removes most of the unwanted parts from the raw grains, and hence, it has a very important role in the quality of the final product.

12.4.2   Processing of Grains

Prior to using barley for fermentation, it should be subjected to different processes, such as malting, malt kilning, dressing, and meshing. Malting is the process by which plain barley is converted to malted barley. First, the grains are immersed or steeped in water for 48 to 72 h until the moisture content of the grain reaches approximately 46%. Later, the grains are transferred to a germination floor and allowed to air-dry and germinate for the next 5 days with regular shuffling. Alternatively, mechanical rakes (Saladin box) or a revolving drum is also used for the same purpose.[4] The temperature of the germination floor should be controlled at approximately 16°C to keep the germinating barley alive; it may otherwise be killed by the heat released during the germination process. Fully germinated barley is called "green malt," and it is then transferred to a kiln. A kiln is a thermally insulated chamber with a controlled temperature where materials can be hardened, burnt, or dried. Inside a kiln, green malt is put on a mesh over a fire containing a certain amount of peat, which gives a peaty taste to the final product. The kilning process imparts a desired color and specification to the green malt, and the resulting malt may be pale, black, chocolate, amber, or crystal (Lewis and Young 2002). After kilning undesirable combings, the rootlets are separated to be used as cattle feed. This process is called "dressing" of the malt, and the resulting malt is ground to form a product called "malt grist." Mashing is a process where ground malt or malt grist is mixed with hot water using a large vessel called a "mash tun." This process helps to dissolve the starch and other molecules present in the malt and provides a controlled condition for the enzymes to act on the starch to produce a fermentable sugar, such as maltose. The fermentable liquid solution collected at the end of the mashing process is called "wort," and it

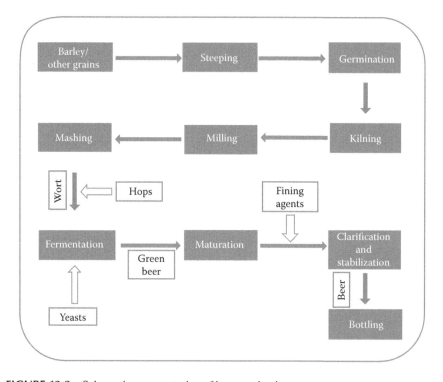

FIGURE 12.2 Schematic representation of beer production process.

should be filtered and cooled close to 0°C prior to fermentation. Figure 12.2 illustrates the different consecutive steps followed for the production of beer.

12.4.3 FERMENTATION

Fermentation of beer can be classified into lager and ale fermentation based on the type of yeast used and the temperature profile followed during fermentation. Bottom-fermenting yeasts are used for lager beer fermentation, and top-fermenting yeasts are used for ale beer fermentation.

12.4.3.1 Lager Fermentation

Primary fermentation, maturation, and cold stabilization are three phases in lager beer fermentation. However, a distinct boundary may not always exist in between these phases. Primary fermentation of lager beer is carried out at low temperatures (4°C–13°C) using special lager yeast, such as *Saccharomyces carlsbergensis* (*S. pastorianus*), which was first isolated by the famous Carlsberg brewery. Higher-temperature fermentation of lager beer generally leads to off-flavors, a possible reason for which is an accelerated rate of metabolism at high temperatures without allowing the yeast to reabsorb its byproducts.[5] As mentioned earlier, the wort initially needs to be cooled close to 0°C to achieve a maximum cold break. Before the onset of fermentation, the cold break should be removed from the wort. There are different methods for cold-break removal, such as by using separate sedimentation tanks, flotation tanks, centrifuges, or by filtration. However, among these methods, filtration seems to be the only method capable of complete removal of the cold break; the other methods are capable of removing only about two-thirds of the same. After removal of the cold break, the resulting wort is warmed up to pitching temperature, which may be somewhere between 5°C and 8°C, and aerated with sterile air to achieve a wort oxygen content of 8 to 10 ppm. Pitching is a process where approximately 500 mL of thick yeast slurry is added to 100 L of wort, and after homogenous distribution of yeast, each milliliter of wort contains approximately 15×10^6 cells. Topping up is an alternative technique for yeast inoculation, common among German lager brewers, in which fresh cooled wort is infused into a tank with strongly fermenting young beer. Based on the pitching and main fermentation temperatures, lager fermentation can be again divided into cold and warm fermentation. In the case of cold fermentation, the pitching temperature is 5°C, and the maximum fermentation temperature is around 9°C. The warm fermentation applies a pitching temperature of 8°C and a maximum fermentation temperature of 10°C to 12°C. Traditional European lager brewers use starter tanks, which are open vessels where yeast is added to the cooled wort, and the contents are then allowed a standing period of up to 24 h (Goldammer 1999). In the fermentation tank, on the third day of fermentation, the maximum temperature is reached, and it continues until the fifth day. At this point, yeast growth slows down, flocculation of the cells starts, and the beer is slowly cooled; however, a sharp decrease in temperature is strictly avoided in order to avoid the probable shocking of yeast cells. Once beer is cooled down, the primary fermentation of beer is completed, and it covers nearly 40% to 60% of the entire brewing process.

Nowadays, mostly vertical cylindroconical fermenters are used for lager production, achieving similar flavor profiles compared to traditional fermentation systems; however, open, square fermenters are still commonly used in traditional lager fermentation (Goldammer 1999). In a modern process, the yeast may be pitched at higher initial temperatures between 7°C and 8°C, and after 48 h, the temperature may be increased to 10°C to 11°C. After nearly 4 days of fermentation, the fermentation temperature is allowed to increase in order to facilitate the yeast rapidly reabsorbing the diacetyl that has been produced during fermentation. At low concentration, diacetyl gives a slippery feeling of the beer in the mouth, and at increased concentration, it gives a buttery or butterscotch flavor. In certain types of beer, the presence of diacetyl can be desirable at low or moderate levels; however, in some other types, the presence of diacetyl is considered undesirable. Therefore, diacetyl concentration in the beer needs to be reduced, and this technique of reabsorbing it from the beer at an elevated temperature is called diacetyl rest. Some modern brewers use the same starting temperature for pitching as mentioned above but then increase the temperature to between 14°C and 15°C. Other breweries prefer a pitching temperature between 10°C and 13°C, and they then increase the temperature to 17°C, and diacetyl rest is shorter in such cases.[6] Once primary fermentation is completed, the yeasts need to be separated from the beer before subjecting it to conditioning, and the separated yeast can be stored for reuse in subsequent fermentations. In traditional lager fermentation using open beer fermenters, after completion of fermentation, the yeast settles down into the bottom of the fermenter. The unconditioned or "green" beer is first separated from the top of the fermenter, and at the end, yeasts are collected from the middle layer of the bottom sediment.

12.4.3.2 Ale Fermentation

S. cerevisiae, also known as top-fermenting yeast, is used for ale fermentation. Ale fermentation is usually carried out between 18°C and 22°C, which is higher than that of lager fermentation. A high fermentation temperature favors the production of esters and other secondary flavor and aroma products by the yeast, and the resulting beer is often slightly fruity (Goldammer 1999). Based on processing conditions and the color and flavor of the final product, ale may be of different varieties, such as brown ale, pale ale, scotch ale, mild ale, and old ale. Similar to lager, ale fermentation can also be initiated using starter tanks with 3 to 36 h of residence time (Goldammer 1999). Traditionally, a fermentation vessel for ale production may be a shallow vessel with a round, square, or rectangular shape; however, modern ale fermentation commonly uses vertical cylindroconical fermenters. Once fermentation is completed, a short diacetyl rest may be applied, following which the beer is cooled, and yeasts are separated by settling down into the bottom of the fermenter.

12.4.4 Beer Maturation

In general, maturation of beer is a secondary or extended fermentation of beer at a relatively lower temperature. Once primary fermentation is completed, the beer still contains fermentation byproducts, such as diacetyl and acetaldehyde. At high concentrations, such compounds can confer undesirable flavor to the final product, and hence, they need to be reduced by the yeast in a subsequent process. Microbiologically, the precise process of reducing such byproducts is known as maturation, and it may be categorized as secondary fermentation, accelerated maturation, lagering, kraeusening, or cask or bottle conditioning (Goldammer 1999). Secondary fermentation involves the slow bioconversion of residual fermentable materials at low temperatures by a relatively small amount of residual biomass of the primary fermentation. The process eliminates residual oxygen from the headspace, reduces undesirable byproducts, incorporates CO_2, and improves the overall quality of the beer. Accelerated maturation is a commercial process, which allows nearly complete fermentation of the beer before it is placed into cold storage. In this method, an extended time profile is used in primary fermentation for complete utilization of fermentable sugars, and a low temperature is precisely maintained toward the end of fermentation. Similarly, lagering is also a maturation process, which involves a long cold storage at low temperatures, and a secondary fermentation of the beer during this period improves the overall quality of the beer. Typically, this method is found in lager fermentation (bottom fermentation) of beers. Kraeusening is a process where secondary fermentation is initiated by adding a small volume of young beer to the beer obtained after primary fermentation. Casking is a process in which, upon completion of primary fermentation, the beer is directly racked from the fermenting vessels into casks, and secondary fermentation is initiated. The process may involve a rough filtration of the beer before filling the casks. Alternatively, bottle conditioning is a beer maturation process where wort or sugar and yeast are primed or infused with the beer after bottling. This technique generally helps in improving the carbonation of the beer. However, a possible problem with beer maturation is that the yeast may shock it into dormancy because of the low temperature used. In such cases, complete reduction of fermentable sugar and byproducts may not be achieved, resulting in under-attenuated beer. Alternatively, cutting maturation can increase profits and production capacity; as a result, most modern commercial lagers hardly undergo any maturation.[7]

12.4.5 Beer Clarification and Stabilization

Beer clarification is the process of removing suspended yeast biomass either by physical processes or by the addition of certain fining agents. Generally, during storage, much of the biomass is settled down in the storage tank, and this simple natural sedimentation is very effective in improving beer quality; however, it is a time-consuming process. Alternatively, centrifugation can be used for precise removal of suspended solids and residual biomass. It is found, in the literature, that clarification by centrifugation is effective at low temperatures, and the optimal yeast separation is achieved at temperatures between 3°C and 5°C (Goldammer 1999). Cross-flow microfiltration is also a recently investigated physical method of beer clarification (Gan et al. 1997). The addition of a fining agent is another commercially used method for beer clarification. Isinglass, gelatin, and bentonite are the few commercially used fining agents. Isinglass is a gelatinous substance collected from the internal membranes of the swim bladders of certain fish. It is believed that positively charged isinglass interacts with negatively charged beer yeast and resulting flocs precipitate. However, fish products are serious allergens, and people having an allergic response to fish must not consume foods made with fish or fish products. Therefore, recently, two alternatives to isinglass—avian collagen and a pea protein extract—have been evaluated as beer clarifying agents (Walker et al. 2007). Similarly, the beer clarification potential of calcined silica gel and zingibain has been described in some other recent reports (Armstead and Quinn 1985; Xiaozhen 2002).

Stabilization of beer is a method of limiting colloidal or nonbiological haze formation, which otherwise can limit the shelf life of beer. This process is different from the clarification of beer where microbiological haze is eliminated. Colloidal haze is mostly formed by the proteins, polyphenols, and carbohydrates present in beer; however, it may also contain some other minor constituents (Anger 1996; McMurrough et al. 1999; Rehmanji et al. 2005). The mechanism of haze formation is an interesting topic of research, and the researchers have proposed that the specific structure of haze-forming proteins and polyphenols has facilitated an interaction between them to form such colloidal particles (McMurrough et al. 1999; Rehmanji et al. 2005; Siebert and Lynn 1998; Siebert et al. 1996). Proteins and polypeptides causing haze formation originate mainly from barley, whereas polyphenols come from barley and hops. Therefore, the principle behind beer stabilization is to reduce the levels of one or both of these haze precursors by suitable physicochemical treatments called chillproofing. There are a number of chillproofing techniques available for reducing chill haze, and one such method is the combined application of silica gel and magnesium silicate (Berg 1989). Silica gel is probably the most widely used chillproofing agent for protein stabilization, whereas polyvinylpolypyrrolidone is known for polyphenol stabilization. The other important chillproofing agents are tannic acid and papain (Rehmanji et al. 2005). Each of these agents has its advantages and drawbacks, and hence, sometimes they are used in combination to fulfill the requirements. However, close examination of a number of key process steps in beer fermentation certainly has the potential to improve beer stability and appearance. For example, polyphenols are mostly lost during mashing, boiling, wort cooling, and maturation. Therefore, Rehmanji et al. (2005) have recommended following precautionary measures for the production of beer with reduced nonbiological haze: (1) very weak wort should be avoided; (2) minimum oxygen contact should be ensured throughout the brewing process; (3) zero or subzero temperatures should be maintained during cold storage, transfer, and filtration of beer; (4) optimum finings, centrifugation, and filtration should be entertained; (5) tannoids and polymerized flavonols should be removed from fresh beer, and polyvinylpolypyrrolidone should be used as a chillproofing agent for such polyphenols; (6) silica gel or a combined stabilizer, having both protein and polyphenol stabilization potential, such as Polyclar Plus 730 (International Specialty Products, Inc.), should be used in a well-balanced manner; and (7) beer transportation and storage conditions should be carefully considered.

Filtration is another important part of the beer fermentation process. It helps to remove the yeast and beer-spoiling bacteria from the fermented beer, which, in turn, helps to maintain the quality of the final product.[8] The beer filtration process can be discussed under three different headings, namely, depth filtration, surface filtration, and double-pass filtration.[9] In the case of depth filtration, particulate materials present in freshly fermented beer are removed by the pore of the filtration media, either by absorption or by mechanical entrapment. Similarly, in the case of surface filtration, a membrane is used, and solid particles are removed by mechanical entrapment.[9] Further, double-pass filtration can also be used for the filtration of beer. In this case, major particulate materials, including yeasts, are removed by a primary filtration, which is followed by a secondary filtration to obtain highly polished beer.[9]

Thus, beer fermentation is a combination of different, well-defined processes, and the final quality of the beer depends on the success of all such processes.

12.5 INDUSTRIAL FERMENTATION OF WINE

12.5.1 RAW MATERIALS

The raw material used for wine production may be different based on the type of the final product desired. Grapes are the main raw material for the production of wine. However, some other fruits, such as apples, berries, cherries, palm, and elderberries, are also used for wine production. *Vitis labrusca* and *Vitis vinifera* are the two grape species mostly used in the wine production process.[10] Grapes contain sugars, acids, tannins, and minerals, which are the necessary ingredients of wine production.

12.5.2 PROCESSING OF FRUITS (DE-STEMMING AND CRUSHING)

As mentioned earlier, for wine production, fruits are mainly used, and grapes are the major fruit used for this purpose. Grape stems contain tannins and impart bitter or unwanted vegetable flavor; therefore, stems should be first removed. Mechanical de-stemmers capable of selectively removing the stems are available for this purpose. Crushing (pressing) of the grapes to release the juice and pulp is the next step of wine processing. Gentle crushing of the grapes helps to release the juice from the pulp located between the seed and skin of the grape, which is smaller in amount but better in quality. In turn, hard crushing will release a large quantity of less-desirable juice from the pulp located near the seeds and skin.[11] Crushing the grapes by human feet is the most suitable method, in which case, the chance of crushing the grape seed and mixing it with the juice is low, and hence a bitter taste in the final product can be avoided. Presently, combined de-stemmer–crushers are available for grape processing purposes.[12] After grape crushing, the resulting juice is called "must," which contains pulp, skin, seed, and 10% to 15% (w/w) glucose.

12.5.3 FERMENTATION

Wine fermentation may be carried out in stainless-steel tanks, in an open wooden vat, or inside a wine barrel. Certain types of wines, such as sparkling wines, are fermented inside the wine bottle itself. First, the grape juice or grape must is transferred to the fermentation vessel, and then yeasts are added. For white wine production, before transferring the grape juice to a vat or fermentation tank, all the grape skins are separated by centrifugation or filtration; however, for red wine, grape must containing both juice and skin is transferred. *S. cerevisiae* is the yeast species used for wine production. It is a domesticated species, which is acclimatized in such a way that it can ferment wine even in the presence of free sulfur dioxide used for controlling the growth of *Acetobacter*. Wild yeasts may be present in the grape juice used; however, for production of wine with predictable quality, the domestic yeast added externally should dominate the fermentation. Further, during wine fermentation, the fermentation vat should be free of oxygen, so growth of unwanted bacteria can be minimized.[13] Traditionally, wine fermentation is carried out at approximately 25°C, and the whole process needs 10 to 30 days. During fermentation, the metabolic activity of yeast is affected by the fermentation temperature; if it exceeds 32°C or falls below 3°C, the rate of fermentation may slow down or stop, and hence, careful control of temperature should be ensured. It can be found in the literature that low-temperature fermentation results in higher-quality wine (Kourkoutas et al. 2001); however, the amount of wine produced is less (Bakoyianis et al. 1992). During low-temperature winemaking, higher alcohols, such as propanol-1 and isobutyl alcohol, are produced at a reduced rate, and this may be the reason for its improved quality (Kourkoutas et al. 2001). Warm or hot fermentation adds toast and vanillin aromas and flavors while diminishing fruity aromas and flavors. Traditionally, during wine fermentation, the barrels or vats are only filled to approximately 75%, so any possible loss of wine resulting from foaming can be avoided. In the course of fermentation, especially with red wine, CO_2 is produced, and it pushes the grape skins to the top of the liquid to produce a layer known as cap. The grape skins present in the cap still contain a sufficient amount of tannin and color constituents. Therefore, manual mixing of the cap and liquid may be necessary to allow these compounds to partition into the liquid fraction. Once fermentation is completed, the wine is separated from the grape skins by pressing and transferred to a holding tank. For the following few days, the wine is kept in the holding tank where residual solid materials and yeast biomass settle down. Actually, before keeping for aging, fermented wine should undergo a number of processes, such as racking, filtration, centrifugation, and sedimentation; and it may take several months.

12.5.4 WINE AGING

After alcoholic fermentation, aging is the next most important stage of the wine production process. In general, it is simply storing the wine in a wooden or stainless-steel vessel at a controlled temperature. Based on the style of wine, aging time may vary, and it may be a few months to several years. White wine is generally aged for 1 to 4 years; however, the aging period for red wine may be 7 to 10 years. Further, some wines never improve with age and should be drunk immediately. Moreover, it may also be possible that if the wines are aged for too long, the flavor-enhancing compounds, such as tannins, may be precipitated. Traditionally, aging is carried out in oak barrels that are stored in damp wine cellars where humidity protects the barrels from drying. However, modern large wineries generally use stainless-steel tanks having a temperature control facility. The aging process improves the overall quality of wine, including aroma, color, clarity, taste, and mouthfeel. In the traditional aging process, the quality of aged wine mainly depends on the wood composition, wine composition, and aging time. Further, the grape variety, cultivation technique, grape-growing region, and winemaking process are some other factors that can influence the quality of aged wine. Aging is a complex biochemical process where yeast, bacteria, sugars, acids, and phenolic compounds, such as tannins, are involved. Malolactic fermentation mainly occurs during the aging of wine, and it contributes to reducing the acidity and mellowing of the wine. Different flavor- and aroma-enhancing compounds are formed during aging, and in the traditional aging process, the wooden barrel used for storing the wine also contributes to improving the flavor; it especially, gives an oaky flavor to aged wine. Actually, during toasting of the wood, numerous specific compounds are formed, and during the aging process, such compounds are slowly, partitioned to the wine. Similarly, highly porous oak wood also favors low oxidation of wine (Vivas and Glories 1993; Garde-Cerdán and Ancín-Azpilicueta 2006). During the aging process, different volatile compounds are transferred from the oak barrels to the wine by means of extraction. The efficiency of such an extraction process is controlled by different factors, such as the quantity of extractable compounds, contact time between the wine and the oak wood surface, and wine composition, especially the alcohol concentration. However, during the course of aging, extracted compounds may undergo microbial transformations, which may alter the final concentration of such substances in the wine (Garde-Cerdán and Ancín-Azpilicueta 2006; Spillman et al. 1998). Further, the porous wood surface may absorb some amount of volatile compounds and, thus, controls their concentration in aged wine. Alternatively, undesirable compounds, such as ethylphenols, with an unpleasant odor may be formed if the wine is aged in used barrels (Garde-Cerdán and Ancín-Azpilicueta 2006). Wood-contaminating yeast is responsible for the formation of such compounds, which can decarboxylate the cinnamic acids and form these phenols (Garde-Cerdán and Ancín-Azpilicueta 2006; Chatonnet et al. 1992). Once aging is completed, the wine is filtered for the last time to remove unwanted sediments and prepare it for bottling. Thus, aging of wine is a complex biochemical process, and its success depends on a number of physical, chemical, and biological factors. Therefore, commercial wine producers always prefer to go for the traditional aging process, which imparts a fairly balanced quality to the wine. However, these conventional processes are extremely slow and require lots of skill. Hence, research work should be focused on developing some quick and economical techniques for wine processing with a maximum capability of preserving the traditional taste.

12.5.5 MALOLACTIC FERMENTATION

Malolactic fermentation is a microbial bioconversion process that converts malic acid into CO_2 and lactic acid. During this process, 1 g of malic acid is converted into approximately 0.67 g of lactic acid and 0.33 g of CO_2. Its commercial significance is that it reduces the acid concentration in red wines and certain white wines.[14] However, it also involves the transformation of flavor compounds, such as esters, aldehydes, and volatile phenols, and the overall improvement of beverage quality. Most importantly, if, during alcoholic fermentation or wine aging, the malolactic fermentation is not carried out under controlled conditions, it may start spontaneously in bottled wine and alter the wine quality. In the winemaking process, malolactic fermentation can occur naturally and can be seen either during alcoholic fermentation or in subsequent stages, and lactic acid bacteria, such as *Oenococcus oeni* and

Lactobacillus brevis, are involved in this process. However, in the industrial production of wine, malolactic fermentation is initiated by introducing selected species of bacteria so that a predictable flavor profile can be ensured. Research work is still on going for optimum utilization of malolactic fermentation in the winemaking process, and recently, Lovitt et al. (2006) reported the successful application of a pilot-scale membrane bioreactor for rapid malolactic fermentation in beverage production.

12.6   PRODUCTION OF DISTILLED BEVERAGES

The raw material preparation for the fermentation of distilled beverages, in particular, spirits, such as whiskey, is more or less similar to that of beer fermentation. It involves the steeping of grain, malting, milling, and mashing to obtain a fermentable liquid called wort. After cooling, the wort is mixed with yeast and fermented for 2 to 3 days. Fermentation results in a weak, alcoholic, clear liquid called "wash," which contains 8% to 9% alcohol by volume and is subjected to distillation. However, wine or fruit juice may be used for distillation in the case of spirits such as brandy. Distillation is carried out in a "pot still," which is a special copper vessel connected from its top to a condenser by a horizontal pipe called the "Lyne arm." The distillation process is generally repeated to achieve a final desired alcohol concentration. Once distillation is complete, a raw or unmatured spirit is produced, and it is transferred to wooden casks for maturation. Maturation is a process similar to wine aging, as mentioned previously, which gives a characteristic taste to the distilled beverage. Traditionally, maturation is carried out in controlled temperature and humidity by leaving the casks in a cool, earth-floored warehouse. In the case of whiskey, it takes a minimum of 3 years and usually 8 to 25 years for maturation. Commonly, whiskey of the same age from different casks are mixed and left for a few months more for further maturation. This process is known as marrying, after which bottling is done. Similarly, industrial production of liqueur also follows the same process as spirits; additionally, it may be blended with sugar and certain other things, such as egg yolks, anise seeds, caraway seeds, coffee beans, spices, and herbs or plant root extracts.

12.7   RECENT ADVANCES IN ALCOHOLIC BEVERAGE FERMENTATION

12.7.1   Genetic and Metabolic Engineering of Yeast

Configuration of highly desirable or more productive microorganisms by genetic and metabolic engineering is a common strategy of modern biotechnology. Similarly, some important research efforts have resulted in the development of novel yeast strains and/or improvement of the performance of *S. cerevisiae* for the production of alcoholic beverages of a desired quality. Easy availability of expression vectors, immunotags, and genetically selectable markers suitable for *S. cerevisiae* as well as the development of very efficient transformation methods have played a role in the genetic engineering of *S. cerevisiae* (Nevoigt 2008; Gietz and Woods 2001; Gueldener et al. 2002). The prevention of wine spoilage and an improved biopreservation ability; improvement of sensory quality, fermentation performance, and wine-processing efficiency; as well as a reduction of the compounds with a negative effect on human health, such as biogenic amines and ethyl carbamate, are the prime targets of yeast genetic engineering (Nevoigt 2008; Pretorius and Bauer 2002). Traditionally, classic genetic techniques, such as protoplast fusion, hybridization, mutagenesis, and cytoduction, were followed for genetic modification of alcoholic beverage-producing industrial yeast strains (Dequin 2001). Such methods were followed by selection of strains with desirable qualities, such as enhanced fermentation efficiency, enhanced carbohydrate utilization, improved ethanol tolerance, and absence of off-flavors in the final product. These traditional techniques were successful in some cases; however, they did not have the option of precisely adding or removing a trait without changing the overall performance of a certain strain (Dequin 2001). Alternatively, the recent advances in molecular biology have significantly changed the concept of genetic engineering, and this has added a new dimension to *S. cerevisiae* genetic engineering as well.

Flocculation is a phenomenon of asexual aggregation of microorganisms into flocs. During alcoholic beverage fermentation, flocculation of yeast cells is observed, and spontaneous sedimentation of such flocs helps in removing the biomass from the fermentation media. Proper flocculation of yeast is desirable for bright beer with good sensory characteristics (Dequin 2001). Physiologically, flocculation is a calcium-dependent, protein–sugar interaction process where specific cell-surface lectin interacts with cell-wall mannan (Dequin 2001; Stratford 1992). In some successful traditional investigations, electrofusion, hybridization, and cytoduction were used to convert nonflocculant strains to flocculant yeast (Dequin 2001; Urano et al. 1993a,b). In other recent works on yeast flocculation, FLO1, the dominant flocculation gene of flocculant *S. cerevisiae*, has been successfully transferred to an ordinary wine yeast to convert it into a flocculant one (Bidard et al. 1994). During white wine production, a fining agent called bentonite (adsorbent) should be used for clarification of the final product and to prevent it from protein haze. However, a considerable amount of wine and aromatic compounds are lost in this process. Alternatively, yeast cell wall mannoproteins have the property of stabilizing protein haze, and deletion of the KNR4 gene results in increased release of mannoproteins in fermentation media (Gonzalez-Ramos et al. 2009). In a recent report, Gonzalez-Ramos et al. (2009) showed that producing wine by using recombinant yeast strains with a deleted KNR4 gene requires 20% to 40% less bentonite than nonrecombinant strains. During the red wine fermentation process, amines are produced, which may be responsible for off-flavors in wine, and they can trigger headaches, hypertension, and migraines upon consumption of such wine. However, by using genetic engineering techniques, recently, one yeast strain has been developed that can reduce such amine production during fermentation.[15] This strain has been commercially used for production of wine in the United States and Canada. Similarly, hydrogen sulfide, with a rotten egg smell, is produced during the fermentation of wine, and it damages the wine's sensory characteristics, and spoilage of wine may result. However, recently, three novel yeast strains have been developed that cannot produce hydrogen sulfide during fermentation.[16] Similarly, numbers of other reports on yeast strain improvement strategy are available and summarized in Table 12.1 (Javadekar et al. 1995; Nevoigt et al. 2002; Fujii et al. 1990; Husnik et al. 2006; Riou et al. 1997).

TABLE 12.1

Different Successful Yeast Strain Improvement Strategies and Alcoholic Beverage Fermentation

	Strain Improvement Strategy	Outcome of the Strategy	Commercial Applications	References
1	FLO1, the dominant flocculation gene transferred to nonflocculant yeast	Flocculant yeast	Not known	Bidard et al. (1994)
2	Electrofusion	Flocculant yeast	Not known	Urano et al. (1993a)
3	Deletion of KNR4 gene from yeast	Increased release of mannoproteins, high stability against protein haze	Not known	Gonzalez-Ramos et al. (2009)
4	Gene from malolactic bacteria transferred to wine yeast to get malolactic yeast (ML01)	Capable of eliminating amines from wine. Produced wine has neither off-flavors nor does it trigger headaches, hypertension, and migraines upon consumption	Yes	Husnik et al. (2006)
5	Not known	Three yeast strains developed that can reduce hydrogen sulfide in wine; hence, resulting wine has no rotten-egg smell of hydrogen sulfide	Yes	See endnote 16
6	Polyethylene glycol (PEG) induced protoplast fusion for transferring killer character to flocculant yeast	Highly flocculant yeast that can kill wild yeasts during fermentation	Not known	Javadekar et al. (1995)
7	GPD1 gene was overexpressed in an industrial lager brewing yeast	Excess production of glycerol-3-phosphate dehydrogenase enzyme, beer with low alcohol content	Not known	Nevoigt et al. (2002)
8	Gene for α-acetolactate decarboxylase has been transferred to brewer's yeast by an integration plasmid, pIARL28	Diacetyl concentration was reduced in fermented wort	Not known	Fujii et al. (1990)
9	Recombinant yeast with HSP30 gene has been developed	The gene can be induced to express in stationary phase and remains active till the end of wine fermentation	Not known	Riou et al. (1997)

Thus, recent advances in genetic engineering have contributed effectively toward the fermentation of alcoholic beverages. However, successful commercial application of genetically modified yeast is not widespread. Insufficient genetic stability of the recombinant yeast strains may be a reason for their rare industrial application. Additionally, different legal restrictions over the application of genetically modified organisms could be seen as the other challenge.

12.7.2 CELL AND ENZYME IMMOBILIZATION IN ALCOHOL FERMENTATION

Whole cell immobilization for alcoholic beverage fermentation has technical advantages and economic benefits over free-cell systems (Kourkoutas et al. 2001; Vasconcelos et al. 2004). The higher cost of microbial recycling, biomass loss in continuous fermentation, a high risk of contamination, as well as an adverse response for environmental variation are some of the disadvantages of a free-cell system. However, these problems can be resolved by a suitable immobilized cell system (Vasconcelos et al. 2004). Further, an immobilized yeast system is known to have better operational stability, a high cell viability, as well as an enhanced alcohol production potential (Razmovski and Vucurovic 2011). Moreover, there is a difference of metabolic activity in between free and immobilized yeast, although the specific reasons behind such differences are not known (Navarro and Durand 1977; Doran and Bailey 1986; Bardi et al. 1997). To be suitable for alcoholic beverage fermentation, the cell-immobilizing materials in an immobilized cell system should be hygienic, cheap, and abundant in nature, preferably of food-grade quality and must improve the taste of the final product (Kourkoutas et al. 2001; Bardi et al. 1997). Entrapment of the microbial cell within the polymeric matrices, such as calcium alginate, polyvinyl alcohol, k-carrageenan, gelatin, and agar agar, are some of the widely studied cell-immobilization techniques (Razmovski and Vucurovic 2011; Sarma and Pakshirajan 2011). Passive adhesion of cells to support materials, such as apple pieces, sugarcane pieces, raisin berries, spent grain, pear pieces, or guava pieces, is another whole cell immobilization technique widely used for alcoholic beverage production (Kourkoutas et al. 2001; Vasconcelos et al. 2004; Razmovski and Vucurovic 2011; Kopsahelis et al. 2007; Tsakiris et al. 2004; Reddy et al. 2006; Mallios et al. 2004). Bardi et al. (1997) reported the immobilization of an alcohol-resistant *S. cerevisiae* strain in delignified cellulosic material and gluten pellets. According to the authors, the materials used for immobilization are of food grade and suitable for low-temperature winemaking. Kourkoutas et al. (2001) investigated the wine-fermentation potential of *S. cerevisiae* immobilized on apple pieces. According to the authors, the immobilized yeast showed good operational stability, and wine production was quicker than with the natural fermentation of grape must. The produced wine was of excellent taste, possibly as a result of a very high amount of ethyl acetate produced during the process. Similarly, sugarcane stalks have also been shown to be very effective as a support material for the immobilization of yeast (Vasconcelos et al. 2004). Reports are also available on sparkling wine production using double immobilized yeast (Tataridis et al. 2005). In another report, van Iersal et al. (2000) have elaborated on the immobilization of yeast to diethylaminoethyl (DEAE)-cellulose, improvement of enzyme activity, and production of alcohol-free beer by limited fermentation. Similarly, spent grains have been used successfully as supporting material for the immobilization of yeast, and continuous fermentation of beer with a highly balanced flavor profile has been reported (Brányik et al. 2005). In another study, the suitability of guava pieces for the immobilization of *S. cerevisiae* has been evaluated, and an overall improvement of the quality of the produced wines has been reported (Reddy et al. 2006). Further, an alginate–maize stem ground tissue matrix used for the immobilization of *S. cerevisiae* has been reported to have very good matrix porosity, better mass transfer characteristics, and better physical stability during alcohol production. Moreover, application of such materials for fermentation was found to reduce the unwanted acetaldehyde and acetic acid content without affecting the alcohol and ethyl acetate content of the distillate (Razmovski and Vucurovic 2011). The presence of malic acid at high concentrations gives a sour taste to the wine produced, and sometimes it is unfavorable for wine quality. *Issatchenkia orientalis* KMBL 5774, a malic acid-degrading yeast, has been successfully immobilized using a mixture of oriental oak (*Quercus variabilis*) charcoal with sodium alginate. By applying such immobilized yeast cells, a 91.6% reduction of malic acid has been achieved without any significant increase in other acids in tested wine (Hong et al. 2010).

Similar to the whole-cell immobilization, the application of immobilized enzymes has also been widely investigated for industrial fermentation of alcoholic beverages. Meena and Raja (2004) have described the feasibility of immobilization of yeast invertase in strontium, barium, calcium–strontium, calcium–barium, and strontium–barium alginates for alcoholic beverage production. Similarly, *S. cerevisiae* invertase has been successfully immobilized on polyethyleneimine (PEI)-coated Sepabeads by using ionic adsorption. The enzyme preparation was fully active even after 15 days of incubation under standard industrial conditions, and optimum activity was observed at temperatures higher than those of soluble enzymes (Torres et al. 2002). Treatment of muscat wine with β-glucosidase can increase the flavor compounds, such as nerol, geraniol, linalool, γ-terpinene, 2-phenylethanol, and benzyl alcohol. Gueguen et al. (1997) have successfully immobilized β-glucosidase from *Candida molischiana* 35M5N and demonstrated its potential for the flavor enrichment of alcoholic beverages using a fluidized bed reactor. Thus, numbers of successful approaches have demonstrated the suitability of immobilized biocatalysts for improved fermentation of alcoholic beverages. Food-grade materials have the potential of improving the sensory characteristics of the final product, and hence, they are more preferred support materials for the immobilization of yeast.

12.7.3 OTHERS

Besides the two mostly focused research areas discussed above, research work is also going on in some other critical areas of the alcoholic beverage fermentation process, such as the designing and controlling of the fermentation tower. Sipsas et al. (2009) evaluated the performance of a multistage fixed-bed tower (MFBT) bioreactor for the production of wine. They have demonstrated both batch and continuous winemaking with the tower fermenter and reported high alcohol productivity even at a low temperature (5°C) as well as an overall improvement in product quality. Similarly, Koutinas et al. (1997) have evaluated the feasibility of the industrial application of a catalytic MFBT bioreactor with mineral kissiris as a promoting material.

They have investigated the alcohol production efficiency of two large-scale MFBT bioreactors of 7000 and 100,000 L. Tata et al. (1999) have studied the continuous fermentation of high-gravity worts using fluidized bed and loop reactors. They have concluded that when operated as a two-stage process, nearly 100% attenuation of the wort sugars was achieved, and the time profile required for fermentation was half of a conventional batch process. Similarly, Yamauchi et al. (1995) demonstrated rapid beer fermentation using a multistage bioreactor system, composed of a continuous stirred-tank reactor (CSTR) and a packed-bed reactor (PBR). Thus, it can be concluded that although alcoholic beverage fermentation is a very old process, researchers are still finding out new possibilities for this technology and have been implementing them successfully on an industrial scale.

12.8 CONCLUSIONS

For a microbiologist, fermentation is a process where organic materials are converted by microbial activity to produce valuable products, such as foods or organic chemicals. Industrial production of alcoholic beverages is a good example of the successful commercial application of fermentation technology. The technology of alcoholic beverage fermentation has been known to human beings since the Neolithic age and traditionally propagated among generations. In the middle of the nineteenth century, Louis Pasteur described the role of yeast in alcoholic beverage fermentation, and since then, research has been focused on yeast to develop novel strains for improvement of beverage quality. Nowadays, different producers have been using genetically engineered yeast for the commercial production of alcoholic beverages because of different advantages offered by such strains. Similarly, the raw materials used for fermentation also influence the quality of the produced beverages. For example, the protein and phenol content of barley is directly related to the development of protein haze during brewing. Spoilage of beverages and production of mycotoxin by contaminating fungi are some serious concerns in the industrial production of alcoholic beverages. However, suitable solutions for such problems have been devised and successfully implemented on an industrial scale. Alternatively, aging is a process that is very vital in improving the overall quality of wine. Traditionally, wine is stored in wooden barrels usually for 8 to 25 years for aging purposes. Slow oxidation and secondary malolactic fermentation of wine in the wooden vessel, partitioning flavor- and color-enhancing compounds to wine from the wood surface, as well as the absorption of undesirable constituents from the wine by the porous wood surface are some of the certain well-known mechanisms behind the wine-aging process. However, traditional aging techniques are still preferably used for industrial production of alcoholic beverages, which is very laborious and time-consuming. Therefore, further research is still needed for the development of fast and economical aging techniques for industrial production of alcoholic beverages.

ACKNOWLEDGMENTS

The authors would like to greatly acknowledge the financial support of the Natural Sciences and Engineering Research Council of Canada (Discovery Grant 355254),

Le Centre de Recherche Industrielle du Québec (CRIQ), MAPAQ (No. 809051), and Ministère des Relations Internationales du Québec (coopération Paraná-Québec 2010-2011). The views or opinions expressed in this work are those of the authors.

NOTES

1. College of Education, The University of Georgia. http://www.coe.uga.edu/ttie/documents/biotech.pdf (accessed May 3, 2012).
2. Scribd.com. http://www.scribd.com/doc/47170205/Alcoholic-Beverages-Industry-report (accessed May 3, 2012).
3. International Center for Alcohol Policies. http://www.icap.org/LinkClick.aspx?fileticket =DZ9ittvJ%2FZs%3D&tabid=75 (accessed May 3, 2012).
4. Simpsons Malt, UK. http://www.simpsonsmalt.co.uk (accessed May 3, 2012).
5. Hubpages, USA. http://tresero.hubpages.com/hub/Ale-and-Lager (accessed May 3, 2012).
6. Braukaiser.com. http://braukaiser.com/wiki/index.php?title=Fermenting_Lagers (accessed May 3, 2012).
7. Meantime Brewing Company, UK. http://www.meantimebrewing.com/our-brewery/maturation-pasteurization (accessed May 3, 2012).
8. Pall Corporation. http://www.pall.com/main/Food-and-Beverage/Final-Filtration-in-Beer-Making-53839.page (accessed May 2, 2012).
9. Beer-brewing.com http://www.beer-brewing.com/beer-brewing/beer_filtration/beer_filtration_methods.htm (accessed May 2, 2012).
10. Madehow.com. http://www.madehow.com/Volume-1/Wine.html (accessed May 3, 2012).
11. Theworldwidewine.com. http://www.theworldwidewine.com/Wine_trivia/How_wine_production.php (accessed May 3, 2012).
12. Bucher Industries, Germany. http://www.bucherind.com/html/en/1891.html (accessed May 3, 2012).
13. Whitman College, USA. http://www.whitman.edu/environmental_studies/WWRB/winemaking.htm (accessed May 2, 2012).
14. The British Columbia Amateur Winemakers Association. http://www.bcawa.ca/winemaking/ml.htm (accessed May 3, 2012).
15. The Vancouver Sun, Canada. http://blogs.vancouversun.com/2011/02/15/genetically-modified-wine-yeast-takes-the-headache-out-of-red-wine-and-chardonnay (accessed May 3, 2012).
16. Beveragedaily.com. New-yeast-strains-promise-to-protect-wine-from-hydrogen-sulfide, available from: http://www.beveragedaily.com (accessed November 30, 2011).

REFERENCES

Anger, H.M. (1996) Assuring nonbiological stability of beer as an important factor for guaranteeing minimum shelf-life, *Brauwelt International*, 14 (2): 142–150.

Armstead, B.H. and Quinn, J.P. (1985) Calcined silicas and their use in beer clarification, US patent, 4515821.

Bakoyianis, V., Kanellaki, M., Kaliafas, A. and Koutinas, A.A. (1992) Low-temperature wine making by immobilized cells on mineral kissiris, *Journal of Agricultural and Food Chemistry*, 40 (7): 1293–1296.

Bardi, E., Koutinas, A.A., Psarianos, C. and Kanellaki, M. (1997) Volatile by-products formed in low-temperature wine-making using immobilized yeast cells, *Process Biochemistry*, 32 (7): 579–584.

Berg, K.A. (1989) Chillproofing with magnesium silicate/silica gel agents, US patent, 4797294.

Bidard, F., Blondin, B., Dequin, S., Vezinhet, F. and Barre, P. (1994) Cloning and analysis of a FLO5 flocculation gene from *S. cerevisiae*, *Current Genetics*, 25 (3): 196–201.

Borneman, A.R., Desany, B.A., Riches, D., Affourtit, J.P., Forgan, A.H., Pretorius, I.S. et al. (2011) Whole-genome comparison reveals novel genetic elements that characterize the genome of industrial strains of *Saccharomyces cerevisiae*, *PLoS Genetics*, 7 (2): e1001287.

Brányik, T., Silva, D.P., Vicente, A.A., Silva, J.B.A. and Teixeira, J.A. (2005) Continuous beer fermentation with yeast immobilized on alternative cheap carriers and sensorial evaluation of the final product, *Proceedings of the 2nd Mercosur Congress on Chemical Engineering*, Rio de Janeiro.

Chatonnet, P., Dubourdie, D., Boidron, J. and Pons, M. (1992) The origin of ethylphenols in wines, *Journal of the Science of Food and Agriculture*, 60 (2): 165–178.

Cudlínová, M., Lukáš, D., Mikeš, P. and Kejzlar, P. (2011) Incorporation of brewer's yeast into the nanofibrous layer by electrospinning, *Proceedings of Nanocon (21–23)*, Brno, Czech Republic.

Dequin, S. (2001) The potential of genetic engineering for improving brewing, wine-making and baking yeasts, *Applied Microbiology and Biotechnology*, 56 (5): 577–588.

Doran, P.M. and Bailey, J.E. (1986) Effects of immobilization on growth, fermentation properties, and macromolecular composition of *Saccharomyces cerevisiae* attached to gelatin, *Biotechnology and Bioengineering*, 28 (1): 73–87.

Dunn, B., Levine, R.P. and Sherlock, G. (2005) Microarray karyotyping of commercial wine yeast strains reveals shared, as well as unique, genomic signatures, *BMC Genomics*, 6 (1): 53.

Fleet, G. (1998) The microbiology of alcoholic beverages, in B.J.B. Wood (ed.) *Microbiology of Fermented Foods*, pp. 217–262. London: Blackie Academic Professional.

Fujii, T., Kondo, K., Shimizu, F., Sone, H., Tanaka, J. and Inoue, T. (1990) Application of a ribosomal DNA integration vector in the construction of a brewer's yeast having alpha-acetolactate decarboxylase activity, *Applied and Environmental Microbiology*, 56 (4): 997–1003.

Gan, Q., Field, R.W., Bird, M.R., England, R., Howell, J.A., McKechnie, M.T. et al. (1997) Beer clarification by cross-flow microfiltration: Fouling mechanisms and flux enhancement, *Chemical Engineering Research and Design*, 75 (1): 3–8.

Garde-Cerdán, T. and Ancín-Azpilicueta, C. (2006) Review of quality factors on wine ageing in oak barrels, *Trends in Food Science and Technology*, 17 (8): 438–447.

Gietz, R.D. and Woods, R.A. (2001) Genetic transformation of yeast, *BioTechniques*, 30 (4): 816–820.

Goldammer, T. (ed.) (1999) *The brewers' handbook*, 2nd edn, KVP Publishers.

Gonzalez-Ramos, D., Quiros, M. and Gonzalez, R. (2009) Three different targets for the genetic modification of wine yeast strains resulting in improved effectiveness of bentonite fining, *Journal of Agricultural and Food Chemistry*, 57 (18): 8373–8378.

Gueguen, Y., Chemardin, P., Pien, S., Arnaud, A. and Galzy, P. (1997) Enhancement of aromatic quality of Muscat wine by the use of immobilized β-glucosidase, *Journal of Biotechnology*, 55 (3): 151–156.

Gueldener, U., Heinisch, J., Koehler, G., Voss, D. and Hegemann, J. (2002) A second set of loxP marker cassettes for Cre-mediated multiple gene knockouts in budding yeast, *Nucleic Acids Research*, 30 (6): e23.

Hong, S.K., Lee, H.J., Park, H.J., Hong, Y.A., Rhee, I.K., Lee, W.H. et al. (2010) Degradation of malic acid in wine by immobilized *Issatchenkia orientalis* cells with oriental oak charcoal and alginate, *Letters in Applied Microbiology*, 50 (5): 522–529.

Husnik, J.I., Volschenk, H., Bauer, J., Colavizza, D., Luo, Z., van Vuuren, H.J.J. (2006) Metabolic engineering of malolactic wine yeast, *Metabolic Engineering*, 8 (4): 315–323.

Javadekar, V.S., Sivaraman, H. and Gokhale, D.V. (1995) Industrial yeast strain improvement: construction of a highly flocculant yeast with a killer character by protoplast fusion, *Journal of Industrial Microbiology and Biotechnology*, 15 (2): 94–102.

Jin, Y.L. and Alex, S.R. (1998) Flocculation of *Saccharomyces cerevisiae*, *Food Research International*, 31 (6–7): 421–440.

Kopsahelis, N., Agouridis, N., Bekatorou, A. and Kanellaki, M. (2007) Comparative study of spent grains and delignified spent grains as yeast supports for alcohol production from molasses, *Bioresource Technology*, 98 (7): 1440–1447.

Kourkoutas, Y., Komaitis, M., Koutinas, A.A. and Kanellaki, M. (2001) Wine production using yeast immobilized on apple pieces at low and room temperatures, *Journal of Agricultural and Food Chemistry*, 49 (3): 1417–1425.

Koutinas, A., Bakoyianis, V., Argiriou, T., Kanellaki, M. and Voliotis, S. (1997) A qualitative outline to industrialize alcohol production by catalytic multistage fixed bed tower (MFBT) bioreactor, *Applied Biochemistry and Biotechnology*, 66 (2): 121–131.

Lewis, M. and Young, T.W. (eds.) (2002) *Brewing*, 2nd edn, Springer.

Linko, M., Haikara, A., Ritala, A. and Penttilä M. (1998) Recent advances in the malting and brewing industry, *Journal of Biotechnology*, 65 (2–3): 85–98.

Lovitt, R., Jung, I. and Jones, M. (2006) The performance of a membrane bioreactor for the malolactic fermentation of media containing ethanol, *Desalination*, 199 (1–3): 435–437.

Mallios, P., Kourkoutas, Y., Iconomopoulou, M., Koutinas, A., Psarianos, C., Marchant, R. et al. (2004) Low-temperature wine-making using yeast immobilized on pear pieces, *Journal of the Science of Food and Agriculture*, 84 (12): 1615–1623.

Mbugua, S.K. and Gathumbi, J. (2004) The contamination of Kenyan lager beers with Fusarium mycotoxins, *Journal of the Institute of Brewing*, 110 (3): 227–229.

McKenzie, H., Main, J., Pennington, C. and Parratt, D. (1990) Antibody to selected strains of *Saccharomyces cerevisiae* (baker's and brewer's yeast) and *Candida albicans* in Crohn's disease, *Gut*, 31 (5): 536–538.

McMurrough, I., Madigan, D., Kelly, R. and O'Rourke, T. (1999) Haze formation shelf-life prediction for lager beer, *Food Technology*, 53 (1): 58–62.

Meena, K. and Raja, T. (2004) Immobilization of yeast invertase by gel entrapment, *Indian Journal of Biotechnology*, 3 (4): 606–608.

Mohamudha, P.R. and Begum, A. (2005) Production and effect of killer toxin by *Saccharomyces cerevisiae* on sensitive yeast and fungal pathogens, *Indian Journal of Biotechnology*, 4: 290–292.

Navarro, J.M. and Durand, G. (1977) Modification of yeast metabolism by immobilization onto porous glass, *Applied Microbiology and Biotechnology*, 4 (4): 243–254.

Nevoigt, E., Pilger, R., Mast-Gerlach, E., Schmidt, U., Freihammer, S., Eschenbrenner, M. et al. (2002) Genetic engineering of brewing yeast to reduce the content of ethanol in beer, *FEMS Yeast Research*, 2 (2): 225–232.

Nevoigt, E. (2008) Progress in metabolic engineering of *Saccharomyces cerevisiae*, *Microbiology and Molecular Biology Reviews*, 72 (3): 379–412.

Pretorius, I.S. (2000) Tailoring wine yeast for the new millennium: Novel approaches to the ancient art of winemaking, *Yeast*, 16 (8): 675–729.

Pretorius, I.S. and Bauer, F.F. (2002) Meeting the consumer challenge through genetically customized wine-yeast strains, *Trends in Biotechnology*, 20 (10): 426–432.

Razmovski, R. and Vucurovic, V. (2011) Ethanol production from sugar beet molasses by *S. cerevisiae* entrapped in an alginate-maize stem ground tissue matrix, *Enzyme and Microbial Technology*, 48 (4–5): 378–385.

Reddy, L., Reddy, Y. and Reddy, O. (2006) Wine production by guava piece immobilized yeast from Indian cultivar grapes and its volatile composition, *Biotechnology*, 5 (4): 449–454.

Rehmanji, M., Gopal, C. and Mola, A. (2005) Beer stabilization technology—Clearly a matter of choice, *Technical Quarterly*, 42 (4): 332–338.

Riou, C., Nicaud, J.M., Barre, P. and Gaillardin, C. (1997) Stationary-phase gene expression in *Saccharomyces cerevisiae* during wine fermentation, *Yeast*, 13 (10): 903–915.

Rodrigues, F., Corte-Real, M., Leao, C., Van-Dijken, J.P. and Pronk, J.T. (2001) Oxygen requirements of the food spoilage yeast *Zygosaccharomyces bailii* in synthetic and complex media, *Applied and Environmental Microbiology*, 67 (5): 2123–2128.

Sarma, S.J. and Pakshirajan, K. (2011) Surfactant aided biodegradation of pyrene using immobilized cells of *Mycobacterium frederiksbergense*, *International Biodeterioration and Biodegradation*, 65 (1): 73–77.

Schwarz, P.B., Casper, H.H. and Beattie, S. (1995) Fate and development of naturally occurring Fusarium mycotoxins during malting and brewing, *Journal of American Society of Brewing Chemists*, 53 (3): 121–127.

Schwarz, P. and Horsley, R. (1996) A comparison of North American two-row and six-row malting barley, Online. Available at http://www.brewingtechniques.com/bmg/schwarz .html (accessed May 3, 2012).

Siebert, K.J., Carrasco, A. and Lynn, P.Y. (1996) Formation of protein-polyphenol haze in beverages, *Journal of Agricultural and Food Chemistry*, 44 (8): 1997–2005.

Siebert, K.J. and Lynn, P.Y. (1998) Comparison of polyphenol interactions with polyvinylpolypyrrolidone and haze-active protein, *Journal of the American Society of Brewing Chemists*, 56 (1): 24–31.

Sipsas, V., Kolokythas, G., Kourkoutas, Y., Plessas, S., Nedovic, V.A. and Kanellaki, M. (2009) Comparative study of batch and continuous multi-stage fixed-bed tower (MFBT) bioreactor during wine-making using freeze-dried immobilized cells, *Journal of Food Engineering,* 90 (4): 495–503.

Spillman, P.J., Iland, P.G. and Sefton, M.A. (1998) Accumulation of volatile oak compounds in a model wine stored in American and Limousin oak barrels, *Australian Journal of Grape and Wine Research*, 4 (2): 67–73.

Stanbury, P.F. (1988) Fermentation technology, in J.M. Walker and R. Rapley (eds) *Molecular Biology and Biotechnology*, pp. 1–24. London: The Royal Society of Chemistry.

Stratford, M. (1992) Yeast flocculation: Reconciliation of physiological and genetic viewpoints, *Yeast*, 8 (1): 25–38.

Tata, M., Bower, P., Bromberg, S., Duncombe, D., Fehring, J., Lau, V. et al. (1999) Immobilized yeast bioreactor systems for continuous beer fermentation, *Biotechnology Progress*, 15 (1): 105–113.

Tataridis, P., Ntagas, P., Voulgaris, I. and Nerantzis, E.T. (2005) Production of sparkling wine with immobilized yeast fermentation, Online. Available at http://e-jst.teiath.gr/eng/ proto_teuxos_eng.htm (accessed May 3, 2012).

Toit, M. and Pretorius, I.S. (2000) Microbial spoilage and preservation of wine: Using weapons from nature's own arsenal-a review, *South African Journal of Enology and Viticulture*, 21: 74–96.

Torres, R., Mateo, C., Fuentes, M., Palomo, J.M., Ortiz, C., Fernández-Lafuente, R. et al. (2002) Reversible immobilization of invertase on sepabeads coated with polyethyleneimine: optimization of the biocatalyst's stability, *Biotechnology Progress*, 18 (6): 1221–1226.

Tsakiris, A., Bekatorou, A., Psarianos, C., Koutinas, A., Marchant, R. and Banat, I. (2004) Immobilization of yeast on dried raisin berries for use in dry white wine-making, *Food Chemistry*, 87 (1): 11–15.

Urano, N., Sahara, H., and Koshino, S. (1993a) Conversion of a non-flocculant brewer's yeast to flocculant ones by electrofusion: Identification and characterization of the fusants by pulsed field gel electrophoresis, *Journal of Biotechnology*, 28 (2–3): 237–247.

Urano, N., Sahara, H., and Koshino, S. (1993b) Conversion of a non-flocculant brewer's yeast to flocculant ones by electrofusion: Small-scale brewing by fusants, *Journal of Biotechnology*, 28 (2–3): 249–261.

Usleber, E., Straka, M. and Terplan, G. (1994) Enzyme immunoassay for fumonisin B1 applied to corn-based food, *Journal of Agricultural and Food Chemistry*, 42 (6): 1392–1396.

van Iersel, M.F.M., Brouwer-Post, E., Rombouts, F.M. and Abee, T. (2000) Influence of yeast immobilization on fermentation and aldehyde reduction during the production of alcohol-free beer, *Enzyme and Microbial Technology*, 26 (8): 602–607.

Vasconcelos, J.N., Lopes, C. and França, F.P. (2004) Continuous ethanol production using yeast immobilized on sugar-cane stalks, *Brazilian Journal of Chemical Engineering*, 21 (3): 357–365.

Vivas, N. and Glories, Y. (1993) Study of mycoflora of oak (*Quercus* sp.) characteristic of wood undergoing natural air-drying, *Cryptogamie Mycologie*, 14 (2): 127–148.

Walker, S.L., Camarena, M.C.D. and Freeman, G. (2007) Alternatives to isinglass for beer clarification, *Journal of the Institute of Brewing*, 113 (4): 347–354.

Wolf-Hall, C.E. and Schwarz, P.B. (2002) Mycotoxins and fermentation: Beer production, *Advances in Experimental Medicine and Biology*, 504: 217–226.

Xiaozhen, T. (2002) Effect of zingibain on beer clarification, *Science and Technology of Food Industry*, 8: 12–14.

Yamauchi, Y., Okamoto, T., Murayama, H., Nagara, A., Kashihara, T., Yoshida, M. et al. (1995) Rapid fermentation of beer using an immobilized yeast multistage bioreactor system, *Applied Biochemistry and Biotechnology*, 53 (3): 261–276.

13 Production of Dairy Products

Luciana Porto de Souza Vandenberghe,
Caroline Tiemi Yamaguishi, Cristine Rodrigues,
Maria Rosa Machado Prado, Michele Rigon Spier,
Adriane Bianchi Pedroni Medeiros,
and Júlio César de Carvalho

CONTENTS

13.1 INTRODUCTION

The manufacture of fermented dairy foods has persisted over centuries and nowadays represents one of the most diverse segments of the food industry. At first, the fermentation of dairy products was performed in order to prolong the shelf life of milk.

Some technological advances, such as the use of starter cultures in fermentation processes, contributed to the development of the dairy chain, leading to standard products that have been directed at specific public sectors. In addition, a better control of milk fermentation improves the taste, texture, color, flavor, and nutritional properties and still may benefit human health.

Milk can be consumed in its fluid form, in a more solid form (such as yogurt), as cheese, or as a major ingredient that is added to other foods. Dairy cases now abound with milk-based products and their reduced-fat and nonfat versions.

Lactic acid bacteria (LAB) are commonly present in fermented milk products. Some of the known LAB are *Lactobacillus*, *Lactococcus*, *Leuconostoc*, *Pediococcus*, and *Streptococcus*. In addition, *Bifidobacterium* and some yeasts and molds are also found, and they give some special characteristics to fermented products. Probiotic cultures have been incorporated into dairy products (e.g., fermented milk, yogurt, cheese, frozen desserts, kefir, etc.) in the form of freeze-dried, spray-dried, and microencapsulated concentrated powders. The development of this market segment has emerged as a result of the increased interest in health-promoting foods.

Artisanal dairy products still exist and are commercialized in some regions. The products are usually sour cream or cheese consisting of milk coming from cows, sheep, goats, or buffalo. Traditionally, farmers and shepherds make cheese using LAB cultures, which naturally occur in milk from different origins.

Basically, the LAB that are used in dairy fermentation can be divided into two groups, according to their optimum growth conditions. The first is mesophilic bacteria that present optimum growth between 20°C and 30°C, and the second is thermophilic bacteria, which have their optimum growth in the range between 30°C and 45°C (Wouters et al. 2002: 91).

In this chapter, some general aspects of the manufacturing of dairy fermented products, as well as some technological challenges of the fermentative process, and the economic and functional importance of these products, are discussed.

13.2 YOGURT

Yogurt is a conventional food known for its therapeutic, nutritional, and sensory properties (Lourens-Hattingh and Viljoen 2001: 1). In most countries, yogurts are produced by the fermentation of pasteurized milk by prosymbiotic cultures of *Streptococcus thermophilus* and *Lactobacillus delbrueckii* ssp. *bulgaricus*. Probiotics are not required in the production but are often added to carry out co-fermentation with the starter cultures.

The production of yogurt begins with the selection of raw materials, such as milk, milk powder, and sugar. Thereafter, the processing is divided into several stages, such as milk standardization, homogenization, pasteurization, inoculation, fermentation, cooling, addition of flavors and other substances, packaging, and storage.

The fermentation process of yogurt is generally conducted as mixed culture of *S. thermophilus* and *L. bulgaricus,* which is balanced (1:1). *S. thermophilus* starts its development with great intensity, giving a favorable environment for *L. bulgaricus,* which then starts its development. Thus, these two cultures are complementary.

Yogurt is still the main vehicle of probiotic cultures preferred by consumers (Hailu et al. 2009: 257; Siegrist et al. 2008: 526). This has resulted in the development of numerous products (Almeida et al. 2008: 178; Antunes et al. 2005: 169; Arayana et al. 2007: 295; Dave and Shah 1997: 31; Ramasubramanian et al. 2008: 4164). It is extremely important to evaluate the interaction and stability of the different probiotic strains (Korbekandi et al. 2009: 75; Sacarro et al. 2009: 397) related to the operational parameters (Mortazavian et al. 2006: 123, 2007: 8; Sodini et al. 2002: 2479) during their processing. In this way, it is crucial to have viable counts of probiotic bacteria that may not decrease along 10^6 CFU/g in order to have a sufficient number of these microorganisms, which may be able to exert the desired therapeutic effects (Shah 2000: 894).

Some factors influence the viability of probiotics in yogurts: strain variation, acid accumulation, interaction with starter cultures, level of dissolved oxygen and hydrogen peroxide (H_2O_2), and storage conditions (Donkor et al. 2006: 1181; Gilliland and Speck 1977: 1394; Nighswonger et al. 1996: 212; Talwalkar and Kailasapathy 2003b: 2537).

Many previous studies focused on devising strategies to improve the viability of particular *Lactobacillus acidophilus* strains that showed suboptimal survival in yogurts. These strategies include the reduction of oxygen content in food by adding ascorbic acid (Dave and Shah 1997: 31) and protecting probiotics by means of encapsulation or addition of cryoprotectants (Capela et al. 2006: 203).

Fermented dairy products, such as yogurt and yogurt drinks, have been marketed and modified successfully to fit target populations with separated formulations for babies and children and for adults concerned about sugar and fat intake. Yogurt drinks, mainly the low-fat and nonfat varieties, are increasing in popularity, targeting the adult female population. The addition of an indigestible carbohydrate and a prebiotic, such as inulin, to yogurt drinks has been linked to improved colon health and increased absorption of calcium and minerals (Hurley and Liebman 2001: 13; Kolida et al. 2002: S193; Kumar et al. 2005: 163; Scholz-Ahrens et al. 2007: 838S), which promotes the marketability of such products.

The oral administration of *Lactobacillus casei* and *Lactobacillus plantarum* showed a prolonged survival time after influenza virus infection and enhanced innate immunity (Yasui et al. 2004: 675; Maeda et al. 2009: 1122; Takeda et al. 2011). *L. bulgaricus* OLL1073R-1 originated from traditional Bulgarian yogurt and was reported to produce immunostimulatory exopolysaccharide (EPS) (Makino et al. 2006: 2873; Nishimura-Uemura et al. 2003: 267).

13.3   FERMENTED MILKS

Fermented milks are defined as the product obtained from milk or regenerated milk and/or milk-based products, regenerated or not, with or without the addition of other ingredients, which, in the dairy base, represents at least 51% of the total of the product ingredients. They are fermented through the action of specific microorganism cultivation and/or the addition of fermented milk and/or other fermented dairy products. The total count of viable dairy bacteria must be, at minimum, 10^6 CFU/g in the final product (FAO/WHO 2011; Codex Alimentarius 2011; Robinson and Tamine 2006: 25).

The classification of fermented dairy beverages with additions refers to the product added to fermented milk in its formulation; such a condition must present a protein content of 1.6 g per 100 g of dairy beverage (Codex Alimentarius 2011; Robinson and Tamine 2006: 25).

A great variety of whey dairy beverages can be found, depending on the processing and the different range of products that are added to them. The manufacturing technology of fermented dairy beverages has been based on the mixing of fermented milk and whey in suitable proportions, followed by ingredients, such as flavoring, food coloring, sweetener, and others, according to the manufacturer's formulation (Codex Alimentarius 2011; Robinson and Tamine 2006: 25; Pescuma et al. 2010: 73).

It is possible to consider that in recent years the consumption of fermented dairy beverages, which have low viscosity and are consumed as mild and fresh beverages, has increased notably.

13.3.1 Characteristics of Fermented Dairy Beverages

The addition of microorganisms is associated with the characteristics that the product will present and with the formation of metabolic components, such as lactic acid, propionic acid, diacetyl, and antagonistic substances that inhibit gram-negative bacteria, responsible for the beverage deterioration. In general, mixed cultures are used, mainly composed by *Lactobacillus* and *Streptococcus* (Robinson and Tamine 2006: 25; Pescuma et al. 2010: 73). Lactic acid, which is formed during the fermentation process, is a natural preservative that makes the product biologically secure. The use of LAB promotes acidification, which improves product conservation and facilitates the beverage's flavor formation and characteristic taste. The production of EPS also occurs, and it has an important function related to beverage texture and stability (Antunes et al. 2007: 173; Athanasiadis et al. 2004: 1091). The acidity increase is determined by the transformation of lactose by microbial enzymes, with the formation of lactic acid, giving the characteristic acidity to each kind of fermented dairy beverages (Güler 2007: 235; Hadadji and Bensoltane 2006: 505; Wszolek et al. 2006: 174).

The fermentative process by dairy bacteria is considered as a regulator agent of the digestive functions. The acidity generated in the process incites the digestive enzymes by the salivary glands. Besides that, some characteristics of dairy fermentation are beneficial for those with lactose intolerance and tendency toward postprandial hyperglycemia (Djurić et al. 2004: 321; Bozanić 2004: 337; Pescuma et al. 2008: 442).

Fermented dairy products generally contain the same nutrient composition as milk. However, the fermentative process makes the nutrients more available to absorption and facilitates the increase of B-complex vitamins, milk proteins, digestibility, and calcium assimilation (Robinson and Tamine 2006; Almeida et al. 2009: 672; Djurić et al. 2004: 321). This fact is a result of the partial hydrolysis of proteins, fats, and lactose.

It is also possible to relate other beneficial properties to these beverages, including anticholesterolemic and anticarcinogenic effects and others. These beneficial effects are highly valorized by consumers, mainly when there is probiotic addition (Pescuma et al. 2008: 442, 2010: 73; Antunes et al. 2007: 173).

Fermented dairy products can present alterations during the storage period related to the change of pH values and acidity, lactose consumption, aroma modifications, and consistency (Fuller et al. 2006: 310; Duboc and Mollet 2001: 19). A pH decrease and acidity increase are directly linked to the type of starter culture used, time and temperature of storage, the microorganism's residual metabolism, and the applied technology in the manufacturing process (Duboc and Mollet 2001: 19; Güler 2007: 235).

13.4 KEFIR

Kefir is an acidic–alcoholic fermented milk product that was originated in the Balkans, Eastern Europe, and the Caucasus (Fontán et al. 2006: 762). The starter culture used to produce this beverage is an irregularly shaped, gelatinous, yellow-white grain (Guzel-Seydim et al. 2000: 35). Traditionally, kefir grains have been used for centuries in many countries, for example, in Tibet and China as the natural starter in the production of this unique, self-carbonated, dairy beverage (Saloff-Coste 1996: 1). Kefir is a consortium of microbes that is mainly used in the production of the low-alcoholic, traditional Russian drink, kefir, where milk constitutes the initial fermenting substrate. It can be produced by fermenting milk with commercial, freeze-dried, kefir starter cultures; traditional kefir grains; and the product that remains after the removal of kefir grains (Bensmira et al. 2010: 1180).

In Soviet countries, kefir has been recommended for consumption by healthy people in order to lower their risk of chronic diseases and has also been provided to certain patients for the clinical treatment of a number of gastrointestinal and metabolic diseases, hypertension, ischemic heart disease, and allergies (Farnworth and Mainville 2003: 77; St-Onge et al. 2002: 1).

Kefir is a refreshing, naturally carbonated, fermented dairy beverage with a slightly acidic taste, a yeasty flavor, and a creamy consistency (Powell et al. 2007: 190). These kefir grains have a varying and complex microbial composition that includes species of yeasts, LAB, acetic acid bacteria (AAB), and mycelial fungi. The grain is a symbiotic association of microorganisms (*Lactobacilli*, *Lactococci*, *Kluyveromyces*, *Candida*, etc.) embedded in a polysaccharide matrix (kefiran), which is composed of glucose and galactose produced by *Lactobacillus kefiranofaciens* (Angulo et al. 1993: 263; Rea et al. 1996: 83). However, cell growth and kefiran production rates have been shown to be increased when *L. kefiranofaciens* is grown in a batch mixed together with *Saccharomyces cerevisiae* (Cheirslip et al. 2003: 43), demonstrating the importance of symbiosis between bacteria and yeast in kefir.

It has been shown that there are specific species that always occur in the grains. In contrast, other microbes may be present, depending on the origin of the grains as well as the method of culturing and substrates added (Pintado et al. 1996: 15). A vast variety of different species of organisms comprising yeast and bacteria have been isolated and identified in kefir grains: *L. acidophilus*, *Lactobacillus brevis*, *L. casei*, *Lactobacillus fermentum*, *Lactobacillus helveticus*, *Lactobacillus kefiri*,

Lactobacillus parakefiri, *Lactococcus lactis*, and *Leuconostoc mesenteroides* (Witthuhn et al. 2004: 33; Simova et al. 2002: 1; Assadi et al. 2000: 541; Garrote et al. 2001: 639). Yeasts isolated from kefir grains include *Kluyveromyces marxianus*, *Torula kefir*, *Saccharomyces exiguous*, and *Candida lambica* (Witthuhn et al. 2004: 33; Assadi et al. 2000: 541). *Acetobacter aceti* and *Acetobacter rasens*, as well as the mycelial fungus *Geotrichum candidum*, have also been isolated from kefir grains (Pintado et al. 1996: 15).

Because of its origin and the way it has been passed onto generations, the beneficial attributes of kefir seem to have been underestimated by the scientific community. Although a great number of studies have shown its benefits, the lack of standardized protocols for clinical trials makes the interpretation of results difficult. In addition, the production of kefir in great amounts in industries using its grains is not standardized, making big batches with the same problematic characteristics (Roberts and Yarunin 2000: 22). A method for mass cultivation of the kefir grains was developed (Schoevers and Britz 2003: 183), which may result in their commercialization, rather than the kefir beverage.

The traditional production of kefir is initiated by the addition of small (0.3–3.5 cm in diameter), irregularly shaped, yellow-white kefir grains to fresh milk that is poured into a clean, suitable container (Güzel-Seydim et al. 2000: 35). The content is left to stand at room temperature for approximately 24 h. The cultured milk is filtered in order to separate and retrieve the kefir grains from the liquid kefir. This fermented milk is appropriate for consumption. The grains are added to more fresh milk, and the process is simply repeated. This simple process can be performed on an indefinite basis because kefir grains are a living ecosystem complex that can be preserved forever as long as it is reused or reinoculated. As active kefir grains are continually cultured in fresh milk to prepare kefir, the grains increase in volume or in biological mass.

Commercial kefir is traditionally manufactured from cow, ewe, goat, or buffalo milk. However, in some countries, animal milk is scarce, expensive, or minimally consumed because of dietary constraints, preferences, or religious customs. Therefore, there have been many attempts to produce kefir from a variety of food sources (Kuo and Lin 1999: 19).

The industrial or commercial process uses direct-to-vat inoculation (DVI) or direct-to-vat set (DVS) kefir starter culture; that is, the microbial blend is well defined. In addition, *Bifidobacterium* spp., *Lactobacillus* spp., and probiotic yeast (*Saccharomyces boulardii*) may be used as adjunct cultures blended with the kefir grains or kefir DVI cultures (Wszolek et al. 2006: 174) for probiotic kefir production.

Because kefir grains are able to metabolize lactose, they can be used to ferment cheese whey, a lactose-rich waste of negligible cost (Papapostolou et al. 2008: 6949). However, scientific information about chemical changes occurring during cheese whey (mainly deproteinized cheese whey) fermentation by kefir grains is still scarce.

There are some reports on single cell protein production using kefir yeasts (Koutinas et al. 2005: 788). The kefir culture can be successfully used for sourdough-type bread-making, leading to bread of good quality, increased shelf life, and better flavor, and it degrades at lower rates compared to breads made with conventional baker's yeast (Plessas et al. 2005: 585, 2008: 40). Recently, freeze-dried kefir culture has been evaluated as a starter in cheese production (Kourkoutas et al. 2006: 6124; Dimitrellou et al. 2007: 1170) and in cheese ripening (Koutinas et al. 2009: 3734). Whey may be a practical base for kefir culture production, and fermented whey is a suitable cryoprotective medium during freeze-drying. The freeze-dried culture retained a high survival rate and showed good metabolic activity and fermentation efficiency, indicating good potential as a value-added starter culture in dairy technology. All these studies have shown promising perspectives for the application of kefir grains in the sector of whey valorization strategies.

Kefiran, the EPS produced by the microflora of kefir grains, improves the viscosity and viscoelastic properties of acid milk gels (Rimada and Abraham 2006: 33), and it is also able to form gels at low temperatures with interesting viscoelastic properties.

In the literature, several health-promoting properties were associated with kefir consumption (Farnwoth 2005: 1). The beneficial action of this fermented milk can be partially attributed to the inhibition of pathogenic microorganisms by metabolic products, such as organic acids, produced by the kefir microflora (Garrote et al. 2000: 364). Kefir cultures are also reported to possess the ability to assimilate cholesterol in milk (Vujičič et al. 1992: 847). In addition, consumption of kefir has been related to a variety of health benefits (McCue and Shetty 2005: 1791; Rodrigues et al. 2005: 404). Some other properties of kefiran, such as immunomodulation or epithelium protection, have been reported (Medrano et al. 2008: 1; Vinderola et al. 2006: 254).

13.5 CHEESES

The first cheese was said to have developed by accident when milk was allowed to ferment. Whether the first cheese was derived from Mongolian yak's milk, the African camel's milk, or the Middle Eastern ewe's milk is unknown and still debated. Cheese can be made from various milks. Milk from cows is typically used in the United States, but milk from sheep, goats, camels, and other animals is used worldwide. In fact, some of the world's finest gourmet cheeses are made from sheep's milk. No matter what type of milk is used, the process is essentially the same (Dole Food Company et al. 2002: 345).

Cheese can be classified according to the animal whose milk is used to make the cheese, geographic origin, texture, type of rind, or production process. According to its texture, cheeses can be classified as fresh and unripened cheeses (cream cheese, ricotta, and cottage cheese), soft cheeses (feta, brinza, lori, chanack), and hard cheeses (Cheddar, Derby, Gouda, Colby, Swiss, Emmental, Monterey Jack). Fresh cheeses are consumed fresh and do not undergo a maturing process. They have a high moisture content. Hard cheeses are firm or hard in texture. They are pressed and require aging, depending on the cheese type, and are usually waxed or vacuum-packaged.

Like any dairy product, cheese is perishable. A general rule is that the harder the cheese, the longer it keeps. Categories of cheese are determined by the method used to make it, the type of milk used, the texture, or even the appearance of the rind. This classification system groups cheeses with common characteristics.

13.5.1 CHEESE PRODUCTION

The pasteurized–standardized milk for cheese production remains in silos (with continuous stirring at 66°C). A part of starter culture is added to the silos for ripening purposes. The milk, prior to its transfer to a coagulator, is heated to 31°C–34°C. As soon as the milk enters the coagulator, the addition of rennet takes place. Rennet is the crude preparation or extract of the abomasum (stomach of a young calf) consisting of two main enzymes: rennin and pepsin. Rennin is an extremely powerful clotting enzyme, which causes rapid clotting without much proteolysis, whereas pepsin induces proteolysis resulting in bitterness in cheese (De 1980: 224). Then, a period of 40 to 50 min is required for coagulation to occur. The coagulated mass is cut into cubes using devices referred to as cheese knives (rectangular stainless-steel frames) (Sandrou and Arvanitoyannis 2000: 327). The cutting of the curd into cubes greatly increases the surface area to facilitate whey expulsion (40–50 min) and concurrent shrinkage of the curd particles, a process known as syneresis (Vedamuthu and Washam 1980: 231). The expulsed whey is collected for different purposes. The curd is washed after all the whey has been removed. This treatment makes the curd firmer and removes acid whey from around it, thereby imparting the desired mildness in flavor (Tsarouhas et al. 2009: 233). Then, the curds are pressed into shapes. Salt may be added at this point. The freshly made cheese is then allowed to ripen, a process that promotes the development of its flavor. Other ingredients also may be added at this point. In general, 11 lbs. of milk is needed to make 1 lb. of cheese (Dole Food Company et al. 2002: 345).

13.5.2 MICROORGANISMS IN CHEESE PRODUCTION

In milk production and cheese manufacture, there is a need for knowledge of the natural biodiversity of microorganisms and their role. Over the last two decades, much experimental work has been carried out, mainly in Europe, in order to demonstrate the specific characteristics of different cheeses (Beuvier and Buchin 2004: 319).

Cheese is a complex ecosystem that is in continuous flux in terms of both external factors, such as cheese-making techniques and ripening conditions, and intrinsic factors, such as physical–chemical composition and interactions among the different microbial communities. The wide range of biochemical reactions and microbial interactions that take place during cheese ripening is the basis of desirable product characteristics, such as good taste and aroma, spoilage prevention, inhibition of foodborne pathogens, and, more recently, modulation of health (Ahola et al. 2002: 799).

Pelaez and Requena (2005: 831) reported that the microbial communities in cheese are very complex. They are more complex in raw milk cheeses than in pasteurized milk cheeses, leading to differences in sensory quality between these two types of cheese (Grappin and Beuvier 1997: 761). The composition of the cheese's microbiota has generally been studied by enumerating certain microbial groups, using discriminating culture media, and isolating and identifying a number of strains using biochemical or molecular tests (Fitzsimons et al. 1999: 3418; de Angelis et al. 2001: 2011). Most of these methods are based on direct analysis of DNA extracted from the cheese, amplification of genes encoding 16S rRNA, and analysis of polymerase chain reaction (PCR) products using a genetic fingerprinting technique.

LAB constitute a heterogeneous group of genera that shares many physiological features. LAB owe their designation to their capacity to ferment sugars primarily into lactic acid via homofermentative or heterofermentative metabolism (Salminen and von Wright 1998). LAB are characterized by being gram-positive, catalase-negative, nonspore-forming, and facultative anaerobic (Kandler and Weiss 1986: 1209). LAB are also usually known to be nonmotile organisms with the exception of *Lactobacillus agilis* and the recently described species *Lactobacillus ghanensis* (Nielsen et al. 2007: 1468) and *Lactobacillus capillatus* (Chao et al. 2008: 2555).

LAB are generally employed because they significantly contribute to the flavor, texture, nutritional value, and microbial safety of fermented foods (Settanni and Corsetti 2008: 121). For these reasons, LAB find numerous applications in the dairy industry; the most commonly selected genera belong to *Lactococcus*, *Lactobacillus*, *Leuconostoc*, and *Streptococcus* (milk) (De Vuyst and Tsakalidou 2008: 476). LAB also provide competitiveness against pathogenic bacteria colonizing the gastrointestinal tracts (Corfield et al. 2001: 1321); thus, several food applications depend on their probiotic effects.

Settani and Moschetti (2010: 691) described that LAB may play different roles in cheese-making: Some species participate in the fermentation process, whereas some others are implicated in the maturation of cheese. In the first case, LAB rapidly ferment lactose-producing high concentrations of lactic acid and are designated as starter LAB (SLAB), and LAB responsible for the ripening process are indicated as nonstarter LAB (NSLAB). The group of SLAB mainly includes *L. lactis* and *Leuconostoc* spp. among mesophilic species and *S. thermophilus*, *L. delbrueckii*, and *L. helveticus* among thermophilic species (Settani and Moschetti 2010: 691).

The group of NSLAB is represented by the genus *Lactobacilli*: *Lactobacillus farciminis*, *L. casei*, *Lactobacillus paracasei*, *L. plantarum*, *Lactobacillus pentosus*, *Lactobacillus curvatus*, *Lactobacillus rhamnosus*, *L. fermentum*, *Lactobacillus buchneri*, *Lactobacillus parabuchneri*, and *L. brevis* (Gobbetti et al. 2002: 511; Coeuret et al. 2004: 451; Svec et al. 2005: 223). The non-*Lactobacillus* species of NSLAB commonly isolated during cheese ripening are *Pediococcus acidilactici*, *Pediococcus pentosaceus*, *Enterococcus durans*, *Enterococcus faecalis*, *Enterococcus faecium*, and also *Ln. mesenteroides* subsp. *cremoris* or *Leuconostoc lactis* with the same species that act as starter cultures (Chamba and Irlinger 2004: 191; Settani and Moschetti 2010: 691).

13.5.3 CHEESE SECURITY

According to Beuvier and Buchin (2004: 319), cheese manufacture is constantly evolving, and there is a tendency to consolidate numerous small units into larger units for most varieties of cheese. This implies changes in milk production with consequences for the quality of the milk. Milk is collected over a wide area, resulting in comingled milks and increased transportation and storage time before processing. This fact induces the development of microbial populations, which are different from those present in milk at the farm. So there is a need to improve hygiene on the farms, which, in turn, is enforced particularly by European microbial standards. In order to destroy pathogens and standardize the milk microflora, pasteurization of milk has become widespread. The use of raw milk and "wild" starters requires constant adaptation of the technological conditions to ensure a good-quality product. In contrast, the use of selected starters, however unspecific, is now in general use. The combination of milk standardization and the general use of a secondary microflora leads to the production of cheeses with a more constant and uniform quality (Beuvier and Buchin 2004: 319). Overall, dairy products and cheese have a remarkably good safety record. The occurrence of pathogenic bacteria has been reported in certain soft cheeses (high moisture and high pH) made from raw milk. Generally, surface-ripened cheeses represent a greater risk for the transmission of pathogens than other cheeses (Beuvier and Buchin 2004: 319).

13.5.4 CHEESE RIPENING

Cheese ripening is a time-consuming process involving complex and well-balanced reactions between glycolysis, proteolysis, and lipolysis of the milk components. In most cheeses, bacterial enzymes play a major role in this process. The proteases, peptidases, esterases, and the enzymes of amino acid catabolism in the LAB have been extensively studied. Those other starters, such as propionibacteria, and surface flora have been studied less extensively. It is well known that changing the bacterial enzyme content in cheese directly affects the rate of cheese ripening and its final flavor (Klein and Lortal 1999: 751).

One of the most efficient ways to increase the bacterial enzyme pool in cheese curd without altering the primary starter and the cheese make is to add whole LAB, which are unable to grow and to produce significant levels of lactic acid but still deliver active ripening enzymes during cheese aging. These starters are normally weakened and referred to as "attenuated." Their preparation can be achieved by various treatments, such as heating, freezing, spray-drying, freeze-drying, fragilization using lysozymes or solvents, or the selection of lactose-negative mutants.

Klein and Lortal (1999: 751) reported that a good attenuated starter can be defined as follows: a strain (or mixture of strains) that is easy to produce and to attenuate in a repeatable and cost-effective way and that is able to positively and significantly influence the ripening process (reduce time and/or increase flavor) at the lowest inoculation level in milk. This implies good retention of the starter in cheese curd, a significant enzyme content, and adequate lysis of the attenuated starter. No doubt, the answer to these various requirements will allow the development of various efficient and safe cheese additives.

13.6 PROBIOTIC MILK PRODUCTS

Probiotic foods are mostly milk-based and have been gaining popularity worldwide because they contribute to intestinal microbial balance and play an important role in health maintenance. The regular consumption of food containing probiotic microorganisms is recommended as a preventive measure against diseases for consumers of all ages.

The incorporation of probiotic strains in milk products has led to growth and development in the dairy chain, and the diversity of products released is quite extensive and has been shown to be economically advantageous.

13.6.1 HISTORY AND DEFINITION OF PROBIOTICS

The first theories about the action of probiotics emerged in approximately 1910 when Ilya Ilyich Metchnikoff (2004) proposed a connection between health and longevity after the ingestion of bacteria present in yogurt. He postulated that *L. bulgaricus* and *S. thermophilus* would suppress the putrefactive-type fermentations of the intestinal flora. In the early 1920s, the therapeutic effects of *L. acidophilus* milk on digestion were documented (Cheplin and Rettger 1922: 24). In the 1930s, Dr. Minoru Shirota selected the strains of intestinal bacteria that could survive passage through the gut. Years later, his product containing *L. casei* Shirota was the basis for the establishment of the Yakult Honsha Company (2011).

There are many definitions for the term "probiotic." One of them is that probiotics are "biopreparations containing living cells or metabolites that stabilize the autochthonous microbiota that colonize and make up the microflora of the animal and human gastrointestinal tract. They exert a stimulating effect on both digestive processes and immune system of the host" (Fuller 1989: 365). But the more-accepted definition is that "probiotics are live microorganisms that, when administered in appropriate amounts, confer a series of health benefits" (FAO/WHO 2001).

The benefits of probiotic strains can be associated with metabolic, trophic, and protective functions (Soccol et al. 2010: 413). Metabolic functions are characterized by the fermentation of nondigestible dietary residue and endogenous mucus, production of vitamin K, and absorption of ions. Trophic functions are based on the control of epithelial cell proliferation and differentiation and enhancement and homeostasis of the immune system. Finally, protective functions are related to a barrier effect and protection against pathogens (Del Piano et al. 2006: S248).

13.6.2 DAIRY PRODUCTS FOR DELIVERING PROBIOTICS

Dairy foods are the main vehicles for probiotic delivery. The preference could be explained as a result of the fact that the manufacturing processes of these products have been continuously optimized to assure and improve the survival of microorganisms. In this way, only minor changes have to be introduced into already consolidated manufacturing technologies and/or processes (Heller 2001: 374). Probiotic bacteria have been used frequently as adjunct cultures for fermented milks and cheeses. The usual practice is the fermentation with both starter microorganisms (*S. thermophilus* and *L. delbrueckii* spp. *bulgaricus*) and one or more species of probiotic bacteria in order to obtain different products.

Microbial cultures used in dairy fermented products are usually originally isolated from the human intestine because these species are recognized as appropriate because of the physiological needs of the host and their easy and high capacity to colonize the intestine. These species include the GRAS (generally recognized as safe) bacteria, such as *Bifidobacterium adolescentis*, *Bifidobacterium bifidum*, *Bifidobacterium breve*, *Bifidobacterium infantis*, *Bifidobacterium longum*, *L. acidophilus*, *L. casei* subsp. *rhamnosus*, and *E. faecium* (Da Cruz et al. 2007: 951). However, the isolation of probiotic strains of other origin has already been related. Recently, a *Lactobacillus* species has been industrialized and launched in the

Spanish market as component of the first dried milk containing probiotic bacteria isolated from breast milk (Zacarías et al. 2011: 548).

In order to assure the biological and therapeutic effects of probiotic products, several organizations established standard regulations for minimum viable counts of probiotic microorganisms. Most of the international organizations have required minimum levels of 10^7 CFU/mL for *L. acidophilus* and 10^6 CFU/mL for Bifidobacteria in fermented milk products at the time of sale (Talwalkar and Kailasapathy 2003a: 117). Already in Japan, the Fermented Milk and Lactic Beverages Association has established at least 10^7 CFU/mL of viable Bifidobacteria in fermented milk drinks. In probiotic yogurts, counts of viable probiotic bacteria should be at least 10^8 CFU/mL to achieve the desired effects (Lourens-Hattingh and Viljoen 2001: 1).

13.6.3   Types of Probiotic Dairy Products

Several probiotic products have been offered in the market and are classified into three categories: (1) conventional food with the addition of probiotic microorganisms, (2) food supplements or fermented milk with food formulations used for delivering probiotic bacteria, and (3) dietary supplements in the form of capsules or pills (Fasoli et al. 2003: 59). Table 13.1 presents some types of probiotic dairy fermented products.

TABLE 13.1
Probiotic Dairy Products

Commercial Name	Probiotic Microorganism	Company	Type of Product
Yakult	*L. casei* Shirota	Yakult Honsha Co. Ltd.	Fermented milk
SVELTY Gastro Protect	*Lactobacillus johnsonii* La1	Nestlé	Fermented milk
Actimel	*L. casei* (Defensis)	Danone	Fermented milk
LC1 Yogurt	*L. johnsonii* La1 and *L. acidophilus*	Nestlé	Yogurt
Activia	*Bifidus Regularis* (*Bifidobacterium animalis*), *L. bulgaricus*, *S. thermophilus*	Danone	Yogurt
Heini's Yogurt Cultured Cheese	*L. acidophilus, L. casei, B. lactis*	Bunker Hill Cheese Company	Cheese
Biorich	*L. acidophilus* LA-5 and *Bifidobacterium* BB-12	Chr. Hansen A/S	Dried bacteria for yogurt manufacture
HOWARU Premium Probiotics	*L. acidophilus* NCFM	Danisco A/S	Dried bacteria for application in dairy products and frozen desserts

Source: Database from companies' websites.

13.6.4   Requirements for Use of Probiotic Strains

Considering that the beneficial effects of probiotics are strain-specific and dose-dependent, the strain-selection step is fundamental to a product's final quality. Microorganisms can be considered probiotic when they are normal inhabitants of the gastrointestinal tract. They must also survive after the passage through the stomach and maintain their viability and metabolic activity in the intestine (Hyun and Shin 1998: 34). In addition, they should not affect the sensory properties of the dairy food, and they should be stable through processing and storage of the final product (Mattila-Sandholm et al. 2002: 173).

Many factors can affect the survival of probiotics. Among them are low pH, the influence of preservatives (Charteris et al. 1998: 123), the presence of some potential microbial growth inhibitors, as well as some products of the lactic acid starter metabolism (e.g., diacetyl, acetaldehyde, lactic acid) (Post 1996: 78). However, the use of additives in the dairy industry (e.g., salt, sugar, sweeteners, pigments, flavors, fruits, and preservatives, such as nisin, natamycin, and lysozyme) has demonstrated to be more damaging to the starter bacteria than probiotic bacteria (Vinderola et al. 2002: 579).

In addition, probiotic microorganisms of the genera *Lactobacillus* and Bifidobacteria have complex nutritional needs (e.g., the assimilation of free amino acids and peptides, vitamins, nucleotides, and certain minerals), and they also require a low oxygen tension environment for optimum growth (Gomes and Malcata 1999: 139).

For probiotic culture suppliers, the challenges are to obtain a high biomass yield on a large scale, which can be concentrated and reactivated after freezing or drying processes. The quality of the products containing probiotic microorganisms is hardly influenced by intrinsic and extrinsic factors, which are associated with processing steps and storage conditions. The technology of encapsulation materials could be employed to improve the stability of the probiotic strains and may confer the health benefits claimed on the label.

13.6.5   Market for Probiotic Dairy Products

The market for probiotics is established mainly for dairy products; nevertheless, other products have shown relative importance (e.g., dietary supplements, probiotic baked products, probiotic ice creams, probiotic chocolates), leading to diversification and popularization of these products. The highlight of this market segment was marked by launching of the fermented milk Yakult Honsha in 1955 whose formulation is based on probiotic bacteria *L. casei* Shirota (Menrad 2003: 181).

Particularly, Europe represents the largest and fastest-growing segment of the probiotic market, followed by Japan (Global Industry Analysis 2010). There is a wide range of probiotic products produced by companies such as BioGaia Biologics AB, Christian Hansen A/S, ConAgra Functional Foods, Danisco, Groupe Danone, Lifeway Foods Inc., Nestlé S.A., Seven Seas Ltd., Valio, and Yakult Honsha Co. Ltd.

More recently, the market for probiotic products has generated US$15.9 billion in 2008 and is forecast to reach US$28.8 billion in 2015 (Markets and Markets 2009; Global Industry Analysis 2010). According to Leatherhead Food Research, the participation of the dairy products in the probiotic market is still the most significant, representing 38% of sales in this sector (Duggan 2010).

13.7 PREBIOTIC MILK PRODUCTS

Prebiotic carbohydrates are defined as nondigestible food ingredient(s) that beneficially affect host health by selectively stimulating the growth, colonization, and/ or activity of one or a limited number of bacteria in the colon (Roberfroid 2002: S105). The prebiotics are dietary components that are resistant to hydrolyzation by the digestive enzymes and/or are not absorbed in the upper part of the gastrointestinal tract, including the small intestine. Indeed, these compounds must pass into the large bowel where most of the indigenous intestinal microbiota is located. A wide variety of dietary carbohydrates that are especially resistant to starch, dietary fibers, and nondigestible oligosaccharides have such characteristics, and they provide available substrates for bacterial fermentation in the colon (Roberfroid 2002: S105).

In addition, prebiotics are increasingly added to yogurt and other fermented dairy products as well as a wide array of other foods (Wang and Gibson 1993: 373). Besides, the market for prebiotics is growing rapidly. They are mainly associated with breakfast cereals, baked goods, cereal bars, baby foods, and some dairy products. The probiotics market has become better established, based primarily on the launch of special yogurt and fermented milk drinks.

Oligosaccharide molecules that are considered prebiotics contain three to approximately 10 monosaccharide residues connected by glycosidic linkages and are considered the main units among prebiotics. Other compounds with potential for development but not yet well known are xylooligosaccharides, acidic oligosaccharides, soy oligosaccharides, pectic oligosaccharides, gluco-oligosaccharides, and isomalto-oligosaccharides (Gibson and Rastall 2000: 25) according to Table 13.2.

The market value of prebiotics as an industrial food ingredient is increasing in the following segments: yogurt and desserts, fast cereals, cereal bars, juices, soft drinks, carbonated drinks, bread and morning goods, pet foods, confectionary, milk and milk drinks, ice creams and frozen desserts, and butter and yellow fats (RTS Resource Ltd. 2012).

Milk can be a good medium for delivery of prebiotics and probiotics, resulting in a significant reduction of gastrointestinal disorders, although, nowadays, prebiotics are also added into other food products, such as bread, cereals, cakes, biscuits, bars, and soups. This chapter focuses on prebiotics and symbiotic milk-based products.

Prebiotics can be baked into bread, cereals, cakes, and biscuits and even added to soups. Symbiotics have also generated interest with some food manufacturers who are exploiting the effects of combining a prebiotic with a probiotic (Sandholm and Saarela 2003: 86).

Prebiotics have been reported as nondigestible oligosaccharides, which contain mixtures of oligomers of different chain lengths and are characterized by the average number of osyl moieties referred to as the degree of polymerization. These nondigestible oligosaccharides are oligomeric carbohydrates, the osidic bond of which is in a spatial configuration that allows resistance to the hydrolytic activities of

TABLE 13.2
Natural Sources of Prebiotics

Prebiotic	Structure and Sources	References
β-glucans	Polymer of β-(1-4)-D-glycopyranosyl units separated by single β-(1-3)-D-glycopyranosyl units Found in barley; oats; onions; wheat; rye; cell walls of *S. cerevisiae*; and mushrooms, such as reishi (*Ganoderma* sp.) and shiitake (*Agaricus blazei*)	Sako et al. (1999); Wasser and Weis (1999: 65)
Cyclodextrins	Polymer of (Glucose)n Produced from starch by cyclodextrin glycosyltransferase and α-amylase	Park (2009: 300)
Fructooligosaccharides	Polymer of gluα1-2-[βfru-2-1]$_n$ Produced from sucrose by transfructosylation using β-fructofuranosidase or inulin hydrolysate	Park (2009: 300)
Inulin	Mixture of glucose, fructose, and a number of fructosyl moieties linked by a β(2-1) bond Chicory root, artichokes, leeks, wheat, bananas	Coussement (1999: 190); Watson and Preedy (2010: 81)
Galacto-oligosaccharides	Polymer of (Galactose)n–Glucose Produced from lactose by transgalactosylation using β-galactosidase	Park (2009: 299)
Gentio-oligosaccharides	Polymer of (Glucose)n Produced from starch by transgalactosylation using β-glucosidase	Park (2009: 300)
Glycosylsucrose	Polymer of (Glucose)n–Fructose Transglucosylation of sucrose and lactose by cyclomaltodextrin glucanotransferase	Park (2009: 300)
Lactosucrose	A functional trisaccharide (lactose and sucrose) Produced from lactose as an acceptor and sucrose as a fructosyl donor by the transfructosylation activity of levansucrase	Choi et al. (2004: 1876)
Isomalto-oligosaccharides	From starch; consists mainly of oligomers with two to four degrees of polymerization, such as isomaltose, panose, and isomaltotriose. Contains α 1→6 glucosidic linkage that resists endogenous digestion	Park (2009: 300); Sako et al. (1999: 69); Mussatto and Mancilha (2007: 587); Chen et al. (2001: 44)
Malto-isomalto-oligosaccharides	Polymer of (Glucose)n Produced by starch by pullanase, isoamylase, α-amylases/transglucosidase	Park (2009: 300); Sako et al. (1999: 69); Mussatto and Mancilha (2007: 587)

(continued)

TABLE 13.2 (Continued)
Natural Sources of Prebiotics

Prebiotic	Structure and Sources	References
Isomaltulose (or palatinose)	Polymer of (glucose–fructose)n Produced from sucrose by transglucosidase	Park (2009: 300); Sako et al. (1999: 69); Mussatto and Mancilha (2007: 587)
Lactulose	Polymer of (galactose–fructose)n Produced by isomerization of lactose by alkali	Park (2009: 300); Sako et al. (1999: 69); Mussatto and Mancilha (2007: 587)
Raffinose	Polymer of (galactose–glucose–fructose) Extraction of plant materials by water or alcohol Direct extraction from beans, soybeans	Park (2009: 300); Sako et al. (1999: 69); Mussatto and Mancilha (2007: 587)
Soybean oligosaccharides	Polymer of (galactose)n–glucose–fructose Such as raffinose and stachyose From soybeans	Park (2009); Sako et al. (1999: 69); Mussatto and Mancilha (2007: 587)
Xylooligosaccharides	Polymer of (xylose)n From sugarcane bagasse, corn bagasse, wheat bran, and peanut husks by enzymatic	Park (2009); Sako et al. (1999: 69); Mussatto and Mancilha (2007: 587)
Galacto-oligosaccharides	Polymer of arabinose and galactose From soybeans by endogactanases	Rastall and Maitin (2002: 490)
Arabino-oligosaccharides	Polymer of arabinose From sugar beet by endoarabinanases	Rastall and Maitin (2002: 490)
Rhamnogalacturono-oligosaccharides	Polymer of rhamnose and galactose From apples by rhamnogalacturonases	Rastall and Maitin (2002: 490)
Arabinoxylooligosaccharides	Arabinose and xylose polymer From wheat by xylanases	Rastall and Maitin (2002: 490)
Galacturono-oligosaccharides	Arabinose and xylose polymer From polygalacturonic acid by endogalacturonases	Rastall and Maitin (2002: 490)
Oligofructose	Oligofructose may contain both a fructose number of fructosyl moieties and glucose and a fructose number of fructosyl moieties	Coussement (1999: 190)
Epilactose	Disaccharide 4-O-β-D-galactopyranosyl-D-mannose From lactose by cellobiose 2-epimerase	Watanabe et al. (2008: 4516)

intestinal digestive enzymes. But they are fermented by at least some of the colonic bacteria. This fermentation produces short-chain fatty acids and gases as well as increased metabolic energy, growth, and proliferation of these bacteria (Roberfroid 2002: S105). Table 13.3 presents some commercial sources of prebiotic milk products, their commercial name, brand, and the respective prebiotic substance used.

TABLE 13.3
Some Commercial Sources of Prebiotic Milk Products

Prebiotic	Prebiotic Composition	Brand/References
Synergy1	Combination of 30% oligofructose and 70% inulin HP (named oligofructose-enriched inulin)	Beneo-Orafti
Aptamil milk Nutriservice	Contains 0.8% of prebiotics (mixture of galacto-oligosaccharides and fructooligosaccharides)	Danone
Nan Confor1	Contains 0.4% of prebiotic (90% of galacto-oligosaccharides and 10% fructooligosaccharides)	Nestlé
Nutrilon (Nutricia)	Blend of galacto-oligosaccharides and fructooligosaccharides	Danone
Vitamilk Prebiotik	Prebiotic type not informed	OT (Indonesia)
Nurture Toodler 3	Inulin	Heinz
Bimuno	Galacto-oligosaccharide	Clasado Ltd.
NutraFlora P-95	FOS, degree polymerization: 2–4	GTC Nutrition Huebnera et al. (2008: 287)
Raftilose P95 (Orafti)	Oligofructose and FOS $Glu\alpha1\text{-}2\text{-}[\beta Fru\text{-}2\text{-}1]_n$ and $[\beta Fru2\text{-}1]_n$	Orafti Huebnera et al. (2008: 287)
Inulin-S	Inulin $Glu\alpha1\text{-}2\text{-}[\beta Fru\text{-}2\text{-}1]_n$	Sigma-Aldrich Huebnera et al. (2008: 287)
Oligomate 55	Galacto-oligosaccharides	Yakult Pharmaceutical
Hiline	Polydextrose	Yakult

13.8   SYMBIOTIC MILK PRODUCTS

Prebiotic substances and symbiotic combinations with probiotics are increasingly attracting consumers and biotechnology industries that facilitate development in the area of functional products. There is a great interest in using prebiotic oligosaccharides as functional food ingredients to contribute to the colonic and microflora growth of nonpathogenic organisms in order to improve health (Rastall and Maitin 2002: 409). The theory is that symbiotics load the colon with good bacteria while ensuring a plentiful supply of the right food on which to thrive (RTS Resource Ltd. 2012).

Table 13.4 shows the composition of symbiotic milk products and related products considering probiotic and prebiotic components and their manufacturers.

13.9   AROMA OF DAIRY PRODUCTS

The aroma of dairy products is composed of a vast array of volatile compounds, such as alcohols, aldehydes, esters, dicarbonyls, short- to medium-chain free fatty acids, methyl ketones, lactones, phenolic compounds, and sulfur compounds. These

TABLE 13.4
Composition of Symbiotic Milk Products and Related Products

Product	Composition	References/Brand
Hiline	Polydextrose	Yakult S.A.
Nido Kinder	Powdered milk containing a prebiotic named Prebio1 not specified	Nestlé
Symbiotic	*Bifidobacterium lactis* HNO19, prebiotic, skim milk, corn syrup solids, cream, sucrose, vegetable oils (soya and sunflower), lactose, fish oil, lecithin, vanillin, vitamins, minerals	Sazawal et al. (2010: e12164)
BioPRO	Fructooligosaccharides, *B. bifidum*, *B. longum*, *B. infantis*, *Bacillus coagulans*, and *L. acidophilus*	Vitamin Research Products (2012)
Probiotic and Prebiotic	FOS (fructooligosaccharides), *L. rhamnosus*, *L. acidophilus*, *B. longum*, *B. bifidum*, *L. bulgaricus*, *S. thermophilus*	AG Adrien Gagnon
MásVital	Fermented dairy drink containing prebiotics from chicory, onion, artichoke, and asparagus (inulin)	Grupo Leche Pascual
Actiflora Bioplus	Probiotic and prebiotic capsules containing *L. bulgaricus*, *L. acidophilus*, *Bifidobacterium* sp., *S. thermophilus*, inulin, oligopeptides, dipeptides, free amino acids, vitamins, trace elements	Kendy USA
Ayrshire Prebiotic Plain Yogurt	Selected cultures and probiotic Howaru cultures with Beneo, containing FOS and inulin	Woolworths
Pro Advantage	Contains Howaru culture and inulin	President's Choice; Loblaws Inc.
Horizon Organic Yogurt	Five active probiotic cultures, Fructan or Nutraflora	Whitewave Foods
Lifeway ProBugs	Organic whole milk kefir for kids has inulin and 10 live active probiotic cultures	Lifeway Foods, Inc.
Yo-Plus	Probiotics (active live cultures) and prebiotics (inulin)	Yoplait
Nutraelle DigestiveCare	*L. acidophilus*, *B. bifidum*, inulin (capsules)	Wellness Nutrition, LLC
Culturelle	*L. rhamnosus* GG and inulin (capsules)	I-Health, Inc.
Ultimate Flora Women's Care	*L. rhamnosus*, *L. casei*, *L. plantarum*, *L. acidophilus*, *L. brevis*, *B. lactis*, *B. longum*, *L. paracasei*, *Lactobacillus salivarius*, *L. bulgaricus*, and FOS (capsules)	Renew Life Formula's, Inc.
Flora Bear for Kids Renew Life	Probiotic blend (*L. acidophilus*, *B. bifidum*, *B. infantis*) and FOS	Renew Life Formula's Inc.
Buddy Bear Probiotic	*L. acidophilus*, *B. bifidus*, *B. infantis*, and FOS 100 mg	Renew Life Formula's Inc.
Julie's Low-Fat Frozen Yogurt	Fructooligosaccharide derived from beet or cane sugar and six active probiotic yogurt cultures	Oregon

volatile compounds contribute to both the aroma and taste of dairy products, characterizing the flavor of the dairy product. Many components of the aroma of dairy products come from the raw milk itself. Others, however, as in the case of fermented dairy products, are the result of a number of chemical and biochemical reactions. The aroma of raw milk is smooth and quite characteristic, although it is made up of more than 200 components at low concentrations. Milk can be used as a flavoring agent itself and serves as a precursor for flavor compound development by providing sugar and amino acids for browning or fermented products (Reineccius 1999: 89).

One must also distinguish the aromas produced naturally during processing of dairy products from those nonnatural aromas used as additives, such as chocolate, strawberry, peach, and vanilla. As an example, the aroma compounds of cheese are used in snack foods, sauces, baked goods, and several other products. Yogurt and buttermilk are also useful as natural aromas (Schutte 1999: 155).

13.9.1 FLAVOR DEVELOPMENT IN DAIRY PRODUCTS

During dairy product processing, chemical and biochemical reactions lead to the formation of characteristic aromas. Flavor compounds are produced from three major milk constituents: lactose, lipids, and proteins (Marilley and Casey 2004: 153).

Chemical and biochemical reactions take place sequentially or simultaneously. The reactions are attributed to the actions of starter and nonstarter bacteria, secondary microflora (from raw milk or postcontamination of pasteurized milk), and their enzymes. Indigenous milk enzymes as lipase, peroxidase, protease, esterase, catalase, and others are also present (Reineccius 1999: 90). Dairy flavor formation can also be attributed to the caramelization of milk sugars and the cow's feed. Development of Maillard reactions between sugars and amino acids or proteins occurs even at low temperatures (60°C). The presence of dimethyl sulfide and vanillin in raw milk is associated with cow's feed (Schutte 1999: 157).

LAB present in the milk convert lactose into lactic acid and into volatile compounds, such as acetoin, diacetyl, and acetaldehyde (Smit et al. 2005: 591). Yeasts, such as *Kluyveromyces*, *Debaromyces*, *Candida*, or *Trichosporon*, are present in many manufactured milk products. These microorganisms modify the sensory characteristics of the products by synthesizing or assimilating volatile nitrogen and sulfur compounds.

13.9.2 CHEESE FLAVORS

Cheese flavor results from the action of microorganisms and enzymes on milk's carbohydrates, fats, and proteins. Numerous breakdown products are formed; among them are short-chain fatty acids, acetic and lactic acids, alcohols, aldehydes, ketones, esters, and sulfur and nitrogen compounds (Sharpell 1985: 967).

Lactose is hydrolyzed by starter cultures, which produce glucose and galactose. Glucose is then oxidized to pyruvate, and galactose is converted by starter bacteria, *Leuconostoc*, and *Lactobacillus* species. Pyruvate is a starting material for the formation of short-chain flavor compounds, such as diacetyl, acetoin, acetate, acetaldehyde, and ethanol. Citrate is metabolized to produce acetolactate, diacetyl, and acetoin. Semihard and hard cheeses are usually manufactured with a combination of starter bacteria *S. thermophilus* and a thermophilic *Lactobacillus*. However, the presence of starter cultures is not sufficient to explain the flavor formation in raw milk cheeses. The native microflora may play an important role (Marilley and Casey 2004: 139).

Fat is another essential milk component to flavor development in cheese. Lipolysis of milk triglycerides releases high concentrations of free fatty acids, which can be precursors of flavor compounds, such as methyl ketones, secondary alcohols, esters, and lactones.

Short-chain fatty acids have a considerable flavor impact, but intensive lipolysis is undesirable in most cheese varieties because of the development of rancidity. Free fatty acids must be counterbalanced with other flavor compounds to develop an appreciated aroma.

LAB, in general, contribute relatively little to lipolysis, but additional cultures, for example, molds, in the case of surface-ripened cheeses (Molimard and Spinnler 1996: 169), often have high activities in fat conversion. Flavors derived from the conversion of fat are particularly important in soft cheeses, such as Camembert and Roquefort (Smit et al. 2005: 591). In Gruyere-type cheeses, the bacteria *Propionibacter shermanii* produces high levels of propionic acid, giving specific flavor to the cheese.

Cheddar cheese is a mixture of protein, fat, carbohydrates, vitamins, and minerals. More than 200 volatile chemical compounds have been identified in Cheddar cheese. The fatty acids seem to contribute a great deal to the flavor of Cheddar cheese. According to the concentration of fatty acids in cheese, it can be indicative of the level of ripening. At very high concentrations, fatty acids are perceived as off-flavors. The right levels of fatty acids contribute to the characteristic flavor that is associated with Cheddar cheese (House and Acree 2002: 481).

Proteolysis directly contributes to cheese flavors by releasing peptides and amino acids. Amino acids are substrates for transamination, dehydrogenation, decarboxylation, and reduction, producing a wide variety of flavor compounds, such as phenyl acetic acid, phenyl ethanol, p-cresol, methanethiol, dimethyl disulfide, 3-methyl butyrate, 3-methyl butanal, 3-methyl butanol, 3-methyl-2-butanone, 2-methyl propionate, 2-methyl-1-propanal, 2-methyl butyrate, and 2-methyl butanal (Marilley and Casey 2004: 139).

Thage et al. (2005: 795) tested three *L. paracasei* subsp. *paracasei* single-strain adjuncts with different aminotransferase activity profiles in reduced-fat, semihard, round-eyed cheese. They found different flavor profiles for cheeses made with the different *Lactobacillus* strains. Use of the adjunct CHCC 4256 significantly increased the content of flavor compounds that were produced from the branched-chain amino acids (leucine, isoleucine, and valine) and asparagine. The concentrations of 3-methylbutanal, 3-methylbutanol, 3-methylbutanoic acid, 2-methylbutanal, 2-methylbutanol, 2-methylpropanol, and diacetyl were at least twice as high as in the cheeses with other *Lactobacillus* adjuncts and four times higher than in control cheeses.

In cheese ripening, the formation of flavors is a rather slow process, involving various chemical and biochemical conversions of milk components (Smit et al. 2005: 591). Several chemical reactions take place on the surface of ripened cheeses, such as Camembert and Brie during maturation as a result of fungi growth. Fungal mycelium of *Penicillium roqueforti* could grow rapidly, and the resulting products are used directly for flavoring foodstuffs with a blue cheese-type flavor.

Marilley and Casey (2004: 139) reported the use of some bacteria strains for cheese ripening with enhanced flavor production. They also mentioned that the catabolism of amino acids is presumably the origin of some major flavor compounds. Nonstarter and adjunct *Lactobacilli* that grow at high numbers in cheese during ripening have been shown to play an important role in flavor formation, especially for long-ripened cheese varieties (Antonsson et al. 2003: 159).

The presence of sulfur compounds originating from the metabolism of methionine on cheeses is responsible for the garlic (methanethiol, dimethyl disulfide, dimethyl trisulfide, S-methyl thioacetate), boiled potato-like (methional), and cooked cabbage (methanethiol) flavors (Marilley and Casey 2004: 139). Sulfur compounds are major contributors to Cheddar cheese aroma and contribute to the garlic notes of well-ripened Camembert cheese. Hydrogen sulfide was the first volatile implicated in Cheddar cheese flavor. Many authors have studied the influence of hydrogen sulfide on Cheddar flavor, although its presence was not essential (Reineccius 1999: 89).

13.9.3   KEFIR FLAVORS

Many strain cultures are responsible for the production of aroma compounds in kefir. The unique flavor and aroma of traditional kefir result from the symbiotic metabolic activity of a number of bacterial and yeast species that form part of the original natural starter-kefir grains. Kefir is a self-carbonated beverage that owes its distinctive flavor to the products of the ongoing lactic acid and alcoholic fermentation (lactic acid, CO_2, ethanol, and other flavor-forming products). The use of starter cultures facilitates kefir manufacture and allows the product to be controlled and standardized. A starter culture for kefir was formed that included LAB and yeasts isolated from kefir grains (Beshkova et al. 2003: 529).

Güzel-Seydim et al. (2000: 35) investigated the production of organic acids and volatile flavor components during kefir starter culture fermentation. Acetaldehyde, ethanol, acetoin, and diacetyl production were monitored using a gas chromatograph equipped with a headspace autosampler. Acetic, propionic, and butyric acids and diacetyl were not detected. Production of ethanol began only after 5 h of incubation, whereas acetaldehyde and acetoin increased during fermentation.

The presence of 2-butanone and acetone during fermentation and storage of kefir was related by Beshkova et al. (2003: 529). They also found *L. lactis* subsp. *lactis* C15 synthesized ethyl acetate more actively than the other single-strain cultures included in the starter. *S. cerevisiae* A13 produced ethanol and CO_2 in amounts (3975 µg g^{-1}; 1.80 g L^{-1}) that lent cultured kefir the distinctive flavor and aroma characteristic of authentic kefir.

13.9.4   AROMA IN YOGURT

Several factors, such as the chemical composition of raw milk, the type of milk, the processing conditions, and the types of starter cultures used to produce the aroma compounds, have influence on the manufacture of yogurt. The final flavor of yogurt is associated with the presence of nonvolatile acids (lactic or pyruvic), volatile acids (butyric to acetic), carbonyl compounds (acetaldehyde to diacetyl), and miscellaneous compounds (from amino acids to products formed by thermal degradation). The major compounds for the final aroma of the yogurt are lactic acid, acetaldehyde, and diacetyl. Their concentrations and relative proportions seem to be essential (Güler and Gursoy-Balc 2011: 1065). Acetaldehyde is considered a key compound for yogurt aroma. Both carbonyl compounds and free fatty acids in yogurt are influenced by the type of starter culture, type and quality of raw milk, incubation, cooling, and storage (Ott et al. 2000: 724).

Ott et al. (1999: 2379) compared the volatile compounds of traditional acidic and mild, less acidic yogurts. Important differences were found for acetaldehyde, 2,3-butanedione, and 2,3-pentanedione. Concentrations of 2,3-butanedione and 2,3-pentanedione increased two- to threefold in mild, less acidic yogurts compared with traditional acidic ones. This is a result of accumulation of the precursors of the diketones, 2-acetolactate, and 2-acetohydroxybutyrate during fermentation in mild, less acidic yogurt. These precursors are subsequently converted to the corresponding diketones during storage. On the contrary, acetaldehyde formation was reduced in the mild yogurt because of growth differences between the LAB used for fermentation of the milk.

13.10 CONCLUSIONS

Important advances were made in the sector of dairy products. Dairy cases now abound with milk-based products and their reduced-fat and nonfat versions. Some technological tools are being employed to optimize the production of a great number of probiotic, prebiotic, and symbiotic products that are available in the market. In this way, new species are always being studied. However, a typical poor growth of these species is observed. Therefore, some efforts must be done to improve growth factors and establish the effective microorganism's survival.

13.11 FUTURE DIRECTIONS

Currently, probiotics have been developed for geriatrics in order to promote the balance of intestinal microflora, which in most cases is poor due to age, medication use, thereby reducing the associated digestive tract disorders. These products have in their composition the addition of prebiotics and can be used as dietary supplements. There is also a long row of children's products that have probiotics and prebiotics in their formulation, which are developed for children with lactose intolerance where the presence of probiotics is essential to prevent certain diseases. A new trend is the development of hypoallergenic products for children who have allergies to milk proteins that are generally not milk-based and have in their composition the combination of probiotics and prebiotics.

Other new products are also constantly appearing in the market, which contain probiotics and prebiotics, but are not derived from milk such as cereal bars, chocolate, fruit juices, and meat products. These products are intended to reach all age groups, always aiming to associate interesting products' characteristics, with health benefits.

REFERENCES

Ahola, A.J., Yli-Knuuttila, H., Suomalainen, T., Poussa, T., Ahlstrom, A., Meurman, J.H. and Korpela, R. (2002) Short-term consumption of probiotic-containing cheese and its effect on dental caries risk factors, *Archives of Oral Biology*, 47: 799–804.

Almeida, K.E., Tamine, A.Y. and Oliveira, M.N. (2009) Influence of total solids contents of milk whey on the acidifying profile and viability of various lactic acid bacteria, *Food Science and Technology*, 42: 672–678.

Almeida, M.H.B., Zoellner, S.S., Cruz, A.G., Moura, M.R.L., Carvalho, L.M.J. and Sant'Ana, A.S. (2008) Potentially probiotic açaí yogurt, *International Journal of Dairy Technology*, 61(2): 178–182.

Angulo, L., Lopez, E. and Lema, C. (1993) Microflora present in kefir grains of the Galician region (North-West of Spain), *Journal of Dairy Research*, 60: 263–267.

Antonsson, M., Molin, G. and Ardö, Y. (2003) *Lactobacillus* strains isolated from Danbo cheese and their function as adjunct cultures in a cheese model system, *International Journal of Food Microbiology*, 85: 159–169.

Antunes, A.E.C., Cazetto, T.F. and Bolini, H.M.A.B. (2005) Viability of probiotic microorganism during storage, post-acidification and sensory analysis of fat-free yogurts with added whey protein concentrate, *International Journal of Dairy Technology*, 58(3): 169–173.

Antunes, A.E.C., Cazetto, T.F. and Bolini, H.M.A.B. (2007) Selective enumeration and viability of *Bifidobacterium animalis* subsp. *lactis* in a new fermented milk product, *Brazilian Journal of Microbiology*, 38: 173–177.

Arayana, K.J., Plauche, S. and Nia, T. (2007) Prebiotic and probiotic fat free yogurt, *Milchwissenschaft*, 62(3): 295–298.

Assadi, M.M., Pourahmad, R. and Moazami, N. (2000) Use of isolated kefir starter cultures in kefir production, *World Journal of Microbiology and Biotechnology*, 16: 541–543.

Athanasiadis, I., Paraskevopoulou, A., Blekas, G. and Kiosseoglou, V. (2004) Development of a novel whey beverage by fermentation with kefir granules: Effect of various treatments, *Biotechnology Progress*, 20: 1091–1095.

Bensmira, M., Nsabimana, C. and Jiang, B. (2010) Effects of fermentation conditions and homogenization pressure on the rheological properties of kefir, *LWT—Food Science and Technology*, 43: 1180–1184.

Beshkova, D.M., Simova, E.D., Frengova, G.I., Simov, Z.I. and Dimitrov, P. (2003) Production of volatile aroma compounds by kefir starter cultures, *International Dairy Journal*, 13: 529–535.

Beuvier, E. and Buchin, S. (2004) *Cheese: Chemistry, Physics and Microbiology: General Aspects*, 3rd edn, vol. 1, London: Elsevier.

Bozanić, R., Tratnik, L., Herceg, Z. and Marić, O. (2004) The influence of milk powder, whey protein concentrate and inulin on the quality of goat and cow acidophilus milk, *Acta Alimentaria*, 33: 337–346.

Capela, P., Hay, T.K.C. and Shah, N.P. (2006) Effect of cryoprotectants, prebiotics and micro-encapsulation on survival of probiotic organisms in yoghurt and freeze-dried yoghurt, *Food Research International*, 39: 203–211.

Chamba, J.-F. and Irlinger, F. (2004) Secondary and adjunct cultures, in P.F. Fox, P.L.H. McSweeney, T.M. Cogan and T.P. Guinee (eds.). *Cheese: Chemistry, Physics and Microbiology*, London: Elsevier, 191–206.

Chao, S.-H., Tomii, Y., Sasamoto, M., Fujimoto, J., Tsai, Y.-C. and Watanabe, K. (2008) *Lactobacillus capillatus* sp. nov., a motile *Lactobacillus* species isolated from stinky tofu brine, *International Journal of Systematic Evolutionary Microbiology*, 58: 2555–2559.

Charteris, W.P., Kelly, P.M., Morelli, L. and Collins, J.K. (1998) Ingredient selection criteria for probiotic microorganisms in functional dairy foods, *International Journal of Dairy Technology*, 51: 123–136.

Cheirsilp, B., Shimizu, H. and Shioya, S. (2003) Enhanced kefiran production by mixed culture of *Lactobacillus kefiranofaciens* and *Saccharomyces cerevisiae*, *Journal of Biotechnology*, 100: 43–53.

Chen, H.L., Lu, Y.-H., Lin, J.-J. and Ko, L.Y. (2001) Effects of isomalto-oligosaccharides on bowel functions and indicators of nutritional status in constipated elderly men, *Journal of the American College of Nutrition*, 20(1): 44–49.

Cheplin, H. and Rettger, L. (1922) The therapeutic application of *Lactobacillus acidophilus*, *Abstracts of Bacteriology*, 6: 24.

Choi, H.J., Kim, C.S., Kim, P., Jung, H.C. and On, D.K. (2004) Lactosucrose bioconversion from lactose and sucrose by whole cells of *Paenibacillus polymyxa* harboring levansucrase activity, *Biotechnology Progress*, 20(4): 1876–1879.

Codex Alimentarius (2011) *Milk and Milk Products*, 2nd edn, Rome: FAO/WHO. Online. Available at ftp://ftp.fao.org/codex/Publications/Booklets/Milk/Milk_2011_EN.pdf (accessed October 21, 2011).

Coeuret, V., Gueguen, M. and Vernoux, J.P. (2004) In vitro screening of potential probiotic activities of selected lactobacilli isolated from unpasteurized milk products for incorporation into soft cheese, *Journal of Dairy Research*, 71: 451–460.

Corfield, A.P., Carroll, D., Myerscough, N. and Probert, C.S. (2001) Mucins in the gastrointestinal tract in health and disease, *Frontier Bioscience*, 6: 1321–1357.

Coussement, P. (1999) Inulin and oligofructose as dietary fiber: Analytical, nutritional and legal aspects, in S.S. Cho, L. Prosky and M.L. Dreher (eds.). *Complex Carbohydrates in Foods*, New York: Marcel Decker Inc.

Da Cruz, A.G., Faria, J.A.F. and Van Dender, A.G.F. (2007) Packaging system and probiotic dairy foods, *Food Research International*, 40: 951–956.

Dave, R.I. and Shah, N.P. (1997) Viability of yoghurt and probiotic bacteria in yoghurt made from commercial starter cultures, *International Dairy Journal*, 7(1): 31–41.

De, S. (1980) *Outlines of Dairy Technology*, New Delhi, India: Oxford University Press.

De Angelis, M., Corsetti, A., Tosti, N., Rossi, J., Corbo, M.R. and Gobbetti, M. (2001) Characterization of non-starter lactic acid bacteria from Italian ewe cheeses based on phenotypic, genotypic, and cell wall protein analyses, *Applied and Environmental Microbiology*, 67: 2011–2020.

Del Piano, M., Morelli, L., Strozzi, G.P., Allesina, S., Barba, M., Deidda, F. et al. (2006) Probiotics: from research to consumer, *Digestive and Liver Disease*, 38: S248–255.

De Vuyst, L. and Tsakalidou, E. (2008) *Streptococcus macedonicus*, a multi-functional and promising species for dairy fermentations, *International Dairy Journal*, 18: 476–485.

Dimitrellou, D., Kourkoutas, Y., Banat, I.M., Marchant, R. and Koutinas, A.A. (2007) Whey cheese production using freeze-dried kefir culture as a starter, *Journal of Applied Microbiology*, 103: 1170–1183.

Djurić, M., Carić, M., Milanović, S., Tekić, M. and Panić, M. (2004) Development of whey based beverages, *European Food Research and Technology*, 219: 321–328.

Dole Food Company, the Mayo Clinic and UCLA Center for Human Nutrition (2002) *Encyclopedia of Foods: A Guide to Healthy Nutrition*, Los Angeles: Academic Press, 345.

Donkor, O.N., Henriksson, A., Vasiljevic, T. and Shah, N.P. (2006) Effect of acidification on the activity of probiotics in yoghurt during cold storage, *International Dairy Journal*, 16: 1181–1189.

Duboc, P. and Mollet, B. (2001) Applications of exopolysaccharides in the dairy industry, *International Dairy Journal*, 11: 19–25.

Duggan P. (2010) *Global Probiotics Market Assessed*. Online. Available at http://www .bordbia.ie/industryservices/information/alerts/Pages/GlobalProbioticsmarketassessed .aspx (accessed November 1, 2011).

Farnwoth, E.R. (2005) Kefir—a complex probiotic, *Food Science and Technology, Bulletin: Functional Food*, 2: 1–17.

Farnworth, E.R. and Mainville, I. (2003) Kefir: a fermented milk product, in E.R. Farnworth (ed.) *Handbook of Fermented Functional Foods*, Boca Raton: CRC Press.

Fasoli, S., Marzotto, M., Rizzotti, L., Rossi, F., Dellaglio, F. and Torriani, S. (2003) Bacterial composition of commercial probiotic products as evaluated by PCR-DGGE analysis, *International Journal of Food Microbiology*, 82: 59–70.

Fitzsimons, N.A., Cogan, T.M., Condon, S. and Beresford, T. (1999) Phenotypic and genotypic characterisation on non-starter lactic acid bacteria in mature cheese, *Applied and Environmental Microbiology*, 65: 3418–3426.

Fontán, M.C.G., Martínez, S., Franco, I. and Carballo, J. (2006) Microbiological and chemical changes during the manufacture of Kefir made from cows' milk, using a commercial starter culture, *International Dairy Journal*, 16: 762–767.

Food and Agriculture Organization/World Health Organization (FAO/WHO) (2001) *Health and Nutritional Properties of Probiotics in Food including Powder Milk with Live Lactic Acid Bacteria*, Report of a Joint FAO/WHO Expert Consultation on Evaluation of Health and Nutritional Properties of Probiotics in Food including Powder Milk with Live Lactic Acid Bacteria, Córdoba: Argentina. Online. Available at http://www.who.int/foodsafety/publications/fs_management/en/probiotics.pdf (accessed October 24, 2011).

Food and Agriculture Organization/World Health Organization (FAO/WHO) (2011) *Milk fermented*, Online. Available at http://www.fao.org/docrep/003/t0251e/T0251E04.htm (accessed October 31, 2011).

Fuller, R. (1989) Probiotics in man and animals, *Journal of Applied Bacteriology*, 66: 365–378.

Fuller, F., Huang, J.K., Ma, H.Y. and Rozelle, S. (2006) Got milk? The rapid rise of China's dairy sector and its future prospects, *Food Policy*, 31: 201–215.

Garrote, G.L., Abraham, A.G. and De Antoni, G.L. (2000) Inhibitory power of kefir: The role of organic acids, *Journal of Food Protection*, 63: 364–369.

Garrote, G.L., Abraham, A.G., and De Antoni, G.L. (2001) Chemical and microbiological characterisation of kefir grains, *Journal of Dairy Research*, 68: 639–652.

Gibson, G.R. and Rastall, R. (2006) *Prebiotics: Developments and Application*, Oxford: John Wiley & Sons, Ltd.

Gilliland, S.E. and Speck, M.L. (1977) Instability of *Lactobacillus acidophilus* in yogurt, *Journal of Dairy Science*, 6: 1394–1398.

Global Industry Analysis (2010) *Global Probiotics Market to Exceed US$28.8 billion by 2015*, New Report by Global Industry Analysis, Inc. San Jose, CA. Online. Available at http://www.strategyr.com/Probiotics_Market_Report.asp (accessed October 21, 2010).

Gobbetti, M., Morea, M., Baruzzi, F., Corbo, M.R., Matarante, A., Considine, T., Di Cagno, R., Guinee, T. and Fox, P.F. (2002) Microbiological, compositional, biochemical and textural characterisation of Caciocavallo Pugliese cheese during ripening, *International Dairy Journal*, 12: 511–523.

Gomes, A.M.P. and Malcata, F.X. (1999) *Bifidobacterium* spp. and *Lactobacillus acidophilus*: Biological, biochemical, technological and therapeutical properties relevant for use as probiotics, *Trends in Food Science and Technology*, 10: 139–157.

Grappin, R. and Beuvier, E. (1997) Possible implications of milk pasteurization on the manufacture and sensory quality of ripened cheese, *International Dairy Journal*, 7: 751–761.

Güler, Z. (2007) Changes in salted yoghurt during storage period, *International Journal of Food Science and Technology*, 42: 235–245.

Güler, Z. and Gursoy-Balc, A.C. (2011) Evaluation of volatile compounds and free fatty acids in set types yogurts made of ewes', goats' milk and their mixture using two different commercial starter cultures during refrigerated storage, *Food Chemistry*, 127: 1065–1071.

Güzel-Seydim, Z.B., Seydim, A.C., Greene, A.K. and Bodine, A.B. (2000) Determination of organic acids and volatile flavor substances in kefir during fermentation, *Journal of Food Composition and Analysis*, 13(1): 35–43.

Hadadji, M. and Bensoltane, A. (2006) Growth and lactic acid production by *Bifidobacterium longum* and *Lactobacillus acidophilus* in goat's milk, *African Journal of Biotechnology*, 5: 505–509.

Hailu, G., Boecker, A., Henson, S. and Cranfield, J. (2009) Consumer valuation of functional foods and nutraceuticals in Canada: A conjoint study using probiotics, *Appetite*, 52(2): 257–265.

Heller, K.J. (2001) Probiotic bacteria in fermented foods: Product characteristics and starter organisms, *American Journal of Clinical Nutrition*, 73: S374–S379.

House, K.A. and Acree, T.E. (2002) Sensory impact of free fatty acids on the aroma of a model Cheddar cheese, *Food Quality and Preference*, 13: 481–488.

Huebnera, J., Wehling, R.L., Parkhurst, A. and Hutkins, R.W. (2008) Effect of processing conditions on the prebiotic activity of commercial prebiotics, *International Dairy Journal*, 18: 287–293.

Hurley, J. and Liebman, B. (2001) Yogurt: diving for cultured pearls, *Nutrition Action Newsletter*, 28(9): 13–15.

Hyun, C. and Shin, H. (1998) Utilization of bovine plasma obtained from a slaughterhouse for economic production of probiotics, *Journal of Fermentation and Bioengineering*, 86: 34–37.

Kandler, O. and Weiss, N. (1986) Genus *Lactobacillus* Beijerinck 1901, in P.H.A. Sneath, N.S. Mair and M.E. Sharpe (eds.). *Bergey's Manual of Systematic Bacteriology*, Baltimore, MD: Williams and Wilkins.

Klein, K.N. and Lortal, S. (1999) Attenuated starters: An efficient means to influence cheeser ripening: A review, *International Dairy Journal*, 9: 751–762.

Kolida, S., Tuohy, K. and Gibson, G.R. (2002) Prebiotic effects of inulin and oligofructose, *British Journal of Nutrition*, 87: S193–197.

Korbekandi, K., Jahdi, M., Maracy, M., Abedi, D. and Jalali, M. (2009) Production and evaluation of a probiotic yogurt using *Lactobacillus casei* ssp., *International Journal of Dairy Technology*, 62(1): 75–79.

Kourkoutas, Y., Kandylis, P., Panas, T., Dooley, J., Poonam, S. and Koutinas, A.A. (2006) Evaluation of freeze-dried kefir co-culture as starter in Greek Feta-type cheese production, *Applied Environment Microbiology*, 72(9): 6124–6135.

Koutinas, A.A., Athanasiadis, I., Bekatorou, A., Iconomopoulou, M. and Blekas, G. (2005) Kefir yeast technology: Scale-up in SCP production using milk whey, *Biotechnology Bioengineering*, 89: 788–796.

Koutinas, A.A., Papapostolou, H., Dimitrellou, D., Kopsahelis, N., Katechaki, E., Bekatorou, A. and Bosnea, L.A. (2009) Whey valorisation: A complete and novel technology development for dairy industry starter culture production, *Bioresource Technology*, 100: 3734–3739.

Kumar, S.G.V., Singh, S.K., Goyal, P., Dilbaghi, N. and Mishra, D.N. (2005) Beneficial effects of probiotics and prebiotics on human health, *Pharmazie*, 60: 163–171.

Kuo, C.Y. and Lin, C.W. (1999) Taiwanese kefir grains: Their growth, microbial and chemical composition of fermented milk, *Australian Journal of Dairy Technology*, 54: 19–23.

Lourens-Hattingh, A. and Viljoen, B.C. (2001) Yoghurt as probiotic carrier food, *International Dairy Journal*, 11: 1–17.

Maeda, N., Nakamura, R., Hirose, Y., Murosaki, S., Yamamoto, Y. and Kase, T. (2009) Oral administration of heat-killed *Lactobacillus plantarum* L-137 enhances protection against influenza virus infection by stimulation of type I interferon production in mice, *International Immunopharmacology*, 9: 1122–1125.

Makino, S., Ikegami, S., Kano, H., Sashihara, T., Sugano, H. and Horiuchi, H. (2006) Immunomodulatory effects of polysaccharides produced by *Lactobacillus delbrueckii* ssp. *bulgaricus* OLL1073R-1, *Journal of Dairy Science*, 89: 2873–2881.

Marilley, L. and Casey, M.G. (2004) Flavours of cheese products: Metabolic pathways, analytical tools and identification of producing strains, *International Journal of Food Microbiology*, 90: 139–159.

Markets and Markets (2009) *Probiotic Market-Advanced Technologies and Global Market (2009–2014)*. Online. Available at http://www.marketsandmarkets.com/Market-Reports/probiotic-market advancedtechnologies-and-global-market-69.html (accessed October 21, 2010).

Mattila-Sandholm, T., Myllarinen, P.M., Crittenden, R., Mogensen, G., Fonden, R. and Saarela, M. (2002) Technological challenges for future probiotic foods, *International Dairy Journal*, 12: 173–182.

McCue, P.P. and Shetty, K. (2005) Phenolic antioxidant mobilization during yogurt production from soymilk using Kefir cultures, *Process Biochemistry*, 40: 1791–1797.

Medrano, M., Pérez, P.F. and Abraham, A.G. (2008) Kefiran antagonizes cytopathic effects of *Bacillus cereus* extracellular factors, *International Journal of Food Microbiology*, 122: 1–7.

Menrad, K. (2003) Market and marketing of functional food in Europe, *Journal of Food Engineering*, 56: 181–188.

Metchnikoff, I.I. (2004) Introduction to reprinted edition, 1977, in I.I. Metchnikoff (ed.) *The Prolongation of Life: Optimistic Studies*, New York: Springer Publishing Company, Inc., xxvi.

Mocquot, 1986 – ver com prof. Luciana (falta também a página).

Molimard, P. and Spinnler, H.E. (1996) Compounds involved in the flavor of surface mold-ripened cheeses: Origins and properties, *Journal of Dairy Science*, 79: 169–184.

Mortazavian, A.M., Ehsani, M.R., Mousavi, S.M., Rezael, K., Sohrabvandi, S. and Reinheimer, J.A. (2007) Effect of refrigerated storage temperature on the viability of probiotic microorganisms in yogurt, *International Journal of Dairy Technology*, 60(1): 123–127.

Mortazavian, A.M., Ehsani, M.R., Reiheimer, J.A., Emamdjomeh, Z., Sohrabvandi, S. and Rezaei, K. (2006) Preliminary investigation of the combined effect of heat treatment and incubation temperature on the viability of the probiotic microorganisms in freshly made yogurt, *International Journal of Dairy Technology*, 59(1): 8–11.

Mussatto, S.I. and Mancilha, I.M. (2007) Non-digestible oligosaccharides: A review, *Carbohydrate Polymers*, 68: 587–597.

Nielsen, D.S., Schillinger, U., Franz, C.M.A.P., Bresciani, J., Amoa-Awua, W., Holzapfel, W.H. and Jakobsen, M. (2007) *Lactobacillus ghanensis* sp. nov., a motile lactic acid bacterium isolated from Ghanaian cocoa fermentations, *International Journal of Systematic and Evolutionary Microbiology*, 57: 1468–1472.

Nighswonger, B.D., Brashears, M.M. and Gilliland, S.E. (1996) Viability of *Lactobacillus acidophilus* and *Lactobacillus casei* in fermented milk products during refrigerated storage, *Journal of Dairy Science*, 79: 212–219.

Nishimura-Uemura, J., Kitazawa, H., Kawai, Y., Itoh, T., Oda, M. and Saito, T. (2003) Functional alternation of murine macrophage stimulated with extracellular polysaccharides from *Lactobacillus delbrueckii* ssp. *bulgaricus* OLL1073R-1, *Food Microbiology*, 20: 267–273.

Ott, A., Germond, J.E., Baumgartner, M. and Chaintreau, A. (1999) Aroma comparisons of traditional and mild yogurts: Headspace gas chromatography quantification of volatiles and origin of α-diketones, *Journal of Agriculture and Food Chemistry*, 47: 2379–2385.

Ott, A., Germond, J.E. and Chaintreau, A. (2000) Vicinal diketone formation in yogurt: 13C precursors and effect of branched-chain amino acids, *Journal of Agricultural and Food Chemistry*, 48: 724–731.

Papapostolou, H., Bosnea, L.A., Koutinas, A.A. and Kanellaki, M. (2008) Fermentation efficiency of thermally dried kefir, *Bioresource Technology*, 99(15): 6949–6956.

Park, Y.W. (2009) *Bioactive Components in Milk and Dairy Products*, Iowa: Wiley BlackWell.

Pelaez, C. and Requena, T. (2005) Exploiting the potential of bacteria in the cheese ecosystem, *International Dairy Journal*, 15: 831–844.

Pescuma, M., Hébert, E.M., Mozzi, F. and Valdez, G.F. (2008) Whey fermentation by thermophilic lactic acid bacteria: Evolution of carbohydrates and protein content, *Food Microbiology*, 25: 442–451.

Pescuma, M., Hébert, E.M., Mozzi, F. and Valdez, G.F. (2010) Functional fermented whey-based beverage using lactic acid bacteria, *International Journal of Food Microbiology*, 141: 73–81.

Pintado, M.E., Da Silva, J.A.L., Fernandes, P.B., Malcata, F.X. and Hogg, T.A. (1996) Microbiological and rheological studies on Portuguese kefir grains, *International Journal of Food Science and Technology*, 31(1): 15–26.

Plessas, S., Koliopoulos, D., Kourkoutas, Y., Psarianos, C., Alexopoulos, A., Marchant, R., Banat, I.M. and Koutinas, A.A. (2008) Upgrading of discarded oranges through fermentation using kefir in food industry, *Food Chemistry*, 106: 40–49.

Plessas, S., Pherson, L., Bekatorou, A., Nigam, P. and Koutinas, A.A. (2005) Bread making using kefir grains as baker's yeast, *Food Chemistry*, 93: 585–589.

Post, R.C. (1996) Regulatory perspective of the USDA on the use of antimicrobials and inhibitors in foods, *Journal of Food Protection* (Supplement): 78–81.

Powell, J.E., Witthuhn, R.C., Todorov, S.D. and Dicks, L.M.T. (2007) Characterization of bacteriocin ST8KF produced by a kefir isolate *Lactobacillus plantarum* ST8KF, *International Dairy Journal*, 17(3): 190–198.

Ramasubramanian, L., Restuccia, C. and Deeth, H.C. (2008) Effect of calcium on the physical properties of stirred probiotic yogurt, *Journal of Dairy Science*, 91(11): 4164–4175.

Rastall, R.A. and Maitin, V. (2002) Prebiotics and symbiotics: Towards the next generation, *Current Opinion in Biotechnology*, 13: 490–496.

Rea, M.C., Lennartsson, T., Dillon, P., Drinan, F.C., Reville, W.J., Heapes, M. and Cogan, T.M. (1996) Irish kefir-like grains: their structure, microbial composition and fermentation kinetics, *Journal of Applied Bacteriology*, 81: 83–94.

Reineccius, G. (1999) *Source Book of Flavors*, 2nd edn, Maryland, USA: Aspen Publishers Inc.

Rimada, P.S. and Abraham, A.G. (2006) Kefiran improves rheological properties of glucono-d-lactone induced skim milk gels, *International Dairy Journal*, 16: 33–39.

Roberfroid, M. (2002) Functional food concept and its application to prebiotics, *Digestive and Liver Disease*, 341: S105–110.

Roberts, M. and Yarunin, S. (2000) Danone moves into Russian kefir market, *New Nutrition Business*, 6: 22–24.

Robinson, R.K. and Tamine, A.Y. (2006) *Types of fermented milks*, UK: Oxford.–ver referência complete com M. Rosa.

Rodrigues, K.L., Gaudino Caputo, L.R., Tavares Carvalho, J.C., Evangelista, J. and Schneedorf, J.M. (2005) Antimicrobial and healing activity of kefir and kefiran extract, *International Journal of Antimicrobial Agents*, 25: 404–408.

RTS Resource Ltd. (2012) *The European Market for Probiotics and Prebiotics*. Online. Available at http://www.docstoc.com/docs/99272698 (accessed March 29, 2012).

Sacarro, D.M., Tamine, A.Y., Pilleggi, A.L.O.P.S. and Oliveira, M.N. (2009) The viability of three probiotic organisms grown with yoghurt starter cultures during storage for 21 days at 4°C, *International Journal of Dairy Technology*, 62(3): 397–404.

Sako, T., Matsumoto, K. and Tanaka, R. (1999) Recent progress on research and applications of non-digestible galacto-oligosaccharides, *International Dairy Journal*, 9: 69–80.

Salminen, S. and Von Wright, A. (1998) *Lactic Acid Bacteria: Microbiology and Functional Aspects*, 2nd edn, New York: Marcel Dekker Inc.

Saloff-Coste, C.J. (1996) Kefir nutritional and health benefits of yoghurt and fermented milks, *Danone World Newsletter*, 11: 1–7.

Sandholm, T.M. and Saarela, M. (2003) *Functional Dairy Products*, vol. 1, England: Woodhead Publishing.

Sandrou, D.K. and Arvanitoyannis, I. (2000) Implementation of hazard analysis critical control point (HACCP) to the cheese making industry, *Food Reviews International*, 16(3): 327–369.

Sazawal, S., Dhingra, U., Hiremath, G., Sarkar, A. and Dhingra, P. (2010) Prebiotic and probiotic fortified milk in prevention of morbidities among children: Community-based, randomized, double-blind, controlled trial, *Plos One*, 5(8): e12164.

Schoevers, A. and Britz, T.J. (2003) Influence of different culturing conditions on kefir grain increase, *International Journal of Dairy Technology*, 56: 183–187.

Scholz-Ahrens, K.E., Ade, P., Marten, B., Weber, P., Timm, W. and Açil, Y. (2007) Prebiotics, probiotics, and symbiotics affect mineral absorption, bone mineral content, and bone structure, *Journal of Nutrition*, 137: 838S–846S.

Schutte, L. (1999) Development and applications of dairy flavors, in R. Teranishi, E.L. Wick and I. Hornstein (eds.). *Flavor Chemistry: Thirty Years of Progress*, New York: Kluwer Academic/Plenum Publishers, 155–157.

Settanni, L. and Corsetti, A. (2008) Application of bacteriocins in vegetable food biopreservation, *International Journal of Food Microbiology*, 121: 123–138.

Settanni, L. and Moschetti, G. (2010) Non-starter lactic acid bacteria used to improve cheese quality, *Food Microbiology*, 27: 691–697.

Shah, N.P. (2000) Probiotic bacteria: Selective enumeration and survival in dairy foods, *Journal of Dairy Science*, 83(4): 894–907.

Sharpell Jr., F.H. (1985) The principles, applications and regulations of biotechnology in industry, agriculture and medicine, in M. Moo-Young (ed.) *Comprehensive Biotechnology*, no. 3, Canada, Toronto: Pergamon Press, 967.

Siegrist, M., Stampfli, N. and Kastenholz, H. (2008) Consumers' willingness to buy functional foods: The influence of carrier, benefit and trust, *Appetite*, 51(3): 526–529.

Simova, E., Beshkova, D., Angelov, A., Hristozova, T., Frengova, G. and Spasov, Z. (2002) Lactic acid bacteria and yeasts in kefir grains and kefir made from them, *Journal of Industrial Microbiology and Biotechno*logy, 28: 1–6.

Smit, G., Smit, B.A. and Engels, W.J.M. (2005) Flavor formation by lactic acid bacteria and biochemical flavor profiling of cheese products, *FEMS Microbiology Reviews*, 29: 591–610.

Soccol, C.R., Vandenberghe, L.P.S., Spier, M.R., Medeiros, A.B.P., Yamaguishi, C.T., Lindner, J.D.D, Pandey, A. and Soccol, V.T. (2010) The potential of probiotics: A review, *Food Technology and Biotechnology*, 48: 413–434.

Sodini, I., Lucas, A., Oliveira, M.N., Remeuf, F. and Corrieu, G. (2002) Effect of milk base and starter culture on acidification, texture, and probiotic cell counts in fermented milk processing, *Journal of Dairy Science*, 85(10): 2479–2488.

St-Onge, M.P., Farnworth, E.R., Savard, T., Chabot, D., Mafu, A. and Jones, P.J. (2002) Kefir consumption does not alter plasma lipid levels or cholesterol fractional synthesis rates relative to milk in hyperlipidemic men: A randomized controlled trial, *BMC Complementary and Alternative Medicine*, 2: 1–7.

Svec, P., Dráb, V. and Sedlácek, I. (2005) Ribotyping of *Lactobacillus casei* group strains isolated from dairy products, *Folia Microbiologica*, 50: 223–228.

Takeda, S., Takeshita, M., Kikuchi, Y., Dashnyam, B., Kawahara, S. and Yoshida, H. (2011) Efficacy of oral administration of heat-killed probiotics from Mongolian dairy products against influenza infection in mice: Alleviation of influenza infection by its immunomodulatory activity through intestinal immunity, *International Immunopharmacology*, doi: 10.1016/j.intimp.2011.08.007.

Talwalkar, A. and Kailasapathy, K. (2003a) A review of oxygen toxicity in probiotic yogurts: Influence on the survival of probiotic bacteria and protective techniques, *Comprehensive Reviews in Food Science and Food Safety*, 3: 117–124.

Talwalkar, A. and Kailasapathy, K. (2003b) Metabolic and biochemical responses of probiotics bacteria in oxygen, *Journal of Dairy Science,* 86: 2537–2546.

Thage, B.V., Broe, M.L., Petersen, M.H., Petersen, M.A., Bennedsen, M. and Ardö, Y. (2005) Aroma development in semi-hard reduced-fat cheese inoculated with *Lactobacillus paracasei* strains with different aminotransferase profiles, *International Dairy Journal*, 15: 795–805.

Tsarouhas, P.H., Arvanitoyannis, I.S. and Varzakas, T.H. (2009) Reliability and maintainability analysis of cheese (feta) production line in a Greek medium-size company: A case study, *Journal of Food Engineering*, 94: 233–240.

Vedamuthu, E.R. and Washam, C. (1980) *Cheese in "Biotechnology,"* Weinheim, Germany: Verlag Chemie.

Vinderola, C.G., Costa, G.A., Regenhardt, S. and Reinheimer, J.A. (2002) Influence of compounds associated with fermented dairy products on the growth of lactic acid starter and probiotic bacteria, *International Dairy Journal*, 12: 579–589.

Vinderola, G., Perdigón, G., Duarte, J., Farnworth, E. and Matar, C. (2006) Effects of the oral administration of the exopolysaccharide produced by *Lactobacillus kefiranofaciens* on the gut mucosal immunity, *Cytokine*, 36: 254–260.

Vitamin Research Products (2012) VRP South Africa S.A. BioPro. Online. Available at http://www.vrp.co.za/products/BioPRO™.aspx (accessed March 16, 2012).

Vujičič, I.F., Vulić, M. and Könyves, T. (1992) Assimilation of cholesterol in milk by kefir cultures, *Biotechnology Letters*, 14: 847–850.

Wang, X. and Gibson, G.R. (1993) Effects of the in vitro fermentation of oligofructose and inulin by bacteria growing in the human large intestine, *Journal of Applied Bacteriology*, 75: 373–380.

Wasser, S.P. and Weis, A.L. (1999) Therapeutic effects of substances occurring in higher *Basidiomycetes* mushrooms: A modern perspective, *Critical Reviews in Immunology*, 19(1): 65–96.

Watanabe, J., Nishimukai, M., Taguchi, H., Senoura, T., Hamada, S., Matsui, H., Yamamoto, T., Wasaki, J., Hara, H. and Ito, S. (2008) Prebiotic properties of epilactose, *Journal of Dairy Science*, 91(12): 4518–4526.

Watson, R.R. and Preedy, V.R. (2010) *Bioactive Foods in Promoting Health: Probiotics and Prebiotics*, London: Elsevier Academic Press.

Witthuhn, R.C., Schoeman, T. and Britz, T.J. (2004) Isolation and characterisation of the microbial population of different South African Kefir grains, *International Journal of Dairy Technology*, 57: 33–37.

Wouters, J.T.M., Ayad, E.H.E., Hugenholtz, J. and Smit, G. (2002) Microbes from raw milk for fermented dairy products, *International Dairy Journal*, 12: 91–109.

Wszolek, M., Kupiec-Teahan, B., Skov Guldager, H. and Tamime, A.Y. (2006) Production of kefir, koumiss and other related products, in A.Y. Tamime (ed.) *Fermented Milks*, Oxford, United Kingdom: Blackwell Publishing.

Yakult Honsha Co. Ltd (2011) *History*. Online. Available at http://www.yakultusa.com/Company/History.php (accessed November 1, 2011).

Yasui, H., Kiyoshima, J. and Hori, T. (2004) Reduction of influenza virus titer and protection against influenza virus infection in infant mice fed *Lactobacillus casei* Shirota, *Clinical and Diagnostic Laboratory Immunology*, 11: 675–679.

Zacarías, M.F., Binetti, A., Laco, M., Reinheimer, J. and Vinderola, G. (2011) Preliminary technological and potential probiotic characterization of bifidobacteria isolated from breast milk for use in dairy products, *International Dairy Journal*, 21: 548–555.

14 Dairy and Nondairy Probiotic Products and Beverages

Jean-Luc Tholozan and Jean-Luc Cayol

CONTENTS

14.1 INTRODUCTION

New trends in food intake by consumers now include nutritional needs for a state of well-being and the fact that diet may modulate some physiological functions and could promote some positive effects on consumers' health. Numerous publications in recent decades have indicated a need for health care through food in fully active young people, a search for a potential increase in life expectancy by means of selected diets or functional foods, and an improved quality of life for older people. Beyond the numerous advertising campaigns and government policies tending toward better consumption practices in the population, mediated by vegetables, fruits, and regular dairy product intake, consumers also look for new, more technological, and healthy foods. The food industry, many years ago, analyzed this quest for new foods and surfed on this wave for consumers' satisfaction. The proposed products have to be safe; have a good marketing aspect, that is, looking as natural as possible, which often means looking as if they could be home-cooked by the consumer himself from raw materials; have to be appetizing; and as often as possible, have to give an additional range of health benefits in addition to mere nutritional needs. Among the numerous proposals of new foods, probiotics have, during the last three decades, taken an important percentage of industrial efforts in the fields of nutritional aspects, technological aspects of food making, and potential positive effects on consumers' health. Numerous types of foods have been proposed to the consumers, corresponding to different ways of research and development of these functional foods in the food industry: Some ancestral traditional foods and beverages have been described, and traditional knowhow has been adapted for industrial processes of production to preserve the original typical aspect, smell, taste, and texture of the product and to modulate and derive products that are well adapted to the consumers' demand. Another way of development has been to promote and to develop new recipes corresponding to consumers' habits and demand by adding functional value through the use of new fermenting bacteria, such as probiotics, simultaneously promoting a better organoleptic quality of the product and health benefits. The third way of food research and development has been to produce functional food additives, participating to the final products, and adding a potentially high added value to the food. Probiotics belong to all three of these directions of new developed food.

14.2 PROBIOTICS, DAIRY PRODUCTS, NEW TENDENCIES

The latest definition of probiotics is "live microorganisms which when administrated in adequate amounts confer a health benefit on the host," replacing the definition given in 2002 and 2003: "live microorganisms, administrated in amounts that positively affect the health of the host" (Sanders 2003; FAO/WHO 2002). Previous definitions included the concept of improving the intestinal microbial balance (Fuller 1989). Further studies added the "proper amount" of microorganisms ingested by the consumer to confer a beneficial effect on the host (Kalliomäki et al. 2001; Rivera-Espinoza and Gallardo-Navarro 2010). The first description of this concept probably issued from the observations of Elie Metchnikoff: He noticed that fermented milk products containing *Lactobacillus* and *Streptococcus* prolonged life, and he attributed these physiological benefits to lactic acid bacteria (Metchnikoff 1908; Bibel 1988; Rasic 2003). In the 1930s, the species *Lactobacillus acidophilus* was incorporated into milk for better survival in the human gastrointestinal system (Rettger et al. 1935). It is often proposed nowadays that a regular intake of probiotics improves and maintains well-balanced intestinal flora and prevents gastrointestinal disorders (Lavermicocca 2006). Specific steps of characterization have thus been proposed for a "probiotic" label attribution for food use: precise taxonomic position of the microorganism, functional characterization, safety assessment, double specific independent protocols of phase 2 human trials, and phase 3 study for attribution of a "probiotic food" (FAO/WHO 2002).

During these numerous decades of probiotic study and probiotic effects quantifications, the gut microbiota was revealed as providing an important functional contribution to its host's physiology and in maintenance of health (Turnbaugh et al. 2007); numerous recent studies emphasized the multiple interactions in the human gut (Lakhdari et al. 2010; Saulnier et al. 2011). Genera *Lactobacillus* and *Bifidobacterium* have been included in probiotic products for a long time with some specific strains claimed to display a positive influence on the consumer's health (Vaughan and Mollet 1999). Probiotics are now known to have a better effect when the treatment is started as early as possible from the onset of diarrhea (Rautanen et al. 1998; Lei et al. 2010). Genus *Lactobacillus* and other lactic acid bacteria also produce different bacteriocins active against closely related microorganisms, often food spoilage microorganisms (Klaenhammer 1995). Probiotics also display a particular ability to survive in the severe conditions encountered in human intestinal tracts, partly because of the production of stress chaperones, such as *groES/groEL* and *dnaK/dnaJ*; this mechanism of survival has been widely used in the production of probiotic microorganisms added to different foods (Kok et al. 1996; Walker et al. 1999; Auty et al. 2001) as developed in the following sections. Other health benefits have been also proposed: a potential effect on detoxification of bile acids inhibitory to bacteria, some potential bio-therapeutic effects on candidiasis as demonstrated in immunodeficient mice (Wagner et al. 1997), reduction of the concentration of cancer-promoting enzymes and/or putrefactive metabolites in the gut, beneficial effects on microbial aberrancies, inflammation and other complaints in connection with inflammatory diseases, normalization of passing stool and stool consistency in subjects suffering from obstipation or irritable colon, prevention or alleviation of allergies and atropic diseases in infants, prevention of respiratory tract infections and other infectious diseases, as well as treatment of urogenital infections (de Vrese and Schrezenmeir 2008). Various genera and species of lactic acid bacteria have been described and used during recent years, including numerous *Bifidobacterium* and *Lactobacillus* species (Boyle and Tang 2006), with survival and benefits efficiency depending partly on the physicochemical environment of the food containing the probiotic microorganism (Ranadheera et al. 2010).

The origins and story of probiotic bacteria explain that a lot of probiotic foods are dairy products. A great variety of yogurts, solid or liquid milk, or products with an appearance close to yogurt or liquid yogurt are omnipresent in the market; additional fermented milk products, such as sour milk, sour whey, sour cream, buttermilk, kefir, and ice cream, are less frequent. For technological reasons, traditional stirred yogurt and drink yogurt are also well suited to the addition of probiotics after fermentation during the stirring (Heller 2001). Probiotic fresh or ripened cheeses are also available in small amounts, although fresh cheese has been rarely described in the food market (de Vrese and Schrezenmeir 2008). Some efforts also tried to improve probiotic culture yields, adding compounds such as maple sap (Al-Hawari et al. 2007). New tendencies include lyophilized probiotics, spray-dried product additives, and, of course, numerous nondairy probiotic products. Various developments are proposed to increase the viability of probiotics in functional foods: They are used in dehydrated form to increase stability, as direct starter cultures, as functional supplements, or as biocontrol agents. The composition of the suspension medium, the temperature regulation during drying, and the addition of cell protectors (skim milk powder, trehalose, glycerol, polysaccharides) efficiently increased cell survival during dehydration or cell protection against oxidation during long periods of storage. Drying cells by atomization, which is considered a long-term preservation method for lactic acid bacteria, demonstrated an optimum moisture content of approximately 4% for 99% of survival even after complementary drying. The use of microencapsulation of *Lactobacillus casei* by spray drying also gave high viable cell recovery during 90 days of storage at 25°C for probiotics included in chocolate (Guergoletto 2012). Another example is given by the patent deposited by Rudolph (2010), enumerating components to be added to the bacterial culture to preserve cell viability during freeze-drying and spray-drying steps and the way to convert the different ingredients together in a frozen dessert. We will now focus on the different types of nondairy probiotic beverages.

14.3 PROBIOTIC NONDAIRY CEREAL BEVERAGES

Besides the dairy probiotic products, the consumer also has a large choice of probiotic nondairy products (Table 14.1). One of the main interests of nondairy probiotic products is that dairy products cannot be consumed by humans allergic to milk proteins or persons displaying severe lactose intolerance: This led to huge research and development work in new directions using a nondairy matrix for probiotic food. An enormous part of the market is represented by cereal-based fermented beverages and foods. Traditional fermentation has been routinely used as a natural means for food preservation. A tentative review of cereals and cereal components' potential role was presented in by Charalampopoulos et al. (2002). Investigation of traditional foods produced in some countries and upgrading of traditional fermented foods to functional foods has been developed during the last decade with the aim of adding value to these fermented products. Other functional foods containing probiotics are represented by the addition of probiotics to a nonprobiotic food during processing, turning it into a probiotic-containing food, too.

TABLE 14.1
Nondairy Probiotic Beverages Composition

Type	Main Compounds	Additional Compounds	Probiotic	Benefits Claimed
Cereals	Maize	Fibers	Bacteria	Antimicrobial
	Millet	Soybeans	Yeasts	Metabolic
	Sorghum	Cooked	Molds	disorders
	Wheat	vegetables		
Soymilk	Soymilk	Yogurts	Bacteria	Antimicrobial
		B vitamins		Antioxidant
				Nutritional
				increase
				Hormone-
				dependent
				disorders
Fruit and vegetable juices	Fruit juices	Sugar	Bacteria	Nutritional value
	Puree	Probiotic		increase
	Pieces of	addition		Antibacterial
	fruits			Antioxidant
Miscellaneous	Tea	Milk	Mushrooms	Antimicrobial
– Kombucha	Vegetables	Yogurts	Yeasts	Antioxidant
– Ice cream			Bacteria	
– Derived beverages/sauces			Spontaneous	
			flora	

A considerable effort has been performed to stimulate fermentation technologies in developing countries, especially for small-scale food industries in the field of sustainable development. This effort started on a large scale by means of analysis of the local context and technical and logistical assessments, followed by institutional assessments, leading to the selection of an appropriate scale for demonstrations and dissemination of the fermentation technology. The FAO's experience has been used as a support for technology transfer with the aim of reducing food losses while increasing food security and food safety, at the same time upgrading nutritional standards of traditional foods (Rolle and Satin 2002). A large simultaneous effort toward a better definition of these products, characterization of microbial strains used for fermentation, and the construction of guidelines for local labeling of these products has accompanied these developments (Senok 2009). The field of cereal-based beverages and foods has been one main focus of these works with a precise characterization and taxonomic position description of the microorganisms responsible for fermentation. Extra roles of these microorganisms have also been frequently demonstrated, conferring to the food the simultaneous status of functional foods (Kalui et al. 2010). Emmer grain, flour, and malt fermented beverages have been thoroughly screened for probiotic fermentation, using numerous different pure lactic acid bacteria cultures as inoculum. The fermented beverages demonstrated a pH value approximately equal to 4 after 5 h of fermentation, an amino acid concentration ranging from 68 to 336 mg/L, and an increase in viscosity resulting from exopolysaccharide accumulation, accompanied by the production of trace amounts of more than 40 volatile compounds. An exhaustive analysis of sensory and nutritional characteristics and the survival of potential probiotic strains revealed a stable panel of qualities during 30 days of storage at 4°C, demonstrating the extreme suitability of emmer grains for making functional beverages (Coda et al. 2011). Oat-derived beverages, fermented by *Lactobacillus plantarum*, demonstrated the main influence of the sugar content and oats on the growth of the probiotic strain; the product remained stable for 21 days when stored at 4°C (Gupta et al. 2010). Maize products led, in general, to nonallergenic, mild-tasting beverages described as *mawe* (Hounhouigan et al. 1994), *kenke*, and *ogi* (Olasupo et al. 1997). In Mexico, *Pozol* is obtained by cooking maize, washed with water, in a 1% lime solution, then grinding it to a dough, shaping it into balls wrapped in banana leaves, and leaving it to ferment spontaneously for up to 4 days; then it is resuspended in water (Wacher et al. 2000). Thorough studies demonstrated with these products show that some local strains used developed antimicrobial products leading to safer and longer-storing foods, such as the maize-fermented porridge *ikii* in Kenya with *Lactobacillus rhamnosus* and *L. plantarum*; in addition, these products also demonstrated an increase in exopolysaccharide production, improved health properties, and technological and taste qualities of the product (Kalui et al. 2009). A similar indigenous cereal-fermented beverage, called *mahewu*, is produced in South Africa from maize and *Streptococcus lactis* and *Lactobacillus lactis* subsp. *lactis* (Blandino et al. 2003) with the addition of some other cereals: sorghum, millet malt, or wheat flour. Previous research on a spontaneously fermented porridge made from sorghum and millet called *bushera*, a Ugandan traditional fermented beverage, demonstrated similar features with numerous strains of *Lactobacillus*, *Streptococcus*, and *Leuconostoc* described as probiotic species (Muyanja et al. 2003). Very similar properties were also demonstrated previously in another traditional beverage from the north of Ghana, *koko*, a lunch or in-between meals beverage made from overnight steeping of pearl millet; discarding of steeping water; and wet-milling of the millet grains together with spices, water, 3 h of fermentation, and the addition of a sedimented bottom layer until it is the desired consistency (Lei and Jacobsen 2004). *Bozo* is a traditional beverage from Bulgaria, Albania, Turkey, and Romania made from different cereals mixed with sugar or saccharine with a flora mainly constituted of yeasts and lactic acid bacteria in a ratio between yeast and lactic acid bacteria of 40% of the global flora being yeasts, and 60% of the global flora being lactic acid bacteria (Prado et al. 2008). In each case, the taxonomic position of these natural starters led to most of the routinely described lactic acid bacteria. They presented probiotic properties with all of the functional properties associated with the definition of a probiotic strain and the corresponding health benefits already described in each case of dairy products: resistance to the gut environment, shortening of diarrhea, production of lantibiotic substances, and antimicrobial activity against some pathogenic microorganisms, such as *Listeria* or *Bacillus* genera. For additional references to traditional African fermented maize products, the interested reader should read the well-documented review from Kalui et al. (2010). Additional local probiotic-containing products obtained by the fermentation of different cereals have been extensively described in Prado et al. (2008). Novel cereal-based drinks have been recently proposed, from the association of different cereals and the development of increased organic acid contents, up to 1 g/L, and higher nutritional properties with final free amino acids and reduced sugar concentrations of 20 mg/L and 2.8 g/L, respectively (Rathore et al. 2012). Not to mention also, the different versions of soy sauce prepared from a blend of cooked soybeans and coarse wheat flour, inoculated by molds, and fermented for 3 days, then stopped with approximately 25% brine solution (Yokotsuka and Sasaki 2006). Industrial developments of new fermented nondairy products have followed these studies with some examples given by a maize porridge with added malted barley used as weaning food consumed in addition to breast milk (Helland et al. 2004; Angelov et al. 2006) produced within a few hours of fermentation, a symbiotic functional drink from oats made by combining a probiotic starter and a whole-grain oat substrate.

More-consistent semisolid foods consumed as breakfast or a snack are also currently described from fermented rice: *idli*, a solid starchy nutritious food, is currently produced and distributed on a large scale in India after a *Lactobacillus*, *Leuconostoc*, and *Pediococcus* fermentation, often associated with some yeasts belonging to *Geotrichum* and *Torulopsis* genera. Another fermented product from rice is *dosa*, produced by grinding wet rice and black gram separately with water, then mixing them and allowing for natural fermentation before cooking in a pan. *Tarhana* in Turkey and in Greece is prepared by mixing wheat flour, sheep-milk yogurt, yeasts, and a variety of cooked vegetables and spices followed by a fermentation of 1 to 7 days; then it is dried and stored to form biscuits (Daglioglu et al. 2002); *kishk* is almost similarly prepared and routinely consumed in the Middle East. Here again with solid foods, new cereal-based probiotic fermented foods are offered to the consumer: *yosa* is a snack food made from oat bran pudding cooked in water and fermented with lactic acid bacteria belonging to *Bifidobacterium*, sustainable for vegetarians, and mainly consumed in Finland and Scandinavian countries (Blandino et al. 2003).

14.4   PROBIOTIC SOYMILK

The second group of developed nondairy probiotic beverages is represented by products derived from soy milk. The use of soy milk was strongly recommended to consumers developing lactose intolerance. From this basic physiological need, many developments have been proposed in the field of better fermentation with lactic acid bacteria, improvement of process production, and/or enrichment of the product with components having some benefits to the health of the consumer. This market offer is very large nowadays, starting from nonfermented soy milk, which is "fortified," as described in the market offer, by probiotic microorganisms thriving in the beverage, mainly belonging to genus *Lactobacillus*. Pioneering experiments demonstrated the drastic decrease in *Helicobacter pylori* survival in corn meal and soy milk yogurts containing probiotics (Oh et al. 2002). The high viability of probiotic *Lactobacillus* has been then assessed and demonstrated in soy milk before proposing such new functional foods (Farnworth et al. 2007; Wang et al. 2009). An extensive review of the different physicochemical properties of soy milk fermented with probiotic yeasts (*Saccharomyces boulardii*) and bacteria (*Lactobacillus acidophilus, L. bulgaricus, L. casei, L. plantarum, L. helveticus*) has been proposed by Rekha and Vijayalakshmi (2008), who stressed the importance of fermentation to the significant increase in the potential health benefits of the probiotic presence: an increase in antioxidant activity of radical scavenging potential of 19.76%, an increase in the bioactive aglycone of up to 13.66 mg/100 mL, and a decrease of up to 1.36 mg/100 mL of poorly digestible glycosides. The high antioxidant activity of soy milk fermented by lactic acid bacteria and bifidobacteria was also demonstrated by Wang et al. (2006). Other developments were performed in the field of process development with the increased growth of *Lactobacilli* upon the immobilization on agro-wastes; rinds of different local plants (mangosteen, durian, cempedak) were ground to form a powder used as an immobilizer for *L. bulgaricus* and *L. acidophilus* growth in soy milk, resulting in a shorter conversion time of soy milk sugars to lactic and acetic acids (The et al. 2010). Other process developments are aimed at improving the nutritional content of soy milk by enhancing the conversion percentage of isoflavone glycosides to the corresponding isoflavone aglycones, highly bioactive compounds easily absorbed in the small intestine. In some cases, a supplementation with lactose resulted in an additional biotransformation of isoflavones to aglycones (Otieno and Shah 2007; Donkor and Shah 2008; Shah et al. 2010; Rekha and Vijayalakshmi 2011). Enhancement of the proteolytic activity in soy milk containing probiotics demonstrated a promising effect on the enhancement of angiotensin-I inhibitory activity with potential antihypertensive effects; the aglycone increase in milk also might provide a potential application for the treatment of hormone-dependent disorders. An extensive hydrolysis of daidzein led to notable quantities of daidzein and dihydro daidzein, acting as precursors for the production of equol, a structural analog of human estradiol, which could bind as estrogen receptors and regulate hormone-dependent diseases, such as osteoporosis and breast cancer (Yeo and Liong 2010a). These authors also demonstrated that a supplementation of soy milk with fructooligosaccharides, usually considered as prebiotic substances stimulating the growth of probiotic bacteria, resulted in a huge increase in the β-glucosidase activity level in the milk with *Bifidobacterium* strains and different species of *Lactobacillus*. This resulted in high conversion rates of isoflavones and in increased rates of organic acid production, directly affecting the growth of gastrointestinal pathogens (Gibson 2004; Yeo and Liong 2010b). Ewe et al. (2010) also demonstrated that the presence of a small initial amount of B vitamins (thiamine, riboflavin, niacin-amide, calcium pantothenate, folic acid, and biotin) added at an initial concentration of 1 mg/L each resulted in an increased production of riboflavin and folic acid of 243% and 175%, respectively, by different strains of the probiotic *Lactobacillus acidophilus* and *L. gasseri*. Other technological treatments are aimed at the preservation of a high level of probiotic survival in soy milk after drying, rehydration, and storage of the starters *Streptococcus*, *Lactobacillus*, and *Bifidobacterium* strains (Wang et al. 2004).

14.5 PROBIOTIC FRUIT AND VEGETABLE JUICES

Fruit and vegetable juices could be an alternative vehicle for the incorporation of probiotics for consumers, provided they facilitate probiotic survival during the shelf and storage life of the products (Table 14.2). In 1994, in Sweden, a lactic acid fermented oatmeal gruel was mixed in a fruit drink. The *L. plantarum* strain 299v used for the fermentation demonstrated a number of benefits to consumer health and remained alive in the food carrier, which had to remain palatable (Molin 2001). Carrot juice was also a good matrix for numerous *Bifidobacterium bifidum* strains' development throughout the fermentation time (Kun et al. 2008). More recently, the effect of *Lactobacillus casei* introduced in *Garcinia mangostana* fruit extract demonstrated an increase in body weight and no mortality in animals infected by the Newcastle disease virus when compared to the reference under placebo in fighting roosters after 90 days of consumption (Bautista-Garfias et al. 2011). A pomegranate juice has been fermented by probiotic lactic acid bacteria *Lactobacillus plantarum*, *L. delbruekii*, *L. paracasei*, and *L. acidophilus* during a 72 h fermentation at 30°C, followed by a storage time of 4 weeks at 4°C. The first two species of *Lactobacillus* demonstrated a notable growth potential and survival in the juice during the first 2 weeks and a drastic decrease in the cell number at the end of the period of storage (Mousavi et al. 2011). Similar results were obtained with *L. casei* strains in fermented cashew apple juice with a cell density of more than 8 log cfu/mL during the 42 days of storage (Pereira et al. 2011). From the marketing point of view, probiotics have been identified as a potential cause of perceptible off-flavors in foods proposed to the consumers. Different strategies have been developed for flavor masking in probiotic juices. Experiments were performed with orange juices as reference, with orange juices containing *L. paracasei* ssp. *paracasei* as a probiotic, and with probiotic orange juices combined with an addition of tropical juices. Tropical juices were effective in masking the probiotic flavor, but surprisingly, interactions between information on the presence or the absence of probiotics in orange juice were clearly demonstrated in the consumer's appreciation of the overall liking of these two latter orange juices

TABLE 14.2
Survival Potential of Probiotics in Nondairy Beverages

Probiotic Microorganism	Type of Food	Survival Potential	References
Zygosaccharomyces bailii	Kombucha	Throughout fermentation	Teoh et al. (2004)
L. plantarum and *L. delbruekii*	Pomegranate juice	Two weeks of storage at 4°C	Mousavi et al. (2011)
L. casei	Cashew apple juice	More than 28 days at 4°C	Pereira et al. (2011)
L. plantarum B28	Oat-based drink	More than 24 days at 4°C	Angelov et al. (2006)
Lactic acid bacteria	Mestizo	More than 8 days	Wacher et al. (2000)
Lactobacillus, *Pediococcus*, and *Weisella*	Koko sour water	More than 20 days	Lei and Jacobsen (2004)
Lactobacillus, *Leuconostoc*, and *Weissella*	Bushera	More than fermentation	Muyanja et al. (2003)
L. acidophilus and *B. bifidum*	Ice cream	More than 17 weeks of storage at −29°C	Cruz et al. (2009)
L. plantarum and *L. delbrueckii*	Cabbage juice	More than 4 weeks at 4°C	Ranadheera et al. (2010)
L. casei Zhang	Soy milk	28 days at 4°C	Wang et al. (2009)
L. acidophilus B94 and *L. casei* L26	Soy milk	More than 28 days at 4°C	Donkor and Shah (2008)
Bifidobacterium longum B6	Dried fermented soy milk	More than 16 weeks at 4°C	Wang et al. (2004)
L. rhamnosus HN001, *L. paracasei* Lpc 37, and *B. lactis* HN001	Fruit juices	More than 6 weeks at 4°C	Shah et al. (2010)

(Luckow et al. 2006). The technical aspect of probiotic juice production was also investigated in the direction of protective or antagonistic effects on probiotic survival or decrease during the storage period of the juices. Mathematical approaches demonstrated the positive effect of some components on probiotic survival during storage, such as high pH values or citric acid in vitro, and similar positive effects on bacteria survival in orange, grapefruit, black currant, and lemon juice, even after 6 weeks of storage; in pomegranate juice, survival time was much shorter, indicating antagonistic effects of some components present in the juice, possibly phenolic compounds, displaying antibacterial properties (Nualkaekul and Charalampopoulos 2011). In a similar way, a process of probiotic beverage production using coconut water as the main raw material was patented by Soccol (2009); the probiotic fermentation occurred after inoculation of coconut water containing an additional carbon and nitrogen source, and was followed by the addition of sugar and additives before packaging. An extensive review of the effect of the fruity food matrices on probiotic survival has been proposed to readers by do Espirito Santo et al. (2011), including carrots, onions, celery, coconuts, acerola, and cacao fruits used in pieces of fruits with probiotic immobilization, puree, chocolate, juice, or ice cream. The good viability of *L. acidophilus* strains and *Bifidobacterium* strains was preserved after 1 month of storage with no decrease of antioxidant activity of β-carotene in the vegetable juices; a decrease in antimicrobial probiotic properties was noticed in apple wedges, and a general increase in polyphenols was demonstrated in juices inoculated with probiotics. In a similar manner, the impact of selected fruits, vegetables, grains, herbs, spices, and extracts was checked on the growth of selected probiotic and pathogenic bacteria. Aqueous extracts of banana, apple, and orange enhanced the growth of each probiotic microorganism (*Lactobacillus reuteri, L. rhamnosus, Bifidobacterium lactis*) and strongly inhibited the growth of different strains of pathogenic *Escherichia coli* (Sutherland et al. 2009). Shah et al. (2010) also demonstrated that the presence of vitamins and antioxidants dramatically influenced, in a positive manner, the survival rate of *L. rhamnosus* after a 6-week storage period. Additional compounds also enhanced lactic acid probiotic bacteria survival as demonstrated for *Bifidobacterium* with the presence of oat fibers added to low pH-value fruit juices (Saarela et al. 2011). The general case of probiotic juices proposed to the consumers consists of *Lactobacillus* bacteria cultured in oatmeal or other nondairy media, then added to fruit juices to provide this beneficial microbe to consumers who cannot tolerate milk or yogurt. Health benefits of pear juices fermented by *L. acidophilus*, during a time of up to 72 h, were assayed for their potential effect on hypoglycemia management and for a potential effect on hypertension. A dramatic decrease in the angiotensin-I-converting enzyme with fermentation time was demonstrated in addition to an accumulation of antioxidant substances, such as epicatechin, an antioxidant flavonol with a highly potent α-glucosidase inhibitory activity; some of the functional properties of the fermented juice were attributed to the presence of epicatechin in combination with other phenolic compounds (Ankolekar et al. 2011). In the Caucasus region, fruit juices are prepared from kefir grains. The resulting drink formed by a microbial consortium useful in the intestine is a probiotic drink; a direct effect on the inhibition of pathogen development was quantified, and the higher inhibiting effects were obtained against *Candida albicans* with brown sugar used as an additional carbon source during the fermentation of the drink (Silva et al. 2009). *L. rhamnosus* survival studies with a whey-based fruit juice matrix given during 2 weeks to healthy adults demonstrated high rates of survival in the human gut and no adverse effects on intestinal function or immunological parameters of the consumers (Suomalainen et al. 2006). An extensive review of fruit-based potential functional foods and related patents has been recently proposed by Sun-Waterhouse (2011). Table olives have also been determined to be a good vector for the delivery of probiotics to humans. Lactic acid bacteria found in olives belong to different species of *Lactobacillus, L. plantarum, L. casei, L. mesenteroides*, and *Pediococcus pentosaceus*. These bacteria display typical numerous probiotic properties, adding functional properties to the nutritional natural value of table olives. The selected probiotic strain should be added into the brine at the onset of fermentation; the current challenge is to maintain high-enough levels of probiotic bacteria in olive-based foods to ensure favorable effects upon consumer health (Peres et al. 2012). High levels of *Lactobacillus* survival, over 10^6 cfu/mL, were counted in orange and pineapple juices after 12 weeks of storage. Acid tolerance of these bacteria and lack of survival after pasteurization treatments of probiotic-containing juices led the authors to propose probiotics as functional supplements in fruit juices (Sheehan et al. 2007).

14.6 OTHER PROBIOTIC BEVERAGES

Some other different nondairy probiotic beverages have been assayed for probiotic properties. Kombucha seems to be a good candidate, produced from green and black tea fermented by probiotic strains, "mushrooms" described as a mixture of yeasts and bacteria. Use of specific culture media allowed purification and isolation of *Candida*, *Rhodotorula*, *Schizosaccharomyces*, *Brettanomyces*, *Torulaspora*, and *Zygosaccharomyces* yeast genera. The highest counts of viable cells were determined in the beginning of tea fermentation; the three last genera remained present throughout the 14 days of fermentation, and *Zygosaccharomyces* was dominant (Teoh et al. 2004). Numerous derived products, including kombucha on the basis of derived recipes, are proposed today on the food market. Other probiotic supports tested were ice creams; low storage temperatures were not in favor of probiotic survival. The slow kinetics of temperature decrease may also decrease the survival rates by causing severe injury to cells. *L. rhamnosus* was nevertheless present at counts as high as 10^8 cfu/g after 1 year of storage of ice creams at $-29°C$ (Cruz et al. 2009). Additional scarce probiotic products are proposed to the consumers on websites. There is the case for example of a fermented "salsa verde" made of unripe tomatoes and of sweet peppers, allowed to ferment with spices for 4 months in the refrigerator; the starter is the natural spontaneous microflora present in the raw materials. This product displays all of the characteristics of a probiotic food. A new neutraceutical probiotic beverage containing herbal mate extract and honey has also recently been proposed: *L. acidophilus* strain ATCC 4356 was the microbial candidate for fermentation, producing 2.0 and 0.75 g/L of lactic and acetic acids, respectively, with a final pH value of 3.67. Final glucose and fructose contents were 1.0 and 1.8 g/L, respectively; antioxidant properties were high as a result of polyphenol present in the herbal mate; and sensorial analyses demonstrated good consumer acceptance when compared to other, similar, commercial beverages (Ferrari Pereira Lima et al. 2012).

14.7 CONCLUSIONS AND PERSPECTIVES

First studies on nondairy probiotic products have demonstrated that cereals are suitable substrates for the growth of human-derived probiotic strains. These studies also demonstrated the need for a systematic approach to a better knowledge of the growth and survival potential of probiotic strains on these raw materials, possibly mediated by the input of functional compounds to food, such as fiber or vitamins. An additional direction for the future potential of probiotics was also suggested with the use of genetically engineered bacteria to shift some physiological or metabolic disorders of the consumers to wellbeing. Ten years later, with a better knowledge of the original indigenous probiotic foods, with some large-scale productions routinely commercialized worldwide, and with a much deeper description of positive and negative interactions between probiotic strains, food compounds, and health benefits, the consumers' knowledge and desire for functional foods are still not well understood. Challenging ongoing research works demonstrate some contradictions in the description of positive and negative interactions between probiotics and food compounds, food compounds and the consumer, and the real need for a clear message to the consumers, who are highly interested but facing the actual large market presenting an enormous diversity of functional foods. More studies are needed to provide and improve the scientific principles and rational information on the efficacy of functional foods and the underlying mechanisms, as well as the industrial aspects of functional food production: selection or choice of probiotic starters, development or associations for optimum interactions between bioactive compounds and food components, type of interactions and microbial genes interfering with the host physiology, and process adjustments to maximize production conditions while preserving the internal quality and potential of these new foods.

REFERENCES

Al-Hawari, J., Cochu, A. and Fournier, D. (2007) Production de bactéries probiotiques à l'aide de sève d'érable, *Conseil National de Recherches Canada*, patent NRC11914.

Angelov, A., Gotcheva, V., Kuncheva, R. and Hristozova, T. (2006) Development of a new oat-based probiotic drink, *International Journal of Food Microbiology*, 112:75–80.

Ankolekar, C., Pinto, M., Greene, D. and Shetty, K. (2011) *In vitro* bioassay based screening of antihyperglycemia and antihypertensive activities of *Lactobacillus acidophilus* fermented pear juice, *Innovative Food Science and Emerging Technologies*, 13:221–230.

Auty, M.A.E., Gardinier, G.E., McBrearty, S.J., O'Sullivan, E.O., Mulvihill, D.M., Collins, J.K., Fitzgerald, G.F., Stanton, C. and Ross, R.P. (2001) Direct in situ viability assessment of bacteria in probiotic dairy products using viability staining in conjunction with confocal scanning laser microscopy, *Applied and Environmental Microbiology*, 67:420–5.

Bautista-Garfias, C.R., Rios-Flores, E. and Garcia-Rubio, V.G. (2011) Comparative effects of *Lactobacillus casei* and a commercial mangosteen dietary supplement on body weight gain and antibody response to Newcastle disease virus vaccine in fighting roosters, *Journal of Medical Food*, 14:828–33.

Bibel, D.J. (1988) Elie Metchnikoff's bacillus of long life, *ASM News*, 54:661–5.

Blandino, A., Al-Aseeri, M.E., Pandiella, S.S., Cantero, D. and Webb, C. (2003) Cereal-based fermented foods and beverages, *Food Research International*, 36:527–43.

Boyle, R.J. and Tang, M.L. (2006) 'The role of probiotics in the management of allergic disease,' *Clinical and Experimental Allergy*, 36:568–76.

Charalampopoulos, D., Wang, R., Pandiella, S.S. and Webb, C. (2002) Application of cereals and cereal components in functional foods: a review, *International Journal of Food Microbiology*, 79:131–41.

Coda, R., Rizzello, C.G., Trani, A. and Gobbetti, M. (2011) Manufacture and characterization of functional emmer beverages fermented by selected lactic acid bacteria, *Food Microbiology*, 28:526–36.

Cruz, A.G., Antunes, A.E.C., Sousa, A.L.Q.P., Faria, J.A.F. and Saad, S.M.I. (2009) Ice-cream as a probiotic food carrier, *Food Research International*, 42:1233–9.

Daglioglu, O., Arici, M., Konyali, M. and Gumus, T. (2002) Effects of tarhana fermentation and drying methods on the fate of *Escherichia coli* O157:H7 and *Staphylococcus aureus*, *European Food Research and Technology*, 215:515–9.

de Vrese, M. and Schrezenmeir, J. (2008) Probiotics, prebiotics, and symbiotics, *Advances in Biochemical Engineering/Biotechnology*, 111:1–66.

do Espirito Santo, A.P., Perego, P., Converti, A. and Oliveira, M.N. (2011) Influence of food matrices on probiotic viability—A review focusing on the fruity bases, *Trends in Food Science and Technology*, 22:377–85.

Donkor, O.N. and Shah, N.P. (2008) Production of β-glucosidase and hydrolysis of isoflavone phytoestrogens by *Lactobacillus acidophilus*, *Bifidobacterium lactis*, and *Lactobacillus casei* in soymilk, *Food Microbiology and Safety*, 73:M15–20.

Ewe, J.A., Wan-Abdullah, W.N. and Liong, M.T. (2010) Viability and growth characteristics of *Lactobacillus* in soy milk supplemented with B-vitamins, *International Journal of Food Sciences and Nutrition*, 61:87–107.

FAO/WHO (2002) Guidelines for the evaluation of probiotics in food, Food and Agriculture Organization of the United Nations/World Health Organization Working group report, London, Ontario, Canada, 30 April–1 May.

Farnworth, E.R., Mainville, I., Desjardins, M.P., Gardner, N., Fliss, I. and Champagne, C. (2007) Growth of probiotic bacteria and bifidobacteria in soy yogurt formulation, *International Journal of Food Microbiology*, 116:174–81.

Ferrari Pereira Lima, I., De Dea Linder, J., Thomaz-Soccol, V., Parada, J.L. and Soccol, C.R. (2012) Development of an innovative nutraceutical fermented beverage from herbal mate (*Illex paraguariensis* A.St.-Hil.) extract, *International Journal of Molecular Sciences*, 13:788–800.

Fuller, R. (1989) Probiotics in man and animal, *Journal of Applied Bacteriology*, 66: 365–78.

Gibson, G.R. (2004) Fibre and effects on probiotics (the prebiotic concept), *Clinical Nutrition Supplements*, 1:25–31.

Guergoletto, K.B. (2012) Dried probiotics for use in functional food applications, in B. Valdez (ed.) *Food industrial processes—Methods and equipment*, Rijeka: InTech.

Gupta, S., Cox, S. and Abu-Ghannam, N. (2010) Process optimization for the development of a functional beverage based on lactic acid fermentation of oats, *Biochemical Engineering Journal*, 52:199–204.

Helland, M.H., Wocklund, T. and Narvhus, J.A. (2004) Growth and metabolism of selected strains of probiotic bacteria, in maize porridge with added malted barley, *International Journal of Food Microbiology*, 91:305–13.

Heller, K.J. (2001) Probiotic bacteria in fermented foods: Product characteristics and starter organisms, *American Journal of Clinical Nutrition*, 73:374S–9S.

Hounhouigan, D.J., Nout, M.J.R., Nago, C.M., Houben, J.H. and Rombouts, F.M. (1994) Microbiological changes in mawe during natural fermentation, *World Journal of Microbiology and Biotechnology*, 10:3959–93.

Kalliomäki, M., Salminen, S., Arvilommi, H., Kero, P., Koskinen, P. and Isolauri, E. (2001) Probiotics in primary prevention of atopic disease: A randomized placebo-controlled trial, *Lancet*, 357:1076–9.

Kalui, C.M., Mathara, J.M. and Kutima, P.M. (2010) Probiotic potential of spontaneously fermented cereal based foods—A review, *African Journal of Biotechnology*, 9:2490–8.

Kalui, C.M., Mathara, J.M., Kutima, P.M., Kliyakia, C. and Wonga, L.E. (2009) Functional characteristics of *Lactobacillus plantarum* and *Lactobacillus rhamnosus* from ikii, a Kenyan traditional fermented maize porridge, *African Journal of Biotechnology*, 8:4363–73.

Klaenhammer, T.R. (1995) Genetics of intestinal lactobacilli, *International Dairy Journal*, 5:1019–58.

Kok, R.G., de Waal, A., Schut, F., Welling, G.W., Weenk, G. and Hellingwerf, K.J. (1996) Specific detection and analysis of a probiotic *Bifidobacterium* strain in infant feces, *Applied and Environmental Microbiology*, 62:3668–72.

Kun, S., Rezessy-Szabo, J.M., Nguyen, Q.D. and Hoschke, A. (2008) Changes of microbial population and some components in carrot juice during fermentation with selected *Bifidobacterium* strains, *Process Biochemistry*, 43:816–21.

Lakhdari, O., Cultrone, A., Tap, J., Gloux, K., Bernard, F., Ehrlich, S.D., Lefèvre, F., Doré, J. and Blottière, H.M. (2010) Functional metagenomics: A high throughput screening methods to decipher microbiota-driven NF-kB modulation in the human gut, *PlosOne* 5:e13092.

Lavermicocca, P. (2006) Highlights on new food research, *Digestive and Liver Disease*, 38:S295–9.

Lei, V., Friis, H. and Fleischer Michaelsen, K. (2010) Spontaneously fermented millet product as a natural probiotic treatment for diarrhea in young children: An intervention study in Northern Ghana, *International Journal of Food Microbiology*, 110:246–53.

Lei, V. and Jacobsen, M. (2004) Microbiological characterization and probiotic potential of *koko* and *koko sour water*, African spontaneously fermented millet porridge and drink, *Journal of Applied Microbiology*, 96:384–97.

Luckow, T., Sheehan, V., Fitzgerald, G. and Delahunty, C. (2006) Exposure, health information and flavor-masking strategies for improving the sensory quality of probiotic juice, *Appetite*, 47:315–23.

Metchnikoff, E. (1908) *The prolongation of life. Optimistic studies*, The English translation, P. Chalmers Mitchell (ed.) New York: G.P. Putnam Sons, the Knickerbocker Press.

Molin, G. (2001) Probiotics in foods not containing milk or milk constituents, with special reference to *Lactobacillus plantarum* 299v, *American Journal of Clinical Nutrition*, 73:380S–5S.

Mousavi, Z.E., Mousavi, S.M., Razavi, S.H., Emam-Djomeh, Z. and Kiani, H. (2011) Fermentation of pomegranate juice by probiotic lactic acid bacteria, *World Journal of Microbiology and Biotechnology*, 27:123–8.

Muyanja, C.M.B.K., Narvhus, J.A., Treimo, J. and Langsrud, T. (2003) Isolation, characterization and identification of lactic acid bacteria from *bushera*: A Ugandan traditional fermented beverage, *International Journal of Food Microbiology*, 80:201–10.

Nualkaekul, S. and Charalampopoulos, D. (2011) Survival of *Lactobacillus plantarum* in model solutions and fruit juices, *International Journal of Food Microbiology*, 146:111–7.

Oh, Y., Osato, M.S., Han, X., Bennett, G. and Hong, W.K. (2002) Folk yogurt kills *Helicobacter pylori*, *Journal of Applied Microbiology*, 93:1083–8.

Olasupo, N.A., Olukoya, D.K. and Odunfa, S.A. (1997) Assessment of a bacteriocin-producing *Lactobacillus* strain in the control of spoilage of cereal-based African fermented food,' *Folia Microbiologica*, 42:31–4.

Otieno, D.O. and Shah, N.P. (2007) Endogenous beta-glucosidase and beta-galactosidase activities from selected probiotic micro-organisms and their role in isoflavone biotransformation in soymilk, *Journal of Applied Microbiology*, 103:910–7.

Pereira, A.L.F., Maciel, T.C. and Rodrigues, S. (2011) Probiotic beverage from cashew apple juice fermented with *Lactobacillus casei*, *Food Research International*, 44:1276–83.

Peres, C.M., Peres, C., Hernandez-Mentosa, A. and Malcata, F.X. (2012) Review on fermented plant materials as carriers and sources of potentially probiotic lactic acid bacteria—with an emphasis on table olives, *Trends in Food Science and Technology*, 26:31–42.

Prado, F.C., Parada, J.L., Pandey, A. and Soccol, C.R. (2008) Trends in non-dairy probiotics beverages, *Food Research International*, 41:111–23.

Ranadheera, R.D.C.S., Baines, S.K. and Adams, M.C. (2010) Importance of food in probiotic efficacy, *Food Research International*, 43:1–7.

Rasic, J.L. (2003) Microflora of the intestine probiotics, in B. Caballero, L. Trugo and P. Finglas (eds.) *Encyclopedia of food sciences and nutrition*, Oxford: Academic Press.

Rathore, S., Salmerón, I. and Pandiella, S.S. (2012) Production of potentially probiotic beverages using single and mixed cereal substrates fermented with lactic acid bacteria cultures, *Food Microbiology*, 30:239–44.

Rautanen, T., Isolauri, E., Salo, E. and Vesikari, T. (1998) Management of acute diarrhoea with low osmolarity oral rehydration solutions and Lactobacillus strain GG, *Archives of Disease in Childhood*, 79:157–60.

Rekha, C.R. and Vijayalakshmi, G. (2008) Biomolecules and nutritional quality of soymilk fermented with probiotic yeasts and bacteria, *Applied Biochemistry and Biotechnology*, 151:452–63.

Rekha, C.R. and Vijayalakshmi, G. (2011) Isoflavone phytoestrogens in soymilk fermented with ββ-glucosidase producing probiotic lactic acid bacteria, *International Journal of Food Sciences and Nutrition*, 62:111–22.

Rettger, L.G., Levy, M.N., Weinstein, L. and Weiss, J.E. (1935) *Lactobacillus acidophilus and its therapeutic applications*. New Haven, CT: Yale University Press.

Rivera-Espinoza, Y. and Gallardo-Navarro, Y. (2010) Non-dairy probiotic products, *Food Microbiology*, 27:1–11.

Rolle, R. and Satin, M. (2002) Basic requirements for the transfer of fermentation technologies to developing countries, *International Journal of Food Microbiology*, 75:181–7.

Rudolph, M.J. (2010) Dormant ferment containing product or live microorganism containing product or ongoing fermenting product, process of preparation or treatment thereof, United States of America, patent USA20100247712.

Saarela, M., Alakomi, H.L., Mättö, J., Ahonen, A.M., Puhakka, A. and Tynkkynen, S. (2011) Improving the storage stability of *Bifidobacterium breve* in low pH fruit juice, *International Journal of Food Microbiology*, 149:106–10.

Sanders, M.E. (2003) Probiotics: considerations for human health, *Nutrition Review*, 61:91–9.

Saulnier, D.M., Santaos, F., Roos, S., Mistrette, T.A., Spinler, J.K., Molenaar, D., Teusink, B. and Versalovic, J. (2011) Exploring metabolic pathway reconstruction and genome-wide expression profiling in *Lactobacillus reuteri* to define functional probiotic features, *PlosOne*, 5:e18783.

Senok, A.C. (2009) Probiotics in the Arabian Gulf region, *Food and Nutrition Research*, 53, DOI: 10.3402/fnr.v53i0.1842.

Shah, N.P., Ding, W.K., Fallourd, M.J. and Leyer, G. (2010) Improving the stability of probiotic bacteria in model fruit juices using vitamins and antioxidants, *Journal of Food Science*, 75:M278–82.

Sheehan, V.M., Ross, P. and Fitzgerald, G.F. (2007) Assessing the acid tolerance and the technological robustness of probiotic cultures for fortification in fruit juices, *Innovative Food Science and Emerging Technologies*, 8:279–84.

Silva, K.R., Rodrigues, S.A., Filho, L.X. and Lima, A.S. (2009) Antimicrobial activity of broth fermented with kefir grains, *Applied Biochemistry and Biotechnology*, 152:316–25.

Soccol, C.R. (2009) Processo tecnológico para produção de uma bebida fermentada a base de água de côco com propriedades probióticas, *República Federativa do Brasil*, patent PI0703244-7 A2.

Sun-Waterhouse, D. (2011) The development of fruit-based functional foods targeting the health and wellness market: A review, *International Journal of Food Science and Technology*, 46:899–920.

Suomalainen, T., Lagström, H., Mättö, J., Saarela, M., Arvilommi, H., Laitinen, I., Ouwehand, A.C. and Salminen, S. (2006) Influence of whey-based fruit juice containing *Lactobacillus rhamnosus* on intestinal well-being and humoral immune response in healthy adults, *LWT-Food Science and Technology*, 39:788–95.

Sutherland, J., Miles, M., Hedderley, D., Li, J., Devoy, S., Sutton, K. and Lauren, D. (2009) In vitro effects of food extracts on selected probiotic and pathogenic bacteria, *International Journal of Food Sciences and Nutrition*, 60:717–27.

Teoh, A.L., Heard, G. and Cox, J. (2004) Yeast ecology of Kombucha fermentation, *International Journal of Food Microbiology*, 95:119–26.

The, S.S., Ahmad, R., Wan-Abdullah, W.N. and Liong, M.T. (2010) Enhanced growth of Lactobacilli in soymilk upon immobilization on agrowastes,' *Food Microbiology and Safety*, 75:M155–64.

Turnbaugh, P.J., Ley, R.E., Hamady, M., Fraser-Liggert, C.M., Knight, R. and Gordon, J.I. (2007) The human microbiome project, *Nature*, 449:804–10.

Vaughan, E.E. and Mollet, B. (1999) Probiotics in the new millennium, *Nahrung*, 3:148–53.

Wacher, C., Canas, A., Barzana, E., Lappe, P., Ulloa, M. and Owens, J.D. (2000) Microbiology of Indian and Mestizo pozol fermentations, *Food Microbiology*, 17:251–6.

Wagner, R.D., Person, C., Warner, T., Dohnalek, M., Farmer, J., Roberts, L., Hilty, M. and Balish, E. (1997) Biotherapeutic effects of probiotic bacteria on candidiasis in immuno-deficient mice, *Infection and Immunity*, 65:4165–72.

Walker, D.C., Girgis, H.S. and Kleanhammer, T.R. (1999) The *groESL* chaperone operon of *Lactobacillus johnsonii*, *Applied and Environmental Microbiology*, 65:3033–41.

Wang, J., Guo, Z., Zhang, Q., Yan, L., Chen, W., Liu, X.M. and Zhang, H.P. (2009) Fermentation characteristics and transit tolerance of probiotic *Lactobacillus casei* Zhang in soy milk and bovine milk during storage, *Journal of Dairy Science*, 92:2468–76.

Wang, Y.C., Yu, R.C. and Chou, C.C. (2004) Viability of lactic acid bacteria and bifidobacteria in fermented soymilk after drying, subsequent rehydration and storage, *International Journal of Food Microbiology*, 93:209–17.

Wang, Y.C., Yu, R.C. and Chou, C.C. (2006) Antioxidative activities of soymilk fermented with lactic acid bacteria and bifidobacteria, *Food Microbiology*, 23:128–35.

Yeo, S.K. and Liong, M.T. (2010a) Angiotensin I-converting enzyme inhibitory activity and bioconversion of isoflavones by probiotics in soymilk supplemented with prebiotics, *International Journal of Food Sciences and Nutrition*, 61:161–81.

Yeo, S.K. and Liong, M.T. (2010b) Effect of prebiotics on viability and growth characteristics of probiotics in soy milk, *Journal of the Science of Food and Agriculture*, 30:267–75.

Yokotsuka, T. and Sasaki, M. (2006) Fermentation of foods in the Orient, in R.H. Hutkins (ed.) *Microbiology and technology of fermented foods*, Ames, Iowa: Blackwell Publishing Professional, 419–55.

15 Bioadditives Produced by Fermentation

*Juliano Lemos Bicas, Mário Roberto Marostica, Junior,
Francisco Fábio Cavalcante Barros,
Gustavo Molina, and Gláucia Maria Pastore*

CONTENTS

15.1 INTRODUCTION

The impact of human activity on the environment represents a great challenge to modern society, and many debates involving sustainability have emerged. In parallel, there is an increasing demand for natural ingredients and additives in the food industry. In this context, fermentative processes have emerged as an attractive alternative for the production of natural industrial ingredients and additives because these processes have been recognized as using renewable, virtually inextinguishable resources and as being environmentally friendly (Hatti-Kaul et al. 2007). This fact has attracted the interest of many scientists worldwide, leading to important activity on research and development in this field.

Food biotechnology has been developed empirically since ancient times when humanity began to dominate the techniques for manufacturing fermented products, for example, beer, wine, cheese, bread. Ever since then, fermentation technology has evolved as one of the main tools for the food industry, and it is applied aiming at either the preservation of food products or the modification of their sensory attributes (aroma, flavor, texture, etc.). Food science and technology have also begun to understand the phenomenology behind these processes, and today, the use of microbial and enzymatic processes for the production of food ingredients is increasingly being developed.

Therefore, biotechnology offers unique opportunities to produce natural food ingredients. However, although it may appear to be an exhausted subject because of its long history, the biotechnological production of food and food additives remains an active field of research, and many promising investigations have still to be developed in this area. In this chapter, the main aspects involving properties and biotechnological production methods of food additives, focusing on bio-aromas, bio-colorants, nondigestible oligosaccharides (NDOs), and biosurfactants, will be discussed. Other relevant products produced by fermentation, such as organic acids and amino acids, have already been reviewed (Krämer 2005; Leuchtenberger et al. 2005; Sauer et al. 2008).

15.2　BIO-AROMAS

Aroma compounds greatly influence the flavor of food products and govern their acceptance by consumers and their market success. The increasing consumer preference for natural products has encouraged remarkable efforts toward the development of biotechnological processes for the production of flavor compounds, which nowadays play an important role in the food, feed, cosmetic, chemical, and pharmaceutical industries (Bicas et al. 2009).

These compounds present a wide variety of chemical functions, hydrocarbons, alcohols, aldehydes, ketones, acids, esters, and lactones that are present in low concentrations in food products (Bicas et al. 2010b). Moreover, terpenes may be considered as the main group of chemicals involved in natural aromas, especially in essential oils (de Carvalho and da Fonseca 2006a).

Aroma compounds consist of volatile molecules (with a molecular weight usually lower than 400 Da) capable of activating specific cells in the olfactory cavity. The sensation provoked by different molecules may be very complex because the aroma perception of a specific volatile may vary according to the matrix (e.g., the food product), the presence of other volatiles, and the concentration of these molecules, among others (German et al. 2007). This directly reflects the great diversity of possible applications of such compounds as food additives.

The methods for obtaining aroma compounds include their direct extraction from nature, chemical transformations, and biotechnological transformations (which include microbial and enzymatic biotransformations, *de novo* synthesis, and the use of genetic engineering tools) (Franco 2004). Flavors have usually been produced by direct recovery from nature, although many disadvantages are encountered, such as (1) low concentrations of the product of interest, which increases the cost of extraction and purification procedures; (2) dependency on seasonal, climatic, and political features; and (3) possible ecological problems involved with the extraction process (Bicas et al. 2009).

Most compounds found on the market are produced via chemical synthesis because of the satisfactory production yields. However, this strategy is associated with a number of environmental challenges and hardly presents adequate regioselectivity and enantioselectivity to the substrate, resulting in a mixture of molecules that may significantly change the aroma perception of the product. It is worth noting, for example, that simple enantiomers may have completely different aroma descriptors, such as *R*-(–)-carvone (sweet spearmint odor) and *S*-(+)-carvone (caraway/dill odor) (de Carvalho and da Fonseca 2006b; Başer and Demirci 2007). Moreover, compounds produced by chemical synthesis are labeled as "artificial" or "nature identical," decreasing their economic interest (Feron and Waché 2006).

In this sense, biotechnology offers interesting tools for the production of so-called bio-aromas, compounds that are defined as "natural" or "naturally produced" flavors. Reactions occur in mild conditions, exhibit high regioselectivity and enantioselectivity, and generally do not generate toxic waste. Additionally, some compounds can be produced exclusively via biotechnology (Demyttenaere et al. 2001).

This section focuses on the production of aroma compounds based on (1) *de novo* microbial processes (fermentation) and (2) bioconversions of natural precursors with microbial cells or enzymes (biotransformation). *De novo* ("from the new") synthesis refers to the production of complex substances from simple molecules through complex metabolic pathways. In general, microorganisms are capable of producing an amazingly broad array of flavor compounds by *de novo* synthesis. However, production levels are usually low, limiting their industrial use. The other strategy (biotransformation or bioconversion) may offer economic advantages because high yields may be achieved. These processes consist of, respectively, a single or a few reactions catalyzed enzymatically (including the use of whole cells) to result in a product structurally similar to the substrate (Feron and Waché 2006). The main biotechnological processes employed for the production of bioflavors by means of *de novo* synthesis and biotransformations are discussed in Section 15.2.1.

15.2.1 *De Novo* Synthesis as Source of Novel Aroma Compounds

The use of *de novo* synthesis allows the production of chemically different volatile flavors, such as short-chain alcohols, esters, aldehydes, ketones, methyl ketones, and acids, as well as pyrazines and lactones that could be formed concurrently (Krings and Berger 1998).

The process of producing aroma compounds by *de novo* synthesis can be summarized as shown in Figure 15.1. Briefly, the microorganism is inoculated into a complex medium, rich in nutrients (Figure 15.1a); the flask is incubated under controlled conditions (Figure 15.1b); and during its growth, some aroma compounds

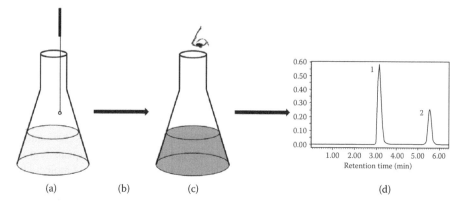

(a)   (b)   (c)   (d)

FIGURE 15.1 Simplified scheme for production of volatile compounds by *de novo* synthesis: (a) inoculation of microorganism in complex medium; (b) incubation on specified conditions; (c) after growth, aroma compounds are produced by fermentation and might be perceived by analysis; (d) identification and quantification of aroma compounds (gas chromatography and mass spectrometry).

can be developed in the fermentation media. These substances can be perceived by the olfactory system of an analyst (Figure 15.1c). After the development of volatile compounds, they can be identified and quantified following appropriate analytical methodologies (Figure 15.1d), such as gas chromatography and mass spectrometry.

There are several publications dealing with the production of volatile fruity aromas during the growth of microbial species in a culture medium, such as *Saccharomyces* sp., *Hansenula* sp., and *Candida utilis* (Longo and Sanromán 2006). Among the compounds involved in fruity flavor, esters are the most important chemicals; their industrial interest is aroused because of their pleasant fruity notes (e.g., ethyl hexanoate), which are highly desirable in fruit-flavored and dairy products (Bauer et al. 2001).

Lactones (cyclic esters) represent another class of compounds with an interesting flavor profile with sweet and pleasant notes, for example, coconut and peach-like (Bauer et al. 2001). Despite the possibility of their production by chemical synthesis, the biogeneration of volatile lactones using fungi and yeast strains has proved to be industrially successful (Cardillo et al. 1990), and some examples include the production of γ-nonalactone, γ-decalactone, δ-decalactone, γ-undecalactone, β-methyl-γ-octalactone (whiskey lactone), and others (Surburg and Panten 2006).

The biotechnological production of aroma alcohols for commercial applications can also be performed. In appropriate concentrations, volatile alcohols can display important aroma sensations in several products. For example, 1-Octen-3-ol has a powerful mushroom aroma character, and its production was reported for *Neurospora* sp. (Pastore et al. 1994), *Pleurotus pulmonarius* (Assaf et al. 1997), and *Penicillium camemberti* (Husson et al. 2002).

Some studies also describe the production of aldehydes using this technique. Benzaldehyde, reported as a bitter almond odor and applied as a starting material for a large number of aliphatic fragrance and flavor materials, has been produced in this way. In a screening of several white-rot fungi, it has been shown that numerous species, such as *Pleurotus sapidus*, *Polyporus* sp., and others, are able to produce benzaldehyde by *de novo* synthesis (Lomascolo et al. 1999).

The production of some other aroma compounds has also been described using *de novo* synthesis. Among them are the terpenoids citronellol, linalool, and geraniol by *Kluyveromyces lactis* and *Ceratocystis moniliformis* (Bluemke and Schrader 2001); pyrazines by *Bacillus subtilis* (Kosuge and Kamiya 1962) or *Corynebacterium glutamicum* (Demain et al. 1967); and sulfur compounds, for example, methional, which are important for baked goods, cheeses, imitation dairy, meat products, condiments, and others, using strains of *Hanseniaspora uvarum*, *Hanseniaspora guilliermondii*, and *Saccharomyces cerevisiae* (Moreira et al. 2008).

Nowadays, advances in industrial biotechnology offer economic alternatives with the use of agro-industrial waste as an alternative culture medium or substrate (Pandey et al. 2000). Considering the importance of using low-cost sources of nutrients and the application of nonconventional media, the biotechnological synthesis of aroma compounds using waste arises as a promising alternative for the building of economically viable processes (Bicas et al. 2010a). Several research groups have made research efforts to find alternative culture media (wastes and residues) capable of supporting microorganism growth with the generation of high-value aroma compounds (Bicas et al. 2010a). One relevant example is the production of fruity banana and pineapple flavor by *Ceratocystis fimbriata* grown on steam-treated coffee husks (Soares et al. 2000) or sugarcane bagasse (Christen et al. 1997).

15.2.2 PRODUCTION OF AROMA COMPOUNDS BY BIOTRANSFORMATION

Biotransformation can be briefly described as chemical reactions catalyzed by microorganisms or enzyme systems and is usually carried out with growing cultures, previously grown cells, immobilized cells, purified enzymes, or multiphase systems in order to result in a product structurally similar to the substrate molecule (Bicas et al. 2009). In some cases, the precursors employed can be considered inexpensive and readily available, such as fatty acids or amino acids, and can be converted to more highly valued chemicals (Krings and Berger 1998).

Briefly, this technique differs from *de novo* synthesis particularly with regard to the culture medium used where the substrates to be transformed are added to the culture media to obtain new compounds from the original molecule. Usually, this substrate is added as the sole carbon source in a mineral medium or buffer (Maróstica and Pastore 2007a). In the field of aroma production, biotransformations have increasingly been investigated in recent years and appear to be one of the most promising manufacturing techniques for the future (Bicas et al. 2009, 2010b).

The biotransformation reactions are directly influenced by a wide range of factors, such as the nature of the biocatalyst employed in the process, the composition of the culture medium, the experimental conditions (temperature, agitation, aeration, etc.), the substrate feeding strategy, and others (Bicas et al. 2008).

Many studies describing the biotransformation of terpenes using enzymes, cell extracts, and whole cells of bacteria, cyanobacteria, yeasts, microalgae, fungi, and plants have been published (Bicas et al. 2009). These terpenes are preferred as substrates because they occur widely in nature and, in some cases, for example, limonene and α-pinene, are inexpensively available in large quantities (de Carvalho and da Fonseca 2006a). Oxyfunctionalization of terpene hydrocarbons can also yield much more valuable, structurally similar molecules (Maróstica and Pastore 2007b).

R-(+)-limonene is one of the most studied precursors in biotransformation experiments for the production of bioflavors. Its chemical structure is similar to valuable oxygenated aroma compounds and allows the production of industrially interesting limonene counterparts, such as carveol, carvone, perillyl alcohol, terpineols, and others (Maróstica and Pastore 2007a). The most recent advances in the field have been obtained for the biotransformation of limonene for α-terpineol production, an alcohol widely used in cosmetics, food, and household products and flavor preparations (Bauer et al. 2001). Some authors have proposed the use of immobilized biomass (Tan and Day 1998a), the addition of organic cosolvents (Tan and Day 1998b), and the use of a closed gas loop bioreactor (Pescheck et al. 2009). The use of genetic engineering tools with a *B. subtilis* strain (Savithiry et al. 1997) has also been considered. After several improvements with a *Fusarium oxysporum* strain, a production of 4 g L^{-1} has been reached (Bicas et al. 2010b). Recently, a paper has described one of the most efficient biotransformation processes for aroma production. Using *Sphingobium* sp. as a biocatalyst and a biphasic system with sunflower oil as the organic phase, an α-terpineol concentration of up to 120–130 g L^{-1} in less than 4 days of processing under nonoptimized conditions has been obtained (Bicas et al. 2010c).

The biotransformation of α- and β-pinenes, other broadly distributed monoterpenes in nature, can also result in a wide variety of metabolites with high commercial interest, such as verbenol, verbenone, pinocarveol, sobrerol, carveol, and carvone (Bicas et al. 2009).

Vanillin, one of the main aroma compounds applied worldwide and extensively used in food and pharmaceutical industries, may also be produced in high yields through microbial transformations (Daugsch and Pastore 2005). This compound illustrates the high difference between the price of the natural compound and its chemically synthesized counterpart. Today, synthetic vanillin, produced from kraft lignin, dominates the market, and its price is approximately US$12 kg^{-1} with a market of 12,000 tons/year, and vanillin extracted from vanilla pods costs up to US$4000 kg^{-1} with a market of 50 tons/year. In contrast, biotechnological processes yield "biotech vanillin," which has a market value of approximately US$1000 kg^{-1} (Schrader et al. 2004; Xu et al. 2007). Consequently, extensive studies have been conducted to obtain vanillin by means of microbial processes (Feron and Waché 2006). Patented bioconversion processes report the production of vanillin in high yields, using precursors such as ferulic acid and eugenol. The bioconversion of ferulic acid into vanillin by strains of *Amycolatopsis* sp. or *Streptomyces setonii* in a 10 L bioreactor could reach final concentrations of 11.5 g L^{-1} (Rabenhorst and Hopp 2000) and 13.9 g L^{-1}, respectively (Muheim et al. 2001). The use of agro-industrial residues as sources of substrate has also been considered (Bicas et al. 2010b).

15.2.2.1 Challenges in Biotransformation of Terpenes

Up until now, most studies dealing with biotransformation of terpenes described are not suitable for industrial application because of problems encountered during the process. The main limiting features are (Van der Werf et al. 1997) (1) chemical instability of substrates, (2) low solubility of substrates, (3) high volatility of substrates and products, (4) toxicity of substrates and products, (5) absence of product accumulation and product degradation, (6) multiple metabolic pathways resulting in the formation of a mixture of products, (7) low product concentrations or yields, (8) enzyme activity not detectable in cell extracts, (9) long incubation times, and (10) short biocatalyst lifetimes.

Some attempts have been made in order to avoid or reduce these problems and achieve potentially marketable yields and product concentrations. An interesting example of these alternatives is the use of statistical tools to simultaneously evaluate the effects of different parameters of the process. These techniques minimize costs by maximizing both yield and productivity (Bicas et al. 2008).

Another important strategy consists of using biphasic systems, involving a water-insoluble organic layer and an aqueous phase, which can reduce the problems associated with the low solubility in water of substrate and product while minimizing their toxic effects on microorganisms. One practical example of this technique is the already cited process for the production of α-terpineol by the biotransformation of limonene where concentrations of up to 120–130 g L^{-1} of product have been obtained (Bicas et al. 2010c). These systems are considered valuable tools in whole-cell biocatalysis because the following advantages may be found (Cabral 2001; León et al. 1998; Gargouri 2001):

- Compatibility with the production of poorly soluble products and efficient bioconversion of sparingly water-soluble substrates.
- The possibility of a significant equilibrium shift in the desired direction so thermodynamically unfavorable reactions become possible.
- An aqueous phase remaining with low concentrations of toxic (hydrophobic) compounds (substrates, products, or by-products), which would otherwise disfavor the biocatalytic process.
- In situ product recovery, which is interesting for (1) reducing end-product inhibition, (2) favoring the bioconversion by shifting the thermodynamic equilibrium toward product formation, and (3) facilitating the recovery of both product and biocatalyst.
- Microbial contamination limited in the presence of the organic solvent.
- Increased yields by reducing the losses by volatilization of volatile organic compounds (substrate and product), which is particularly important on aerobic processes.

The use of agro-industrial residues as a substrate has also been considered for the production of aroma compounds using biotransformations. This strategy is adapted to situations where the cost of the substrate is a key parameter in the economic evaluation of a process, which is often the case in microbial transformations. A nice example of this approach has been described in which cassava wastewater is used to support fungal growth. The resulting biomass is then used in biotransformation processes involving byproducts as a source of terpene substrates, for example, orange oil as a source of R-(+)-limonene (Maróstica and Pastore 2007a).

Other strategies are (1) in situ recovery of the product; (2) fed-batch feeding of the substrate; (3) use of special reactors (membrane, solid state, closed loop); and (4) use of nonconventional media, such as gas phases (Berger 2009).

15.3 BIO-COLORANTS

Color is one of the main attributes of foodstuffs. Because it is generally related to the first impression (appearance) in a sensory evaluation, this parameter may rapidly determine the acceptance or rejection of a product (Aberoumand 2011).

Coloration is a result of molecules naturally present in or intentionally added to food with the objective of conferring, maintaining, or intensifying its color. These coloring molecules may present different chemical characteristics and origins. Synthetic ("artificial") colorants, for example, have been widely applied in industry, although the list of permitted substances includes only a few compounds. For this reason, there is a growing interest in identifying natural colorants with potential applications in foodstuff, so new molecules, which are stable in processing and storage conditions and are safe from a toxicological point of view, may integrate technological tools available to the food industry (Aberoumand 2011). Therefore, as mentioned before, biotechnology represents a fundamental strategy for the production of such compounds.

15.3.1 COLOR

Color refers to our ability to perceive electromagnetic radiation with wavelengths ranging from approximately 380 to 770 nm, which corresponds to the so-called "visible spectrum." Therefore, depending on the frequency, the observed color may vary from red to violet, passing through orange, yellow, green, blue, and their nuances. The color of a food product occurs as a result of its capacity to reflect or refract electromagnetic radiations in the visible spectrum, and they are further detected by the retina. This condition results mainly from the composition and proportion of molecules composing that specific product. Therefore, foods may be naturally colored or, yet, may suffer intentional addition of coloring compounds to alter or reinforce their color (Schwartz et al. 2010).

As emphasized before, the color of a particular food is crucial to its acceptance. When confronted with an undesirable color, for example, consumers may assume that a product is spoiled, resulting in its rejection. Additionally, there are predetermined expectations for certain colors of foodstuffs: red for cherry and strawberry products, orange for carrots, purple for grapes, etc. Therefore, it is no exaggeration to say that we "inevitably eat with our eyes" (Dufossé 2006).

15.3.2 NATURAL COLORANTS IN FOOD

There is a basic difference between the terms "pigments" and "dyes": The first refers to a colored particulate organic or inorganic solid, usually insoluble and unaffected by the vehicle or substrate into which it is incorporated, and the last corresponds to soluble molecules used to impart molecules to a material. However, both of them may be grouped into the term "colorant," which has been preferred because of its more generic definition (Delgado-Vargas and Paredes-Lópes 2003). Therefore, the term "colorant" will be mostly used, although "pigment" will be considered a synonym in this chapter as occurs in other scientific reports.

The main natural colorants present in foods present complex structures with different functional groups and molecular masses, being classified as (1) porphyrins, such as hemoglobin and myoglobin (red) and chlorophyll (green); (2) betalains; (3) flavonoids, especially anthocyanins (red to blue); (4) carotenoids, such as lutein (yellow) and β-carotene and lycopene (red-orange); (5) tannins (pale yellow to light brown); and (6) others (quinones, polyphenols, etc.) (Bobbio and Bobbio 2001). Some examples of natural food pigments are bixin, an apocarotenoid found in annatto; capsanthin, found in paprika; and the betalains, betacyanin and betaxanthins, found in beetroots (Schwartz et al. 2010). Therefore, many products have been colored with vegetable extracts, that is, red from paprika, beetroots, or tomatoes; yellow from saffron; and green from vegetable leaves (Dufossé 2006).

15.3.3 BIOTECHNOLOGICAL PRODUCTION OF BIO-COLORANTS

Many natural colorants may be obtained from plants, animals, and other sources. Some well-known examples are curcumine, present in the rhizome of *Curcuma longa*; carmine, obtained from the female *Dactylopius coccus* cochineal insects; and caramel, a coloring preparation resulting the heat treatment of sugars in different conditions (Pintea 2008). Besides these traditional sources, some microorganisms offer a virtually inextinguishable source of natural colorants, produced by means of fermentation processes (Dufossé 2006). Such biotechnological processes represent many advantages over extractions from natural sources: The so-called "microbial pigments" have a natural character; their production is independent of the seasonal, climatic, and geographical conditions; and they may be produced under controlled conditions with predictable and optimized yield (Babitha 2009).

Pigment-producing microorganisms are relatively common. Carotenoids, melanins, flavins, quinines, and more specifically, monascins, violaceins, phycocyanin, and indigo are among the coloring molecules generated (Dufossé et al. 2005).

The functions of pigments in microorganisms are (1) protection against ultraviolet radiation; (2) protection against oxidants; (3) protection against extremes of heat and cold; (4) protection against antimicrobials produced by other microorganisms; (5) antimicrobial activity against other microorganisms; (6) acquisition of nutrients; and (7) participation in photosynthesis, as in cyanobacteria (Liu and Nizet 2009).

The production of microbial pigments is well documented in the scientific literature. Processes involving filamentous fungi, yeasts, bacteria, and microalgae have been reviewed (Chattopadhyay et al. 2008; Dufossé 2006; Dufossé et al. 2005), and some others have been patented (Jones et al. 2004; Long 2004; Sardaryan 2002; Song et al. 2010). Most of the published work deals with the production of yellow-orange and red colorants, such as the traditional carotenoids (astaxanthin, lycopene, zeaxanthin, and β-carotene), whose coloration varies from yellow and orange to red, and the red-orange pigments derived from fungi of the genus *Monascus*, *Penicillium*, and *Fusarium* (see examples in the sequence). Blue colorants, in turn, are less common, but some articles dealing with these compounds may be found. For example, phycocyanin-producing microalgae, such as *Porphyridium* (Dufossé et al. 2005) and

Spirulina (Eriksen 2008; Sarada et al. 1999; Walter et al. 2011); *Streptomyces* sp. producing actinorhodin (Bytrykh et al. 1996; Lu et al. 2002; Zhang et al. 2006); and indigotin-like compounds (Novakova et al. 2010) have been reported.

Despite all the advantages attributed to the microbial production of colorants, it is important to note that these substances may have some impediments to their wide adoption. Virulence factors are related to some microbial pigments (Liu and Nizet 2009), and others may be accompanied by toxins generated during the fermentation process (Carvalho et al. 2005). On the other hand, it is claimed that some coloring substances may bring health benefits to their consumers, being considered "nutraceutical" compounds (Delgado-Vargas and Paredes-Lópes 2003; Amiot-Carlin et al. 2008).

Some microbial pigments are already produced on an industrial scale applied to food products. *Monascus* colorants and β-carotene produced by the fungus *Blakeslea trispora* have been employed by the food industry in Asia and Europe, respectively. There are also other processes on the level of industrial production, for example, riboflavin (yellow) from the fungus *Ashbya gossypii* and Arpink Red from *Penicillium oxalicum*, as well as several examples of products in the development stage or in research projects (Dufossé 2006). According to Dufossé (2008), a good candidate for the large-scale production of microbial pigments in the coming years is the pink-red colorant astaxanthin derived from the yeast *Xanthophyllomyces dendrorhous* (*Phaffia rhodozyma*).

15.3.3.1 Examples of Microbial Pigments Produced by Fermentation Processes

The production of red colorants by *Monascus* sp. is one of the most well-established and studied processes. *Monascus*-grown rice (named "anka" or "ang-kak") has been used for centuries in Asian countries as a food colorant, and today, it is still consumed in oriental countries (Babitha 2009; Dufossé 2006). However, despite its enormous economic potential, the use of this color additive is not approved in the major markets of the western world (Carvalho et al. 2005; Dufossé 2006).

Color substances from *Monascus* are secondary metabolites belonging to the group of azaphilones that are produced during fungal fermentation. Different molecules may be produced, resulting in colonies with varied colors from orange-yellow to purple-red, depending on the type and proportion of the pigments present and also the culture conditions (medium composition, temperature, and mainly pH). The six primary *Monascus* pigments are monascin and ankaflavin (yellow), rubropunctain and monascorubin (orange), and rubropunctamine and monascorubramine (red) (Babitha 2009; Delgado-Vargas and Paredes-Lópes 2003). These red colorants present greater interest because they are possible substitutes for synthetic colors, such as erythrosine (FD & C red no. 3) (Carvalho et al. 2007). Novel pigments with potential functionalities are still being discovered (Mukherjee and Singh 2011).

The most important products found in *Monascus* colorants, the reddish compounds monascorubramine and rubropunctatin (Figure 15.2), may be produced by solid-state fermentation using its traditional substrate (rice or broken rice) (Carvalho et al. 2005; Rosenblitt et al. 2000; Vidyalakshmi et al. 2009), although other substrates, for example, agro-industrial residues, can also be employed (Babitha 2009; Carvalho et al. 2007; Nimnoi and Lumyong 2011). Submerged fermentations may also be considered for the production of these molecules (Kongruang 2011; Silveira et al. 2008).

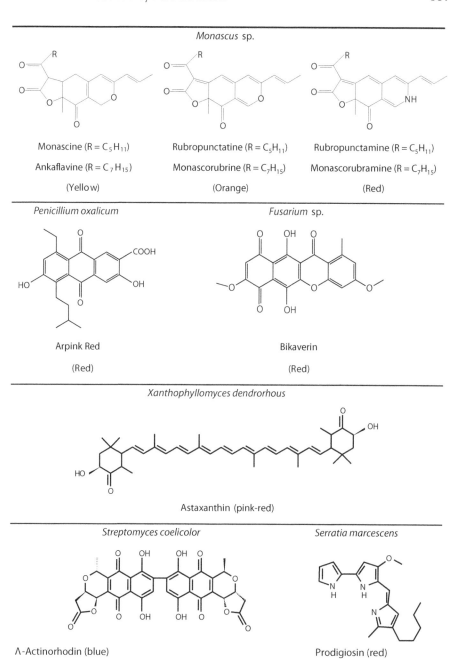

Monascus sp.

Monascine (R = C₅H₁₁)

Ankaflavine (R = C₇H₁₅)

(Yellow)

Rubropunctatine (R = C₅H₁₁)

Monascorubrine (R = C₇H₁₅)

(Orange)

Rubropunctamine (R = C₅H₁₁)

Monascorubramine (R = C₇H₁₅)

(Red)

Penicillium oxalicum

Arpink Red

(Red)

Fusarium sp.

Bikaverin

(Red)

Xanthophyllomyces dendrorhous

Astaxanthin (pink-red)

Streptomyces coelicolor

Λ-Actinorhodin (blue)

Serratia marcescens

Prodigiosin (red)

FIGURE 15.2 Diversity of pigments produced by microorganisms.

It is important to note that the mycotoxin citrinin is usually coproduced with *Monascus* pigments (Wang et al. 2005). Even though there are studies showing that some "red mold rice" presented no toxicity effect in albino rats, even in concentrations of 5 g/kg body weight (Kumari et al. 2009), the possible presence of citrinin is a major bottleneck for the approval of *Monascus* pigments in some countries. One interesting approach to overcome this problem is the selection of nonmycotoxigenic microorganisms producing such molecules. Some species of the genus *Penicillium*, for example, have been shown to be able to produce yellow, orange, and purple-red *Monascus*-like pigments free of citrinin (Mapari et al. 2009).

In fact, microorganisms from the genus *Penicillium* are also reported to be producers of commercial biopigments. The Czech company Ascolor Biotech manufactures a natural food colorant called Arpink Red (Figure 15.2), which is claimed to be produced by *P. oxalicum* var. Armeniaca CCM 8242 by means of a patented process (Sardaryan 2002). Other *Penicillium* strains are also able to generate yellow-to-bluish-green colored molecules (Table 15.1) as reviewed in Mapari et al. (2005).

Fusarium is another fungal genus that may produce microbial pigments by fermentation. One of the main molecules produced by *Gibberella fujikuroi* (an anamorph of *Fusarium fujikuroi*) or other *Fusarium* species is bikaverin (Figure 15.2),

TABLE 15.1
Some Examples of Microbial Pigments Produced by Fermentation

Microorganism	Pigment[a]	Color	References
Monascus sp.	Monascine, ankaflavine; rubropuctatine, monascorubrine; rubropunctamine, monascorubramine	Yellow, orange, red	Babitha 2009, Dufossé 2006, and Pattanagul et al. 2007
P. oxalicum	Arpink Red (anthraquinone-type)	Red	Sardaryan 2002, reviewed by Dufossé 2006
Penicillium sp.	Others	Yellow, red, blue-green	Mapari et al. 2005
Fusarium sp.	Bikaverin	Red	Limón et al. 2010
Bl. trispora, *Dunaliella* sp.	β-Carotene	Yellow-orange	Choudhari and Singhal 2008, Papaioannou and Liakopoulou-Kyriakides 2010, and Ribeiro et al. 2011
X. dendrorhous	Astaxanthin	Pink-red	Rodríguez-Sáiz et al. 2010 and Schmidt et al. 2011
S. marcescens	Prodigiosin	Red	Venil and Lakshmanaperumalsamy 2009
Streptomyces sp.	λ-Actinorhodin	Blue	Zhang et al. 2006
Spirulina sp.	Phycocyanin	Blue	Eriksen 2008

[a] Structures of some pigments may be seen in Figure 15.2.

a reddish pigment with claimed antitumor properties and antibiotic activity against a variety of organisms, including protozoa (especially *Leishmania braziliensis*), nematodes, and filamentous fungi (Limón et al. 2010; Rodríguez-Ortiz et al. 2010). The production of bikaverin by fungal fermentation has already been studied using a fluidized bed (Chávez-Parga et al. 2005) or airlift bioreactors (Chávez-Parga et al. 2008). Genetically modified strains of *Fusarium sporotrichioides* were also reported to effectively produce carotenoids, in this case, lycopene (Jones et al. 2004).

Carotenoids constitute the largest class of natural pigments. They are widely distributed, occurring naturally in large amounts in some fruits (e.g., papayas, guavas), vegetables (e.g., tomatoes, carrots), fungi, flowers, and also in animals (e.g., birds, crustaceans, and salmon), where they impart colors ranging from yellow to red. They are lipid-soluble substances known for their structural diversity; more than 650 different carotenoids have been isolated and identified so far. These molecules are very attractive for the food and beverage industries because they represent natural sources of orange, yellow, and red colorings (Ötles and Çagindi 2008).

As stated previously, carotenoids may be produced by microorganisms. The most distinguishable example is the production of β-carotene by *Bl. trispora* (Choudhari and Singhal 2008; Papaioannou and Liakopoulou-Kyriakides 2010). *X. dendrorhous* producing astaxanthin is also another important example (Rodríguez-Sáiz et al. 2010; Schmidt et al. 2011) (Table 15.1). Microalgae have also demonstrated great potential for the production of these pigments. The genus *Dunaliella* is one of the most reported for the production of carotenoids. This is a halotolerant, unicellular microalgae, which grows in high salt concentrations. Species from this genus can accumulate large amounts of β-carotene (Ribeiro et al. 2011), and many companies are involved in the cultivation of *Dunaliella* for commercial production of this carotene for food, feed, or pharmaceutical formulations (Dufossé et al. 2005). This and many other examples involving fungi, yeasts, bacteria, and microalgae may be found in the literature as reviewed by Dufossé et al. (2005) and Ribeiro et al. (2011). Another important perspective is the development of metabolic engineering approaches from noncarotenogenic microorganisms, such as *Escherichia coli*, for the production of carotenoids by genetically modified organisms (Das et al. 2007).

15.4 NONDIGESTIBLE OLIGOSACCHARIDES

NDOs, generally referred to as prebiotics, are short-chain carbohydrates that can play a role in the composition, or metabolism, of the gut microbiota in a beneficial manner (MacFarlane et al. 2006). The criteria to classify a product or ingredient as prebiotic are (Gibson and Roberfroid 1995; MacFarlane et al. 2006; Roberfroid 2007; Kolida and Gibson 2007) that (1) it resists host digestion and absorption; (2) it is selectively fermented by large intestinal microbiota; and (3) it beneficially affects the host by selectively stimulating the growth and/or activity of one or a limited number of bacteria in the colon and, thus, improves host health. According to MacFarlane et al. (2006), prebiotics receive much attention because of their health properties: (1) they could play a role in the healthy or balanced gut microbiota; (2) they could promote the selective growth of healthy microbiota strains; (3) they could be a better technological alternative to probiotics as the later are more difficult to be used in the food industry; and (4) furthermore, they could be used as stabilizers.

Nowadays, the NDOs recognized as prebiotics (which meet the criteria above) are β-2,1 fructans, which include inulin and fructooligosaccharides (FOSs). Other candidates are galactooligosaccharides (GOSs), gluco-oligosaccharides, isomaltooligosaccharides, lactulose, mannooligosaccharides (MOSs), nigerooligosaccharides (such as β-glucans from oats), raffinose, oligosaccharides from soybeans, transgalactooligosaccharides, and xylooligosaccharides (XOSs) (Roberfroid 2007; Aachary and Prapulla 2011; Torres et al. 2010).

Fructans are found in several vegetables, such as garlic, artichokes, asparagus, onions, and chicory. Long-chain fructans are generally hard to obtain, and there are few plants that can synthesize them. However, short-chain fructans (known as fructooligosaccharides) are produced by the hydrolysis of long-chain fructans or by enzyme synthesis using sucrose as a substrate (Kolida and Gibson 2007). Figure 15.3 shows the general structures of inulin, FOSs, and GOSs. Inulin and FOSs have the general molecular formula GF_n (where G = glucose, F = fructose, and n is the number of fructose moieties). GOSs are galactose oligomers whose generic formula may be represented by $[Gal]_n$-Glc where Glc = glucose, Gal = galactose, and n is the number of galactose moieties. Different glycosidic linkages have already been identified for GOS: β-1,2; β-1,3; β-1,4; and β-1,6 (Barreteau et al. 2006) (see Table 15.2 for examples).

15.4.1 Biotechnological Production of NDO

Prebiotics can be obtained by extraction from vegetables or by enzymatic means. Inulin can be found in several vegetables, such as Jerusalem artichokes, onions, garlic, leeks, wheat, rye, bananas, and yacón. However, the most important source of inulin is chicory, which contains more than 70% of inulin on a dry-weight basis (Roberfroid et al. 2010). The main steps of the conventional methods of producing inulin are tuber washing, milling, hot water extraction, pH adjustment, filtering, precipitation, and concentration. Another inulin production process is shown in Figure 15.4 (Franck 2002). The use of hot water (80°C–90°C) is important to inactivate enzymes and to improve inulin solubilization. High-performance inulin (HP inulin)

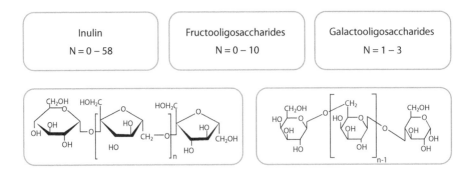

FIGURE 15.3 Structures of some NDOs.

TABLE 15.2

Structure of Some Oligosaccharides Found in Commercial Preparations of GOS (Gal = galactose, Glc = glucose)

Disaccharides

β-D-Gal-(1→6)-D-Glc

β-D-Gal-(1→6)-D-Gal

β-D-Gal-(1→3)-D-Glc

β-D-Gal-(1→2)-D-Glc

β-D-Gal-(1→3)-D-Gal

Trisaccharides

β-D-Gal-(1→6)-β-D-Gal-(1→6)-D-Glc

β-D-Gal-(1→6)-β-D-Gal-(1→4)-D-Glc

β-D-Gal-(1→6)-β-D-Gal-(1→6)-D-Gal

β-D-Gal-(1→3)-β-D-Gal-(1→4)-D-Glc

β-D-Gal-(1→4)-β-D-Gal-(1→4)-D-Glc

Tetrasaccharides

β-D-Gal-(1→6)-β-D-Gal-(1→6)-β-D-Gal-(1→4)-D-Glc

β-D-Gal (1→6)-β-D-Gal-(1→3)-β-D-Gal-(1→4)-D-Glc

β-D-Gal (1→3)-β-D-Gal-(1→6)-β-D-Gal-(1→4)-D-Glc

Pentasaccharides

β-D-Gal-(1→6)-β-D-Gal-(1→6)-β-D-Gal-(1→6)-β-D-Gal-(1→4)-D-Glc

Source: Based on Manucci, F., 'Enzymatic synthesis of galactooligosaccharides from whey permeate.' MSc Thesis, Dublin Institute of Technology, 2009.

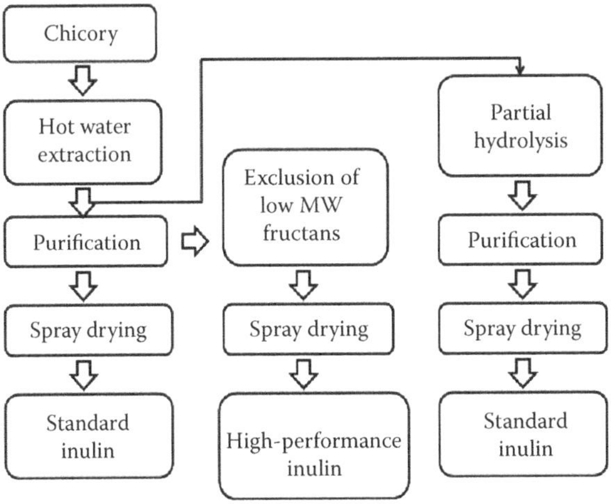

FIGURE 15.4 Industrial production of inulin and its derivatives. (From Franck, A., *British Journal of Nutrition*, 87: S287–S291, 2002. With permission.)

can also be obtained by the removal of low molecular-weight fructans, resulting in a less-sweet product when compared to conventional inulin (Figure 15.4).

FOS is produced on a commercial scale using two techniques; one of them consists of the partial enzymatic hydrolysis of inulin using endo-inulinase, resulting in a mixture of oligosaccharides with two to seven monomers. After purification and concentration, an oligofructose syrup is obtained. FOS can also be produced by the transglycosylation of sucrose using fructofuranosidase (E.C. 3.2.1.26) or fructosyltransferase (E.C. 2.4.1.9) (Dalzenne 2003; Mussato and Mancilha 2007). The resulting compounds (two to four sugar moieties) are commercialized by some Japanese companies (Franck 2002).

GOSs are commercially produced using lactose as a substrate. The enzyme β-galactosidase can be obtained from basidiomycete and filamentous fungi, especially *Sporobolomyces singularis*, *Sterigmatomyces eliviae*, *Aspergillus oryzae*, and *Scopulariopsis* sp. Trisaccharides, tetrasaccharides, pentasaccharides, and hexasaccharides can be produced. Lactose concentration is one of the most important factors for GOS production. On one hand, in low lactose concentrations, hydrolysis reactions take place more extensively, and on the other hand, higher lactose concentrations stimulate transgalactosylation, resulting in GOS production. Temperature is another important factor on the reaction yields, and the thermal stability of enzymes plays an important role in this context (Sako et al. 1999).

Figure 15.5 shows some of the main steps involved in industrial GOS production. The GOSs available on the market are, generally, mixtures of several oligosaccharides (~55%) (Table 15.2), lactose (~20%), sucrose (~20%), and small amounts of galactose (Sako et al. 1999). The kinds of glycosidic bonds depend on the enzymes used and the

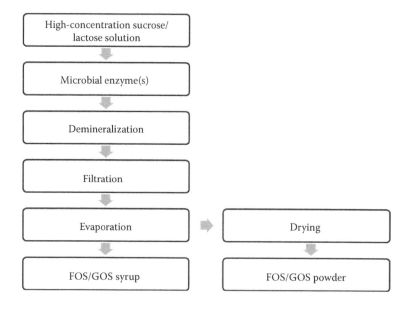

FIGURE 15.5 Industrial production of FOS and GOS. (Adapted from Sako, T., Matsumoto, K. and Tanaka, R., *International Dairy Journal*, 9: 69–80, 1999.)

reaction conditions. β-galactosidases from *Bacillus circulans* or *Cryptococcis laurentii* generally act on β-1,4 bonds between galactose molecules (producing 4′-GOS); in another way, enzymes from *A. oryzae* or *Streptococcus thermophilus* act on β-1,6 linkages (resulting in 6′-GOS) (MacFarlane and MacFarlane 2003).

As described by Nakakuki (2002), some challenges in the production of oligosaccharides are as follows: (1) continuous and efficient processes must be developed to yield high-purity oligosaccharides with a low cost; (2) the need for development of oligosaccharides with higher functional potential; and (3) functional oligosaccharides must be conceived, and a system to evaluate the functional properties must be established. With these advances, the market for oligosaccharides should progress more intensely.

15.4.2 Uses of NDO

Prebiotics, because of their several technological and physiological properties, are extensively used in the food industry. Inulin and FOS are structurally different and are used for diverse purposes in foodstuffs. Inulin has a longer chain and is less soluble than FOS. Inulin is largely used as a fat replacer for milk products, fillers, desserts, bakery products, meat products, sauces, and soups (MacFarlane and MacFarlane 2003).

Initially, inulin was used solely in coffee and similar products. Nowadays, it can be used in several other products as inulin does not strongly affect their flavors. The water retention properties of inulin can contribute to an extended shelf life of foodstuffs. Some precipitates can be formed in products with the intentional addition of inulin. In these cases, starch can be added to prevent such events (MacFarlane and MacFarlane 2003).

FOSs are more soluble than inulin. Besides its relevant physiological properties (Sako et al. 1999), FOSs present some other main technological characteristics (Yun 1996): (1) relatively low sweetening power (about one-third that of sucrose), which is interesting for products where the sweet intensity of sucrose restricts its use; (2) nondigestibility, so they do not contribute to the caloric content, although some of their fermentation products in the colon may be used as source of energy for the body; and (3) they are noncariogenic. These characteristics make FOSs suitable for the formulation of different food and pharmaceutical products. Therefore, FOSs are found in dairy products, confectionery, ice creams, jellies, and others (Sangeetha et al. 2005). FOSs are also used as thickening agents and for the prevention of syneresis in milk products. Furthermore, they are able to reduce the freezing point of desserts and are also used in cereal bars. Oligofructose is combined with sweeteners and applied as a sugar substitute; these products also have the property of masking the aftertaste of aspartame and acesulfame (MacFarlane and MacFarlane 2003).

Because of their pH and heat stabilities (no modifications at 160°C/10 min/pH = 7, 120°C/10 min/pH = 3, or 100°C/10 min/pH = 2) (Sako et al. 1999), GOSs can be used in several foodstuffs, including thermally treated products (Manucci 2009). These products have been used in fermented milk products, breads, jellies, and beverages (Sako et al. 1999). GOSs can also be used as fat replacers as they can increase the viscosity of a solution, impacting mouthfeel (MacFarlane and MacFarlane 2003).

Nowadays, GOSs and FOSs are used in baby food as they can provide some of the health benefits provided by oligosaccharides naturally present in human milk. Sialic acid, N-acetylglucosamine, L-fucose, D-glucose, and D-galactose are some of the compounds found in human milk that can increase the number of bifidobacteria, reduce pathogenic bacteria, improve immune response, and reduce diarrhea and infections (Vandenplas 2002).

15.4.3 Functional Properties of NDO

These oligosaccharides are compounds that present with important physicochemical and physiological properties, which are beneficial to the health of consumers. Noncariogenicity, low caloric value, and the ability to stimulate the growth of beneficial bacteria in the colon are some of their proprieties. In the gastrointestinal tract, prebiotics selectively stimulate indigenous beneficial bacteria, such as *Bifidobacteria* and *Lactobacilli*. The stimulation of these microorganisms has been considered as advantageous because of their action as immunomodulators, the inhibition of the growth potential of pathogens, the reduction of ammonia formation, the lowering of blood cholesterol levels, and the restoration of the normal flora during antibiotic therapy (Roberfroid et al. 2010).

Anaerobic microbial fermentation of these oligosaccharides results in the production of short-chain fatty acids (SCFAs), particularly acetate, propionate, and butyrate, and gases hydrogen (H_2) and carbon dioxide (CO_2). The production of these organic acids is one of the most important physiological processes mediated by colonic microorganisms in the large gut. The vast majority of these metabolites are absorbed from the gut, enabling the host to save energy from food not digested in the upper bowel (Roberfroid et al. 2010). SCFAs are the main mucosal energy source in the colon, stimulating mucosal cell proliferation, mucus production, and mucosal blood flow. Besides being the major source of respiratory fuel for colonic mucosal cells, they are the main source of acetyl-coenzyme A for lipid synthesis and cell membrane assembly, necessary for maintaining the integrity of the mucosal cells (Borchers et al. 2009). Moreover, SCFAs may also play an important role for the optimal functioning of the colonic epithelium and the absorption of various cations, including Ca^{2+}, Mg^{2+}, and Fe^{2+}, because of the decrease in the intestinal pH (Roberfroid et al. 2010). They can also influence the intestinal absorption of calcium by both active and passive transport because it reduces the pH of the medium and, thus, causes the ionization of calcium, favoring the positive balance in the metabolism of this mineral (Bruzzese et al. 2006).

15.5 BIOSURFACTANTS

Biosurfactants are molecules produced by several microorganisms and widely spread in nature, presenting a highly varied molecular structure. A common characteristic of this group of compounds is their amphiphilic properties. This occurs because such molecules contain both hydrophobic and hydrophilic moieties that confer surface-active properties. The hydrophobic part of the molecule usually consists of saturated, unsaturated, and/or hydroxylated fatty acids or fatty alcohols, and the hydrophilic moiety is formed by monosaccharides, oligosaccharides, or polysaccharides, peptides, or proteins (Barros et al. 2007; Nitschke and Costa 2007; Saharan et al. 2011).

Thus, these molecules have the ability to accumulate in the interface of fluid phases such as oil/water or air/water, reducing the surface and interfacial tension and forming emulsions (Barros et al. 2007; Nitschke and Costa 2007; Saharan et al. 2011). They are the primary chemical base of biological membranes and assure the transport and exchange of materials (Bicas et al. 2010). Because of their surface activity properties, surfactants may be applied in a variety of applications in industry and agriculture (Desai and Banat 1997; Murkerjee et al. 2006).

15.5.1 PROPERTIES OF BIOSURFACTANTS

The main functions of emulsifiers, including biosurfactants, are (1) to increase the surface area of hydrophobic water-insoluble substrates, (2) to increase the bioavailability of hydrophobic substrates by increasing their apparent solubility, and (3) to influence accession of microorganisms to surfaces (Barros et al. 2007). The reduction of surface and interface tensions also allows the formation of liquid–liquid dispersion. This property is very important for industrial applications. Currently, the market is basically supplied by synthetic surfactants, which are used in several industrial areas, such as oil, food, cosmetic, detergents, environmental and pharmaceutical industries, and agriculture (Barros et al. 2008a; Bicas et al. 2010; Desai and Banat 1997).

Moreover, a growing number of investigations of the functional properties of biosurfactants are reported. These molecules have been referred to as antibiotics, antivirals, and antitumor agents; immunomodulatory or inhibitors of enzymes and toxins; inhibitors of biofilm formation from other bacteria, including pathogens; administrators of drugs via the lungs; antimycoplasmatic; and effective agents in thrombolytic therapy. The mechanism of action resulting in these properties has not yet been clarified in detail, although it is known that their activities and membrane surfaces play an important role in diverse systems (Barros et al. 2007; Makkar and Cameotra 2002; Singh and Cameotra 2004).

Besides, it is also known that biosurfactants generally present low toxicity and can be applied in food, cosmetics, and pharmaceuticals, especially if the microorganisms producing these molecules are recognized as GRAS (generally recognized as safe) (Nitschke and Costa 2007).

15.5.2 CLASSIFICATION OF BIOSURFACTANTS

Biosurfactants, in general, can be classified into five groups: (1) glycolipids, (2) liposaccharides, (3) lipopeptides, (4) phospholipids, and (5) fatty acids/neutral lipids (Desai and Banat 1997). Because of the complexity involving the structural and microbial origins of biosurfactants, several criteria are proposed for their classification. One of these parameters is the molecular weight: There are high and low molecular-weight biosurfactants. Another possibility is the classification of biosurfactants based on the ionic character of the polar part of their chain or the size of the hydrophobic part (Nitschke and Costa 2007). However, despite the great number of biosurfactant classes, only two of them are widely reported in the literature: the lipopeptides and glycolipids.

15.5.3 Applications of Biosurfactants

Biosurfactants attract great interest in industry because of their noticeable physico-chemical and biological properties. Most work on the application of biosurfactants has been focused on the bioremediation of pollutants. However, these microbial compounds exhibit a variety of useful properties for the food industry because they can act as emulsifiers; solubilizers; and foaming, wetting, antiadhesive, and anti-microbial agents (Nitschke and Costa 2007; Barros et al. 2008a). Several industrial sectors require compounds with these characteristics, including petroleum, petro-chemical, food, beverage, cosmetic, pharmaceutical, mining, metallurgical, agro-chemical, fertilizer, environmental, pulp and paper, among others (Chen et al. 2006; Barros et al. 2008a; Thavasi et al. 2008).

The first reports of the use of these molecules date from the 1940s, but it was only in the mid-1960s that biosurfactant production in hydrophobic media was identified, leading to studies of its application in the treatment of waste oil, in oil recovery, and dispersion in the bioremediation of oil spills (Barros et al. 2007). Till then, the interest in these molecules has considerably increased as they represent an environ-mentally friendly alternative to chemical surfactants, especially in the food, phar-maceutical, and oil industries (Desai and Banat 1997; Nitschke and Pastore 2006; Singh and Cameotra 2004).

Although the oil industry is still the major potential market for biosurfactants, the chemical diversity of these molecules provides a wide variety of specific applications that the classical surfactants may lack (Barros et al. 2007; Nitschke and Costa 2007). Unlike synthetic surfactants, compounds produced by microorganisms are easily biodegradable, have low toxicity, and are therefore suitable for environmental appli-cations. These are the main reasons for the interest in biosurfactants (Murkherjee et al. 2006; Nitschke and Costa 2007).

15.5.4 Biosurfactants in Food

Emulsifiers are very important for the food industry. By definition, an emulsion is a system of at least two immiscible liquid phases with a particle size bigger than true solutions and smaller than suspensions. Emulsions are thermodynamically unstable; however, the emulsifiers reduce the interfacial tension, reducing the surface energy between the two phases and preventing the coalescence of the particles (Barros et al. 2007). These substances (1) promote the stability of the emulsion by controlling the agglomeration of fat globules and stabilizing aerated systems, (2) improve the tex-ture and shelf life of products containing starch by forming complexes with compo-nent starch, (3) modify the rheological properties of wheat flour by interaction with gluten, (4) improve the consistency and texture of fat-based products for the control of polymorphism and the crystal structure of fats, and (5) promote the solubilization of flavors (Barros et al. 2007; Nitschke and Costa 2007).

Some advantages of the application of biosurfactants in food can be listed as follows: Their production is relatively simple, and inexpensive procedures can be applied. New types of surfactants that are not readily synthesized by a chemical process can be obtained, and they are considered environmentally friendly because of their fully biodegradable character and low toxicity (Makkar and Cameotra 2002; Barros et al. 2008a; Nitschke and Costa 2007). In addition, other characteristics can also be cited. They present excellent surfactant power—even at low concentrations, they reduce the surface tension of water and the interfacial tension of the system water/hydrocarbons—and are good emulsifying agents for vegetable oils and hydrocarbons, and they are very stable at high temperatures, extreme pH, high salinity, and also in the presence of enzymes (Singh and Cameotra 2004; Murkherjee et al. 2006). Despite the advantages demonstrated by biosurfactants, few reports are available regarding their use in food products and food processing (Nitschke and Costa 2007). In foods, the use of emulsifiers have potential applications for many types of products, such as bakery products, meat products, margarine and related products, dairy products and ice cream, salad dressings, and instant products, among others (Barros et al. 2007).

15.5.5 Biotechnological Production of Biosurfactants

Although some biosurfactants can be obtained by enzymatic synthesis, the majority are obtained by means of fermentation processes using bacteria (predominant), yeasts, or molds. Biotechnological biosurfactant production is extensively studied, although industrial-scale processes have not yet been established mainly because of economic factors (Bicas et al. 2010a).

It is widely known that it is necessary to have a correct equilibrium of nutrients to promote suitable conditions for microbial development and production of biosurfactants (Makkar and Cameotra 2002). Some of the most important factors to be considered in the production of biosurfactants are the carbon and nitrogen sources besides process conditions, such as pH, agitation, temperature, and dissolved oxygen (Desai and Banat 1997; Saharan et al. 2011). Optimizing the fermentation conditions is essential to improving the overall process economics in biosurfactant production. It has been estimated that raw materials account for about 10% to 30% of the total cost of biosurfactant production. Generally, biosurfactants are produced during growth on hydrocarbons, which are usually expensive and, therefore, increase the overall process cost (Desai and Banat 1997; Nitschke and Pastore 2006). However, industrial and agro-industrial effluents have been considered recently as a good perspective in the search for cheaper materials used in fermentation processes (Bicas et al. 2010a; Saharan et al. 2011). Thus, the use of wastes with high contents of carbohydrates and lipids appears to be an appropriate option for the production of biosurfactants.

Taking into account that millions of tons of environmentally harmful agro-industrial wastes are generated annually, their use in biotechnological processes seems to be a plausible alternative. Thus, a large amount of cheap substrates could be available for use in bioprocesses. Wastes from the processing of cassava, soybean, olive, potato, molasses, and others were reported (Bicas et al. 2010a; Makkar and Cameotra 2002). In addition, a number of attempts have allowed the increase in biosurfactant productivity by manipulating physiological conditions and medium composition. Genetic manipulation has also been another approach to improve yields (Desai and Banat 1997).

Besides, studies for the optimization of fermentation conditions have resulted in a significant increase in production yields (Desai and Banat 1997; Murkherjee et al. 2006). The development of a continuous process (Chen et al. 2006), an upscale process (Barros et al. 2008b), and an efficient and cheap recovery process (Chen et al. 2006; Murkherjee et al. 2006; Barros et al. 2008b; Saharan et al. 2011) provides promising prospects in this way.

Thus, considering the economic aspects of biosurfactant production, three strategies have been adopted worldwide to make this process cost-competitive: (1) the use of cheaper and waste substrates to lower the initial raw material costs involved in the process; (2) the development of efficient bioprocesses, including optimization of the culture conditions and a cost-effective separation process for maximum biosurfactant production and recovery; and (3) the development and use of overproducing mutant or recombinant strains for enhanced biosurfactant yields (Bicas et al. 2010a; Chen et al. 2006; Murkherjee et al. 2006; Nitschke and Pastore 2006; Thavasi et al. 2008).

Undesirable foaming is one of the major problems associated with many bioprocesses. Foams are created by the vigorous stirring and the aeration, which are required to supply a sufficient amount of oxygen in aerobic fermentations, and are stabilized by the presence of surface-active solutes in the culture broth (Davis et al. 2001). However, in biosurfactant production, foam generation may be very useful. As previously discussed, biosurfactants accumulate in the liquid–gas interface, which means that their concentration is much higher in the foam. Thus, withdrawal of foam is used as a method for the primary recovery of biosurfactants in these procedures (Davis et al. 2001; Barros et al. 2008b). Other processes may use antifoams that do not interfere with the stability of the compounds (Kim et al. 2006).

15.5.5.1 Production of Lipopeptides

This group of biosurfactants is extensively mentioned in the literature. Their general structure consists of amino acids, generally organized in a cyclical chain, with a fatty acid residue resulting in the apolar part of the molecule. *Bacillus* is the main genus described as a producer of this class of biosurfactant (Desai and Banat 1997; Barros et al. 2007; Thavasi et al. 2008).

There is great diversity among compositions and types of culture media for the production of lipopeptides. However, in general, it is possible to categorize these substrates into two classes: (1) hydrophilic media with high levels of carbohydrates and (2) hydrophobic media, supplemented with vegetable oils and hydrocarbons (Makkar and Cameotra 2002; Mukherjee et al. 2006). Usually, the media are rich in carbohydrates because they are the best carbon sources for the synthesis of lipopeptides by *Bacillus* (Bicas et al. 2010a) and because the concentration of this C source strongly influences yield (Chen et al. 2006). Although glucose is the main constituent of this type of medium, other carbohydrates are also used. This occurs because these microorganisms have the ability to use not only low molecular-weight sugars, as glucose, fructose, and sucrose, but polysaccharides such as starch as well, because the genus *Bacillus* presents amylolytic bacteria. The amylolytic enzymes degrade the sugar polymers to oligosaccharides and monosaccharides (Nitschke and Pastore 2006); hence, waste starchy materials are alternative raw materials for the production of biosurfactants (Barros et al. 2008b; Nitschke and Pastore 2006). Thus, a large number of agro-industrial residues with this characteristic (potato effluent, cassava wastewater, cashew juice, whey, molasses, etc.) have been reported in the literature (Barros et al. 2008b; Bicas et al. 2010a; Makkar and Cameotra 2002; Mukherjee et al. 2006).

The use of hydrophobic media, on the other hand, is not so usual for the production of this group of biosurfactants (Bicas et al. 2010a), even though some reports may be found in the literature. An example is the use of a soybean oil residue to produce lipopeptides from *B. subtilis*, which has small surface and emulsification activities against crude oil. Vegetable oils, such as soybean, olive, castor, sunflower, and coconut fat, added to minimal culture medium were also tested with *Serratia marcescens* to stimulate biosurfactant production (Mukherjee et al. 2006). Another example is the production of iturin A by *B. subtilis* in solid-state fermentation using rice bran and peanut oil (Shih et al. 2008). Hydrocarbons have also been used as carbon sources. One example is the use of lubrication oil for the production of biosurfactants in a culture of *B. subtilis* (Makkar and Cameotra 2002).

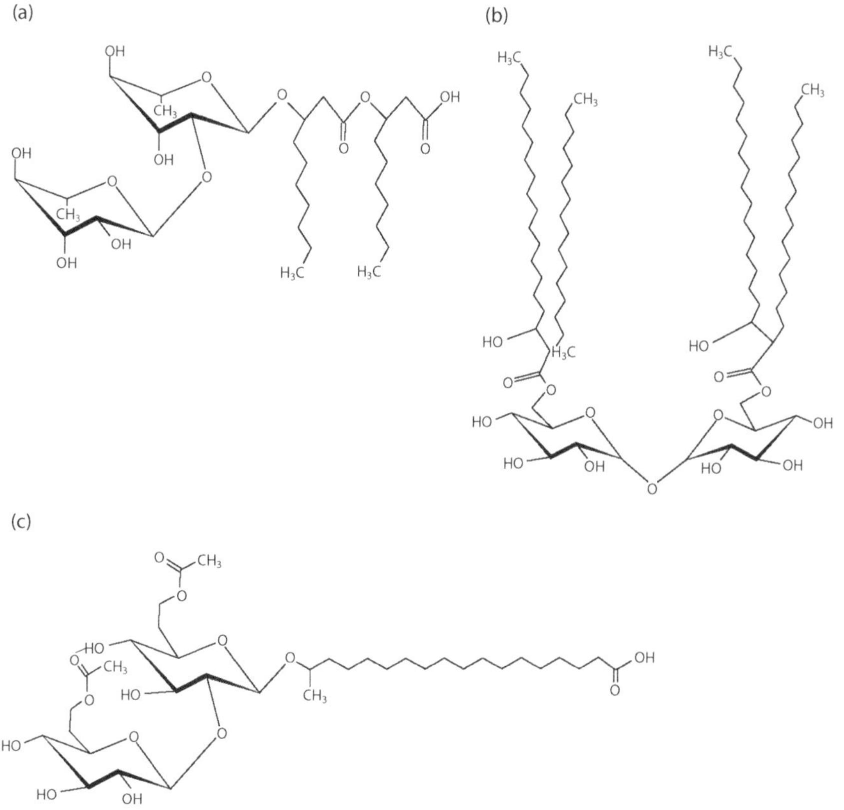

FIGURE 15.6 Example of structure of some common glycolipid biosurfactants. (a) Rhamnolipid: two rhamnose subunits linked to two hydroxy acids. (b) Trehalolipid: trehalose linked to two long-chain α-branched β-hydroxy fatty acids. (c) Sophorolipid: sophorose linked to long-chain hydroxy fatty acid. (Based on Desai, J.D. and Banat, I.M., *Microbiology and Molecular Biology Reviews*, 61: 47–64, 1997.)

TABLE 15.3
Biosurfactants and Their Microbial Sources

	Glycolipids
Rhamnolipids	*P. aeruginosa*
	Pseudomonas sp.
	Pseudomonas chlororaphis,
	Serratia rubidea
Trehalolipids	*Rhodococcus erythropolis*
	Nocardia erythropolis
	Mycobacterium sp.
	Arthrobacter sp.
	Corynebacterium sp.
Sophorolipids	*Candida bombicola*
	Candida apicola
	Torulopsis petrophilum
	Candida riodocensis
	Candida stellata
	Candida antartica
	Candida botistae
	Candida bogoriensis
	Candida lipolytica
Cellobiolipids	*Ustilago zeae*
	Ustilago maydis
Monnosylerythritol lipids	*C. antartica*
	Kurtzmanomyces sp.
	Pseudozyma siamensis
	Candida sp.
Lipomannan	*Candida tropicalis*
	Lipopeptides and Lipoproteins
Peptide–lipid	*Bacillus licheniformis*
Serrawettin	*S. marcescens*
Viscosin	*Pseudomonas fluorescens*
	Leuconostoc mesenteroides
Surfactin	*B. subtilis*
	Bacillus amyloliquefaciens
Iturin	*B. subtilis*
	B. amyloliquefaciens
Subtilisin	*B. subtilis*
Lichenysin	*B. subtilis*
	B. licheniformis
Fengycin	*B. subtilis*
Gramicidins	*Bacillus brevis*
Polymyxins	*Bacillus polymyxa*

(*continued*)

TABLE 15.3 (Continued)
Biosurfactants and Their Microbial Sources

Ornithine lipids	*Pseudomonas* sp.
	Thiobacillus thiooxidans
	Agrobacterium sp.

Fatty Acids, Neutral Lipids, and Phospholipids

Fatty acids	*Corynebacterium lepus*
	Clavibacter michiganensis subsp. *insidiosus*
Neutral lipids	*N. erythropolis*
Phospholipids	*Turulopsis thiooxidans*
	Acinetobacter sp.
	C. lepus
Corynomicolic acids	*Corynebacterium insidibasseosum*

Polymeric Surfactants

Emulsan	*A. calcoaceticus*
Alasan	*Acinetobacter radioresistens*
Biodispersan	*A. calcoaceticus*
Mannan–lipid–protein	*C. tropicalis*
Liposan	*C. lipolytica*
	C. tropicalis
Carbohydrate–protein– lipid	*P. fluorescens*
	Debaryomyces polymorphis
Protein PA	*P. aeruginosa*

Particulate Biosurfactants

Vesicles and fimbriae	*A. calcoaceticus*
	Pseudomonas marginilis
	Pseudomonas maltophila
Whole cells	*Variety of bacteria*
	Cyanobacteria

Source: Based on Desai, J.D. and Banat, I.M., *Microbiology and Molecular Biology Reviews*, 61: 47–64, 1997; Nitschke, M. and Costa, S.G.V.A.O., *Trends in Food Science and Technology*, 18: 252–259, 2007; Saharan, B.S., Sahu, R.K. and Sharma, D., *Genetic Engineering and Biotechnology Journal*, 2011: GEBJ–29, 2011.

15.5.5.2 Glycolipids

Glycolipids are the most commonly known biosurfactants. Their molecular structure is based on carbohydrates linked to long-chain aliphatic acids or hydroxyaliphatic acids (Desai and Banat 1997). Among the glycolipids, the best known are rhamnolipids, trehalolipids, and sophorolipids (see examples in Figure 15.6). There are several microbial genera described as producers of this kind of surfactant (Table 15.3): *Pseudomonas* and *Serratia*, which produce rhamnolipids (Makkar and Cameotra 2002; Mukherjee et al. 2006); *Torulopsis* and *Pseudozyma* (*Candida*), which produce sophorolipids (Desai and Banat 1997; Sarubbo et al. 2007; Saharan et al. 2011); *Mycobacterium, Nocardia, Corynebacterium, Rhodococcus,* and *Arthrobacter,* which produce trehalolipids (Desai and Banat 1997; Saharan et al. 2011); and *Bacillus megaterium* (Thavasi et al. 2008) and *Candida* sp., which produce mannosylerythritol lipid (Kim et al. 2006).

As for lipopeptides, culture media suitable for the cultivation of glycolipid-producing microorganisms can also be classified into two groups: those exhibiting high levels of carbohydrates and those with hydrophobic composition. However, in the case of production of glycolipids, the use of culture media with hydrophobic components is preponderant. In most cases, hydrocarbons or vegetable oils are added to mineral media (Bicas et al. 2010a; Saharan et al. 2011). Although glucose, glycerol, and triglycerides can be used in the fermentative process of *Pseudomonas* to produce rhamnolipids, the yields are low when compared to the use of hydrocarbon media. However, some culture media rich in carbohydrates should be used as a substrate because of its low cost. Another alternative can be the use of vegetable oils, such as soybean oil; wastes from rapeseed, soybean, and corn oil refineries; frying oil; and sunflower oil (Bicas et al. 2010a; Mukherjee et al. 2006).

Processes involving the production of glycolipids in hydrophilic media, on the other hand, are not easily found in the literature. However, some examples of this application may be cited: the production of rhamnolipids by *Pseudomonas aeruginosa* using distillery waste, curd whey, cashew apple juice, or sugar industry effluent, or the production of a biosurfactant from *Acinetobacter calcoaceticus* in cashew apple juice (Bicas et al. 2010a; Saharan et al. 2011).

15.6 CONCLUDING REMARKS

The food industry has entered into a new era. Consumers are no longer accepting food as simple means to satisfy their basic energetic needs. Besides the increase in the demand for "natural" additives, there is also a tendency to search for functional food ingredients, that is, those that can contribute favorably to the promotion of a consumer's health and wellbeing. This scenario is even more evident for countries with well-established economies, such as the United States, Japan, Germany, France and others. In this chapter, some examples of the production of food additives by biotechnological means have been presented. This is one of the most promising techniques to substitute the traditional methods for obtaining such compounds (chemical synthesis or direct extraction from nature). The advantage of biotechnological-based products, as presented here, is that they can be labeled as "natural" and, in some cases, may also present desirable functional properties. However, developments in bioprocess engineering and microbial genetic improvements should be considered to overcome low yields and high manufacturing costs that may hamper the commercial adoption of some of these processes.

REFERENCES

Aachary, A.A. and Prapulla, S.G. (2011) Xylooligosaccharides (XOS) as an emerging prebiotic: Microbial synthesis, utilization, structural characterization, bioactive properties, and applications. *Comprehensive Reviews in Food Science and Food Safety,* 10: 2–16.

Aberoumand, A. (2011) A review article on edible pigments properties and sources as natural biocolorants in foodstuff and food industry. *World Journal of Dairy and Food Sciences*, 6: 71–78.

Amiot-Carlin, M.J., Babot-Laurent, C. and Tourniaire, F. (2008) Plant pigments as bioactive substances, in: C. Socaciu (ed.) *Food colorants: Chemical and functional properties*, Boca Raton: CRC Press, pp. 127–146.

Assaf, S., Hadar, Y. and Dosoretz, C.G. (1997) 1-octen-3-ol and 13-hydroperoxylinoleate are products of distinct pathways in the oxidative breakdown of linoleic acid by *Pleurotus pulmonarius*. *Enzyme and Microbial Technology*, 21: 484–490.

Babitha, S. (2009) Microbial pigments, in: P.S. Nigam and A. Pandey (eds.) *Biotechnology for agro-industrial residues utilisation*, Springer.

Barreteau, H., Delattre, C. and Michaud, P. (2006) Production of oligosaccharides as promising new food additive generation. *Food Technology and Biotechnology*, 44: 323–333.

Barros, F.F.C., Quadros, C.P., Maróstica Jr., M.R. and Pastore, G.M. (2007) Surfactina: Propriedades químicas, tecnológicas e funcionais para aplicações em alimentos. *Química Nova*, 30: 409–414.

Barros, F.F.C., Quadros, C.P. and Pastore, G.M. (2008a) Propriedades emulsificantes e estabilidade do biossurfactante produzido por *Bacillus subtilis* em manipueira. *Ciência e Tecnologia de Alimentos*, 28: 979–985.

Barros, F.F.C., Ponezi, A.N. and Pastore, G.M. (2008b) Production of biosurfactant by *Bacillus subtilis* LB5a on a pilot scale using cassava wastewater as substrate. *Journal of Industrial Microbiology and Biotechnology*, 35: 1071–1078.

Başer, K.H.C. and Demirci, F. (2007) Chemistry of essential oils, in: R.G. Berger (ed.) *Flavours and fragrances: Chemistry, bioprocessing and sustainability*, Berlin: Springer, pp. 42–86.

Bauer, K., Garbe, D. and Surburg, H. (2001) *Common fragrance and flavor materials preparation, properties and uses*, 4 ed. Wiley-VCH, Weinheim, Germany, 293 pp.

Berger, R. (2009) Biotechnology of flavors—the next generation. *Biotechnology Letters*, 31: 1651–1659.

Bicas, J.L., Barros, F.F.C., Wagner, R., Godoy, H.T. and Pastore, G.M. (2008) Optimization of *R*-(+)-α-terpineol production by the biotransformation of *R*-(+)-limonene. *Journal of Industrial Microbiology and Biotechnology*, 35: 1061–1070.

Bicas, J.L., Dionísio, A.P., Silva, J.C., Barros, F.F.C. and Pastore, G.M. (2010a) Agro-industrial residues in biotechnological processes, in: J. Krause and O. Fleischer (eds.) *Industrial fermentation: Food processes, nutrient sources and production strategies*, Hauppauge: Nova Science Publishers, Inc., pp. 275–295.

Bicas, J.L., Dionısio, A.P. and Pastore, G.M. (2009) Bio-oxidation of terpenes: An approach for the flavor industry. *Chemical Reviews*, 109: 4518–4531.

Bicas, J.L., Fontanille, P., Pastore, G.M. and Larroche, C. (2010c) A bioprocess for the production of high concentrations of *R*-(+)-α-terpineol from *R*-(+)-limonene. *Process Biochemistry*, 45: 481–486.

Bicas, J.L., Silva, J.C., Dionisio, A.P. and Pastore, G.M. (2010b) Biotechnological production of bioflavors and functional sugars. *Ciência e Tecnologia de Alimentos*, 30: 7–18.

Bluemke, W. and Schrader, B.J. (2001) Integrated bioprocess for enhanced production of natural flavors and fragrances by *Ceratocystis moniliformis*. *Biomolecular Engineering*, 17: 137–142.

Bobbio, P.A. and Bobbio, F.O. (2001) *Química do processamento de alimentos*, 3 ed. Varella, São Paulo.

Borchers, A.T., Selmi, C., Meyers, F.J., Keen, C.L. and Gershwin, M.E. (2009) Probiotics and immunity. *Journal of Gastroenterology*, 44: 26–46.

Bruzzese, E., Volpicelli, M., Squaglia, M., Tartaglione, A. and Guarino, A. (2006) Impact of prebiotics on human health. *Digestive and Liver Disease*, 38: S283–S287.

Bystrykh, L.V., Fernández-Moreno, M.A., Herrema, J.K., Malpartida, F., Hopwood, D.A. and Dijkhuizen, L. (1996) Production of actinorhodin-related "blue pigments" by *Streptomyces coelicolor* A3(2). *Journal of Bacteriology*, 178: 2238–2244.

Cabral, J.M.S. (2001) Biotransformations, in: C. Ratledge and B. Kristiansen (eds.) *Basic biotechnology*, 2 ed., Cambridge: Cambridge University Press, pp. 471–501.

Cardillo, R., Fungarnti, C., Sacerdote, G., Barbeni, M., Cabella, P. and Squarcia, F. (1990) Eur. Pat. No 356,291.

Carvalho, J.C., Oishi, B.O., Pandey, A. and Soccol, C.R. (2005) Biopigments from *Monascus*: Strains selection, citrinin production and color stability. *Brazilian Archives of Biology and Technology*, 48: 885–894.

Carvalho, J.C., Oishi, B.O., Woiciechowski, A.L., Pandey, A., Babitha, S. and Soccol, C.R. (2007) Effect of substrates on the production of *Monascus* biopigments by solid-state fermentation and pigment extraction using different solvents. *Indian Journal of Biotechnology*, 6: 194–199.

Chattopadhyay, P., Chatterjee, S. and Sen, S.K. (2008) Biotechnological potential of natural food grade biocolorants. *African Journal of Biotechnology*, 7: 2972–2985.

Chávez-Parga, M.C., Gonzalez-Ortega, O., Negrete-Rodríguez, M.L.X., Vallarino, I.G., Alatorre, G.G. and Escamilla-Silva, E.M. (2008) Kinetic of the gibberellic acid and bikaverin production in an airlift bioreactor. *Process Biochemistry*, 43: 855–860.

Chávez-Parga, M.C., Gonzalez-Ortega, O., Sánchez-Cornejo, G., Negrete-Rodríguez, M.L.X., Alatorre, G.G. and Escamilla-Silva, E.M. (2005) Mathematical description of bika-verin production in a fluidized bed bioreactor. *World Journal of Microbiology and Biotechnology*, 21: 683–688.

Chen, C.-Y., Baker, S.C. and Darton, R.C. (2006) Continuous production of biosurfactant with foam fractionation. *Journal of Chemical Technology and Biotechnology*, 81: 1915–1922.

Choudhari, S. and Singhal, R. (2008) Media optimization for the production of β-carotene by *Blakeslea trispora*: a statistical approach. *Bioresource Technology*, 99: 722–730.

Christen, P., Meza, J.C. and Revah, S. (1997) Fruity aroma production in solid state fermen-tation by *Ceratocystis frimbriata*: influence of the substrate type and the type and the presence of precursors. *Mycological Research*, 101: 911–919.

Dalzenne, N.M. (2003) Oligosaccharides: state of the art. *Proceedings of the Nutrition Society*, 62: 177–182.

Das, A., Yoon, S.-H., Lee, S.-H., Kim, J.-Y., Oh, D.-K. and Kim, S.-W. (2007) An update on microbial carotenoid production: Application of recent metabolic engineering tools. *Applied Microbiology and Biotechnology*, 77: 505–512.

Daugsch, A. and Pastore, G.M. (2005) Obtenção de vanilina: Oportunidade biotecnológica. *Química Nova*, 28: 642–645.

Davis, D.A., Lynch, H.C. and Varley, J. (2001) The application of foaming for the recov-ery of surfactin from *Bacillus subtilis* ATCC 21332 cultures. *Enzyme Microbiology Technology*, 28: 346–354.

De Carvalho, C.C.C.R. and Da Fonseca, M.M.R. (2006a) Biotransformation of terpenes. *Biotechnology Advances*, 24: 134–142.

De Carvalho, C.C.C.R. and Da Fonseca, M.M.R. (2006b) Carvone: Why and how should one bother to produce this terpene? *Food Chemistry*, 95: 413–422.

Delgado-Vargas, F. and Paredes-Lópes, O. (2003) *Natural colorants for food and nutraceuti-cal uses*, CRC Press, Boca Raton.

Demain, A.L., Jackson, M. and Trenner, N.R. (1967) Thiamine-dependent accumulation of tet-ramethylpyrazine accompanying a mutation in the isoleucine–valine pathway. *Journal of Bacteriology*, 94: 323–326.

Demyttenaere, J.C.R., Adams, A., Vanoverschelde, J. and De Kimpe, N. (2001) Biotransformation of (*S*)-(+)-linalool by *Aspergillus niger*: An investigation of the culture conditions. *Journal of Agricultural and Food Chemistry*, 49: 5895–5901.

Desai, J.D. and Banat, I.M. (1997) Microbial production of surfactants and their commercial potential. *Microbiology and Molecular Biology Reviews*, 61: 47–64.

Dufossé, L. (2006) Microbial production of food grade pigments. *Food Technology and Biotechnology*, 44: 313–321.

Dufossé, L. (2008) Pigments from microalgae and microorganisms: Sources of food colorants, in: C. Socaciu (ed.) *Food colorants: Chemical and functional properties*, Boca Raton: CRC Press, pp. 399–426.

Dufossé, L., Galaup, P., Yaron, A., Arad, S.M., Blanc, P., Murthy, K.N.C. and Ravishankar, K. (2005) Microorganisms and microalgae as sources of pigments for food use: A scientific oddity or an industrial reality? *Trends in Food Science and Technology*, 16: 389–406.

Eriksen, N.T. (2008) Production of phycocyanin—a pigment with applications in biology, biotechnology, foods and medicine. *Applied Microbiology and Biotechnology*, 80: 1–14.

Feron, G. and Waché, Y. (2006) Microbial biotechnology of food flavor production, in: K. Shetty, G. Paliyath, A. Pometto, and R. Levin (eds.) *Food biotechnology*, CRC Press Taylor & Francis.

Franck, A. (2002) Technological functionality of inulin and oligofructose. *British Journal of Nutrition*, 87: S287–S291.

Franco, M.R.B. (2004) *Aroma e sabor dos alimentos: Temas atuais*, Livraria Varela, São Paulo, 246 pp.

Gargouri, M. (2001) Biocatalysis in liquid–liquid biphasic media: coupled mass transfer and chemical reactions, in: A.G. Volkov (ed.) *Liquid interfaces in chemical, biological and pharmaceutical applications*, New York: Marcel Dekker, Inc., pp. 553–584.

German, J.B., Yeritzian, C. and Tolstoguzov, V.B. (2007) Olfaction, where nutrition, memory and immunity intersect, in: R.G. Berger (ed.) *Flavours and fragrances: Chemistry, bioprocessing and sustainability*, Berlin: Springer, pp. 25–41.

Gibson, G.R. and Roberfroid, M.B. (1995) Dietary modulation of the colonic microbiota: Introducing the concept of prebiotics. *Journal of Nutrition*, 125: 1401–1412.

Hatti-Kaul, R., Törnvall, U., Gustafsson, L. and Börjesson, P. (2007) Industrial biotechnology for the production of bio-based chemicals: a cradle-to-grave perspective. *Trends in Biotechnology*, 25: 119–124.

Husson, F., Thomas, M., Kermasha, S. and Belin, J.M. (2002) Effect of linoleic acid induction on the production of 1-octen-3-ol by the lipoxygenase and hydroperoxide lyase activities of *Penicillium camemberti*. *Journal of Molecular Catalysis B: Enzymatic*, 19–20: 363–369.

Jones, J.D., Hohn, T.M. and Leathers, T.D. (2004) *Fusarium sporotrichioides* strains for production of lycopene. US Patent N° 6696282.

Kim, H.-S., Jeon, J.-W., Kim, B.-H., Ahn, C.-Y., Oh, H.-M. and Yoon, B.-D. (2006) Extracellular production of a glycolipid biosurfactant, mannosylerythritol lipid, by *Candida* sp. SY16 using fed-batch fermentation. *Applied Microbiology and Biotechnology*, 70: 391–396.

Kolida, S. and Gibson G.R. (2007) Inulin and oligofructose: Health benefits and claims: A critical review. *Journal of Nutrition*, 137: 2503S–2506S.

Kongruang, S. (2011) Growth kinetics of biopigment production by Thai isolated *Monascus purpureus* in a stirred tank bioreactor. *Journal of Industrial Microbiology and Biotechnology*, 38: 93–99.

Kosuge, T. and Kamiya, H. (1962) Discovery of a pyrazine in a natural product: tetramethyl pyrazine from cultures of a strain of *Bacillus subtilis*. *Nature* (London), 193: 776.

Krämer, R. (2005) Production of amino acids: physiological and genetic approaches. *Food Biotechnology*, 18: 171–216.

Krings, U. and Berger, R.G. (1998) Biotechnological production of flavors and fragrances. *Applied Microbiology and Biotechnology*, 49: 1–8.

Kumari, H.P.M., Naidu, K.A., Vishwanatha, S., Narasimhamurthy, K. and Vijayalakshmi, G. (2009) Safety evaluation of *Monascus purpureus* red mould rice in albino rats. *Food and Chemical Toxicology*, 47: 1739–1746.

León, R., Fernandes, P., Pinheiro, H.M. and Cabral, J.M.S. (1998) Whole-cell biocatalysis in organic media. *Enzyme and Microbial Technology*, 23: 483–500.

Leuchtenberger, W., Huthmacher, K. and Drauz, K. (2005) Biotechnological production of amino acids and derivatives: Current status and prospects. *Applied Microbiology and Biotechnology*, 69: 1–8.

Limón, M.C., Rodríguez-Ortiz, R. and Avalos, J. (2010) Bikaverin production and applications. *Applied Microbiology and Biotechnology*, 87: 21–29.

Liu, G.Y. and Nizet, V. (2009) Color me bad: Microbial pigments as virulence factors. *Trends in Microbiology*, 17: 406–413.

Lomascolo, A., Stentelaire, C., Asther, M. and Lesage-Meessena, L. (1999) Basidiomycetes as new biotechnological tools to generate natural aromatic flavors for the food industry. *Trends in Biotechnology*, 17: 282–289.

Long II, T.V. (2004) Process for production of carotenoids, xanthophylls and apo-carotenoids utilizing eukaryotic microorganisms, US Patent No. 6783951.

Longo, M.A. and Sanromán, M.A. (2006) Production of food aroma compounds: microbial and enzymatic methodologies. *Food Technology and Biotechnology*, 44: 335–353.

Lu, L., Chui, H.-L., Chen, Y.-N. and Yuan, S. (2002) Isolation and identification of *Streptomyces* sp. and assay of its exocellular water-soluble blue pigments. *Folia Microbiologica*, 47: 493–498.

MacFarlane, S. and MacFarlane, G.T. (2003) Regulation of short-chain fatty acid production, *Proceedings of the Nutrition Society*, 62: 67–72.

MacFarlane, S., MacFarlane, G.T. and Cummings, J.H. (2006) Review article: Prebiotics in the gastrointestinal tract. *Alimentary Pharmacology Therapy*, 24: 701–714.

Makkar, R.S. and Cameotra, S.S. (2002) An update on the use of unconventional substrates for biosurfactant production and their new applications. *Applied Microbiology and Biotechnology*, 58: 428–434.

Manucci, F. (2009) Enzymatic synthesis of galactooligosaccharides from whey permeate. MSc Thesis, Dublin Institute of Technology.

Mapari, S.A.S., Meyer, A.S., Thrane, U. and Frisvad, J. (2009) Identification of potentially safe promising fungal cell factories for the production of polyketide natural food colorants using chemotaxonomic rationale. *Microbial Cell Factories*, 8: 24–40.

Maróstica Jr., M.R. and Pastore, G.M. (2007a) Production of *R*-(+)-α-terpineol by the biotransformation of limonene from orange essential oil, using cassava waste water as medium. *Food Chemistry*, 101: 345–350.

Maróstica Jr., M.R. and Pastore, G.M. (2007b) Biotransformation of limonene: A review of the main metabolic pathways. *Química Nova*, 30: 382–387.

Moreira, N., Mendes, F., Pinho, P.G., Hogg, T. and Vasconcelos, I. (2008) Heavy sulphur compounds, higher alcohols and esters production profile of *Hanseniaspora uvarum* and *Hanseniaspora guilliermondii* grown as pure and mixed cultures in grape must. *International Journal of Food Microbiology*, 124: 231–238.

Muheim, A., Muller, B., Munch, T. and Wetli, M. (2001) Microbiological process for producing vanillin. United States Patent No. 6,235,507.

Murkherjee, S., Das, P. and Sen, R. (2006) Towards commercial production of microbial surfactants. *Trends in Biotechnology*, 24: 509–515.

Mukherjee, G. and Singh, S.K. (2011) Purification and characterization of a new red pigment from *Monascus purpureus* in submerged fermentation. *Process Biochemistry*, 46: 188–192.

Mussatto, S.I. and Mancilha, I.M. (2007) Non-digestible oligosaccharides: A review. *Carbohydrate Polymers*, 68: 587–597.

Nakakuki, T. (2002) Present status and future of functional oligosaccharide development in Japan. *Pure and Applied Chemistry*, 74: 1245–1251.

Nimnoi, P. and Lumyong, S. (2011) Improving solid-state fermentation of *Monascus purpureus* on agricultural products for pigment production. *Food Bioprocess Technology*, 4: 1384–1390.

Nitschke, M. and Pastore, G.M. (2006) Production and properties of a surfactant obtained from *Bacillus subtilis* grown on cassava wastewater. *Bioresource Technology*, 97: 336–341.

Nitschke, M. and Costa, S.G.V.A.O. (2007) Biosurfactants in food industry. *Trends in Food Science and Technology*, 18: 252–259.

Novakova, R., Odnogova, Z., Kutas, P., Feckova, L. and Kormanec, J. (2010) Identification and characterization of an indigoidine-like gene for a blue pigment biosynthesis in *Streptomyces aureofaciens* CCM 3239. *Folia Microbiology*, 55: 119–125.

Ötles, S. and Çagindi, Ö. (2008) Carotenoids as natural colorants, in: C. Socaciu (ed.) *Food colorants: Chemical and functional properties*, Boca Raton: CRC Press, pp. 51–70.

Pandey, A., Soccol, C.R. and Mitchell, D. (2000) New developments in solid state fermentation: I-bioprocesses and products. *Process Biochemistry*, 35: 1153–1169.

Papaioannou, E.H. and Liakopoulou-Kyriakides, M. (2010) Substrate contribution on carotenoids production in *Blakeslea trispora* cultivations. *Food and Bioproducts Processing*, 88: 305–311.

Pastore, G.M., Park, Y.K. and Min, D.B. (1994) Production of fruity aroma by *Neurospora* from beiju. *Mycological Research*, 98: 1300–1302.

Pattanagul, P., Pinthong, R., Phianmongkol, A. and Leksawasdi, N. (2007) Review of angkak production (*Monascus purpureus*). *Chiang Mai Journal of Science*, 34: 319–328.

Pescheck, M., Mirata, M.A., Brauer, B., Krings, U., Berger, R.G. and Schrader, J. (2009) Improved monoterpene biotransformation with *Penicillium* sp. by use of a closed gas loop bioreactor. *Journal of Industrial Microbiology and Biotechnology*, 36: 827–836.

Pintea, A.M. (2008) Food colorants derived from natural sources by processing, in: C. Socaciu (ed.) *Food colorants: Chemical and functional properties*, Boca Raton: CRC Press, pp. 101–124.

Rabenhorst, J. and Hopp, R. (2000) Process for the preparation of vanillin and microorganisms suitable therefore, US Patent No. 6,133,003.

Ribeiro, B.D., Barreto, D.W. and Coelho, M.A.Z. (2011) Technological aspects of β-carotene production. *Food and Bioprocess Technology*, 4: 693–701.

Roberfroid, M. (2007) Prebiotics: the concept revisited. *Journal of Nutrition*, 137: 830S–837S.

Roberfroid, M., Gibson, G.R., Hoyles, L., McCartney, A.L., Rastall, R., Rowland, I., Wolvers, D., et al. (2010) Prebiotic effects: metabolic and health benefits. *British Journal of Nutrition*, 104: S1–S72.

Rodríguez-Ortiz, R., Mehta, B.J., Avalos, J. and Limón, M.C. (2010) Stimulation of bikaverin production by sucrose and by salt starvation in *Fusarium fujikuroi*. *Applied Microbiology and Biotechnology*, 85: 1991–2000.

Rodríguez-Sáiz, M., Fuente, J.L. and Barredo, J.L. (2010) Xanthophyllomyces dendrorhous for the industrial production of astaxanthin. *Applied Microbiology and Biotechnology*, 88: 645–658.

Rosenblitt, A., Agosin, E., Delgado, J. and Perez-Correa, R. (2000) Solid substrate fermentation of *Monascus purpureus*: growth, carbon balance, and consistency analysis. *Biotechnology Progress*, 16: 152–162.

Saharan, B.S., Sahu, R.K. and Sharma, D. (2011) A review on biosurfactants: Fermentation, current developments and perspectives. *Genetic Engineering and Biotechnology Journal*, 2011: GEBJ–29.

Sako, T., Matsumoto, K. and Tanaka, R. (1999) Recent progress on research and applications of non-digestible galactooligosaccharides. *International Dairy Journal*, 9: 69–80.

Sangeetha, P.T., Ramesh, M.N. and Prapulla, S.G. (2005) Recent trends in the microbial production, analysis and application of fructooligosaccharides. *Trends in Food Science and Technology*, 16: 442–457.

Sarada, R., Pillai, M.G. and Ravishankar, G.A. (1999) Phycocyanin from *Spirulina* sp.: Influence of processing of biomass on phycocyanin yield, analysis of efficacy of extraction methods and stability studies on phycocyanin. *Process Biochemistry*, 34: 795–801.

Sardaryan, E. (2002) Strain of the microorganism *Penicillium oxalicum* var. Armeniaca and its application. US Patent No. 6,340,586 B1.

Sarubbo, L.A., Farias, C.B.B. and Campos-Takaki, G.M. (2007) Co-utilization of canola oil and glucose on the production of a surfactant by *Candida lipolytica*. *Current Microbiology*, 54: 68–73.

Sauer, M., Porro, D., Mattanovich, D. and Branduardi, P. (2008) Microbial production of organic acids: Expanding the markets. *Trends in Biotechnology*, 26: 100–108.

Savithiry, N., Cheong, T.K. and Oriel, P. (1997) Production of alpha-terpineol from *Escherichia coli* cells expressing thermostable limonene hydratase. *Applied Biochemistry and Biotechnology*, 63: 213–220.

Schmidt, I., Schewe, H., Gassel, S., Jin, C., Buckingham, J., Hümbelin, M., Sandmann, G. and Schrader J. (2011) Biotechnological production of astaxanthin with *Phaffia rhodozyma/Xanthophyllomyces dendrorhous*. *Applied Microbiology and Biotechnology*, 89: 555–571.

Schrader, J., Etschmann, M.M.W., Sell, D., Hilmer, J.-M. and Rabenhorst, J. (2004) Applied biocatalysis for the synthesis of natural flavour compounds: Current industrial processes and future prospects. *Biotechnology Letters*, 26: 463–472.

Schwartz, S.J., Von Elbe, J.H. and Giusti, M.M. (2010) Corantes, in: S. Damodaran, K.L. Parkin and O.R. Fennema (eds.) *Química de alimentos de fennema*, 4 ed., Porto Alegre: Artmed.

Shih, I.-L., Kuo, C.-Y., Hsieh, F.-C., Kao, S.-S. and Hsieh, C. (2008) Use of surface response methodology to optimize culture conditions for iturin A production by *Bacillus subtilis* in solid-state fermentation. *Journal of the Chinese Institute of Chemical Engineers*, 39: 635–643.

Silveira, S.T., Daroit, D.J. and Brandelli, A. (2008) Pigment production by *Monascus purpureus* in grape waste using factorial design. *LWT—Food Science and Technology*, 41: 170–174.

Singh, P. and Cameotra, S.S. (2004) Potential applications of microbial surfactants in biomedical sciences. *Trends in Biotechnology*, 22: 142–146.

Soares, M., Christen, P., Pandey, A. and Soccol, C.R. (2000) Fruity flavor production by *Ceratocystis fimbriata* grown on coffee husk in solid-state fermentation. *Process Biochemistry*, 35: 857–861.

Song, S., Tang, Q., Wang, H. and Zhu, J. (2010) Blue pigment producing bacteria and method for preparing crude preparation by using the same. CN Patent No. 101864380 (A).

Surburg, H. and Panten, J. (2006) *Common fragrance and flavor materials, preparation, properties and uses*, 5 ed. WILEY-VCH Verlag, Weinheim.

Tan, Q. and Day, D.F. (1998a) Bioconversion of limonene to α-terpineol by immobilized *Penicillium digitatum*. *Applied Microbiology and Biotechnology*, 49: 96–101.

Tan, Q. and Day, D.F. (1998b) Organic co-solvent effects on the bioconversion of (R)-(+)-limonene to (R)-(+)-α-terpineol. *Process Biochemistry*, 33(7): 755–761.

Thavasi, R., Nambaru, V.R.M.S., Jayalakshmi, S., Balasubramanian, T. and Banat, I.M. (2008) Biosurfactant production by *Pseudomonas aeruginosa* from renewable resources. *Indian Journal of Microbiology*, 51: 30–36.

Torres, P.M.D., Gonçalves, M.P.F., Teixeira, J.A. and Rodrigues, L.R. (2010) Galacto-oligosaccharides: Production, properties, applications, and significance as prebiotics. *Comprehensive Reviews in Food Science and Food Safety*, 9: 438–454.

Van der Werf, M.J., De Bont, J.A.M. and Leak, D.J. (1997) Opportunities in microbial bio-transformation of monoterpenes. *Advances in Biochemical Engineering Biotechnology*, 55: 147–177.

Vandenplas, Y. (2002) Oligosaccharides in infant formula. *British Journal of Nutrition* 87: S293–S296.

Venil, C.K. and Lakshmanaperumalsamy, P. (2009) An insightful overview on microbial pigment, Prodigiosin. *Electronic Journal of Biology*, 5: 49–61.

Vidyalakshmi, R., Paranthaman, R., Murugesh, S. and Singaravadivel, K. (2009) Microbial bioconversion of rice broken to food grade pigments. *Global Journal of Biotechnology and Biochemistry*, 4: 84–87.

Walter, A., Carvalho, J.C., Soccol, V.T., Faria, A.B.B., Ghiggi, V. and Soccol, C.R. (2011) Study of phycocyanin production from *Spirulina platensis* under different light spectra. *Brazilian Archives of Biology and Technology*, 54: 675–682.

Wang, Y.-Z., Ju, X.-L. and Zhou, Y.-G. (2005) The variability of citrinin production in *Monascus* type cultures. *Food Microbiology*, 22: 145–148.

Wiemann, P., Willmann, A., Straeten, M., Kleigrewe, K., Beyer, M., Humpf, H.-U. and Tudzynski, B. (2009) Biosynthesis of the red pigment bikaverin in *Fusarium fujikuroi*: Genes, their function and regulation. *Molecular Microbiology*, 72: 931–946.

Xu, P., Hua, D. and Ma, C. (2007) Microbial transformation of propenylbenzenes for natural flavour production. *Trends in Biotechnology*, 25: 571–576.

Yun, J.W. (1996) Fructooligosaccharides-occurrence, preparation, and application. *Enzyme and Microbial Technology*, 19: 107–117.

Zhang, H., Zhan, J., Su, K. and Zhang, Y. (2006) A kind of potential food additive produced by *Streptomyces coelicolor*: Characteristics of blue pigment and identification of a novel compound, λ-actinorhodin. *Food Chemistry*, 95: 186–192.

16 Microalgae for Food Production

Jorge Alberto Vieira Costa and
Michele Greque de Morais

CONTENTS

16.1 INTRODUCTION

Food manufacturing is a rapidly changing environment, due to increased competition, globalization of prices and consumer demands for quality, safety, better health and convenience foods. The main focus of research by these industries is on increasing the efficiency of processes, combined with reducing environmental impact, expanding the development of technologies that add value and increasing the development of foods that promote health and wellbeing (Damodaran et al. 2010).

Biotechnology has introduced a new dimension into food production, in addition to innovation. It has the potential to increase food production, quality and the healthiness of products, to reduce dependence of agriculture on chemicals and to lower the costs of raw material, all in an environmentally-sustainable fashion (Damodaran et al. 2010).

Research has been directed toward the development of functional products, by adding vitamins, antioxidants, highly digestible protein, and essential fatty acids to foods. Microalgae can provide several of these nutrients and have the potential to promote health benefits. Microalgae have been used in animal feed because of their protein and lipid content, or as an ingredient in production. Microalgae are used in this way because of their antiviral, antioxidant, antiallergenic and immunomodulatory potential (Cavani et al. 2009; Petracci et al. 2009). Microalgae can contain 60–70 per cent of protein and are rich in vitamins (especially vitamin B12), minerals, chlorophyll, carotenoids, carbohydrates, sterols, pigments and fatty acids (Peiretti and Meineri, 2011).

In addition to their importance in improving the nutritional quality of food, these micro-organisms are also an alternative for the biotechnological production of biopolymers, which can be used for the development of biodegradable packaging. Poly (3-hydroxybutyrate) (PHB) and Poly(3-hydroxybutyrate-co-valerate) (PHB-HV) are polymers from the polyhydroxyalkanoate (PHA) family, which are synthesized and accumulated as reserve substances in various microorganisms. They have thermoplastic properties, as well as physical and mechanical characteristics that are similar to polypropylene. However, the advantage with them is that they are biodegradable when exposed to biologically active environments: this makes them an important alternative for the replacement of traditional plastics (Leary et al. 2009; Sudesh et al. 2000).

The nutritional requirements for cultivation of microalgae may be available in industrial effluents that cause changes to the ecosystem. Hence, what was previously considered to be a problem can become the raw material for obtaining coproducts with high added value. If industries are to become sustainable they need to incorporate biotechnology into numerous processes. Microalgal biotechnology can be used to treat industrial effluents. Molecular oxygen from microalgal photosynthesis is used as an electron acceptor to degrade organic compounds. The use of ammonium, nitrate, and phosphate as nutrients by microalgae has the advantage of removing these compounds during wastewater treatment (Munõz and Guieysse, 2006).

The source of carbon that is required to grow microalgae represents 60 percent of the cost of nutrients. The use of alternative sources of carbon, such as CO_2 emitted from the combustion of coal in power plants, minimizes the problems caused by the emission of this gas, such as global warming. At the same time, there is a reduction in the cost of providing this nutrient and carbon credits that can be generated, which can be negotiated with other countries that need to reduce greenhouse gas emissions.

The objective of this study was to present several applications of microalgae in the food sector, including supplementation of products from biomass, specific biocompounds, obtaining polymers with potential for developing packaging and the treatment of industrial effluents.

16.2 MICROALGAE WITH POTENTIAL IN HUMAN AND ANIMAL FOOD

In the 1950s, the increasing world population and the prediction that there would not be enough protein to feed people led to a search for alternative and unconventional sources of protein. Microalgae emerged as strong candidates for this purpose. Their cultivation on a commercial scale began in 1960 in Japan with the cultivation of *Chlorella* by Nihon Chlorella. In 1970, *Spirulina* was cultivated and harvested in Lake Texcoco by Sosa Texcoco S.A. (Mexico City, Mexico). By 1980, 46 Asian companies were producing more than 1000 kg/month of microalgal biomass on a commercial scale. The production of *Dunaliella salina* to obtain β-carotene was started in 1986 by the companies Western Biotechnology (Hutt Lagoon, Australia) and Betatene (Whyalla, Australia) (Spolaore et al. 2006). It then expanded to the USA (AquaCarotene, Cyanotech), Australia (Cognis Nutrition & Healthy), Japan (Sohonsha Nikken Corporation), and India (Parry Pharmaceuticals) (Rosenberg et al. 2008).

More recently, several plants have been set up to produce *Haematococcus pluvialis* in the USA (Bioreal, Cyanotech, Mera Pharmaceuticals), Israel (Alga Technologies) and India (Parry Pharmaceuticals) for extraction of astaxanthin, (Biotechnological and Environmental Applications of Microalgae) (Spolaore et al. 2006). The microalgae Schizochytrium and Crypthecodinium are grown by BlueBiotech International GmbH (Germany), Spectra Stable Isotopes (Maryland, USA), and Martek Biosciences (Maryland, USA) to obtain poly-unsaturated fatty acids. *Spirulina*, *Chlorella*, and *Chlamydomonas* are used as food supplements in Germany (BlueBiotech International GmbH) and in the USA (Cyanotech, Earthrise Nutritionals, and Phycotransgenics) (Rosenberg et al. 2008). In southern Brazil a commercial plant that makes *Spirulina* capsules as a nutritional supplement is being set up, aiming at annual production of 6000 kg.

Microalgae are currently used in the form of tablets, capsules or liquids. These micro-organisms can be incorporated into pastas, cookies, food, candy, gum, and beverages (Liang et al. 2004). Because of their different chemical properties, microalgae can be applied as a nutritional supplement or as a source of natural dyes, antioxidants and polyunsaturated fatty acids (Spolaore et al. 2006; Soletto et al. 2005)

TABLE 16.1
General Composition of Different Human Sources of Nutrients and Microalgae (% Dry Weight)

Product	Proteins	Carbohydrates	Lipids
Meat	43.0	1.0	34.0
Milk	26.0	38.0	28.0
Rice	8.0	77.0	2.0
Soy	37.0	30.0	20.0
Anabaena cylindrica	43.0–56.0	25.0–30.0	4.0–7.0
C. vulgaris	51.0–58.0	12.0–17.0	14.0–22.0
D. salina	57.0	32.0	6.0
Porphyridium cruentum	28.0–39.0	40.0–57.0	9.0–14.0
S. obliquus	50.0–56.0	10.0–17.0	12.0–14.0
Spirulina máxima	60.0–71.0	13.0–16.0	6.0–7.0
Synechococcus sp.	63.0	15.0	11.0

Source: Becker, W., 'Microalgae in human and animal nutrition,' pp. 312–351. In Richmond, A. (Ed). *Handbook of Microalgal Culture*, Blackwell, Oxford, 2004. With permission.

(Table 16.1). Every year, 5 million pounds of algal biomass are produced worldwide, and have a value around US$ 330/kg (Pulz and Gross, 2004). Approximately one fifth of this biomass is used as fish food (Muller-Feuga, 2004). Scientists have also studied the possibility of using microalgae as a system of expression for recombinant protein and for vaccines for fish and humans (Manuell et al. 2007).

16.2.1 SPIRULINA

Spirulina is a prokaryotic microalga, order Cyanophyceae, division Cyanophyta (Cyanobacteria). *Spirulina* is distinguished by its arrangement of multicellular cylindrical trichomes in an open helix throughout its length. The helical shape of the trichomes is characteristic of the genus, but the length and size of the helix vary by species (Vonshak, 1997).

Spirulina is found in alkaline lakes that contain sodium carbonate or bicarbonate, as well as other minerals and sources for nitrogen fixation. Natural occurrences of microalgae have been recorded in lakes in Chad in Central Africa, Texcoco in Mexico, Nakaru and Elementeita in Kenya, and Aranguadi and Kilotes in Ethiopia. In Brazil, the occurrence of *Spirulina* has been recorded in Lake Mangueira (Morais et al. 2008).

Through photosynthesis, *Spirulina* converts the nutrients in the medium into cellular material and oxygen is released (Henrikson, 1994). Its reproduction occurs through binary fission, where a trichome undergoes elongation, accompanied by an increase in the number of cells, and then breaks into multiple filaments (Vonshak, 1997). Several components are required for cell growth, such as sources of carbon, nitrogen, phosphorus, potassium, magnesium, iron and other micronutrients.

Spirulina has been consumed for decades by the Kanembu people, who live around Lake Chad in Africa. These people harvest a hardened blue-green algae pie from the riverbank, which they dry in the sun, called dihé (Henrikson, 1994).

Both *Chlorella* and *Spirulina* are GRAS certified (Generally Recognized As Safe) and can be used as food without posing a health risk. Since June 23, 1981, *Spirulina* has been legally accepted by the FDA (Food and Drug Administration), which stated "*Spirulina* is a source of protein and contains many vitamins and minerals. It can be legally marketed as food or food supplement as long as it is precisely qualified and free from contamination and adulteration with substances" (Fox, 1996).

Cultivation of *Spirulina* requires highly selective environments, in order for them to remain free of contamination by other algae and protozoa. In a simultaneous cultivation carried out with the toxigenic cyanobacteria *Microcystis aeruginosa*, *Spirulina platensis* did not have any susceptibility to contamination (Costa et al. 2006). However, molecular markers to verify the purity of *Spirulina* have been developed in order to prevent adulterated or contaminated products reaching the consumer (Morais et al. 2008).

Spirulina has a high protein content, which ranges from 50–70 per cent in dry weight. This concentration is rare in micro-organisms, even in the best sources of vegetable protein, such as soybean meal, where this content is less than 35 per cent (Becker, 1994). Its proteins are highly digestible, as this microalga has no cell wall but a layer of soft mucopolysaccharides. The fat content is between 5 to 10 per cent (Vonshak, 1997), mainly fatty acids, and the carbohydrate content is 10 to 20 per cent.

Minerals and trace elements represent about 6 to 9 per cent of the biomass of *Spirulina*, the main ones of which are iron, magnesium, calcium, phosphorus, and potassium.

Clinical studies suggest that *Spirulina* presents certain beneficial therapeutic effects: it reduces blood cholesterol, protects against some cancers, enhances the immune system, increases the number of intestinal lactobacilli and reduces radiation, hyperlipidemia and obesity. Supplementation with 0.5 g.d^{-1} biomass of *Spirulina platensis* in the diet of rabbits resulted in decreased levels of total cholesterol and triglycerides and increased levels of HDL. Jiménez et al. (2003) reported that the aqueous extract of *Spirulina platensis* partially inactivated HIV-1 replication in human cells.

16.2.2 CHLORELLA

Chlorella is a spherical unicellular eukaryotic microalga, and its diameter varies from 5–10 μm (Illman et al. 2000). It is a microalgae from the *Chlorophyta* division, order *Chlorococcales*. This microalga contains 53 per cent protein, 23 per cent carbohydrates, 9 per cent fat, and 5 per cent minerals (Henrikson, 1994). It also contains more than 2 per cent chlorophyll, which enables it to grow rapidly because, in the same way as higher plants, its main metabolism is photosynthesis, where the main energy source is sunlight (Vonshak, 1997).

This microalga was first discovered by the Japanese (Richmond, 1990), who are traditional consumers of algae, and usually enjoy and use algae as a food supplement in salads and sushi, due to the high content of nutrients such as chlorophyll "a" and "b", proteins, vitamins, minerals and essential amino acids. *Chlorella* is also rich in B vitamins, especially B12, which is vital in the formation and regeneration of blood cells. In addition to B12, *Chlorella* contains iron, so can be used for the treatment and prevention of anemia.

According to Borowitzka (1999), the first unialgal culture was carried out in 1890 by Beijerinck with *Chlorella vulgaris*. From 1948 to 1950, research at Stanford, California, focused heavily on the large-scale cultivation of microalgae using the laboratory technique. In order for the nutrients to be fully utilized by the organism, *Chlorella* cells, which are protected by a cell wall, must be disintegrated during the drying process, so that their nutrients can be largely absorbed by metabolism (Henrikson, 1994).

According to Sung et al. (1999), *Chlorella* has an almost constant specific growth rate at pH values greater than 4.2, and therefore it grows easily in tanks and lakes. In addition, one of the great advantages of *Chlorella* compared with the other microalgae is its high tolerance to temperature and CO_2 concentration, while maintaining growth at 42°C and 40 per cent CO_2 (Sakai et al. 1995).

16.2.3 SCENEDESMUS

Scenedesmus is a green alga from the family Scenedesmaceae, order Chlorococcales, division Chlorophyta. *Scenedesmus* species are distributed in lakes and soils. The cylindrical cells are grouped in rows of 4, 8, or 16, forming a colony. The cells are uninuclear, contain chloroplast and pyrenoid and are formed by numerous long filaments that help the movement of cells in liquid medium.

Scenedesmus reproduces by producing autocolonies, which are sister colonies formed by cells from the parent colony. The content of the parent cell divides into numerous daughter cells without a flagellum, which subsequently regroup to form a new daughter colony: an autocolony. In cultivation, the formation of colonies often does not occur and the cells remain separate. *Scenedesmus* species are found in lakes or brackish water (Hoek et al. 1995).

Just like *Spirulina* and *Chlorella*, *Scenedesmus* is a microalgae that is rich in protein. It contains 53 per cent protein, 29 per cent carbohydrates, 15 per cent lipids, and 5 per cent minerals. It is used in biotechnology in biomass production cultures for the development of fish feed and as a food supplement; it is also grown in liquid residues from the orange juice concentrate industry (Herodek et al. 1989).

In 1973, Peruvian-Germain began operating a pilot plant in Trujillo, in northern Peru, to grow the microalga *Scenedesmus acutus*. The pilot plant is equipped with several small tanks in an area of 100 m², which are stirred with blades. The microalgae are collected from suspension with a centrifuge and dried in a cylindrical kiln.

16.2.4 *HAEMATOCOCCUS*

The green microalga *Haematococcus* is known for its ability to accumulate astaxanthin and ketocarotenoids above 0.2 to 2.0 per cent (dry weight). Its products have applications in pharmaceuticals, nutraceuticals and animal nutrition (Lorenz and Cysewski, 2000).

Haematococcus can be grown in autotrophic and heterotrophic conditions. Astaxanthin from the *Haematococcus* microalga is licensed for use by the Food and Drug Administration (FDA), and several European countries have approved it as a dietary supplement or food ingredient for humans. This microalga produces around 1.5 to 3.0 per cent astaxanthin (Lorenz and Cysewski, 2000). Its applications in human health include the prevention of degenerative diseases and cancer (Dufossé et al. 2005). The cultivation of this microalga in open tanks is difficult due to contamination by other species of microalgae. However, it can be grown in closed photobioreactors, such as tubular reactors (Fábregas et al. 2001).

Haematococcus grows at a temperature of 20–28°C, illuminance of 15 klux, and pH between 6.8 and 7.4. Because it has optimum growth at neutral pH, this microalga is very susceptible to contamination by bacteria, fungi and protozoa. Under excess light, cell growth is significantly affected (Lorenz and Cysewski, 2000).

16.2.5 *DUNALIELLA*

The microalga of the genus *Dunaliella* is green, halotolerant, unicellular, and from the Chlorophyceae family. It has a rigid cell wall and large chloroplasts. This microalga develops in salt concentrations of 1.5 ± 0.1 M NaCl. *Dunaliella* requires bicarbonate as a source of carbon and other nutrients such as nitrate, sulfate and phosphate. The initial phase of growth requires 12 to 14 days in medium that is rich in nitrate. Magnesium salt is essential for chlorophyll production (Dufossé et al. 2005).

This microalga accumulates massive amounts of β-carotene in response, especially at high light intensity. When it is exposed to the optimum conditions of growth, the *Dunaliella* biomass yields approximately 400 mg β-carotene per square meter of cultivated area. In order to produce carotenoids, it is essential that there are enough nutrients, salt, and excess light (usually 25–30 klux). In addition to β-carotene, this microalga is rich in protein and essential fatty acids, which can be consumed safely due to its GRAS status (Generally Recognized As Safe) (Dufossé et al. 2005).

The biomass of *Dunaliella* microalga has exhibited various biological activities, such as anti-hypertension, bronchodilator, analgesic, muscle relaxant and anti-edema properties. Natural β-carotene contains several essential nutrients that are not present in the same pigment produced synthetically (Yousry, 2002). The human body converts β-carotene into vitamin A, without forming toxins in the liver. β-carotene has antioxidant activity, and counteracts the effects of free radicals. The biomass of *Dunaliella* can be used directly in the formulation of foods and pharmaceutical formulations.

Dunaliella salina can be used as a food source for fish and crustaceans, both to feed the animals and boost their characteristic color, particularly salmon and shellfish (Cantrell et al. 2003).

16.3   ADVANTAGES OF USING MICROALGAE INSTEAD OF OTHER CULTURES

Microalgae can grow in soil that is unfit for agriculture and livestock, in lakes or in tanks located on inhospitable lands, such as deserts, which are usually unsuitable for growing any kind of food. Microalgae are highly productive and can double their biomass in just three-and-a-half days. Doubling the biomass of land-plants, genetically modified or not, takes months, while doubling the concentration of animal protein can take years (Chisti, 2007).

The production of proteins from microalgae consumes three times less water than the same amount of protein from soy, which is the plant that is richest in this nutrient. When compared to cattle, the difference is much greater: microalgae use 50 times less water to produce the same amount of protein and the space required is 200 times smaller.

Some microalgae, especially *Spirulina* and *Chlorella*, have GRAS status (Generally Recognized As Safe) and can be added to food without any risk to health. Their nutritional and therapeutic properties are used to prevent or inhibit cancers in humans and animals, reduce blood cholesterol levels, reduce the nephrotoxicity of heavy metals and drugs, and to protect against radiation.

Since 1998 the Federal University of Rio Grande (FURG), in southern Brazil, through the Laboratory of Biochemical Engineering (LEB), has developed a project to study the cultivation of Spirulina on a pilot scale on the banks of Lake Mangueira. The Spirulina is added to snack foods for children in the region. The project has developed products that are easy to prepare, store and distribute, and which are highly nutritious and accepted by consumers. These include instant noodles, pudding, cake mix powder, biscuits, chocolate milk powder, instant soup, isotonic drinks, powdered gelatin, and cereal bars. These products are prepared at the Center for Enrichment of Foods with Spirulina (CEAS), which is located at the university.

Microalgae use photosynthesis as their main metabolism for obtaining organic carbon through the inorganic carbon contained in CO_2, and oxygen is releases into the atmosphere. Unlike most conventional energy sources, such as oil and coal, solar power is virtually unlimited, does not cause the emission of waste and its use does not cause harm to the environment (Reinhardt et al. 2008).

The fixation of CO_2 by microalgae employs the photosynthetic abilities of organisms and may be one of the most efficient processes for removal of this gas, without the need for radical changes in the global energy matrix and in productive activities. It has the advantage of reducing global warming, which if not tackled, could cause disastrous consequences: a reduction in food production, species extinction, changes in the supply of fresh water and the occurrence of a greater number of cyclones, droughts, storms and floods (Wang et al. 2008).

Compared to trees, microalgae have many advantages including their greater efficiency for CO_2 fixation and the fact that their cultivation does not depend on soil quality; also the biomass produced has potential applications. The results of fixation by cultures of microalgae are immediate, while trees need months or years. The cultivation of microalgae enables gas to be injected directly into reactors, while trees capture the gas that is dispersed in the atmosphere, where the concentration is about 0.038 per cent.

As for the oceans, studies estimate that they have captured half of the CO_2 generated between 1800 and 1994, and that the gas is saturating the upper layers of the oceans, which could transform the seas from sinks into sources of CO_2. In addition, the pH of water has decreased, which has altered the cycle of $CaCO_3$ that is required to form the shells and skeletons of marine animals.

Microalgal biomass can be used in the formulation of food or feed and it helps to minimize problems such as malnutrition. By-products can also be obtained, such as pigments. Lipids are a major component of microalgae and play an important nutritional role as a source of energy and essential fatty acids. In the extraction of polyunsaturated fatty acids (PUFAs) for food additives, microalgae have advantages over fatty acids extracted from plants and animals, as they do not have the unpleasant odor of fish, reduce the risk of chemical contamination and are easy to refine (Guedes et al. 2011). Microalgae produce biopolymers that are biodegradable when exposed to biologically active environments, so they are important as a replacement for traditional plastics in food packaging (Sudesh et al. 2000).

The gas barrier properties of PHAs may lead to them having applications in food packaging, such as in the replacement of polyethylene terephthalate (PET) in beverage bottles. This property also means that PHAs can be used in long-life type paper cartons for products such as juices, sports drink, and pasteurized milk. They can also be used as packaging or secondary packaging for fruit or biscuits; however, when used with fruits they must be properly prepared to permit gas exchange (Sharma et al. 2007).

Low density polyethylene (LDPE), which is used for coating paper, does not enable degradation of the paper when it is discarded, and also hinders the recycling of packaging, since its components are difficult to separate. Thus, LDPE can be replaced by PHAs in order to reduce the problems with these residues.

PHAs can be used in humans, since they are non-toxic, compatible with living tissue and their degradation leads to R-b-hydroxybutyric acid, a metabolite which is present in mammals in concentrations from 30 to 10 mg/100 ml blood in normal adults. Low molecular weight PHB has been detected bound to albumin and also low density lipoproteins in human blood serum.

16.4   REQUIREMENTS FOR CULTIVATION OF MICROALGAE

The growth of microalgae is a complex phenomena and subject to variables. The increase of a population of microalgae responds to the mutual interaction of abiotic and biotic factors. Photosynthesis, the assimilation of a carbon source, and photobioreactors employed are decisive factors in the rate of multiplication of these micro-organisms.

Photosynthesis can be defined as a physicochemical process by which photosynthetic organisms synthesize organic compounds using solar energy, water, carbon dioxide and minerals (Poli et al. 2004). Photosynthesis can be represented by the following chemical reaction:

$$6CO_2 + 12H_2O + \text{Light energy} \rightarrow C_6H_{12}O_6 + 6O_2 + 6H_2O$$

This complex process involves many physical and chemical reactions that occur in a coordinated way in systems of proteins, pigments and other compounds. In general, the photosynthetic process is analyzed in two simultaneous and interdependent steps: the photochemical step, previously named the light phase and the chemical step, also called the reductive photosynthetic carbon cycle, previously named the dark phase.

In the photochemical phase, radiant energy excites the photosynthetic pigments, and this excited state (energy) is transferred with the aid of water (redox), to the molecules of nicotinamide adenine dinucleotide phosphate (NADP) and adenosine triphosphate (ATP) (chemical energy). The primary products of the photochemical step are ATP and metilenotetraidrofolato redutase ($NADPH_2$). At this stage, oxygen is released as a byproduct of the dissociation of the water molecule. In the chemical phase, the carbon obtained from a CO_2 molecule is assimilated by a series of enzymatic reactions using the energy stored in ATP and $NADPH_2$ molecules and, ultimately, forms the first product of photosynthesis, the carbohydrate (CH_2O) (Poli et al. 2004).

The light that provides energy for the photosynthetic assimilation of CO_2 is extremely important for microalgae. The wavelength and light intensity cause variations in the photosynthetic responses, in the growth and metabolism of microalgae (Dubinsky, 1990). Solar energy is propagated in the form of waves, forming a broad electromagnetic spectrum. Within this broad spectrum, the primary producers (microalgae) capture energy for photosynthesis from a narrow band called visible radiation, which has a wavelength between 400 and 720 nm (Lips & Avissar, 1990).

Many microalgae are exposed to H_2CO_3 and its ions HCO^{3-} and CO_3^{2-} as well as atmospheric CO_2 available for terrestrial plants. When in equilibrium with air at 15°C, water contains 10 μM of dissolved CO_2 (Vyzamal, 1995). CO_2 crosses the boundary between air and water at the water surface. This is possible because of the slight attraction between the hydrogen atoms of water and the carbon atom of CO_2 (due to the dipole nature of the water molecule and the carbon atom). Part of the dissolved CO_2 is removed by microalgae; the rate depends on the photosynthetic activity and proximity of the algal trichome (Fox, 1996).

Other CO_2 molecules react with the water to form carbonic acid (H_2CO_3). This carbonic acid is dissociated into bicarbonate (HCO^{3-}) and H^+ ions, and bicarbonate can be dissociated into CO_2 or carbonate ions (CO_3^{-2}) depending on the pH of the medium. All these reactions are very fast, and while the microalgae remove CO_2 from the medium, OH– ions are formed, which raises the pH (Poli et al. 2004; Fox, 1996).

Below pH 5.0, only free CO_2 is predominant; between pH 7.0 and 9.0, bicarbonate is more significant and above pH 9.5, carbonate prevails. Bicarbonate is used by microalgae through dehydration that occurs within the cytoplasm by the action of the enzyme carbonic anhydrase to form CO_2 and OH– ions. The CO_2 is used in photosynthesis by diffusion in the cells through membranes, and this CO_2 is exclusively used as a substrate by the enzyme Rubisco.

The rate of CO_2 fixation in photosynthesis depends on external and internal factors and interactions between them. Internal factors include the genus and species of the alga, as well as the stage of the cell in the cell division cycle. External factors include the supply of inorganic carbon, pH, intensity and wavelength of light, oxygen levels, nutrient supply and temperature. The interaction between these factors controls the amount and activity of enzymes (Vyzamal, 1995).

During the night, in the heterotrophic growth phase, the cellular concentration of photosynthetic metabolisms decreases and is restored in the subsequent phase of autotrophic growth. According to Ogbonna & Tanaka (2000), the addition of an organic carbon source in the night can be used not only to prevent the loss of biomass, but also to enable continuous cell growth in constant light/dark cycles—a technique known as mixotrophic cultivation.

Efficient transfer of CO_2 mass in the culture medium is desirable since undissolved CO_2 is lost. The O_2 mass transfer in the system is also an important consideration because of the need to remove oxygen derived from photosynthesis before it reaches inhibitory concentrations. The hydrodynamics (characteristics of the mixture) are a function of reactor geometry and operating conditions (flow rate of gas and liquids) and is considered the main determinant of the use of light by the culture. This affects the photosynthetic efficiency, productivity and cellular composition.

Once the biological kinetic reaction and the mass transfer are understood, it is essential to understand how biological reactors work. Open systems can be divided into natural waters (lakes, ponds) or artificial tanks. With respect to complexity, extensive raceway tanks are simpler than closed systems. In raceway tanks, the cultures are agitated by rotating blades and typically operate at depths between 15 and 20 cm. Circular tanks are similar in design, and are used in Asia and Ukraine.

According to Jiménez et al. (2003), the main advantages of cultivation in an open system are the low capital investment and the use of solar energy. However, this is only true for certain species in selected environments, and the yields obtained are lower than theoretically possible. Until recently, open systems were the most commonly used type for cultivation of microalgae; however, the preparation of products with high added value, such as pharmaceuticals and cosmetics, have led to the adoption of closed systems, which enable greater reproducibility (Ogbonna & Tanaka, 2000).

Column reactors have a large length in relation to their diameter. In such reactors, the mixture is supplied by the force of compressed gas in the reactor which rises through the liquid. This type of reactor has been used for many years in the chemical industry because of its various advantages: it is relatively low cost, is mechanically simple to setup, and has low operating costs because of the minimal energy requirements (Table 16.2) (Bailey & Ollis, 1986).

There are more unusual reactors used in biotechnological industries, such as the tower bioreactor, which has been used on a large scale for the brewing of beer and the manufacture of vinegar. The design of the tower is an essential element for large scale processes that are employed in the cultivation of micro-organisms for use as animal feed. In some cases, a single column, which contains internal plates and stirrers, is used in some or all stages (Bailey & Ollis, 1986).

The reactors can be used in both batch and continuous mode. In air-lift reactors or reactors with a cycle from the pressure, an external aid is used to agitate the liquid. The advantage of this type of stirring is its high-efficiency heat exchange. In addition, the circulation increases the definition of the flow and mixing properties of the reactor (Bailey & Ollis, 1986). There are many variants of the airlift-type reactor, in which air is introduced into a chimney inside the reactor through a material with small pores. This mechanism causes intense agitation in the liquid, depending on the flow of air used. These reactors are suitable for the cultivation of cells that are sensitive to shear, such as in the cultivation of animal and plant cells (Schmidell et al. 2001).

Several designs of photobioreactors (most commonly tubular and plates) have been investigated by researchers worldwide. A photobioreactor must provide light, oxygen supply with minimal losses, removal of oxygen produced photosynthetically and efficient temperature control (Weissman et al. 1988). The efficiency of light transmission per unit volume of culture is determined by the geometry and mixing conditions, which are controlled by the design and operation. The precise magnitude of the effect on productivity caused by mass transfer and hydrodynamics are highly specific to the system.

Vonshak (1997) reported that when the accumulation of oxygen in cultures of Spirulina sp. in open tanks was allowed, there was no decrease in the photosynthetic rate with up to 300 per cent saturation (20–22 mg L^{-1}). However, there was a decline

TABLE 16.2
A Comparison Between Cultures Carried out in Tubular and *Raceway* Type Photobioreactors

Condition	Raceway	Tubular
Area that is in contact with light (m^2)	500.0	600.0
Total volume (m^3)	75.0	7.0
Space that is required (m^2)	550.0	110.0
Biomass concentration (g L^{-1})	0.3–0.5	5.0–8.0
Yield (g L^{-1} day^{-1})	0.05–0.1	0.8–1.2

Source: Pulz, O., *Applied Microbiology and Biotechnology* 57: 287–293, 2001. With permission.

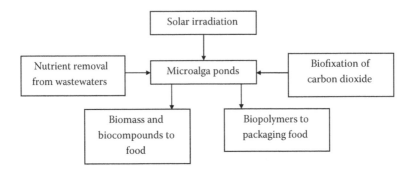

FIGURE 16.1   Flowchart of microalgal production for application in food sector.

in productivity of 15 per cent at 450 per cent of saturation (30–32 mg L^{-1}) and 25 per cent with 800 per cent saturation (50–60 mg L^{-1}).

According to Weissman et al. (1988), an increase in agitation did not improve productivity in open tanks, but resulted in low biomass density. Other researchers have investigated the effects of agitation on the productivity in photobioreactors operating at high cell densities and concluded that there is an optimal degree of agitation, which provides a sufficient cycle of cells between the light and dark zones. Excessive agitation can be unproductive due to the turbulent motion (Richmond, 1990). Figure 16.1 shows the flowchart for production of microalgae for use in the food industry.

16.5   MICROALGAL APPLICATIONS IN FOOD INDUSTRY

Microalgae are able to improve the nutritional content of foods, and have a positive effect on the health of humans and animals. This characteristic is due to their chemical composition. Their biomass can also be used to extract biopolymers that can be used in food packaging. Microalgal cultivation can be coupled with CO_2 biofixation and wastewater treatment, thus reducing environmental problems and the cost of cultivation.

16.5.1   Proteins

Proteins play a central role in biological systems. Although the information on the evolution and biological organization of cells is contained in DNA, the enzymes carry out, in an exclusive fashion, the chemical and biological processes that sustain the life of organisms. Proteins are composed of 21 different amino acids. The numerous biological functions performed by proteins would not be possible without the complexity of their composition, which produces several three-dimensional structural forms with different biological functions (Damodaran et al. 2010).

Proteins contain 50–55 per cent carbon, 6–7 per cent hydrogen, 20–23 per cent oxygen, 12–19 per cent nitrogen, and 0.2 to 3 per cent sulfur. Protein synthesis occurs in ribosomes. After synthesis, cytoplasmic enzymes modify certain constituents of amino acids. This process changes the elemental composition of some proteins.

TABLE 16.3
Content of Amino Acids (g/mg) in Eggs, Milk, Meat (Damodaran et al. 2010), Spirulina, Chlorella, and FAO/WHO Standard for Essential Amino Acids for Children from 2 to 5 Years

Amino Acid	Eggs	Milk	Meat	Spirulina	Chlorella	FAO/WHO
Histidine	22.0	27.0	34.0	27.2	21.8	19.0
Isoleucine	54.0	47.0	48.0	43.6	20.1	28.0
Leucine	86.0	95.0	81.0	80.2	47.1	66.0
Lysine	70.0	78.0	89.0	29.5	39.8	58.0
Methionine + cysteine	57.0	33.0	40.0	21.2	18.4	–
Phenylalanine + tyrosine	93.0	102.0	80.0	89.5	50.9	–
Threonine	47.0	44.0	46.0	48.7	24.6	34.0
Tryptophan	17.0	14.0	12.0	25.3	10.4	11.0
Valine	66.0	64.0	50.0	46.1	30.3	35.0

Source: Morais, M. et al., *Aquaculture* 294: 60–64, 2009. With permission.

Proteins that are not enzymatically modified in the cells are called homoproteins, and those that are modified or complexed with non-protein components are called conjugated proteins or heteroproteins.

The quality of a protein is mainly related to its essential amino acid content and digestibility. High quality proteins are those that contain all essential amino acids at levels higher than the FAO/WHO reference values, which have comparable or better digestibility than proteins from egg whites or milk (Table 16.3).

The high protein content of some species is a major reason to consider them an unconventional source of protein. The amino acid standard of most microalgae is positively comparable with the best vegetable sources of amino acids (Spolaore et al. 2006). Spirulina has protein content (60–70 per cent) with a low concentration of nucleic acids and amino acid content similar to that recommended by the FAO (Pelizer et al. 2003).

The proteins present in Spirulina have a balanced amino acid content, and high digestibility and are free of toxic compounds. The high digestibility of proteins contained in *Spirulina* (75 per cent to 92 per cent absorption in the intestine) is due to the fact that its cell wall is composed of soft mucopolysaccharides, which are easily degraded.

The proteins of the microalgae *Chlorella* and *Scenedesmus* are less digestible than *Spirulina* because they have a cell wall. Thus studies on the enzymatic hydrolysis of cells have been carried out to optimize the digestibility of microalgal protein (Morris et al. 2008).

16.5.2 FATTY ACIDS

The lipids are a large group of chemically diverse compounds that are soluble in organic solvent. The major components of lipids are fatty acids, compounds containing an aliphatic chain and a carboxylic acid group. Most naturally-occurring fatty acids have an even number of carbons in a linear chain, because due to the biological process of chain elongation two carbons are added at a time (Damodaran et al. 2010).

The edible lipid profile has altered dramatically with the evolution of agricultural practices. It is believed that our ancestors consumed diets with approximately equal amounts of ω-3 and ω-6. The development of modern agriculture has increased the availability of refined fats, mainly vegetable oils, modifying the human diet ratio of ω-6 to ω-3 to more than 7:1. This is a very fast change, in evolutionary terms, which is problematic, because humans convert ω-6 fatty acids into ω-3 at low speeds.

The levels of ω-3 fatty acids in the diet are important, because these bioactive lipids play a vital role in membrane fluidity, cell signalling, gene expression, and metabolism of eicosanoids. Therefore, the consumption of ω-3 fatty acids is essential for the promotion and maintenance of health (Damodaran et al. 2010).

Polyunsaturated fatty acids (eicosapentaenoic acid EPA, ω-3, C20:5 and docosahexaenoic acid, DHA, ω-3, C22:6) are important because of their role in human metabolism. These fatty acids have been used for the prevention and treatment of heart and inflammatory disease and also in nutritional supplements for humans and marine organisms (Sijtsma et al. 2005; Chi et al. 2007).

Several species of microalgae can be induced to produce specific fatty acids by manipulating the physical and chemical conditions of the culture, and this byproduct has numerous industrial applications. The accumulation of lipids in microalgae cells depends on factors such as the growth temperature, pH, nutrient balances of carbon, nitrogen, and phosphorus, the growth regime (autotrophic, heterotrophic or mixotrophic) the physiological condition of the inoculum and the species of microalgae selected (Wen and Chen, 2003; Chisti, 2007). The lipid content of heterotrophically cultivated *Chlorella protothecoides* microalgae is 55 per cent higher compared to that produced in an autotrophic culture (Xu et al. 2006). The production of polyunsaturated fatty acids EPA and DHA is greater in cultures of the microalgae *Tetraselmis* spp. (when there is no illumination, Chen et al. 2007).

At temperatures that are below the optimum for microalgae growth, more unsaturated fatty acids are synthesized, because reducing the temperature by 10–15°C decreases the membrane fluidity and thus there is a high activity of desaturase (acyl-CoA desaturase, acyl-ACP desaturase, and acyl lipid desaturase), which promotes the unsaturation of the lipid membrane (Garcia et al. 2011).

16.5.3 Pigments

Pigments are natural substances in the cells and tissues of plants and animals that provide color, while dyes are any chemical, synthetic or natural product that provides color. Color and appearance are fundamental attributes, if not the most important, for food quality. This is due to the human ability to easily perceive these factors, which are the first to be evaluated by consumers when buying food. One can provide consumers with food that is more nutritious, safe and more economical, however, if it is not attractive, they will not buy it (Damodaran et al. 2010).

Consumers also link specific colors to food quality and the perception of flavor. Many compounds that are responsible for the colors of fruits and vegetables have antioxidant activity. Some pigments are unstable during the processing and storage of foods and depending on the pigment, the stability is affected by factors such as presence or absence of light, oxygen, metals, oxidizing and reducing agents, temperature, water activity, and pH. Due to the instability of pigments, dyes are sometimes added to food.

Chlorophylls are the main light-absorbing pigments in green plants, microalgae, and photosynthetic bacteria. They are completely unsaturated macrocyclic structures containing four pyrrole rings linked by a single carbon bond. Several chlorophylls are found in nature. Their structures differ in terms of the substituents of the forbin core.

Chlorophylls a and b are found in green plants in a ratio of approximately 3:1. Chlorophyll a contains a methyl group, whereas chlorophyll b contains a formyl group. Chlorophyll c is found in association with chlorophyll a in marine algae, dinoflagellates and diatoms. Chlorophyll d is a minority constituent and accompanies chlorophyll a in red algae (Damodaran et al. 2010).

Chlorophylls are located in the lamellae of intercellular organelles in microalgae, known as chloroplasts. They are associated with carotenoids, lipids and lipoproteins. Covalent bonds occur between these molecules, which are easily broken. Chlorophylls can be extracted by soaking the biomass in organic solvents. Microalgae have 0.5 to 1.0 per cent (dry weight) of chlorophyll (Kadam & Prabhasankar, 2010).

Attempts to apply chlorophylls in foods, have involved the use of zinc or copper complexes. The copper complexes of phaeophytin and phaeophorbids are available commercially under the names copper chlorophyll and copper chlorophyllin, respectively.

These chlorophyll derivatives cannot be used in the USA, however their use is permitted in European countries under the control of the European Economic Community. The Food and Drug Administration (FDA) has certified their use as secure for food as long as the concentration of free ionized copper does not exceed 200 ppm (Damodaran et al. 2010).

Carotenoids are pigments that are widely distributed in nature. Their annual terrestrial production is 100 million tons, most of which are biosynthesized by algae and microalgae. Carotenoids are sensitive to light, excessive heat and exposure to acids. This sensitivity makes them highly vulnerable during processing and storage, so great care must be taken to minimize their losses (Damodaran et al. 2010).

The main role of carotenoids in the diet of humans is their ability to act as a precursor of vitamin A. A prerequisite for the activity of vitamin A is the existence of a retinoid structure (with the β-ionone ring) of carotenoids. β-carotene is the carotenoid that has the highest activity of provitamin A, because of its two β-ionone rings (Garcia et al. 2005).

β-carotene has several health benefits: it strengthens the immune system, improves activity and changes in the number of immune cells and reduces the risk of degenerative diseases such as cancer, cardiovascular disease and cataract formation (Campos et al. 2003). Carotenoids have anti-cancer properties, although in vitro and in vivo studies in humans and animals have not confirmed this characteristic (Becker, 2004; Guerin et al. 2003; Del Campo et al. 2000; Gouveia & Empis, 2003).

Unlike vitamin A, which is toxic when present in excessive amounts, β-carotene does not appear to cause toxic effects when ingested in large quantities. Excess β-carotene is stored in adipose tissues or merely remains available by circulating in the bloodstream. The maximum recommended intake of vitamin A is 10,000 international units (IU). β-carotene can be taken in daily doses that exceed 300,000 IU (180 mg) with no observable adverse effects (Henriques et al. 1998).

Algae and microalgae are important sources of carotenoids, such as astaxanthin, β-carotene, lutein, canthaxanthin, violaxanthin, lutein and neoxanthin (Bernal et al. 2011). The microalga *Dunaliella salina* produces 14 per cent of β-carotene in dry weight (Garcia et al. 2011). This microalga can be grown in open tanks and in extreme conditions, such as hypersalinity, low nitrogen availability and high levels of solar radiation. The largest producer of *Dunaliella* is Cognis Nutrition and Healthy, which has an 800 hectare plant in Australia. This company produces *Dunaliella* for the extraction of β-carotene, powder for food and animal feed; the prices of these products range from US$ 300 to 3000/kg (Spolaore et al. 2006; Ben-Amotz, 2004).

Another very important carotenoid is astaxanthin, which has a red color and is widely produced as food for salmon. Astaxanthin is considered a super vitamin E, because of its antioxidant activity. Astaxanthin has been investigated with regard to properties that protect against cancer induced by chemicals, boost the immune system, prevent the effects of ultraviolet radiation and treat atherosclerosis (Fan & Chen, 2007).

The production of synthetic astaxanthin is carried out by BASF and Hoffman-La Roche, and it has a value of US$2500/kg. Natural Astaxanthin can be synthesized by the microalga *Haematococcus pluvialis*, reaching concentrations of 1.5–3.0 per cent of dry weight. This cultivation is carried out in two stages: in the first stage, optimization is carried out for the production of microalgae biomass; the microalga is then exposed to excess light to stimulate the synthesis of astaxanthin (Hejazi & Wijffels, 2004; Todd & Cysewski, 2000).

The main commercial phycobiliproteins of interest are phycoerythrin and phycocyanin extracted from *Spirulina* and *Porphyridium* microalgae. The first application of phycocyanin is as a pigment for food. The company Dainippon Ink & Chemicals (Sakura) has produced gums, ice cream, candy and drinks with this pigment (Viskari & Colyer, 2003; Bermejo et al. 2002). This blue pigment can also be extracted from the microalga *G. sulfuraria* under limited conditions of carbon and excess nitrogen in a heterotrophic culture (Schmidt et al. 2005).

16.5.4 ANTIOXIDANTS

The production of free radicals that act on macromolecules, such as lipid membranes, proteins and DNA, can cause many health problems in humans, such as cancer, diabetes mellitus, neurodegenerative and antiinflammatory diseases that cause several alterations in tissues (Butterfield et al. 2002). The two major mechanisms responsible for damage caused by free radicals in living organisms are the overproduction of superoxides by mitochondria and nonenzymatic lipid peroxidation. Parkinson's disease, Alzheimer's and amyotrophic lateral sclerosis are associated with oxidative damage of the mitochondria.

Free radicals are molecules that contain atoms that have at least one unshared electron in their outer orbit. This electronic configuration makes them very unstable and reactive and they are highly capable of combining in a nonspecific way with the various molecules of the cell structure. The deterioration of some foods has been linked to oxidation or rancidity of lipids and formation of secondary products from lipid peroxidation (Kim & Wijesekara, 2010).

In the food and pharmaceutical industries, many synthetic antioxidants, such as butylated hydroxytoluene (BHT), butylated hydroxyanisole (BHA), tertbutylhydroquinone (TBHQ), and propyl gallate (PG), have been used to retard the peroxidation process. However, the use of synthetic antioxidants must be strictly controlled because of potential damage that indiscriminate use can cause in consumers (Park et al. 2001).

Thus, research on natural antioxidants as alternatives to synthetic antioxidants is very important for the food industry (Rajapakse et al. 2005). In a study to determine the antioxidant potential of the biomass of *Spirulina platensis* and phycocyanin pigment extracted from this microalga, it was found that the samples had antioxidant potential in the soybean oil and olive oil lipid systems that had been subjected to the oxidation process. There was a reduction of 31 per cent for *Spirulina platensis* and 81 per cent for the phycocyanin pigment (Bertolin et al. 2006).

Gad et al. (2011) evaluated the effect of soy protein concentrate (WPC), *Spirulina* and their combinations in vitro, against the action of the superoxide anion (O2-) induced by 1,1-diphenyl-2-picrylhydrazyl (DPPH). It was observed that the action against the induction of DPPH was stronger when the sample containing *Spirulina* was used, followed by the combination of Spirulina and WPC and finally by the pure WPC. This result was related to carotenoids, vitamin E, chlorophyll and phycocyanin contained in the biomass of this microalga, which play a role in reducing the activity of DPPH radicals due to their ability to donate hydrogen.

16.5.5 BIOPOLYMERS

Throughout history, mankind has developed a growing need for the use of packaging. Food packaging is used for transportation and conservation, and in many cases, the sale of a product is affected by the attractiveness of packaging.

Companies need a multidisciplinary approach to packaging and have to consult with various sectors, such as engineering, marketing, design, law, quality, production, logistics, sales and environment in order to design products that meet the needs of the entire chain. Packaging is used in 80 per cent of preprepared and preserved foods. However, it is thrown away after the product has been used and creates waste that is discarded into the environment (Bucci, 2003).

The replacement of conventional plastics by biodegradable ones in the manufacturing of packaging is one solution to reduce waste. PHB is completely and rapidly degraded by microbial attack (fungi, bacteria and enzymes). Thus, the use of PHB in the development of packaging can contribute to this new concept of eco-friendly packaging.

PHB and PHB-HV are produced by Monsanto in the United States, under the generic name Biopol. In Brazil and Latin America, the first company to produce PHB and PHB-HV is Copersucar (Cooperative of Sugar and Alcohol Producers of the State of São Paulo), which has been developed jointly with Institute of Technologic Research/Sao Paulo (IPT/SP) (Nascimento, 2001).

According to Bucci (2003), the biodegradation of pots that are developed with PHB in household waste after 60 days was 65.33 per cent; after 90 days, the samples had degraded by 100 per cent. This result is very positive and confirms the hypothesis that waste caused by the disposal of plastic packaging can be reduced by the use of biodegradable plastics.

PHB can be used in injection processes for packages using the same equipment that is currently used for polypropylene (PP), as long as the process conditions are adjusted to the characteristics of this polymer and a specific mold is used (Bucci 2003).

To reduce the cost of PHB production, companies must use highly productive strains of microalgae, develop strategies for cultivation and use low cost carbon sources. Several studies have shown that the synthesis of PHB is directly linked to the growth and nutritional conditions to which the culture is exposed. The growth phase, light–dark cycle, temperature, pH, nutrient limitation (phosphorus and nitrogen), excess nutrients (carbon), mixotrophic and chemotrophic cultures, and limitations of gas exchange (oxygen) have been studied in order to increase the amount of PHB in the cells (Sharma et al. 2007).

Nishioka et al. (2001) obtained 55 per cent of PHB in *Synechococcus* sp. MA19 under limited phosphate conditions. Sharma et al. (2007), cultivating *Nostoc muscorum* with 0.17 per cent sodium acetate, 5 mg L $^{-1}$ K2HPO 4, 0.16 per cent glucose and 95H in the dark, obtained 47.4 per cent of PHB. Supplementation of glucose, fructose, maltose and ethanol did not stimulate the accumulation of PHB in cells of *Synechocystis* sp. PCC 6803 (Panda et al. 2006).

Experiments carried out with *Synechocystis* sp. PCC 6803 in batch cultivation produced maximum yields of PHB in the stationary phase of growth (4.5 per cent w/w) compared to the lag phase (1.8 per cent) and logarithmic phase (2.9 per cent) (Panda et al. 2006). Panda & Mallick (2007) found that phosphorus deficiency caused a 90 per cent reduction in the level of dissolved oxygen, and this oxygen reduction played an important role in the accumulation of PHB. Supplementation of 0.4 per cent acetate and poor gas exchange resulted in the accumulation of 30 per cent of PHB in cells of *Synechocystis* sp. PCC 6803 after 7 d of cultivation. The authors also found that when cultures in the stationary phase are incubated with simultaneous limitation of gas exchange and phosphate in the presence of 0.4 per cent fructose and 0.4 per cent acetate, 38 per cent of PHB was obtained (Panda & Mallick 2007).

Microalgae respond to nitrogen limitation by a process known as chlorosis, which is due to proteolytic degradation of phycobiliproteins (Jau et al. 2005). Jau et al. (2005) found that after 5 to 7 d of cultivation with nitrogen limitation, cells of *Spirulina* sp. had a yellow color. The limitation of phosphorus had the same characteristics of loss of pigmentation observed in cultures with limited nitrogen, and a 1.2 per cent maximum production of PHB.

16.5.6 Treatment of Effluents

The use of microalgae production in wastewater treatment is a promising technology that associates the growth of micro-organisms with biological cleaning of waste. Thus, the microalgae receive nutrition from the organic compounds of nitrogen and phosphorus that is available in the effluents. The microalgae mitigates the effects of the effluent and provides clean water, which can be applied in aquaculture. The removal of nitrogen and carbon from water can help to reduce eutrophication in aquatic environments (Mata et al. 2010).

Microalgae of the genus *Scenedesmus*, *Spirulina*, and *Chlorella* are the most commonly used to remove nitrogen and phosphorus from effluents, followed by those of the gender *Nannochloris*, *Botryococcus* (Órpez et al. 2009). The use of microalgae not only results in the intake of nutrients from the effluent, but also provides oxygen through photosynthesis required by the bacteria responsible for degradation of nutrients via the aerobic pathway, reducing the need for agitation and aeration.

One of the challenges for the use of microalgae in the treatment of effluent is the tolerance of microalgae to toxic compounds in the effluent, such as phenols and heavy metals. To circumvent some of these problems, the use of immobilized cells has been proposed (Mallick, 2002).

Metals can be removed by adsorption on the surface of microalgal cells without being metabolized (biosorption), metabolized inside the cell (bioaccumulation), or chemically transformed. Biosorption depends on the interaction between cell surface molecules (usually proteins, carbohydrates and lipids) with metals. The removal of a metal by bioaccumulation is a specific process that is dependent upon the identification of metal by the cell as a nutrient or a toxic agent, and this information determines whether or not the removal occurs (Chojnacka et al. 2005).

Aslan & Kapdan (2006), used *Chlorella vulgaris* to remove nitrogen and phosphorus from industrial effluent, and obtained a mean removal efficiency of 72 per cent for nitrogen and 28 per cent for phosphorus. *Scenedesmus obliquus* and *Scenedesmus quadricauda* were grown in concentrations of up to 8.0 ppm of zinc. Between 0.5 and 8.0 ppm, the increase in concentration of the metal increased its adsorption but reduced the dry weight of the cells and the content of chlorophyll a and b, carotenoids, and amino acids (Omar, 2002).

The association of microalgae and bacteria has been investigated to improve the stabilization of tanks, removal of organic matter and nutrients from wastewater. The oxidation of organic matter and ammonia was studied with the microalgae *Chlorella sorokiniana* with a bacteria culture in a tubular photobioreactor (Gonzales et al. 2008). This reactor presented a 99 per cent removal of $NH4^+$ and 75% of dissolved organic carbon.

16.6   CONCLUSION

The photosynthetic production of microalgal biomass has received special attention from various companies. These micro-organisms can be grown with environmental CO_2 as a source of carbon, as well as nitrogen and phosphorus sources from industrial effluents. Microalgal biomass that is produced can synthesize biocompounds with the potential for supplementation of human and animal foods, such as proteins, essential fatty acids, pigments, and antioxidants. Biopolymers can also be extracted from biomass, with special interest in the development of food packaging. Thus, microalgal biotechnology is a sustainable alternative that is environmentally correct and has the potential for application in the food sector.

REFERENCES

Aslan, S. and Kapdan, I. (2006) Batch kinetics of nitrogen and phosphorus removal from synthetic wastewater by algae. *Ecological Engineering* 28: 64–70.

Bailey, J. and Ollis, D. (1986) Biochemical Engineering Fundamentals 2nd ed., Singapore: McGraw-Hill, pp. 180–88.

Becker, W. Microalgae in human and animal nutrition, (2004), pp. 312–351. In Richmond, A. (Ed). *Handbook of Microalgal Culture*, Blackwell, Oxford.

Becker, E. W. (1994) Microalgae Cambridge University Press, Cambridge.

Ben-Amotz, A. (2004) Industrial production of microalgal cell-mass and secondary products— major industrial species—*Dunaliella*, pp. 273–80. In Richmond, A. (Ed). *Handbook of Microalgal Culture*. Blackwell, Oxford.

Bermejo, R., Alvarez, J., Acien F. and E. Molina, (2002) Recovery of pure B-phycoerythrin from the microalga. *Porphyridium cruentum. Journal of Biotechnology* 93: 73–85.

Bernal, J., Mendiola, J., Ibanez, E. and Cifuentes, A. (2011) Advanced analysis of nutraceuticals. *Journal of Pharmaceutical and Biomedical Analysis* 55: 758–74.

Bertolin, T., Furlong, E. and Costa, J. (2006). Radicais livres e o processo de envelhecimento. pp. 79–95. UPF Editora, Passo Fundo.

Borowitzka, M. (1999) Commercial production of microalgae: Ponds, tanks, tubes and fermenters. *Journal of Biotechnology* 70: 313–21.

Bucci, D. (2003) Avaliação de embalagens de PHB (poli ácido 3-hidroxibutírico) para alimentos Florianópolis, 166f. Dissertação (Mestrado em Engenharia de Produção), Universidade de Santa Catarina.

Butterfield, D., Castenga, A., Pocernich, C., Drake, J., Scapagnini, G. and Calabrese, V. (2002) Nutritional approaches to combat oxidative stress in Alzheimer's disease. *Journal of Nutritional Biochemistry* 13: 444–61.

Campos, F., Pinheiro-Sant'ana, H., Stringheta, P. and Chaves, J. (2003) Teores de Beta-Caroteno em Vegetais Folhosos Preparados em Restaurantes Comerciais de Viçosa-MG. *Brazilian Journal of Food Technology* 6: 163–69.

Cantrell, A., McGarvey, D., Trustcott, G., Rancan, F. and Bphm, F. (2003) Singlet oxygen quenching by dietary carotenoids in a model membrane environment. *Archives of Biochemistry and Biophysics* 412: 47–54.

Cavani, C., Petracci, M., Trocino, A. and Xiccato, G. (2009) Advances in research on poultry and rabbit meat quality. *Italian Journal Animals Science* 8: 741–50.

Chen, G., Jiang, Y. and Chen, F. (2007) Fatty acid and lipid class composition of the eicosapentaenoic acid producing microalga. *Nitzschia laevis Food Chemistry* 104: 1580–85.

Chi, Z., Pyle, D., Wen, Z., Frear, C. and Chen, S. (2007) A laboratory study of producing docosahexaenoic acid from biodiesel waste glycerol by microalgal fermentation. *Process Biochemistry* 42: 1537–45.

Chisti, Y. (2007) Biodiesel from microalgae. *Biotechnology Advances* 25: 294–06.

Chojnacka, K., Chojnacki, A. and Górecka, H. (2005) Biosorption of Cr^{3+}, Cd^{2+} and Cu^{2+} ions by blue-green algae *Spirulina* sp.: Kinetics, equilibrium and the mechanism of the process. *Chemosphere* 59: 75–84.

Costa, J., Morais, M., Dalcanton, F., Reichert, C. and Durante, A. (2006) Simultaneous cultivation of *Spirulina platensis* and the toxigenic cyanobacteria. *Microcystis aeruginosa Zeitschrift für Naturforschung* 61c: 105–10.

Damodaran, S., Parkin, K. and Fennema, O. (2010) Química de Alimentos 4th ed., Porto Alegre. ISBN 978-85-363-2248-3.

Del Campo, J., Moreno, J., Rodriguez, H., Vargas, M., Rivas, L. and Guerrero, M. (2000) Carotenoid content of chlorophycean microalgae: Factors determining lutein accumulation in *Muriellopsis* sp (Chlorophyta). *Journal of Biotechnology* 76: 51–9.

Dubinsky, Z. Productivity of algae under natural conditions (1990) In: Richmond, A. (Ed). CRC Handbook of microalgas mass cultura. CRC Press, Boca Raton pp. 101–115.

Dufossé, L., Galaup, P., Yaronm, A., Arad, S., Blanc, P., Murthy, M. and Ravishankar, G. (2005) Microorganisms and microalgae as sources of pigments for food use: A scientific oddity or an industrial reality? *Trends in Food Science and Technology* 16: 389–06.

Fábregas, J., Otero, A., Maseda, A. and Dominguez, A. (2001) Two-stage cultures for the production of astaxanthin from *Haematococcus pluvialis*. *Journal of Biotechnology*, 89: 65–71.

Fan, K. and Chen, F. (2007) Chapter 11: Production of high-value products by marine microalgae Thraustochytrids. *Bioprocessing for Value-Added Products from Renewable Resources* pp. 293–322.

Fox, R. (1996) *Spirulina* production and potential, France: Edisud, ISBN 2-84744-883-X.

Gad, A., Khadrawy, Y., Nekeety, A., Mohamed, S., Hassan, N. and Wahhab, M. (2011) Antioxidant activity and hepatoprotective effects of whey protein and *Spirulina*. *Nutrition* 27: 582–89.

Garcia, O., Escalanta, F., Bashan, L. and Bashan, Y. (2011) Heterotrophic cultures of microalgae: Metabolism and potential products. *Water Research* 45: 11–36.

Garcia, M., Moreno, J., Manzano, J., Florencio, F. and Guerrero, M. (2005) Production of *Dunaliella salina* biomass rich in 9-cis-β-carotene and lutein in a closed tubular phtobioreactor. *Journal of Biotechnology* 115: 81–0.

Gonzales, C., Marciniak, J., Villaverde, S., Leon, C., Garcia, P. and Muñoz, R. (2008) Efficient nutrient removal from swine manure in a tubular biofilm photo-bioreactor using algae-bacteria consortia. *Water Science Technology* 58: 95–02.

Gouveia, L. and Empis, J. (2003) Relative stabilities of microalgal carotenoids in microalgal extracts, biomass and fish feed: effect of storage conditions. *Innovation Food Science Emergent Technology* 4: 227–33.

Guedes, A., Amaro, H., Barbosa, C., Pereira, R. and Malcata, F. (2011) Fatty acids composition of several wild microalgae and cyanobacteria, with a focus on eicosapentaenoic, docosahexaenoic and α-linolenic acids for eventual dietary uses. *Food Research International* 44: 2721–29.

Guerin, M., Huntley, M. and Olaizola, M. (2003) *Haematococcus* astaxanthin applications for human health and nutrition. *Trends of Biotechnology* 21: 210–16.

Hejazi, M. and Wijffels, R. (2004) Milking of microalgae. *Trends of Biotechnology* 22: 189–94.

Henrikson, R. (1994) Microalga *Spirulina*: Superalimento del futuro, Barcelona: Ediciones S.A. Urano, ISBN: 84-7953-047-2.

Henriques, N., Navalho, J., Varela, J. and Cancela, M. (1998) *Dunaliella*: uma fonte natural de beta-caroteno com potencialidades de aproveitamento biotecnológico, pp. 12–18, Boletim de Biotecnologia nº 61.

Herodek, S., Tátrai, I., Oláh, J. and Vörös, L. (1989) Feeding experiments with silver carp *Hypophthalmichthys molitrix*. *Aquaculture* 83: 31–3.

Hoek, C., Mann, D. and Jahns, H. (1995) Algae: An introduction to phycology. *Australia*. *Cambridge University Press*. ISBN 0521304169.

Illman, A., Scragg, A. and Shales, S. (2000) Increase in *Chlorella* strains calorific values when in low nitrogen medium. *Enzyme Microbial Technology*. 27: 631–35.

Jau, M., Yew, S., Toh, P., Chong, A., Chu, W., Phang, S., Najimudin, N. and Sudesh, K. (2005) Biosynthesis and mobilization of poly(3-hydroxybutyrate) [P(3HB)] by *Spirulina platensis*. *International Journal of Biological Macromolecular* 36: 144–51.

Jiménez, C., Cossio, B., Labella, D. and Niell, F. (2003) The feasibility of industrial production of *Spirulina* (*Arthrospira*) in Southern Spain. *Aquaculture* 217: 179–90.

Kadam, S. and Prabhasankar, P. (2010) Marine foods as functional ingredients in bakery and pasta products. *Food Research International* 43: 1975–80.

Kim, S. and Wijesekara, I. (2010) Development and biological activities of marine-derived bioactive peptides: A review. *Journal of Functional Foods* 2: 1–9.

Leary, D., Vierros, M., Hamon, G., Arico, S. and Monagle, C. (2009) Marine genetic resources: A review of scientific and commercial interest. *Marine Policy* 33: 183–94.

Liang, S., Xueming, L., Chen, F. and Chen, Z. (2004) Current microalgal health food R&D activities in China. *Hydrobiologia* 512: 45–8.

Lips, S. and Avissar, Y. (1990) Photosynthesis and ultrastructure microalgae. In: A. Richmond, (Ed). CRC Handbook of microalgal mass culture. CRC Press. Boca Raton, pp.49–67.

Lorenz, R. and Cysewski, G. (2000) Commercial potential for *Haematococcus* microalgae as a natural source of astaxanthin. *Reviews* 18: 160–67.

Mallick, N. (2002) Biotechnological potential of immobilized algae for wastewater N, P and metal removal: A review. *BioMetals* 15: 377–90.

Manuell, A., Beligni, M., Elder, J., Siefker, D., Tran, M., Weber, A., McDonald, T. and Mayfield, S. (2007) Robust expression of a bioactive mammalian protein in *Chlamydomonas* chloroplast. *Plant Biotechnology Journal* 5: 402–12.

Mata, T., Martins, A. and Caetano, N. (2010) Renewable and sustainable energy. *Reviews* 14: 217–32.

Morais, M., Radmann, E., Andrade, M., Teixeira, G., Brusch, L. and Costa, J. (2009) Pilot scale semicontinuous production of Spirulina biomass in southern Brazil. *Aquaculture* 294: 60–4.

Morais, M., Reichert, C., Dalcanton, F., Durante, A., Marins, L. and Costa, J. (2008) Isolation and characterization of a new Arthrospira strain. *Zeitschrift für Naturforschung* 63c: 144–50.

Morris, H., Almarales, A., Carrillo, O. and Bermudez, R. (2008) Utilisation of *Chlorella vulgaris* cell biomass for the production of enzymatic protein hydrolysates. *Bioresource Technology* 99: 7723–29.

Muñoz, R. and Guieysse, B. (2006) Algal bacterial processes for the treatment of hazardous contaminants: a review. *Water Resource* 40: 2799–15.

Muller-Feuga, A. (2004) Microalgae for aquaculture: the current global situation and future trends. In Handbook of microalgal culture. Edited by – Richmond A. Blackwell Science 352–64.

Nascimento, J. (2001) Estudo da processabilidade e da caracterização do poli(ácido 3-hidroxibutírico)-PHB obtido a partir da cana-de-açúcar, Campinas, 58f. Dissertação (Mestrado em Engenharia Química), Universidade Estadual de Campinas.

Nishioka, N., Nakai, K., Miyake, M., Asada, Y. and Taya, Y. (2001) Production of poly-β-hydroxybutyrate by thermophilic cyanobacterium *Synechoccocus* sp. MA19 under phosphate-limited conditions. *Biotechnology Letters* 23: 1095–99.

Ogbonna, J. and Tanaka, H. (2000) Light requirement and photosynthetic cell cultivation: development of processes for efficient light utilization in photobioreactors. *Journal Applied Phycology* 12: 201–07.

Omar, O. (2002) Bioremoval of zinc ions by *Scenedesmus obliquus* and *Scenedesmus quadricauda* and its elect on growth and metabolism. *International Biodeterioration and Biodegradation* 50: 95–00.

Órpez, R., Martínez, M., Hodaifa, G., Yousfi, F., Jbari, N. and Sánchez, S. (2009) Growth of the microalga *Botryococcus braunii* in secondarily treated sewage. *Desalination* 246: 625–30.

Panda, B. and Mallick, N. (2007) Enhanced poly-β-hydroxybutyrate accumulation in a unicellular cyanobacterium, *Synechocystis* sp. PCC 6803. *Letter in Applied Microbiology* 44: 194–98.

Panda, B., Jain, P., Sharma, L. and Mallick, N. (2006) Optimization of cultural and nutritional conditions for accumulation of poly-β-hydroxybutyrate in *Synechocystis* sp. PCC 6803. *Bioresource Technology* 97: 1296–01.

Park, P., Jung, W., Nam, K., Shahidi, F. and Kim, S. (2001) Purification and characterization of antioxidative peptides from protein hydrolysate of lecithin-free egg yolk. *Journal of American Oil Chemists Society* 78: 651–56.

Peiretti, P. and Meineri, G. (2011) Effects of diets with increasing levels of *Spirulina platensis* on the carcass characteristics, meat quality and fatty acid composition of growing rabbits. *Livestock Science* 140: 218–24.

Pelizer, P., Danesi, E., Rangel, C., Sassano, C., Carvalho, J., Sato, S. and Moraes, I. (2003) Influence of inoculum age and concentration in *Spirulina platensis* cultivation. *Journal of Food Engineering* 56: 371–75.

Petracci, M., Bianchi, M. and Cavani, C. (2009) Development of rabbit meat products fortified with n-3 polyunsaturated fatty acids. *Nutrients* 1: 111–18.

Poli, C., Poli, A., Andratta, E. and Beltrame, E. (2004) Aqüicultura Experiências Brasileira. Florianópolis, SC. Multitarefa Editora Ltda, CDD: 63930981.

Pulz, O. and Gross, W. (2004) Valuable products from biotechnology of microalgae. *Applied Microbiology and Biotechnology* 65: 635–48.

Pulz, O. (2001) Photobioreactors: Production systems for phototrophic microorganisms. *Applied Microbiology and Biotechnology* 57: 287–93.

Rajapakse, N., Mendis, E., Byun, H. and Kim, S. (2005) Purification and in vitro antioxidative effects of giant squid muscle peptides on free radical mediated oxidative systems. *Journal of Nutritional Biochemistry* 16: S62–9.

Reinhardt, G., Rettenmaier, N. and Koppen, S. (2008) How sustainable are biofuels for transportation? In *Bioenergy: Challenges and opportunities*. International Conference and Exhibition on Bioenergy.

Richmond, A. (1990) Handbook of microalgal mass culture, Boston: CRC Press, ISBN 0-8493-3240-0.

Rosenberg, J., Oyler, G., Wilkinson, J. and Betenbaugh, M. (2008) A green light for engineered algae: Redirecting metabolism to fuel a biotechnology revolution. *Current Opinion in Biotechnology* 19: 430–436.

Sakai, N., Sakamoto, Y., Kishimoto, N., Chihara, M. and Karube, I. (1995) *Chlorella* strains from hot springs tolerant to high temperature and high CO_2. *Energy Conversion and Management* 16: 693–696.

Schmidell, W., Lima, A., Aquarone, E. and Borzani, W. (2001) *Biotecnologia Industrial*, v. 2., São Paulo: Edgard Blücher LTDA.

Schmidt, R., Wiebe, M., Eriksen, N. (2005) Heterotrophic high cell density fed-batch cultures of the phycocyanin-producing red alga *Galdieria sulphuraria*. *Biotechnology and Bioengineering* 90: 77–4.

Sharma, L., Singh, A., Panda, B. and Mallick, N. (2007) Process optimization for poly-β-hydroxybutyrate production in a nitrogen fixing cyanobacterium *Nostoc muscorum* using response surface methodology. *Bioresource Technology* 98: 987–93.

Sijtsma, L., Anderson, A. and Ratledge, C. (2005) Alternative carbon sources for heterotrophic production of docosahexaenoic acid by the marine alga *Crypthecodinium cohnii*. In: Cohen, Z., Ratledge, C. (Eds). Single cell oils. AOCS Press, Champaign, Il, USA, 107–23.

Soletto, D., Binaghi, L., Lodi, A., Carvalho, J. and Converti, A. (2005) Batch and fed-batch cultivations of *Spirulina platensis* using ammonium sulphate and urea as nitrogen sources. *Aquaculture* 243: 217–24.

Spolaore, P., Joanni, C., Duran, E. and Isambert, A. (2006) Commercial application of microalgae. *Journal of Bioscience and Bioengineering* 101: 87–96.

Sudesh, K., Abe, H. and Doi, Y. (2000) Synthesis, structure and properties of polyhydroxyalkanoates: Biological polyesters. *Program Polymer Science* 25: 1503–55.

Sung, K., Lee, J., Park, S. and Choi, M. (1999) CO_2 fixation by *Chlorella* sp KR-1 and its cultural characteristics. *Bioresource Technology* 68: 269–73.

Todd, L. and Cysewski, G. (2000) Commercial potential for *Haematococcus* microalgae as a natural source of astaxanthin. *Trends Biotechnology* 18: 160–7.

Viskari, P. and Colyer, C. (2003) Rapid extraction of phycobiliproteins from culture cyanobacteria samples. *Analysis Biochemistry* 319: 263–71.

Vonshak, A. (1997) *Spirulina platensis (Arthrospira)* physiology, cell-biology and biotechnology, London: Taylor & Francis, ISBN 0-7484-0674-3.

Vyzamal, J. (1995) Algae and element cycling in wetlands. Boca Raton – Florida ISBN 0-87371-899-2.

Wang, B., Li, Y., Wu, N. and Lan, C. (2008) CO_2 bio-mitigation using microalgae. *Applied Microbiology and Biotechnology* 79: 707–18.

Weissman, J., Goebel, R. and Benemann, J. (1988) Photobioreactor design: Mixing, carbon utilization, and oxygen accumulation. *Biotechnology and Bioengineering* 31: 336–44.

Wen, Z. and Chen, F. (2003) Heterotrophic production of eicosapentaenoic acid by microalgae. *Biotechnology Advances* 21: 273–94.

Xu, H., Miao, X. and Wu, Q. (2006) High quality biodiesel production from a microalga *Chlorella protothecoides* by heterotrophic growth in fermenters. *Journal of Biotechnology* 126: 499–07.

Yousry, N. (2002) Color your customer health with carotenoids. Available at http://www.hnherbs.com/carotenoids.pdf.

17 Biorefinery Concept Applied to Valorization of Agro-Food Coproducts and Wastes

Integrated Process for Waste Recycling and Effluent Treatment

Carlos Ricardo Soccol, Susan Grace Karp,
Paula Fernandes de Siqueira,
Cássia Tiemi Nemoto Sanada,
Vanete Thomaz-Soccol, and Ashok Pandey

CONTENTS

17.1 INTRODUCTION

The utilization of agro-residues to produce commercial biomolecules is considered an important element in developing techno-economically feasible technologies that would contribute to the development of sustainable processes and the generation of value-added products. Brazilian and Indian economies are among the most important agriculture-based economies in the world, and the residues generated by the intense agricultural activity represent potential feedstock that could be utilized in diverse production chains instead of being discarded (Singhania et al. 2009; Soccol and Vandenberghe 2003).

Because many predictions indicate that availability of the nonrenewable carbon derived from petroleum, coal, and natural gas will diminish in the next decades, extensive efforts are being directed toward developing effective exploitation of plant biomass resources. The agricultural biomass, rich in renewable green carbon, can be biotransformed in an economical and sustainable way to produce biofuels, food, feed, and other bioproducts for future needs.

In this chapter, some results are presented from an integrated project in which the biorefinery concept was applied for the integral exploitation of residues of soybean industry, for the production of bioethanol, xanthan gum, alpha-galactosidase, and lactic acid on laboratory and pilot scales.

17.2 THE CONCEPT OF A BIOREFINERY

According to IEA Bioenergy Task 42, biorefining is the sustainable processing of biomass into a spectrum of marketable products and energy (Cherubini 2010). The term "biorefinery" is associated with the coproduction of transportation biofuels, bioenergy, and marketable chemicals from renewable biomass sources. The concept is analogous to the petroleum refinery that produces multiple products from petroleum; however, in the case of a biorefinery, all products are bio-based and derived from renewable resources (Cherubini and Ulgiati 2010). According to Cherubini (2010), renewable, carbon-based, raw materials for the biorefinery are provided from five different sectors: agriculture (dedicated crops and residues), forestry, industries (process residues and leftovers), households (municipal solid waste and wastewaters), and aquaculture (algae and seaweeds).

Conceptually, a biorefinery would apply hybrid technologies from different fields, including polymer chemistry, bioengineering, and agriculture (FitzPatrick et al. 2010), with the aim of separating the biomass resources (wood, grasses, corn) into their building blocks (carbohydrates, proteins, triglycerides) to be converted to value-added products, biofuels, and chemicals (Cherubini 2010). This allows for the development of systems in which the term "waste" is not applicable as each production stream has the potential to be converted into a by-product stream rather than waste streams (FitzPatrick et al. 2010).

The biological production of chemicals is not a new technology, although it is one that is central to the sustained development of biorefining technologies because of the high value and relatively low material demands of this industry (FitzPatrick et al. 2010). Concerning the state of the art in the field of bio-based products, they are currently obtained from basic biomass components, such as starch, oil, and cellulose (Cherubini and Ulgiati 2010). Since the first half of the twentieth century, several commodity products are produced by fermentation, including acetic acid, citric acid, lactic acid, and itaconic acid. Between 1945 and 1950, one-tenth of the acetone and two-thirds of the n-butanol in the United States were produced through the fermentation of molasses and starch, respectively (FitzPatrick et al. 2010). However, most of these biofuels and biochemicals are produced in single production chains and not within a biorefinery framework, and they usually require materials that compete with the food and feed industry (Cherubini and Ulgiati 2010). An integrated biorefinery process has the advantage of enhancing the economic profitability of manufacturing and the supplying of environmentally friendly products. The high profit variability of a biorefinery resulting from the price fluctuations in the market could be reduced by diversifying the raw materials and products (Yun et al. 2009). A biorefinery could make a higher profit from the flexible production of, for example, ethanol, which is a relatively cheap product, and biochemicals with a higher market value, such as lactic acid and enzymes.

Yun et al. (2009) described a biorefinery process for the manufacture of multiple products from corn or wheat. The process begins with the liquefaction and saccharification of the starchy raw materials to generate a glucose solution to feed the fermenters. Then the material is divided into two streams, one to produce ethanol and the other to produce biochemicals, such as citric, lactic, and itaconic acids. All products are recovered in a beer column and a rectifier. The broths are filtered successively in a filtration unit, an ion exchanger, and a decolorization unit. The remaining liquid is condensed before the products are crystallized and dried.

Alriols et al. (2010) presented a process for the treatment of nonwoody lignocellulosic biomass (*Miscanthus sinensis*) to obtain cellulose, different molecular weight lignin fractions, and hemicelluloses. Organosolv technology enabled the fractionation of the raw material into different products (cellulose, hemicellulose-derived sugars, and lignin) allowing the subsequent recovery of the solvents by distillation with high yields and low energy consumption. Membrane ultrafiltration was used to obtain specific molecular weight lignin fractions as it proportioned excellent fractionation capability with low chemical consumption and low energy requirements. A cost estimation of the obtained lignin fractions by the ultrafiltration process resulted in €52/ ton of lignin, which was found to be higher than the reported costs of kraft lignin (€33/ton); however, the obtained product presented higher quality related to its high purity and specific characteristics (low polydispersity, determined functional groups).

Oleskowicz-Popiel et al. (2012) investigated the integration of a biorefinery with an organic farm where ethanol was produced from germinated rye grains and whey by utilizing inherent amylase activity from germination of the seed, and the effluent was separated into two streams: the protein-rich solid fraction, to be used as animal feed, and the liquid fraction, which can be codigested with clover grass silage to produce biogas. The effluent from the anaerobic digestion was assumed to serve as a natural fertilizer. According to the techno-economic analysis, the total capital investment was estimated to be approximately US$4 million. Setting a methane selling price according to available incentives for green electricity (US$0.72/m^3) led to a minimum ethanol selling price of US$1.89/L.

17.3 ASPECTS OF SOYBEAN PRODUCTION AND INDUSTRIALIZATION IN BRAZIL

Brazil has a territorial area of 851 million hectares, of which 54% is occupied by preservation areas (including the Amazon forest); 26% is utilized for cattle production; 9% for agricultural production; and 11% represents unoccupied areas that could be used for agricultural production in the future. The country's climate is tropical to subtropical, and one of the most important freshwater reserves on the planet, representing 20% of the world's freshwater, belongs to the Brazilian territory.

All these characteristics make Brazil an important producer and exporter of agricultural products. Currently, the country is the first ranked worldwide as a producer and exporter of coffee, sugarcane, tropical fruits, beans, and meat; the second ranked as a producer of ethanol, soybeans, and cassava; and the third ranked as a producer of corn. The global production of oilseeds, for the harvest of 2011 and 2012, is estimated at 455 million tons, (USDA 2011), and the soybean production is estimated at 83.8 million tons, and 75 million tons for the United States and Brazil, respectively.

The Brazilian production of soybeans was about 68 million tons in 2010, which represented almost 30% of the global production. The major Brazilian producers are the states of Mato Grosso and Paraná, contributing 29% and 20% of the national production, respectively. Soybean cultivation in Brazil corresponds to a harvested area of 23.5 million hectares (IBGE 2011). The major part of the soybean produced and exported by Brazil is in the raw, nonindustrialized form. The industrialization of soybeans is extremely important for the Brazilian economy because it would generate thousands of new jobs as well.

Soybeans are an important vegetable source of proteins, carbohydrates, and lipids with high biological value and low cost (Table 17.1).

Figure 17.1 shows the typical products obtained from industrial processing of soybeans in Brazil, which are soybean meal, oil, and lecithin. The protein concentrate is the soybean meal with approximately 70% protein (dry basis), obtained by the extraction of sugars from defatted soy bran using ethanol as a solvent. An important

TABLE 17.1

Physicochemical Composition of Soybeans

Component	Percentage on Dry Basis
Proteins	42.70
Lipids	21.35
Carbohydrates	25.84
Fibers	4.494
Ashes	5.618

Source: EMBRAPA—Empresa Brasileira de Pesquisa e Agropecuária. Online. Available at http://www.cnpso.embrapa.br/html/compsoja .htm (accessed October 5, 2007).

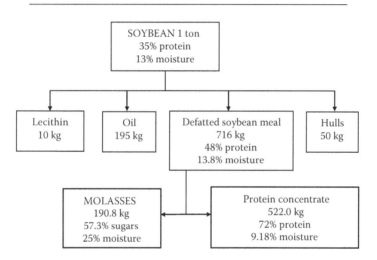

FIGURE 17.1   Industrial products obtained from soybeans.

Brazilian soybean-processing company in the region of Araucária produces the protein concentrate (600 tons/day), generating 220 tons/day of soybean molasses.

Sugars extracted from the defatted soybean meal are the main components of soybean molasses, totaling 57.3% of its dry mass (Table 17.2). The most representative carbohydrates include sucrose (28.4%), a fermentable sugar for many microorganisms, and stachyose (18.6%) and raffinose (9.68%), which are complex sugars that are not easily fermented. The molasses also contains significant amounts of proteins (9.44%) and lipids (21.2%).

Stachyose [gal(α1→6)gal(α1→6)glc(α1↔2β)fru], raffinose [gal(α1→6)glc(α1↔2β)fru] and sucrose [glc(α1↔2β)fru] are the most important carbohydrates present in soybean molasses. The β-1,2 bond is cleaved by the enzyme invertase (β-D-fructofuranoside fructohydrolase, E.C. 3.2.1.26), widely distributed in different microorganisms. The other bond is of the α-1,6 type, cleaved by the enzyme α-galactosidase (α-D-galactoside galactohydrolase, E.C. 3.2.1.22).

TABLE 17.2
Composition of Soybean Molasses

Component	Percentage on Dry Basis
Stachyose	18.6
Raffinose	9.68
Sucrose	28.4
Lactose	–
Glucose	0.243
Fructose	0.127
Galactose	0.254
Total Carbohydrates	57.3
Proteins	9.44
Lipids	21.2
Fibers	5.7
Ash	6.36

Source: Siqueira, P.F. (2006) 'Production of bio-ethanol from soybean molasses by *Saccharomyces cerevisiae*,' *Master's dissertation*, Federal University of Paraná/Universities of Provence and of the Mediterranean, Curitiba, Brazil.

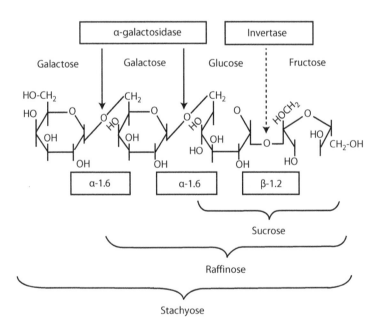

FIGURE 17.2   Structural formulas of stachyose, raffinose, and sucrose. (From LeBlanc, J.G. et al., *Genetics and Molecular Research*, 3: 432–440, 2004. With permission.)

17.4   PRODUCTION OF BIOETHANOL FROM SOYBEAN MOLASSES

17.4.1   INTRODUCTION

Bioethanol is one of the most important renewable fuels contributing to the reduction of negative environmental impacts generated by the worldwide utilization of fossil fuels (Cardona and Sánchez 2007). According to the consulting firm F.O. Licht, the world currently produces approximately 89 billion liters of ethanol per year. Brazil produces 27 billion liters of bioethanol per year from sugarcane, and the United States produces 51 billion liters of bioethanol from maize per year. The Brazilian National Bio-Fuel Program, initiated in 1975, stimulated the substitution of gasoline by sugarcane alcohol for automobile use and intensified the use of a mixture of ethanol (24%) and gasoline (76%) as fuel for common cars (Soccol et al. 2005). Today, all gasoline sold in Brazil contains at least 25% ethanol. Approximately 60% of all motor vehicles produced in Brazil are "flex," that is, they can run on any mixture of alcohol/gasoline as well as on 100% alcohol (Grad 2006).

A process for bioethanol production from soybean molasses was developed at the bioprocess engineering and biotechnology department of the Federal University of Paraná (DEBB/UFPR) in partnership with a Brazilian soybean-processing company on laboratory, pilot, and industrial scales (Siqueira et al. 2008).

17.4.2   MICROORGANISM SELECTION

Yeasts, particularly *Saccharomyces cerevisiae*, are the most commonly and traditionally used microorganisms for ethanol fermentation. Table 17.3 presents the results of a screening test to select the most appropriate strain for bioethanol production from

TABLE 17.3

Screening of *S. cerevisiae* Strains for Bioethanol Production from Soybean Molasses

Strain	Total Sugars (g/L)	Ethanol (g/L)	Yield[a] over Initial Sugars (%)
N. f. m.[b]	222.1	–	–
LPB 1	135.2	38.9	34.3
LPB 2	125.0	35.9	31.6
LPB 3	131.5	37.0	32.6
LPB 4	135.6	42.9	37.8
LPB 5	134.9	37.5	33.0
LPB 6	118.3	33.2	29.3
LPB-SC	**126.9**	**47.1**	**41.5**
LPB-MA	145.0	30.6	27.0
LPB-JP	121.7	30.5	26.9

Note: Yields for different strains after 24 h of fermentation. [a]Percentage of the maximum theoretical yield: 51.1% (w/w) of sugar content; yield's average standard deviation was 1.41. [b]N. f. m.—nonfermented medium.

soybean molasses. Ten *S. cerevisiae* strains, available from the culture collection of DEBB/UFPR, were assayed.

The strain LPB-SC presented the highest ethanol yield (41.5% of the theoretical yield from the total initial sugars). This strain was chosen for the subsequent fermentation tests with soybean molasses.

17.4.3 BIOETHANOL PRODUCTION KINETICS UNDER OPTIMIZED CONDITIONS

Fermentation tests were conducted on a laboratory scale to define the optimal conditions for ethanol production by the strain LPB-SC. The soybean molasses provided the necessary nutrients for growth, and there was no need to supplement the fermentation medium with inorganic salts ($MgSO_4$, NH_4NO_3) or yeast extract. The molasses was fermented with an initial concentration of 30% of soluble solids (or 30°Brix) to provide a minimum ethanol concentration that would be economical for the distillation process. For an initial biomass concentration of 3×10^8 viable cells/mL, fermentation finished in 6 h. Table 17.4 shows the kinetics, under optimized conditions, of the batch fermentation of soybean molasses to produce bioethanol, conducted in a bench-scale bioreactor (8 L of total capacity).

Maximum ethanol productivity (10.8 g/L h) occurred after 3 h of fermentation, and the average productivity was 8.08 g/L h. Evolution of biomass concentration was negligible because there was no oxygen supply.

The yield of bioethanol produced was 45.4% from consumed substrate, which represented 88.8% of the theoretical maximum (51.1% because 48.9% is converted into CO_2). Considering only the initial substrate concentration (182 g/L), ethanol yield was 45.8% of the theoretical maximum. This was equivalent to 134.1 kg or 169.8 L of absolute ethanol per ton of dry molasses because 1 ton of molasses contained 573 kg of sugar on a dry basis (Table 17.2).

TABLE 17.4
Kinetic Parameters of Bioethanol Production from Soybean Molasses

Time (h)	S (g/L)	X (g/L)	P (g/L)	r_S (dS/dt)	r_X (dX/dt)	r_P (dP/dt)	$Y_{X/S}$,% (r_x/r_s)	$Y_{P/S}$,% (r_p/r_s)
0	182.0	6.00	0.400	–	–	–	–	–
1	176.1	6.05	3.20	5.9	0.05	2.8	0.847	47.5
2	163.6	6.18	8.80	12.5	0.13	5.6	1.04	44.8
3	140.5	6.31	19.6	23.1	0.13	10.8	0.563	46.7
4	123.9	6.49	27.0	16.6	0.18	7.4	1.08	44.6
5	102.1	6.59	36.4	21.8	0.10	9.4	0.459	43.1
6	87.6	6.72	43.0	14.5	0.13	6.6	0.897	45.5
	Average			15.7	0.12	8.08	0.815	45.4

Note: Average standard deviations were 4.77% for sugar (S), 2.10% for biomass (X), and 5.24% for ethanol (P) concentrations.

The high residual substrate concentration (87.6 g/L) indicated that nearly 50% of the total sugars were not fermentable by the strain LPB-SC. It is known that *S. cerevisiae* produces intracellular and extracellular invertase. The extracellular invertase resides in the periplasmic space and is responsible for cleaving sucrose into glucose and fructose, monomers that are assimilated and converted into ethanol (Zech and Görisch 1995). In this way, based on the composition analysis of soybean molasses (Table 17.2), it was concluded that the strain LPB-SC did not produce the enzyme α-galactosidase necessary for the assimilation of molasses' complex sugars.

17.4.4 BIOETHANOL PRODUCTION ON A PILOT SCALE (1 M³ MOLASSES PER DAY)

The main objective of scaling up a process is to identify the problems that might not have been significant at laboratory scale and also to check if the fermentation yield can be maintained. The pilot plant was designed according to the parameters defined on a laboratory scale with a capacity to produce 1 m³ of fermented broth per day (Figure 17.3). For the first batch fermentation at pilot scale, a major problem identified was the high formation of foam. The soybean molasses has a significant concentration of proteins and lipids that could contribute to increasing the viscosity and surface tension. Although molasses' viscosity should diminish while ethanol concentration increases, the intense formation of CO_2 bubbles is a factor that strongly favors foam formation (Togrul and Arslan 2004). An alternative to control CO_2 and foam formation is the fed-batch process, which is considered one of the most useful systems for economical ethanol production (Roukas 1996). The main advantage of the fed-batch system is that intermittent feeding of the substrate prevents inhibition

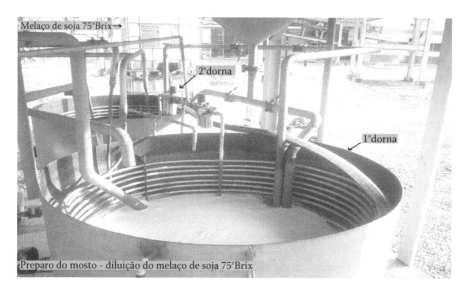

FIGURE 17.3 Pilot plant for bioethanol production.

TABLE 17.5

Results of Pilot-Scale Fermentations for Bioethanol Production with Different Concentrations (°Brix or % of Soluble Solids) of Soybean Molasses

Cycle nr./ op. mode[a]	Brix	Initial Sugar (g/L)	Final Sugar (g/L)	Ethanol (g/L)	Time (h)	Productivity (g/L h)	Yield[b] (%)
1/B[c]	20.0	116.6	54.00	26.10	6	4.350	43.80
2/B	30.0	251.8	118.2	58.60	6	9.767	45.54
3/FB[d]	21.0	166.9	59.1	39.28	7	5.611	46.06
4/FB	22.0	168.9	79.2	41.92	5	8.384	48.57
5/FB	23.0	185.0	89.9	43.72	5	8.744	46.25
6/FB	21.0	182.1	83.3	40.01	4	10.00	43.00
7/FB	21.0	189.5	76.9	40.50	6	6.750	41.82
8/FB	21.5	190.7	85.0	40.81	6	6.802	41.88
9/FB	22.0	180.8	83.2	41.14	5	8.228	44.53
10/FB	23.0	173.2	80.0	38.00	4	9.500	42.93
11/FB	30.0	230.7	74.37	48.41	7	6.916	41.06
		Average productivity/yield				7.882[e]	44.13

Notes: Initial biomass concentration was 3×10^8 cells/mL. [a]Cycle number/operational mode; [b]Yield over total initial sugars; [c]B—batch; [d]FB—fed batch; [e]For the fed batches only.

and catabolite repression, improving the productivity of fermentation by maintaining a low substrate concentration (Prasad et al. 2007).

Table 17.5 shows the results of 11 fermentation cycles conducted at the pilot scale plant. An average yield of 44.13% (standard deviation ± 1.323) over the total initial sugars represented an ethanol yield of 129.2 kg or 163.6 L of absolute ethanol per ton of dry molasses. The average productivity for the fed-batch process was 7.882 g/L h (standard deviation ± 0.8560).

17.4.5   BIOETHANOL PRODUCTION ON AN INDUSTRIAL SCALE (10 M³ ETHANOL PER DAY)

After satisfactory results were obtained at the pilot scale plant and the necessary parameters were defined, the process was transferred to an industrial plant equipped with distillation devices with a production capacity of 10 m³ ethanol per day (Figure 17.4).

A major problem identified after starting up the plant was the contamination with bacteria. Microscopic analysis indicated the presence of Gram-positive bacilli. This problem was expected because the biomass was recycled and might lose activity, allowing the development of opportunistic contaminants. The problem was solved by the addition of an appropriate antibiotic, together with the previous treatment of the medium (removal of insoluble solids by centrifugation), in order to facilitate the separation of yeast biomass at the end of fermentation and increase the efficiency of the inoculum's acid treatment. Previous treatment of soybean molasses may also preserve the equipment (tanks, pipelines, distillation columns).

FIGURE 17.4 Industrial plant for bioethanol production.

The increase in ethanol productivity along fermentation cycles at the industrial plant is shown in Figure 17.5.

According to the mass balance of industrial-scale bioethanol production, 1 ton of soybean molasses (dry basis) yielded 162.7 L of absolute ethanol and 3729 tons of vinasse with 19.5% solids.

Table 17.6 summarizes the results of bioethanol production at different production scales. Bioethanol yields obtained after the scale-up were satisfactory, showing small decreases from 169.8 to 163.6 L (pilot scale) and 162.7 L (industrial scale) of absolute ethanol per ton of dry molasses.

The generation of vinasse—a liquid waste produced by distillation of the fermented broth—in high volumes was a considerable environmental problem. The treatment of this waste has been one of the most challenging issues in the industrial production of bioethanol in Brazil because of its high Biochemical Oxygen Demand (BOD) values ranging from 30 to 60 gO_2/L (Navarro et al. 2000). The BOD of the vinasse produced by fermentation of soybean molasses was 77.2 gO_2/L, making

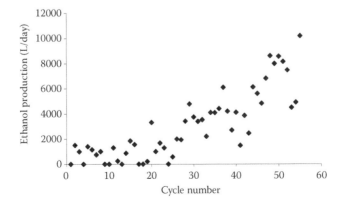

FIGURE 17.5 Increase in ethanol productivity (L/day) at industrial plant.

TABLE 17.6
Comparison between Results Obtained at Laboratory, Pilot, and Industrial Scales

	Capacity	Ethanol Yield (L/ton[a])
Laboratory scale	10 L/day	169.8
Pilot scale	1 m³/day	163.6
Industrial scale	10 m³ ethanol/day	162.7

[a] Liters of absolute ethanol per ton of dry molasses.

TABLE 17.7
Average Physicochemical Composition of Soybean Vinasse from Bioethanol Production

Component	Percentage on Dry Basis
Stachyose	11.09
Raffinose	22.07
Sucrose	0
Glucose	0
Fructose	0
Galactose	1.84
Total Carbohydrates	35.0
Proteins	13.3
Lipids	27.8
Fibers	14.6
Ash	9.24

it impossible to be treated as a common wastewater. The initial alternative was to recover it by concentration through water evaporation and burning of the residue in the industrial boiler, producing steam for industrial activities. Table 17.7 presents the composition of soybean vinasse.

17.5 PRODUCTION OF XANTHAN GUM FROM SOYBEAN MOLASSES

17.5.1 INTRODUCTION

A second alternative for the exploitation and reuse of soybean molasses was studied on a laboratory scale to produce xanthan gum. Xanthan gum, or simply xanthan, is an extracellular heteropolysaccharide produced by bacteria of the genus *Xanthomonas*, particularly the species *X. campestris*. Its molecular structure (Figure 17.6) is composed of D-glucosyl, D-mannosyl, and D-glucuronyl residues in the proportion 2:2:1 besides variable proportions of O-acetyl and pyruvyl (Katzbauer 1998).

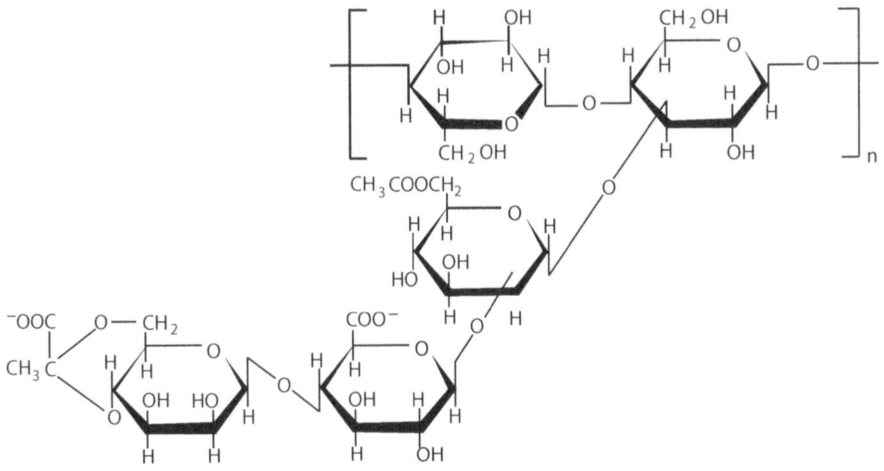

FIGURE 17.6 Molecular structure of xanthan gum. (Adapted from Garcia-Ochoa, F. et al., *Biotechnology Advances*, 18: 549–579, 2000. With permission.)

This composition, together with the tridimensional structure formed by xanthan chains associated with each other, confers to the xanthan gum aqueous solutions certain characteristics that distinguish it from other natural gums of commercial interest (Rosalam and England 2005): high thickening capacities, even at low concentrations (0.06%–0.2% [w/v]); pronounced pseudoplastic behavior (its viscosity in aqueous solutions decreases with the augmentation of applied shear stress); stability at a broad range of temperatures, pH, and salt concentrations; and synergism of properties when combined with other gums.

Since its classification as a GRAS (generally recognized as safe) product issued by the Food and Drug Administration of the United States (FDA) in 1969, xanthan gum has been applied as an additive in the food industry. It is employed as a stabilizing and thickening agent in sauces, juices, soups, dairy products, and desserts and in the pharmaceutical and cosmetics industries, where it is present in the composition of creams, lotions, toothpastes, and shampoos (Woiciechowski 2001).

Xanthan gum is also applied in large amounts in petroleum drilling and extraction procedures where it allows the pumping up of an oil–water mix with consistent viscosity, making oil drilling more efficient. Besides, it is also being employed in the composition of cleaning products, inks, agricultural chemicals, and wherever insoluble ingredients have to be uniformly distributed even during long-term storage.

The following describes, as an example, a process for xanthan gum production, which has been developed on a laboratory scale at the bioprocess engineering and biotechnology department of the UFPR.

17.5.2 MICROORGANISM SELECTION

Five strains of *Xanthomonas campestris* were tested for xanthan gum production from soybean molasses. The results are presented in Table 17.8.

TABLE 17.8

Biomass and Xanthan Gum Production by Different _X. campestris_ Strains in a Medium Containing Soybean Molasses 20°Brix, 1% (w/v) of Yeast Extract, 1% (w/v) of K$_2$HPO$_4$

Strain	Biomass Concentration (g/L)	Xanthan Gum Concentration (g/L)
LPB 001/DEBB[a]	7.2	47.3
CCT 6508/ATF[b]	4.6	7.2
CCT 6507/ATF	4.1	17.4
CCT 5677/ATF	3.0	33.0
CCT 6510/ATF	0.7	0.0

[a] Bioprocess engineering and biotechnology department/UFPR; [b] André Tosello Foundation.

In this initial stage of the process design, the soybean molasses employed was not submitted to any pretreatment, which further showed to be extremely important for the process success. In the experiments described below, the strain _X. campestris_ var. _campestris_ reference LPB001 was used, which gave the best results.

17.5.3 SOYBEAN MOLASSES PRETREATMENT

Soybean molasses presents in its composition substances that may decrease the fermentation yields and adhere to the polysaccharide, affecting rheological quality and other physicochemical properties of the final product. These "contaminants" include heavy metals; high molecular weight compounds, such as proteins; and substances resulting from Maillard reactions, which may render intense color to the xanthan produced. Efficient downstream processes, for example, multiple operations and/ or precipitation methods highly selective for xanthan gum, may solve the problem of rheological quality, but it will not reduce the influence of these contaminants in the fermentation yield (multiple operations may also reduce downstream yields). However, if these contaminants could be removed before the fermentation by substrate pretreatment, there might be no harmful impact on fermentation and downstream yields, and therefore, it might result in good quality xanthan.

Figure 17.7 shows the results for the combination of the best pretreatment methods chosen for further investigation. Molasses at 10°Brix was submitted to pH reduction (from 5.6 to 4.0, 2.0, or 0.0) with or without the addition of active carbon (5 g/L, powder) and, after 10 min, followed by centrifugation for precipitate removal (4500×g, 15 min). Results were recorded for the absorbance measurement at 420 nm.

As shown in Figure 17.7, pH reduction to 2.0 coupled with the addition of 5 g/L of active carbon resulted in a very expressive absorbance reduction. This indicated that it could be a good strategy for obtaining a higher-quality xanthan gum.

Figure 17.8 shows the influence of active carbon treatment on the fermentation yield. This experiment was conducted in a 2-L, bench-scale bioreactor. Molasses at 10°Brix was treated for pH reduction to 2.0 (later corrected to pH 7.0), diluted to 5°Brix, and fermented in the bioreactor.

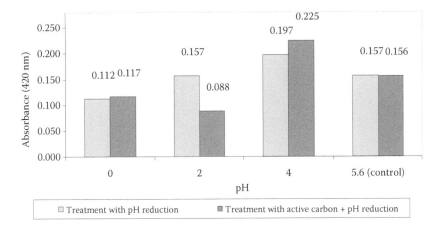

FIGURE 17.7 Soybean molasses' pretreatments evaluation. *Note*: The best result corresponds to the lowest absorbance (0.088).

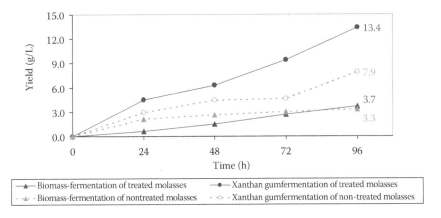

FIGURE 17.8 Effect of pretreatment with active carbon (5 g/L, powder) on fermentation yields.

These results suggested that the pretreatment was highly beneficial for the microbial growth, and there was a slight improvement in the xanthan gum production.

17.5.4 XANTHAN GUM PRODUCTION KINETICS UNDER OPTIMIZED CONDITIONS

Parameter optimization was conducted in 500-mL Erlenmeyer flasks, which showed that the more diluted the molasses used for broth composition, the higher the amount of xanthan obtained (Figure 17.9). The reason for this could be related to the broth's optimal sugar content for growing the *X. campestris* cells. However, some compounds present in the molasses that cannot be removed during the pretreatment phase may interfere in xanthan production.

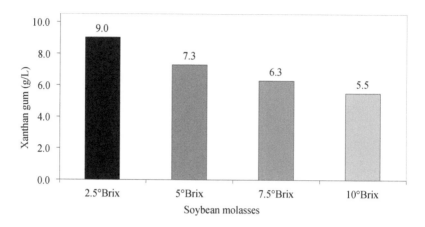

FIGURE 17.9   Comparison of xanthan gum production in different media dilutions.

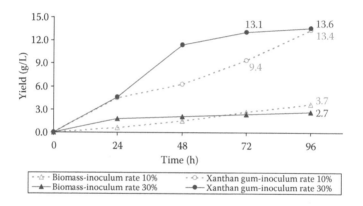

FIGURE 17.10   Xanthan and biomass production under different inocula rates.

An important observation about this study was that the Erlenmeyer flask tests did not give similar yields to those achieved in bioreactor fermentations. This was probably related to a poor oxygen transfer within the medium in these flasks that did not receive direct aeration by sparring like that which occurred in the bioreactor.

Based on a previous work (Shamala and Prasad 2001), the influence of an inoculum's rate in a bioreactor was investigated (Figure 17.10). By changing the inoculum rate from 10% to 30% (w/v), it was possible to reduce the time needed to obtain the maximum concentration of xanthan gum (13 g/L) by 24 h.

17.5.5   Downstream Process

Figure 17.11 shows a comparison between the rheological profile of xanthan gum obtained from the fermentation of soybean molasses and that of two commercial xanthan gums. For these tests, xanthan obtained from molasses was precipitated with three volumes of ethanol, dried (24 h, 70°C), ground, and resuspended in water

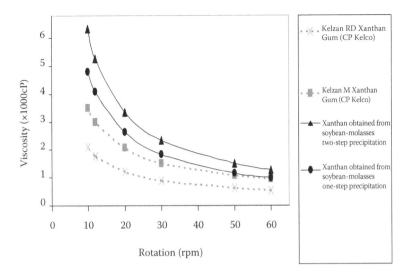

FIGURE 17.11   Rheological behavior of xanthan gum obtained by fermentation of soybean molasses, compared to two commercial xanthan gums.

(1%, w/v). One sample was submitted to the same process twice, resulting in a xanthan precipitated twice. In these downstream procedures, the ethanol produced from the fermentation of soybean molasses by *S. cerevisiae* was used. The samples were then analyzed in a viscometer at 25°C. The industrial-grade xanthan gums Kelzan M and KelzanRD produced by CP Kelco were also analyzed at 1% (w/v) concentration.

Results demonstrated that xanthan obtained from soybean molasses was similar to that of commercial samples, considering viscosity patterns. Although a two-step precipitation rendered a product with higher viscosity, one step was sufficient to obtain a product with an acceptable rheological behavior.

17.6   PRODUCTION OF ALPHA-GALACTOSIDASE FROM SOYBEAN VINASSE

17.6.1   INTRODUCTION

The enzyme α-D-galactoside galactohydrolase (EC 3.2.1.22), also named as α-galactosidase, catalyzes the hydrolysis of α-1,6-galactosidic bonds, releasing α-D-galactose. These bonds are found in oligosaccharides, such as melibiose, raffinose, and stachyose, which are associated with flatulence in monogastric animals and humans (Suarez et al. 1999). Human consumption of soy-derived products has been limited by the presence of these nondigestible oligosaccharides (NDOs) because most mammals, including man, lack pancreatic α-galactosidase. However, such NDOs can be fermented by gas-producing microorganisms present in the cecum and large intestine, thus inducing flatulence and other gastrointestinal disorders in sensitive individuals (LeBlanc et al. 2004). Alpha-galactosidases are applied in biocatalytic processes to improve products destined for human nutrition and also as components in animal diets to increase digestibility and reduce the fermentation of NDOs.

Nutritional studies using commercial preparations of α-galactosidase as supplements in the feed of monogastric animals demonstrated a significant increase in weight gain simultaneously with digestibility in swine and chickens. The food industry, especially the soy derivative sector, has great interest in the reduction of galactooligosaccharides, which are resistant to heat and, therefore, are not eliminated during the conventional processing (Said and Pietro 2004). Mulimani et al. (1997) reported that crude α-galactosidase treatment on soybean flour reduced the raffinose and stachyose contents by 90.4% and 91.9%, respectively.

Lactic acid bacteria, such as *Lactobacillus plantarum*, *Lactobacillus fermentum*, *Lactobacillus brevis*, *Lactobacillus buchneri*, and *Lactobacillus reuteri*, are able to hydrolyze α-galactooligosaccharides into digestible carbohydrates during vegetable fermentations. Recently, the characterization of genes involved in α-galactooligosaccharides hydrolysis by *Lactococcus raffinolactis* has been described (LeBlanc et al. 2005).

In order to give a destination to the soybean vinasse produced in the alcoholic fermentation of soybean molasses, other than burning for energy generation, the possibility of producing α-galactosidase from this residue was evaluated in submerged cultivation on a laboratory scale. The composition of soybean vinasse (Table 17.7) showed considerable amounts of stachyose (11.09%) and raffinose (22.07%), which were the substrates for α-galactosidase. Results of these studies have been published by Sanada et al. (2009).

17.6.2 MICROORGANISM SELECTION

Table 17.9 shows the results of the microorganism screening test. The analysis of variance at the level of 0.05 demonstrated that the response presented by *Lactobacillus*

TABLE 17.9

Extracellular α-Galactosidase Activity for Nine Different Strains of Lactic Acid Bacteria

Strains	α-Galactosidase Activity (U/mL)		
	48 h	72 h	96 h
L. paracasei LPB E1	0.156 (0.008)	0.205 (0.006)	0.209 (0.009)
L. pentosus NRRL B-227	0.081 (0.004)	0.198 (0.003)	0.145 (0.006)
L. plantarum NRRL B-4496	0.063 (0.003)	0.165 (0.007)	0.201 (0.01)
L. delbrueckii FAT 0846	0.019 (0.0007)	0.175 (0.002)	0.186 (0.008)
L. fermentum LPB 7	0.094 (0.006)	0.218 (0.0075)	0.150 (0.008)
L. agilis LPB 56	**0.393** (0.03)	**0.545** (0.025)	**0.498** (0.03)
L. casei rhamnosus LPB H19	0.134 (0.008)	0.106 (0.009)	0.109 (0.004)
L. salivarius NRRL B-1949	0.000 (0)	0.003 (0.0001)	0.090 (0.004)
L. raffinolactis DSM 20443	0.054 (0.004)	0.098 (0.003)	0.074 (0.003)

Note: Values in parentheses represent average standard deviations.

agilis LPB 56 was significantly higher than the other averages; hence, this strain was selected for further studies.

17.6.3 Kinetics of Cellular Growth in MRS and Vinasse Media

Man-Rogosa-Sharpe (MRS) broth and soybean vinasse were compared as media for cellular production, aiming at the substitution of the synthetic medium for inoculum preparation. The kinetics of cellular growth for *L. agilis* LPB 56 is shown in Figure 17.12 for both the media.

Cellular growth in the vinasse presented a satisfactory result of the order of 10^8 colony-forming units per milliliter (CFU/mL); this, for the MRS medium, was around 10^9 CFU/mL. The stationary growth phase was about 4 h longer for the vinasse medium. The sudden decline in cellular viability, without a stationary phase, has already been reported by Karp (2007) for *L. agilis* LPB 56 grown in MRS medium. Even in the vinasse, the reduction in viable cell concentration was very expressive after 16 h and could possibly be attributed to pH reduction as a result of lactic acid production.

Extracellular enzymatic activity presented after 72 h in the fermentation medium was 8% higher for the medium inoculated with 10% (v/v) of vinasse (an average of 0.586 U/mL). Intracellular activity was assayed after cell disruption and presented a value of 1.52 U/mL. So it was concluded that the selected strain *L. agilis* LPB 56 produced intracellular and extracellular α-galactosidase. The utilization of vinasse as an inoculum medium, besides reducing fermentation costs, provided the beginning of strain adaptation in the fermentation medium. Subsequently, the C/N ratio was optimized in order to improve the cellular growth in the vinasse.

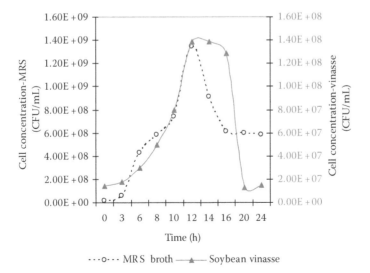

FIGURE 17.12 Cellular concentration (in CFU/mL) of the microorganism *L. agilis* LPB 56 in MRS broth and soybean vinasse at 10°Brix. Average standard deviation was 7%.

17.6.4 SUPPLEMENTATION OF VINASSE-BASED INOCULUM MEDIUM WITH DIFFERENT NITROGEN SOURCES AND C/N RATIOS

Results presented in Table 17.10 demonstrated that, among the inorganic and organic nitrogen sources assayed, yeast extract at the C/N ratio of 6 provided the best cellular production in the inoculum medium. The kinetics of cell growth were evaluated for the addition of the yeast extract. Results in Figure 17.13 confirmed that

TABLE 17.10

Cellular Production (×10⁸ CFU/mL) in the Vinasse after 12 h for Different Nitrogen Sources and C/N Relations

Nitrogen Source	C/N Relation					
	1	3	6	9	11	13
Ammonium sulfate	0.025	1.6	3.5	1.3	1.0	0.20
Ammonium nitrate	0.012	1.2	2.8	1.1	1.2	0.16
Yeast extract	0.037	3.8	5.6	0.96	1.3	0.17
Peptone proteose	0.0088	0.74	2.2	0.92	1.3	0.078
Urea	0.012	0.62	1.2	0.89	0.96	0.085

Note: Average standard deviation was 6.7%.

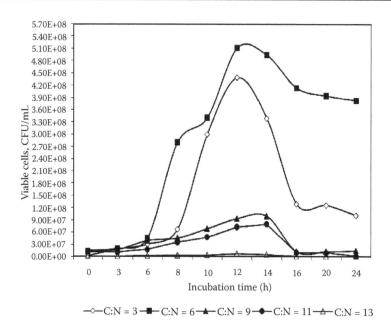

FIGURE 17.13 Growth kinetics of *L. agilis* LPB 56 in vinasse-based inoculum medium supplemented with yeast extract at different C/N relations. Average standard deviation was 9.1%.

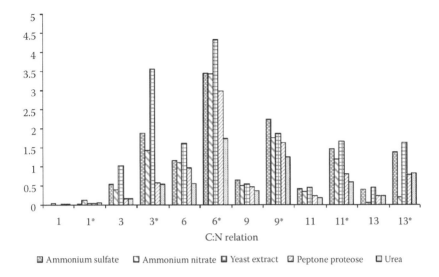

FIGURE 17.14 Alpha-galactosidase activity in fermentation medium after 72 h for different nitrogen sources and C/N relations in the inoculum. *Refers to total activity (intracellular + extracellular). Average standard deviation was 5.4%.

the maximum concentration of viable cells (approximately 5×10^8 CFU/mL) was obtained for a C/N ratio of 6 at around 12 h. The second maximum cellular production was obtained for the C/N relation of 3, but it represented a higher addition of nitrogen source. There was no significant difference among the proportions 9 and 11 (around 10^7 CFU/mL). The C/N ratio of 1 resulted in a concentration of viable cells close to 10^6 CFU/mL (data not shown in Figure 17.13).

According to the Tukey test, α-galactosidase activity (1.60 U/mL extracellular and 4.33 U/mL total) presented by the strain *L. agilis* LPB 56 when using yeast extract and a C/N relation of 6 was significantly higher than the other averages (Figure 17.14). The same condition was found by Gote et al. (2004), who reported α-galactosidase activity of 1.15 U/mL with 0.5% of yeast extract as a nitrogen source.

17.6.5 SCREENING OF CULTURE CONDITIONS AND MEDIUM COMPOSITION

The values of temperature and pH were chosen according to literature reports (Roissart and Luquet 1994). Rotation, soluble solid concentration, and size of inoculum were defined based on the parameters used by Karp (2007). C/N ratios were fixed according to the inoculum preparation test. Preliminary experiments revealed that supplementation of mineral salts in the fermentation medium (K_2HPO_4, KH_2PO_4, $MnSO_4·H_2O$, $FeSO_4·7H_2O$, $MgSO_4·7H_2O$, $CaCl_2$, and $ZnSO_4·7H_2O$) did not significantly affect α-galactosidase production (data not shown). Chosen levels are shown in Table 17.11, and α-galactosidase activities obtained for the different factors and levels are presented in Table 17.12.

According to the Pareto chart of the Plackett–Burman design (Figure 17.15), three variables, including soluble solids percentage (X_4), size of inoculum (X_5), and C/N ratio (X_6), significantly influenced α-galactosidase activity, increasing its value when their levels were increased. These variables were subsequently investigated for their optimal concentrations.

TABLE 17.11

Coded and Real Values of Culture Conditions and Medium Compositions Used for Plackett–Burman Design

Level		−1	+1
Temperature (°C)	X_1	25	35
pH	X_2	6	7
Rotation (rpm)	X_3	110	130
Soluble solids (°Brix)	X_4	10	20
Inoculum (% v/v)	X_5	10	20
Yeast extract (C/N relation)	X_6	3	6

TABLE 17.12

Results Obtained for Plackett–Burman Design: α-Galactosidase Activities for Different Culture Conditions and Medium Compositions after 72 h of Fermentation

Run	Temp. (°C) X_1	pH X_2	Rotation (rpm) X_3	Soluble Solids (°Brix) X_4	Inoculum (% v/v) X_5	Yeast Extract (C:N) X_6	Dummy X_7	Total α-gal Activity (U/mL)
1	−1	−1	−1	−1	−1	−1	1	2.53
2	−1	−1	+1	−1	+1	+1	−1	6.23
3	−1	+1	−1	+1	−1	+1	−1	6.88
4	−1	+1	+1	+1	+1	−1	1	5.97
5	+1	−1	−1	+1	+1	−1	−1	5.91
6	+1	−1	+1	+1	−1	+1	1	6.97
7	+1	+1	−1	−1	+1	+1	1	6.03
8	+1	+1	+1	−1	−1	−1	−1	2.63

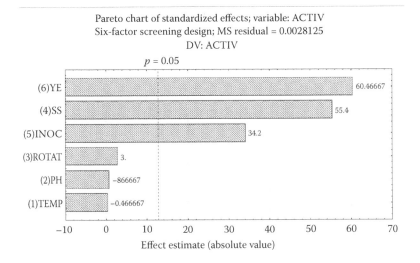

FIGURE 17.15 Pareto chart for screening of significant effects on α-galactosidase production, demonstrating the positive effect of variables 4 (soluble solids), 5 (size of inoculum), and 6 (C/N ratio). Significance of effects is determined by student's *t* test at the confidence level of 95% and represented by cutoff line at $p = 0.05$.

17.6.6 OPTIMIZATION OF CULTURE CONDITIONS AND MEDIUM COMPOSITIONS

The levels of soluble solids concentration and size of inoculum were increased in relation to the previous experiment because the best reported values for lactic acid production from the soybean vinasse by the strain *L. agilis* LPB 56 were 30% and 25%, respectively (Karp 2007). Levels of C/N ratios were chosen in order to compare the optimum relation previously obtained (a C/N of 6) and the C/N ratio in non-supplemented fermentation medium. Values are presented in Table 17.13.

The response surfaces presented in Figure 17.16, described by the second-order polynomial equation:

$$\text{Activity (U/mL)} = 7.589 - 2.95013*[\text{soluble solids}] - 2.76238*[\text{soluble solids}]^2 -$$
$$2.66672*[\text{inoculum\%}]^2 - 2.56397*[\text{C/N ratio}]^2 \qquad (17.1)$$

for coded values, showed the results for total (extracellular + intracellular) α-galactosidase activity (U/mL) when soluble solids concentration, size of inoculum, and C/N relation were varied, using the strain *L. agilis* LPB 56. All response surfaces showed that increasing the levels of the three aforementioned variables was beneficial for α-galactosidase production until a certain point. Subsequent increases caused a negative effect on enzyme production. This was expected for the variable soluble solids as the effect of substrate inhibition has been well known. For the size of inoculum, the result was in accordance with the previous studies on lactic acid production (Karp 2007). It was observed that the maximum α-galactosidase production

TABLE 17.13

Alpha-Galactosidase Activities Obtained after 72 h of Fermentation for Different Culture Conditions and Medium Compositions Used for Central Composite Design

Run	Soluble Solids% (°Brix)	Inoculum (% v/v)	Yeast Extract (C/N Relation)	Total α-gal Activity (U/mL)
1	25	15	6	4.87
2	35	15	6	0.873
3	25	35	6	4.39
4	35	35	6	2.38
5	25	15	12	4.64
6	35	15	12	0.788
7	25	35	12	4.02
8	35	35	12	1.90
9	22	25	9	6.98
10	38	25	9	2.13
11	30	8	9	2.86
12	30	42	9	6.52
13	30	25	2	2.65
14	30	25	14	7.02
15	30	25	9	7.55
16	30	25	9	7.31
17	30	25	9	7.60

(7.60 U/mL) was reached when the soluble solid concentration was 30%, the size of inoculum was 25% (v/v), and the C/N ratio was 9 (nonsupplemented medium).

According to Table 17.14, the significant variable that more affected the response was the soluble solids percentage (°Brix) for presenting the highest effect (absolute value) in the quadratic term (2.95838). The model revealed a correlation coefficient (R^2) of 0.81647. Verification of the model was performed in Erlenmeyer flasks (five experiments, in duplicates) for conditions within the experimental range. Results are presented in Table 17.15. According to the calculated deviations, the average relative deviation of the model in relation to real values was 25%.

17.6.7　PRODUCTION OF α-GALACTOSIDASE IN BIOREACTOR UNDER OPTIMIZED CONDITIONS

The kinetic study of α-galactosidase production in a bioreactor under optimized conditions is presented in Table 17.16. The highest enzymatic activity (11.07 U/mL) was reached after 144 h of fermentation. It was observed that the enzymatic activity did not drop after 72 h as for the previous experiments. This was probably a result of continuous pH control besides $CaCO_3$ addition at the beginning of fermentation. The maximum α-galactosidase activity was obtained during the stationary phase, and the intracellular fraction had considerably higher enzyme activity than the

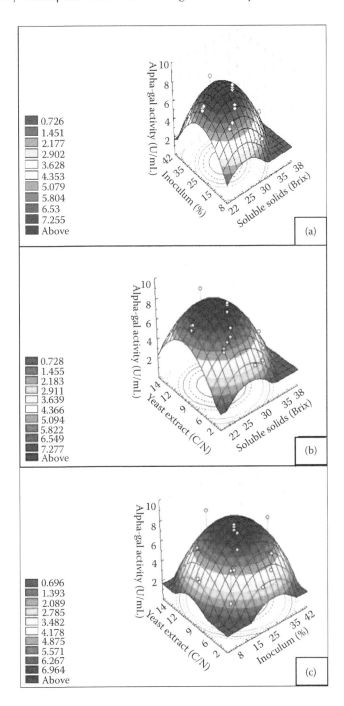

FIGURE 17.16 Response surface plot described by model, showing effect of (a) soluble solids (°Brix) and size of inoculum (% v/v); (b) soluble solids (°Brix) and yeast extract (C/N relation); (c) size of inoculum (% v/v) and yeast extract (C/N ratio) on α-galactosidase production.

TABLE 17.14

Identification of Significant Variables ($p < 0.05$) for α-Galactosidase Production Using Central Composite Design

Factor	Coefficients	Standard Error	t-Value	p-Value
Soluble solids (°Brix) (L)	−1.47507	0.423785	−3.48070	0.010257
Soluble solids (°Brix) (Q)	−1.38119	0.466875	−2.95838	0.021154
Inoculum% (Q)	−1.33336	0.466875	−2.85593	0.024480
Yeast extract (C/N) (Q)	−1.28199	0.466875	−2.74589	0.028674

TABLE 17.15

Experiments for Verification of Predicted Model of α-Galactosidase Production

Run	Soluble Solids% (°Brix)	Inoculum (% v/v)	Yeast Extract (C/N Relation)	Predicted (U/mL)	Experimental (U/mL)
1	25	25	12	5.81	5.67
2	27.5	20	11.5	6.47	6.09
3	30	15	9	4.92	6.65
4	32.5	30	7.5	4.12	3.87
5	35	35	6	0	1.95

Note: Results represent the average of the duplicates with a standard deviation below 10%.

TABLE 17.16

Kinetic Study of α-Galactosidase, Biomass, and Lactic Acid Production and Sugar Consumption

Time (h)	Biomass (g/L)[a]	Total Sugar (g/L)	Lactic Acid (g/L)	Extracellular Activity (U/mL)	Total Activity (U/mL)	Specific Activity (U/mg)
0	0.272	105.6	7.826	0	0.03487	0.2019
24	3.92	53.03	48.44	0.324	4.939	35.05
48	6.83	26.49	66.79	2.111	6.839	46.32
72	7.05	5.704	72.20	3.002	7.374	44.78
96	7.77	4.924	71.29	3.356	7.600	43.93
120	7.46	4.744	72.54	3.652	7.545	41.25
144	8.22	4.432	70.30	5.925	11.07	51.97
168	8.12	4.432	73.57	3.480	9.703	49.41
192	8.26	3.952	73.65	3.301	10.1	44.83

Note: Results represent the average of two experiments with a standard deviation below 10%; [a]Dry weight.

extracellular fraction, which was in accordance with the results obtained by Yoon and Hwang (2008). Improvements in product yields as recorded in this investigation (from 7.60 U/mL in shake flasks to 11.07 U/mL in the fermenter) were expected in the fermenter resulting from better control of process parameters. The enhancement in α-galactosidase activity could be mainly attributed to the maintenance of optimum pH.

Some studies have reported values for α-galactosidase activity as 2.0 U/mL (extracellular) after 24 h in submerged fermentation by *Bacillus stearothermophilus* using soybean meal (Gote et al. 2004) and 5.0 U/mL after 24 h for submerged fermentation by *L. fermentum* in a synthetic raffinose-based medium (LeBlanc et al. 2004).

17.7 PRODUCTION OF L(+)-LACTIC ACID FROM SOYBEAN VINASSE

17.7.1 INTRODUCTION

Lactic acid, the popular name of the 2-hydroxypropanoic acid ($C_3H_6O_3$), is the most widely occurring carboxylic acid in nature (Narayanan et al. 2004). Lactic acid can be produced either by chemical synthesis or fermentation. By the chemical way, the racemic mixture D/L is always produced, and optically pure isomers (D or L) can be obtained by fermentation when the appropriate microorganism is selected. Approximately 90% of lactic acid is produced by the fermentative route (Joglekar et al. 2006).

L-lactic acid is classified as GRAS by the FDA and can be used as a food additive. However, the D-isomer is not metabolized by animals, causing acidosis and decalcification (Wee et al. 2006). There are four major categories for the applications of lactic acid: food, cosmetics, pharmaceuticals, and chemicals. The most recent application of lactic acid is its polymerization to poly-lactic acid (PLA), which is a biocompatible and biodegradable plastic that finds many applications in the medical field (Narayanan et al. 2004; Wee et al. 2006). The worldwide production level of lactic acid is approximately 350,000 tons per year, and the market growth is believed to be approximately 12% to 15% per year (Joglekar et al. 2006).

The soybean vinasse can be a suitable medium for lactic acid fermentation according to its physicochemical composition analysis (Table 17.7), which shows the presence of the complex sugars stachyose (11.09%) and raffinose (22.07%), proteins (13.3%), and lipids (27.8%). Reduction of stachyose and raffinose contents in soymilk fermented with lactic bacteria has been reported (Chien et al. 2006). Several lactic acid bacteria exhibit proteolytic activity, and they predominantly use peptides to meet their demand for complex nitrogen. Assimilation of lipids and their conversion to the corresponding alcohols by lactic acid bacteria has been reported in sourdough fermentation (Gänzle et al. 2007). Lipid oxidation products were identified in deoiled soy lecithin fermentation by *Lactococcus lactis*, *Lactococcus cremoris*, *Streptococcus thermophilus*, and *Lactobacillus helveticus* (Suriyaphan et al. 2001).

The process for L-lactic acid production from the soybean vinasse, as another alternative for the exploitation of this residue, was developed on a laboratory scale at the bioprocess engineering and biotechnology department (UFPR) and on a pilot scale in partnership with a Brazilian soybean-processing company. Results have been published by Karp et al. (2011).

17.7.2 Microorganism Selection

Table 17.17 shows the results of lactic acid production by 10 strains available from the culture collection of DEBB/UFPR, cultivated in the soybean vinasse and molasses. *Lactobacillus pentosus* showed the best lactic acid productivity in the molasses medium (1.224 g/L h or 29.38 g/L after 24 h). As presented in Table 17.2, the molasses contained a significant amount of sucrose (28.4% on a dry basis), which was easily metabolized by many microorganisms. The vinasse, however, contains mainly complex sugars. The only strain that was able to produce a significant amount of lactic acid in the vinasse (27.76 g/L in 72 hours) was *L. agilis* LPB 56. Because the aim of this work was to use the vinasse as the main raw material, the strain *L. agilis* LPB 56 was chosen for the subsequent studies.

TABLE 17.17

Results of Lactic Acid Production after 24, 48, and 72 h and Maximum Productivities for 10 Different Strains Cultivated in Soybean Molasses and Vinasse

Strain	Medium	Lactic Acid 24 h (g/L)	Lactic Acid 48 h (g/L)	Lactic Acid 72 h (g/L)	Productivity (g/L h)
Lactobacillus	SBM[a]	29.38	31.25	31.55	1.224
pentosus	SBV[b]	10.44	9.767	8.667	0.4350
L. delbrueckii 0246	SBM	20.42	22.95	11.51	0.8508
	SBV	0	0	0	0
L. delbrueckii 1344	SBM	2.981	0	0	0.1242
	SBV	0.5930	0	1.067	0.02471
L. delbrueckii 1377	SBM	5.602	1.067	3.351	0.2334
	SBV	1.075	0.4550	1.284	0.04479
L. ruminis	SBM	0	0	0	0
	SBV	0	0	0	0
L. salivarius	SBM	2.276	7.530	22.27	0.3093
	SBV	2.723	2.904	3.074	0.1346
L. plantarum	SBM	17.52	18.35	21.63	0.7300
	SBV	0.2690	10.99	13.83	0.2290
LPB 2	SBM	4.887	3.804	3.580	0.2036
	SBV	0.633	0	0	0.02637
LPB 7	SBM	16.19	26.52	31.44	0.5525
	SBV	5.643	8.273	8.889	0.2351
L. agilis LPB 56	SBM	15.99	22.07	25.73	0.6663
	SBV	9.644	22.29	27.76	0.4644

[a] SBM—Soybean molasses; [b]SBV—Soybean vinasse.

The optical purity of L-lactic acid produced by *L. agilis* LPB 56 was 91%, determined by an enzymatic assay.

17.7.3 LACTIC ACID PRODUCTION KINETICS UNDER OPTIMIZED CONDITIONS

Fermentation tests were conducted on a laboratory scale to define the optimal conditions for lactic acid production by the strain *L. agilis* LPB 56. Supplementation of the medium with inorganic nitrogen sources [$(NH_4)_2HPO_4$ and $(NH_4)_2SO_4$] and yeast extract showed no positive effect on lactic acid production. Fermentation pH had to be controlled by the addition of $CaCO_3$ (6% w/v at the beginning) and $Ca(OH)_2$ (along with fermentation to keep the pH constant).

Table 17.18 shows the kinetics, under optimized conditions, of the lactic fermentation of soybean vinasse, conducted in a bench-scale bioreactor (2 L of total capacity).

A product conversion yield of 0.864 was obtained, which was considered satisfactory, especially when comparing with the literature-reported values, for example, 0.71 for soybean stalk fermentation by *Lactobacillus casei* and *Lactobacillus sake* (Xu et al. 2007) and 0.70 for synthetic MRS medium fermentation by *Lactobacillus delbrueckii* and *L. casei* (Lee 2005). Values of $Y_{P/S}$ higher than 1 at the beginning of fermentation indicated that some quickly assimilable carbon sources (possibly pentoses) for lactic acid production were not considered.

The reduction of stachyose and raffinose concentrations and their conversion to lactic acid is shown in Figure 17.17.

TABLE 17.18
Kinetics of Lactic Acid and Biomass Production and Sugar Consumption in Soybean Vinasse Fermented by *L. agilis* LPB 56 Inoculated at 10% (v/v)

Time (h)	Lactic Acid (g/L)	Total Sugars (g/L)	Biomass (g/L)	r_x (dX/dt)	r_S (dS/dt)	r_P (dP/dt)	$Y_{P/S}$ (r_p/r_s)	$Y_{X/S}$ (r_x/r_s)
0	2.518	68.3	0.2075	–	–	–	–	–
4	3.537	67.54	0.2453	0.00945	0.19	0.255	1.34	0.0500
8	10.85	61.06	0.3010	0.0139	1.62	1.83	1.13	0.00860
12	17.62	55.14	0.3567	0.0139	1.48	1.69	1.14	0.00941
16	25.25	47.69	0.4085	0.0129	1.86	1.91	1.02	0.00695
20	29.96	40.8	0.4901	0.0204	1.72	1.18	0.683	0.0118
24	35.06	33.03	0.5597	0.0174	1.94	1.27	0.656	0.00896
28	42.14	24.7	0.6413	0.0204	2.08	1.77	0.850	0.00980
32	47.84	17.94	0.6950	0.0134	1.69	1.43	0.843	0.00794
36	49.27	16.1	0.7707	0.0189	0.46	0.357	0.777	0.0411
40	51.17	13.8	0.7886	0.00447	0.575	0.475	0.826	0.00778
44	52.01	10.29	0.7866	−0.0005	0.877	0.21	0.239	−0.0006
48	50.28	6.883	0.7965	0.00247	0.852	−0.433	−0.51	0.00291
		Average		0.0145	1.32	1.13	0.864	0.0162

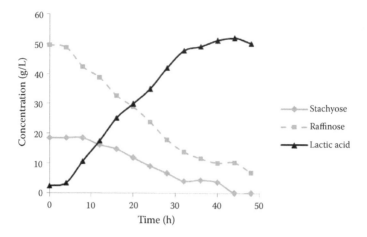

FIGURE 17.17 Consumption of individual sugars in vinasse.

17.7.4 LACTIC ACID PRODUCTION IN CONCENTRATED MEDIUM

The concentration of the vinasse was considered as a possibility to obtain higher lactic acid concentrations in the fermented broth. A vinasse medium, containing 19.5% solids, was supplemented with molasses (75% solids) to reach a final concentration of 30% solids. Because fermentation time increased significantly in the concentrated vinasse, the use of higher inoculum percentages was tested in order to accelerate the fermentation. Results presented in Figure 17.18 showed that the inoculation percentages of 25% and 50% showed practically no difference in fermentation time, productivities being, respectively, 0.930 and 0.899 g/L h in 120 h. Long fermentation times are of common occurrence in lactic acid production processes (John et al. 2006). The highest yields (71.12% from total initial sugars and 81.82% from consumed sugars) were achieved with the inoculum size of 25%, because lactic acid concentration decreased at the final stages when using 50% inoculum.

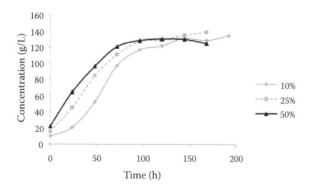

FIGURE 17.18 Lactic acid production kinetics in medium composed of vinasse + molasses at 30°Brix using three different inoculum sizes: 10%, 25%, and 50% (v/v).

17.7.5 LACTIC ACID PRODUCTION AT PILOT SCALE

Eight batches were conducted on a pilot scale according to the conditions defined on a laboratory scale in order to check the reproducibility of results and evaluate the need for adjustments in operational procedures. It was observed that the pH control system was a critical device because the fermentation pH decreased very expressively, and the low solubility of $Ca(OH)_2$ required efficient homogenization and pumping. Effective mixing and cleaning of the bioreactor were also important parameters to be assured. Table 17.18 presents the average values obtained for four batches, conducted under stable operational conditions, after the necessary adjustments were established. In comparison with the laboratory-scale fermentation of concentrated vinasse with 25% inoculum, there was no significant difference in fermentation yield ($Y_{P/S}$), and an increase of 18.1% in productivity (r_P) was observed. This enhancement could be attributed not only to scale-up but also to the different pH control strategy, which was essential to maintain lactic acid productivity (Qin et al. 2010). There was no significant improvement in L-lactate optical purity, which oscillated between 90% and 93%.

The separation of L-lactic acid from the fermented vinasse and molasses would be a difficult task because of the high concentration of "contaminants" remaining after the fermentation (residual sugars, lipids, proteins, ashes) that conferred a dark color to the product. In this way, the obtaining of pure L-lactic acid tends to be economically unfeasible and could generate a significant amount of residues.

The production of a calcium lactate-based feed additive is an alternative that allows the integral use of the fermented broth as a commercial product. Mass and cost balances for this novel process are presented in Figure 17.19 and Table 17.19.

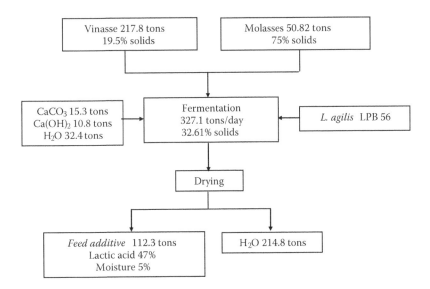

FIGURE 17.19 Mass balance for feed additive production: potential application of calcium lactate.

TABLE 17.19
Cost Balance for Calcium Lactate Production

Variable Costs	Price (US$/kg)	Amount/Month	Cost/Month (US$)
Vinasse + molasses 30°Brix	Equiv. firewood	6420 tons	90,105.26
Calcium carbonate	0.605	250 tons	151,315.79
Calcium hydroxide	0.974	177.3 tons	172,634.21
Water	0.00158	10,000 m³	15,789.47
Steam	0.0246	26.49 tons	652,631.58
Electric energy	0.22/kWh	190,328.00 kWh	41,070.78
Lab. material	–	–	1,578.95
Total			1,125,126.04
Fixed costs			87,134.21
	Production capacity		3045 tons per month of *feed additive*
	Production cost		US$ 0.40/kg

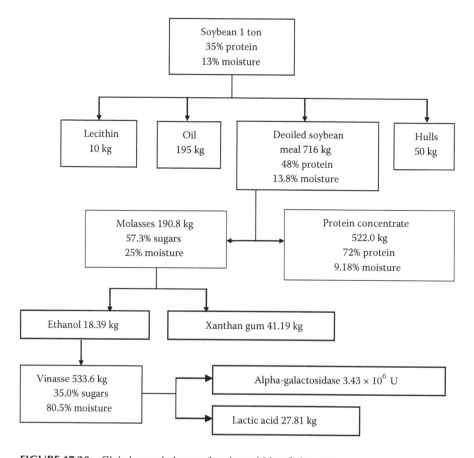

FIGURE 17.20   Global mass balance of soybeans' biorefining process.

17.8   CONCLUSIONS

This work presented the development of a project within the concept of a biorefinery for the integral exploitation of an important agricultural residue produced in Brazil: soybean molasses. Efforts were focused on the implementation of integrand technologies to produce high commercial value products from renewable agricultural materials and their wastes. Results obtained presented promising perspectives for the utilization of the soybean vinasse for value addition through biotechnological interventions. Figure 17.20 presents the global mass balance of soybeans' biorefining process.

REFERENCES

Alriols, M.G., García, A., Llano-ponte, R. and Labidi, J. (2010) 'Combined organosolv and ultrafiltration lignocellulosic biorefinery process,' *Chemical Engineering Journal*, 157: 113–120.

Cardona, C.A. and Sánchez, O.J. (2007) 'Fuel ethanol production: Process design trends and integration opportunities,' *Bioresource Technology*, 98: 2415–2457.

Cherubini, F. (2010) 'The biorefinery concept: Using biomass instead of oil for producing energy and chemicals,' *Energy Conversion and Management*, 51: 1412–1421.

Cherubini, F. and Ulgiati, S. (2010) 'Crop residues as raw materials for biorefinery systems: A LCA case study,' *Applied Energy*, 87: 47–57.

Chien, H.L., Huang, H.Y. and Chou, C.C. (2006) 'Transformation of isoflavone phytoestrogens during the fermentation of soymilk with lactic acid bacteria and bifidobacteria,' *Food Microbiology*, 23: 772–778.

EMBRAPA—Empresa Brasileira de Pesquisa e Agropecuária. Online. Available at http://www.cnpso.embrapa.br/html/compsoja.htm (accessed October 5, 2007).

FitzPatrick, M., Champagne, P., Cunningham, M.F. and Whitney, R.A. (2010) 'A biorefinery processing perspective: Treatment of lignocellulosic materials for the production of value added products,' *Bioresource Technology*, 101: 8915–8922.

Gänzle, M.G., Vermeulen, N. and Vogel, R.F. (2007) 'Carbohydrate, peptide and lipid metabolism of lactic acid bacteria in sourdough,' *Food Microbiology*, 24: 128–138.

Garcia-Ochoa, F., Santos, V.E., Casas, J.A. and Gómez, E. (2000) 'Xanthan gum: Production, recovery and properties,' *Biotechnology Advances*, 18: 549–579.

Gote, M., Umalkar, H., Khan, I. and Khire, J. (2004) 'Thermostable α-galactosidase from *Bacillus stearothermophilus* (NCIM 5146) and its application in the removal of flatulence causing factors from soymilk,' *Process Biochemistry*, 39: 1723–1729.

Grad, P. (2006) 'Biofuelling Brazil: An overview of the bio-ethanol success story in Brazil,' *Refocus*, 7: 56–59.

IBGE—Instituto Brasileiro de Geografia e Estatística. Online. Available at http://www.ibge.gov.br/home/presidencia/noticias/noticia_visualiza.php?id_noticia=832&id_pagina=1 (accessed November 15, 2011).

Joglekar, H.G., Rahman, I., Babu, S., Kulkarni, B.D. and Joshi, A. (2006) 'Comparative assessment of downstream processing options for lactic acid,' *Separation and Purification Technology*, 52: 1–17.

John, R.P., Nampoothiri, K.M. and Pandey, A. (2006) 'Solid state fermentation for L-lactic acid production from agro wastes using *Lactobacillus delbrueckii*,' *Process Biochemistry*, 41: 759–763.

Karp, S.G. (2007) 'Production of L-lactic acid from the soybean vinasse,' *Master's dissertation*, Federal University of Paraná/Universities of Provence and of the Mediterranean, Curitiba, Brazil.

Karp, S.G., Igashiyama, A.H., Siqueira, P.F., Carvalho, J.C., Vandenberghe, L.P.S., Thomaz-Soccol, V., Coral, J., Tholozan, J.L., Pandey, A. and Soccol, C.R. (2011) 'Application of the biorefinery concept to produce L-lactic acid from the soybean vinasse at laboratory and pilot scale,' *Bioresource Technology*, 102: 1765–1772.

Katzbauer, B. (1998) 'Properties and applications of xanthan gum,' *Polymer Degradation and Stability*, 59: 81–84.

LeBlanc, J.G., Silvestroni, A., Connes, C., Juillard, V., Giori, G.S., Piard, J.C. and Sesma, F. (2004) 'Reduction of non-digestible oligosaccharides in soymilk: Application of engineered lactic acid bacteria that produce alpha-galactosidase,' *Genetics and Molecular Research*, 3: 432–440.

LeBlanc, J.G., Piard, J.C., Sesma, F. and Giori, G.S. (2005) '*Lactobacillus fermentum* CRL 722 is able to deliver active alpha-galactosidase activity in the small intestine of rats,' *FEMS Microbiology Letters*, 248: 177–182.

Lee, K.B. (2005) 'Comparison of fermentative capacities of lactobacilli in single and mixed culture in industrial media,' *Process Biochemistry*, 40: 1559–1564.

Mulimani, V.H., Thippeswamy, S. and Ramalingam, S. (1997) 'Enzymatic degradation of oligosaccharides in soybean flours,' *Food Chemistry*, 59 (2): 279–282.

Narayanan, N., Roychoudhury, P.K. and Srivastava, A. (2004) 'L(+) lactic acid fermentation and its product polymerization,' *Electronic Journal of Biotechnology*, 7: 167–178.

Navarro, A.R., Sepúlveda, M.C. and Rubio, M.C. (2000) 'Bio-concentration of vinasse from the alcoholic fermentation of sugar cane molasses,' *Waste Management*, 20: 581–585.

Oleskowicz-Popiel, P., Kádár, Z., Heiske, S., Klein-Marcuschamer, D., Simmons, B.A., Blanch, H.W. and Schmidt, J.E. (2012) 'Co-production of ethanol, biogas, protein fodder and natural fertilizer in organic farming: Evaluation of a concept for a farm scale biorefinery,' *Bioresource Technology*, 104: 440–446.

Prasad, S., Singh, A. and Joshi, H.C. (2007) 'Ethanol as an alternative fuel from agricultural, industrial and urban residues,' *Resources, Conservation and Recycling*, 50: 1–39.

Qin, J., Wang, X., Zheng, Z., Ma, C., Tang, H. and Xu, P. (2010) 'Production of L-lactic acid by a thermophilic *Bacillus* mutant using sodium hydroxide as neutralizing agent,' *Bioresource Technology*, 101 (19): 7570–7576.

Roissart, H. and Luquet, F.M. (1994) *Bactéries lactiques—aspects fondamentaux et technologiques*. Uriage: Lorica.

Rosalam, S. and England, R. (2005) 'Review of xanthan gum production from unmodified starches by *Xanthomonas camprestris* sp.,' *Enzyme and Microbial Technology*, 39: 197–207.

Roukas, T. (1996) 'Ethanol production from non-sterilized beet molasses by free and immobilized *Saccharomyces cerevisiae* cells using fed-batch culture,' *Journal of Food Engineering*, 27: 87–96.

Said, S. and Pietro, R.C.L.R. (2004) *Enzimas como Agentes Biotecnológicos*. Ribeirão Preto: Legis Summa.

Sanada, C.T., Karp, S.G., Spier, M.R., Portella, A.C., Gouvêa, P.M., Yamaguishi, C.T., Vandenberghe, L.P.S., Pandey, A. and Soccol, C.R. (2009) 'Utilization of soybean vinasse for α-galactosidase production,' *Food Research International*, 42: 476–483.

Shamala, T.R. and Prasad, M.S. (2001) 'Fed-batch fermentation for rapid production of xanthan by *Xanthomonas campestris*,' *Food Biotechnology*, 3: 169–177.

Singhania, R.R., Patel, A.K., Soccol, C.R. and Pandey, A. (2009) 'Recent advances in solid-state fermentation,' *Biochemical Engineering Journal*, 44: 13–18.

Siqueira, P.F. (2006) 'Production of bio-ethanol from soybean molasses by *Saccharomyces cerevisiae*,' *Master's dissertation*, Federal University of Paraná/Universities of Provence and of the Mediterranean, Curitiba, Brazil.

Siqueira, P.F., Karp, S.G., Carvalho, J.C., Sturm, W., Rodríguez-León, J.A., Tholozan, J.L., Singhania, R.R., Pandey, A. and Soccol, C.R. (2008) 'Production of bio-ethanol from soybean molasses by *Saccharomyces cerevisiae* at laboratory, pilot and industrial scales,' *Bioresource Technology*, 99: 8156–8163.

Soccol, C.R. and Vandenberghe, L.P.S. (2003) 'Overview of applied solid-state fermentation in Brazil,' *Biochemical Engineering Journal*, 13: 205–218.

Soccol, C.R., Vandenberghe, L.P.S., Costa, B., Woiciechowski, A.L., Carvalho, J.C., Medeiros, A.B.P., Francisco, A.M. and Bonomi, L.J. (2005) 'Brazilian Biofuel Program: An overview,' *Journal of Scientific and Industrial Research*, 64: 897–904.

Suarez, F.L., Springfield, J., Furne, J.K., Lohrmann, T., Kerr, P.S. and Levitt, M.D. (1999) 'Gas production in humans ingesting a soybean flour derived from beans naturally low in oligosaccharides,' *American Journal of Clinical Nutrition*, 69 (1): 135–139.

Suriyaphan, O., Drake, M.A. and Cadwallader, K.R. (2001) 'Lipid oxidation of deoiled soy lecithin by lactic acid bacteria,' *Lebensmittel Wissenschaft und Technologie*, 34: 462–468.

Togrul, H. and Arslan, N. (2004) 'Mathematical model for prediction of apparent viscosity of molasses,' *Journal of Food Engineering*, 62: 281–289.

USDA—United States Department of Agriculture. Online. Available at http://www.usda .gov/oce/weather/pubs/Other/MWCACP/Graphs/Brazil/BrazilSoybean.pdf (accessed November 9, 2011).

Wee, Y.J., Kim, J.N. and Ryu, H.W. (2006) 'Biotechnological production of lactic acid and its recent applications,' *Food Technology and Biotechnology*, 44: 163–172.

Woiciechowski, A.L. (2001) 'Desenvolvimento de bioprocesso para a produção de goma xantana a partir de resíduos agroindustriais de café e de mandioca,' *Doctoral thesis*, Federal University of Paraná, Curitiba, Brazil.

Xu, Z., Wang, Q., Wang, P., Cheng, G., Ji, Y. and Jiang, Z. (2007) 'Production of lactic acid from soybean stalk hydrolysate with *Lactobacillus sake* and *Lactobacillus casei*,' *Process Biochemistry*, 42: 89–92.

Yoon, M.Y. and Hwang, H. (2008) 'Reduction of soybean oligosaccharides and properties of α–D-galactosidase from *Lactobacillus curvatus* R08 and *Leuconostoc mesenteroides* JK55,' *Food Microbiology*, 25: 815–823.

Yun, C., Kim, Y., Park, J. and Park, S. (2009) 'Optimal procurement and operational planning for risk management of an integrated biorefinery process,' *Chemical Engineering Research and Design*, 87: 1184–1190.

Zech, M. and Görisch, H. (1995) 'Invertase from *Saccharomyces cerevisiae*: Reversible inactivation by components of industrial molasses media,' *Enzyme and Microbial Technology*, 17: 41–46.

Index

Page numbers followed by *f* and *t* indicate figures and tables, respectively.